HARVARD ECONOMIC STUDIES

HARVARD ECONOMIC STUDIES

VOLUME LVIII

AWARDED THE DAVID A. WELLS PRIZE FOR THE YEAR 1933–34 AND PUBLISHED FROM THE INCOME OF THE DAVID A. WELLS FUND. THIS PRIZE IS OFFERED ANNUALLY, IN A COMPETITION OPEN TO SENIORS OF HARVARD COLLEGE AND GRADUATES OF ANY DEPARTMENT OF HARVARD UNIVERSITY OF NOT MORE THAN THREE YEARS STANDING, FOR THE BEST ESSAY IN CERTAIN SPECIFIED FIELDS OF ECONOMICS

LONDON : HUMPHREY MILFORD
OXFORD UNIVERSITY PRESS

MARKET CONTROL IN THE ALUMINUM INDUSTRY

BY

DONALD H. WALLACE

ASSISTANT PROFESSOR OF ECONOMICS IN HARVARD UNIVERSITY

CAMBRIDGE

HARVARD UNIVERSITY PRESS

1937

PRINTED AT THE HARVARD UNIVERSITY PRESS

CAMBRIDGE, MASS., U.S.A.

To

MARCIA

PREFACE

THE present study first took partial form as a doctoral dissertation (presented in 1931) upon the aluminum monopoly in the United States. Thereafter, the scope of the inquiry was widened to include market control in Europe and international relations in this industry. It was my hope that study of the industry in Europe, where producers have at times competed and at times united in cartel organization, might enable a surer assessment of the market results here, where there has been but one producer of primary aluminum, as well as furnish some conclusions about the consequences of cartel control and of "oligopoly," to use Professor Chamberlin's expression for the condition where sellers are few, which appear to be the practicable alternatives in this industry. The grant of a traveling fellowship for the year 1931–32 from the Social Science Research Council gave me the opportunity to study in Germany, France, and Switzerland, where much of the material for the analysis of European experience was gathered. Since then nearly all of the original thesis has been completely revised in accordance with changes in the conception of some of the problems involved and in the light of additional material; and an attempt has been made to unify the analysis of both American and European experience relating to the fundamental problems presented by the alternative possible mixtures of competitive and monopolistic elements in this industry.

I have benefited much from conversations with many officials of companies engaged in producing aluminum or its products in various countries. In accordance with their wishes, acknowledgment is made anonymously, and the information and ideas which they have contributed appear without citation. The following gentlemen have provided me with courteous and valuable assistance: Mr. Richard Whitely, attorney for the Federal Trade Commission; Messrs. Furness and McGrath of the Bureau of Foreign and Domestic Commerce; Dr. Martin Doering of the Verein der deutschen Maschinenbauanstalten in Berlin; Dr. Thews and Dr.

Regensburg, Berlin journalists; Dr. Josten of the Reichswirtschaftsministerium; Dr. Apfelstädt of the Statistisches Reichsamt; Dr. Rudolf Schwarzmann of the Statistisches Amt in Bern; and M. Thibaud of the Ministère des Travaux Publics in Paris.

I owe much to Professors Taussig and Ripley, whose encouragement and guidance were of great aid in carrying the project through the troublesome early stages. Professors W. L. Crum and A. P. Usher of the Department of Economics and Professors R. S. Meriam and Samuel S. Stratton of the Graduate School of Business Administration of Harvard University have aided me with valuable criticism. I have also benefited from the suggestions of members of two discussion groups at Harvard to whom I presented in their original form parts of the analysis of Chapters XV, XVI, and XX. My greatest debts are to Professor Edward Chamberlin, as will be apparent to all who are familiar with his *Theory of Monopolistic Competition* (Cambridge, 1933), and to Professor Edward Mason, whose thinking upon the sort of problems treated in this book has exercised an immeasurable influence upon my ideas. Responsibility for the views expressed in this volume is, of course, entirely my own.

I am indebted to the Bureau of International Research of Harvard University and Radcliffe College for permission to reproduce here much of the material contained in my chapter on aluminum contributed to the study entitled *International Control in the Nonferrous Metals,* directed by Professor W. Y. Elliott, which is about to be published by the Bureau as this book goes to press. Much of that chapter represents a brief condensation of several portions of Parts I-III of the present book, and contains several tables and charts which are reproduced here in similar or more elaborate form. As a result of better information obtained since that chapter was completed some time ago, the statistics and other information presented in this book differ in some instances from what was given there.

The major part of the tedious work of preparing the manuscript for the press has been done by my wife, who has also assisted in proofreading. Her criticism of ideas and expression has aided me greatly in many instances. Financial assistance from the A. W. Shaw Fund, administered by the Committee on Research in the Social Sciences of Harvard University, facilitated the latter stages of preparation of the manuscript. I am very grateful to Miss Margaret Ballard for her meticulous care in this part of the work.

Statistics of production and foreign trade in aluminum ordinarily appear in pounds or short tons for the United States and in metric tons or long tons for Europe. In order to facilitate comparison I have reduced the figures for all countries to a metric ton basis. In Part IV, which deals only with relations within the aluminum industry of the United States, the pound unit has been retained.

My original hope of making the book equally understandable to the economist and to the general reader, whose lack of familiarity with the tools of economic analysis is overborne by an interest in the nature of modern market processes, is not, I fear, well realized. The endeavor to put certain parts of the argument in non-technical language has resulted in a style of exposition demanding the forbearance of the trained economist, to whom it will seem unduly tedious and awkward; while a frank resort to technical exposition in those parts dealing with more complicated phases of the problem requires the forgiveness of others.

For the benefit of the general reader I wish to stress the fact that many words and phrases of ordinary language are here used in the special meanings given to them in economics, and, in some instances, with a particular sense of my own. I hope I have avoided misunderstanding by defining terms in cases where ambiguity seemed to threaten. It should be quite clear that, unless otherwise specified, the word "monopoly" in all its forms has been used in an economic sense — signifying complete control of supply, or sufficient control to affect appreciably the fundamental market relations between investment, output, price, earnings, and demand — and not in the legal sense attached to it by the antitrust laws of the United States. Such terms as "discrimination" and "unfair methods of competition" are also used with particular economic meanings rather than their legal meanings, except where the contrary is stated. Questions of discrimination in the aluminum price structure have not been related to the Robinson-Patman Act, which was passed after most of the analysis of discrimination had been concluded.

In several instances terms which sound quite "academic," such as "ideal investment" and "best utilization," have been employed in order to give precision to the analysis. The reader must not draw the mistaken inference that it is implied that ideal conditions can necessarily be obtained under some form of market control. Such phrases are merely technical terms used to designate certain market relationships which represent useful standards by which to measure the results of different sorts of market control.

The present work is an attempt to appraise critically the limited material bearing upon several important economic problems. Inasmuch as the material available has proved in some respects inadequate, the statements in the text must be regarded merely as the considered judgments of the author. I also wish to make it plain that the economic conclusions are not intended to express any moral or ethical judgment on the conduct of persons in this industry. At the present time an understanding of economic processes sufficient to distinguish those business policies which promote general economic welfare from those which tend in the opposite direction is not widespread enough to exert any very salutary influence upon the general scale of moral values and standards of conduct.

One symptom of the tardy appreciation of the range of alternative mixtures of competitive and monopolistic elements in modern markets is the absence of any term for the structure of market forces which does not imply the predominance of either competitive or monopolistic elements. I have used the term "market control" to fill this gap. In this sense market control may be of various types, ranging from pure competition to single-firm monopoly. The phrase selected is not entirely satisfactory, for it is often convenient to employ it, as I have sometimes done, in the older and more limited sense of substantial monopoly control.

Those who are chiefly interested in the reasons for existence in the aluminum industry of a few producers only will find a unified treatment of this topic in Part II. The main conclusions about the actual results of oligopoly in this industry are contained in Chapters XII–XIV. The analysis in these chapters will be better understood if Chapter X is also read. Section 1 of Chapter XV embodies a unified theoretical essay comparing the probable results of oligopoly and single-firm monopoly under certain assumed conditions, the most important of which are a rapid forward movement of demand and great uncertainties about the rate of movement and the sorts of new adaptations of a basic product that can be made. This section can be read apart from the rest of the book by those interested in the pure theory of oligopoly.

As this book goes to press I discover that Professor Arthur R. Burns's new book, *The Decline of Competition* (New York, 1936), contains several ideas about oligopoly under dynamic conditions, price stabilization, and the relations between monopoly and vertical integration which are similar to ideas worked out here.

CAMBRIDGE, MASSACHUSETTS D. H. W.
December, 1936

CONTENTS

INTRODUCTION xxv

PART I

HISTORICAL DEVELOPMENT

I. DEVELOPMENT OF MARKETS, 1888–1914 3

1. Silver from Clay, 3.
2. Early Marketing Difficulties, 8.
3. Transmission Lines and the Automobile, 14.

II. INDUSTRIAL STRUCTURE AND MARKET CONTROL . . . 24

1. The Aluminum Company of America, 24.
2. The Four European Producers, 31.

III. RESEARCH AND MARKETS, 1915–1935 43

1. The War Stimulus, 43.
2. New Alloys and New Markets, 47.

IV. EXPANSION AND POLITICAL RELATIONS, 1915–1935 . . 69

1. The Aluminum Company of America and Aluminium Limited, 69.
2. New and Old National Monopolies in Europe, 82.

PART II

THE NATURE OF MARKET CONTROL — THE BASIC PRODUCT

V. EARLY MARKET CONTROL 101

1. Monopoly in America, 101.
2. Competition and Monopoly in Europe, 118.

VI. POTENTIAL COMPETITION — CONTROL OF ORE AND POWER 129

1. Potential Competition in America — the Uihlein and Duke-Haskell Episodes, 129.
2. Control of Bauxite and Power, 137.

VII. NATIONAL MONOPOLIES AND INTERNATIONAL RELATIONS 149

1. Continuing Monopoly in America, 149.
2. National Monopolies in Europe, 153.
3. International Competition and Coöperation, 157.

PART III

SOME CONSEQUENCES OF VARIOUS KINDS OF MARKET CONTROL

VIII. Structural Effectiveness — Integration 173
1. Issues and Terminology, 173.
2. Theoretical Relations between Integration and Monopoly, 177.
3. Integration in the Aluminum Industry, 179.

IX. Structural Effectiveness — Horizontal Extension and Integrated Balance 189

X. Investment, Price, and Demand — Introductory . . 204
1. Issues, 204.
2. Some Characteristics of Supply and Demand Conditions, 212.
3. Some Features of the Aluminum Price Structure, 216.

XI. Investment, Price, and Demand in the United States 225
1. Earnings, 225.
2. Capacity and Price, 238.
3. Changes in Demand, 252.

XII. Investment, Price, and Demand — The First European Cartel 264
1. Issues and Facts, 264.
2. Conclusions, 271.

XIII. Investment, Price, and Demand in Europe during the Post War Decade 277
1. Earnings, 277.
2. Capacity and Price, 286.
3. Oligopoly and Cartels, 299.

XIV. Rationalization in the Short Run 313

XV. Conclusions and Possibilities 331
1. Theoretical Comparison of Different Types of Market Control, 331.
2. Possible Alternatives for Government Policy, 352.

PART IV

RELATIONS BETWEEN MONOPOLY OF THE BASIC PRODUCT AND INDEPENDENT COMPETITORS AT LATER STAGES — SOME ASPECTS OF AMERICAN EXPERIENCE

XVI. The Nature of Competition in Sheet Production . 369
1. Introductory, 369.
2. The Rolling-Mill Case, 371.
3. Independent Rolling Mills and Ingot Supply, 374.
4. The Ingot-Sheet Price Differential, 379.

XVII. Competitive Methods and Government Control . . 396
1. Consent Decree of 1912, 396.
2. Government by Investigation, 398.
3. Criteria of Unfair Methods of Competition, 402.
4. Injunctions of the Consent Decree of 1912, 404.

XVIII. Competitive Methods in the Utensil Industry . . 408
1. Introductory, 408.
2. Miscellaneous Practices, 415.
3. Price Discrimination, 417.
4. Delivery Delays, 421.
5. Attitude toward Potential Competition, 437.
6. Summary, 440.

XIX. Scrap and Sand Castings 443
1. Complaint of Unfair Methods, 443.
2. Preliminary Analysis, 449.
3. Scrap, 456.
4. Pricing of Castings, 462.
5. Summary, 472.

XX. The Effectiveness of Government Control . . . 474
1. The NRA Code, 474.
2. Results of Attempts at Government Supervision, 479.
3. Private Litigation, 480.
4. Possibilities for Economic Reform, 485.

Appendices 501
A. Aluminum and the Electrochemical Revolution, 503.
B. Patent Litigation in the United States, 527.
C. Investment and Earnings of the Aluminum Company of America, 1909–1935, 538.
D. Consent Decree of 1912, 547.
E. United States of America before Federal Trade Commission — in the Matter of the Aluminum Company of America. Complaint, Answer, Order of Dismissal, 556.
F. Statistical Tables, 570.

Bibliography 575

Index 587

TABLES

1. Average Yearly Prices of Aluminum and Price Ratios of Aluminum and Other Nonferrous Metals in the United States, 1900–1907 . . 17
2. Per Cent of Aluminum Sold in Different Forms in the United States in Certain Years 21
3. Estimated Capacities of Aluminum Producers of the World in 1914 40
4. Ratios of World Production of Aluminum in Certain Years to World Production of Other Nonferrous Metals by Volume 46
5. Estimated Annual Consumption of Aluminum in the United States by the Automobile Industry, 1921–1934 61
6. Estimated Consumption of Aluminum in Various Markets in the United States in 1930 Expressed as Percentages of Annual Output . 64
7. Aluminum Conductors in High-Tension Transmission Lines in Certain Countries in 1933 67
8. Estimated Consumption of Aluminum in Various Markets of Germany in 1929 Expressed as Percentages of Total Consumption . . 67
9. Estimated Capacities of Aluminum Companies of the World in 1936 96
10. Estimated Capacities for Aluminum Production in 1936 by Countries 98
11. Estimated Capacities of Old and New European Companies in Certain Years 123
12. Estimated Ratios of Earnings to Investment of the Aluminum Company of America, 1909–1935 226
13. Price Data for 98–99 per cent Virgin Aluminum Ingot, 1920–1932 . 240
14. Monthly Average Prices of 98–99 per cent Virgin Aluminum Ingot in the New York Open Market 242
15. Imports of Aluminum Ingot into the United States, 1919–1933 . . 243
16. Ratios of Yearly Average List Prices of Aluminum Ingot to Yearly Average Prices of Copper and a Weighted Nonferrous Metals Index in the United States, 1919–1934 245
17. Sales of Various Products by the Aluminum Company of America, 1923–1929 251
18. Ratios of Earnings to Investment of Three European Aluminum Companies, 1922–1935 278
19. Approximate Average Yearly Prices of 98–99 per cent Aluminum Ingot in Europe, 1922–1930 287

20. Ratios of Average Yearly Prices of Aluminum Ingot to Average Yearly Prices of Copper in Europe in Certain Years 288

21. Estimated Capacities of Aluminum Plants of the World in Certain Years 291

22. Division of Cartel Output of Ingot Aluminum between Leading Companies, 1922–1930 301

23. Statistics of Foreign Trade in Aluminum Ingot of Cartel Countries, 1923–1929 303

24. Division of Aggregate Ingot Capacity of Leading European Aluminum Companies in Certain Years 305

25. Estimated Production of Ingot Aluminum by American and European Companies, 1918–1930 308

26. Relative Positions of the American and European Companies in Foreign Trade in Aluminum Ingot, 1922–1930 309

27. Statistics of Output and Sales of Aluminum in the United States, 1926–1935 325

28. List Prices of Aluminum Company of America for 99 per cent Ingot and Certain Classes of Sheet and Resultant Spreads, 1918–1931 . . 382

29. Sales of Similar Duralumin Alloy Sheet in Competitive Sizes by the Baush Machine Tool Company and the Aluminum Company of America, 1925–1931 385

30. Aluminum Company of America — Price and Cost Data for Aluminum and Aluminum-Alloy Sheet of All Classes and 99 per cent Ingot, 1925–1930 388

31. Statistics of Growth of Aluminum-Ware Industry 409

32. Discounts from Schedule Prices of Raw and Semifabricated Aluminum of the Aluminum Company of America Accorded Various Cooking-Utensil Firms and Users of Aluminum in Other Industries during the Period November 15, 1921–October 3, 1922 419

33. Percentages of Obligations Shipped by the Aluminum Company of America in Certain Periods to Various Cooking-Utensil Companies . 424

34. Percentages of Obligations at Different Seasons Shipped by the Aluminum Company of America to Four Manufacturers as Compared to Shipments to the Aluminum Goods Manufacturing Company and the Aluminum Cooking Utensil Company Based on a Fiscal rather than a Calendar Month 429

35. Sales of Aluminum Castings by the Aluminum Company of America and Two Independents in Certain Years 470

36. Earnings of the Aluminum Company of America on Sand Castings 471

37. Investment, Earnings, and Rate of Return of the Aluminum Company of America, 1909–1935 544
38. Estimated World Production of Primary Aluminum by Countries, 1890–1935 570
39. Estimated Production and Foreign Trade in Aluminum of the United States, 1900–1935 572

CHARTS

I. Estimated World Production of Primary Aluminum, 1900–1935 . 35
II. Price Data for 98–99 per cent Primary Aluminum Ingot in the United States and Europe, 1920–1932 239
III. Estimated European Production of Primary Aluminum by Leading Companies, 1910–1930 302

ABBREVIATIONS

AG = *Les Assemblées générales*

BMTC appellant v. ACOA = Baush Machine Tool Company, appellant, v. Aluminum Company of America

BMTC v. ACOA appellant = Baush Machine Tool Company v. Aluminum Company of America, appellant

BR = Benham Report, Department of Justice

EMJ = *Engineering and Mining Journal*

FTC = Federal Trade Commission

HR = Henderson Report, National Recovery Administration

JFE = *Journal du four électrique*

KFR = Kitchen Furnishings Report, Federal Trade Commission

MI = *The Mineral Industry . . . in the United States*

MR = *The Mineral Resources of the United States*

MW = *Metallwirtschaft*

NRA = National Recovery Administration

INTRODUCTION

For the student of monopolistic and competitive forces the aluminum industry presents an unusually interesting specimen. The Aluminum Company of America has remained the only producer of virgin aluminum ingot in the United States since its organization in 1888 to work the patent which gave birth to this new industry. The activities of the company affecting the domestic market for virgin ingot have never been held, in a final proceeding, to violate the antitrust laws, nor does it appear that they have infringed accepted notions of business ethics. How explain the lack of domestic competitors? Except during periods of depression imports over the tariff duty have usually been rather small relative to the sales of the domestic firm. No marked separation of ownership and control has existed in the Aluminum Company to vitiate direct motivation. Here seems to be an opportunity to test at once the pessimism of the older monopoly theory and the optimism infused into the discussion of monopoly by the rationalizers. In Europe some degree of competition has existed at times in national and international markets, while national monopolies and international cartel control have prevailed for periods of several years. Analysis of the consequences of oligopoly and of cartel control abroad in this industry may be of interest in this country, where the violent death of NRA is not likely to allay the appeals for permanent revision of the antitrust laws to permit cooperative self-government of business for "planned" control of production and marketing in each industry.

This study is an inquiry into the nature of monopoly and oligopoly, with and without agreements, under dynamic conditions. Its purpose is to explain the continued existence of single-firm monopoly[1] of the basic product in the United States and of strong

[1] Owing to the presence of foreign producers in the market for virgin aluminum in the United States the condition is, strictly speaking, one of oligopoly. Apparently, however, the foreign companies have not set up capacity for large exports to this country; and much of the time their sales here have been quite small. Since we have no simple phrase to describe this sort of oligopoly, it will be convenient to use the term "single-firm monopoly."

monopolistic elements abroad (Part II), to evaluate the consequences and assess the relative merits of alternative mixtures of competitive and monopolistic elements (Part III), and to examine some of the intricate problems created by the existence in some fabricating stages of independents competing with the firm or firms from whom they obtain their raw materials (Part IV). No attempt is made to consider the effects of monopolistic elements upon labor, bankers, different classes of investors; no questions are raised concerning working conditions or the division of gross earnings among all those who have claims to a part. The chapters in Part I are designed to give sufficient knowledge of technology, markets, public relations, and international relations for the analysis which follows. Two chapters upon the early history of scientific discovery and inventive activity centering about aluminum are printed as Appendices A and B in the hope that others may share my lively interest in this hitherto neglected chapter of economic history.

The study was undertaken with the aim of adding to our knowledge of the workings and results of competitive and monopolistic forces in present-day economic organization. I share with many others the belief that we need more studies of the different mixtures of competitive and monopolistic forces in particular industries, of the various sorts of markets which make up that part of the economy in which the conditions of pure competition—that is, complete absence of monopoly elements—are not even approximated. Unfortunately it has been impossible in the present study to reach assured conclusions upon several points. In large degree this has been due to inability to obtain adequate and accurate information; but the limitations of economic method and the unavoidable employment of hypothesis must bear part of the responsibility. With the exception of some financial and technical data, little information of basic economic importance has been voluntarily published by the aluminum companies. In the last ten years there has appeared a tendency towards more publicity on the part of the Aluminum Company of America and some of the other firms. They do not yet, however, ordinarily give out the sort of information required for analysis of the kind of questions raised in this book. Government reports and the records of private litigation contain a large amount of information, but much of it is inadequate for the treatment of such questions. The material available in trade and scientific journals and the yearbooks of private organizations falls far short of remedying this lack. For statistical data on production and capacity it has been necessary to rely largely upon the estimates appearing in such sources. Requests addressed directly to the aluminum companies have yielded some important information; but the companies have been unwilling to give me the sort of data desired for many parts of this study, doubtless partly through fear that disclosure of such information might hurt their business. Even the sketch of the historical development of the industry has suffered from the lack of information of certain sorts. While there is a vast literature in trade and scientific journals dealing with the technical aspects of aluminum, few articles attempt to survey the changing industrial importance of this metal or to provide the quantitative materials for such study. Finally, it is particularly unfortunate that the government investigations of the position and competitive methods of the Aluminum Company of America in the fabricating stages of the industry do not provide sufficient information of the sort required to resolve the true economic issues.

Furthermore, it must be recognized that economic method has not yet designed tools of analysis keen enough to dissect neatly the results of various mixtures of competitive and monopolistic forces operating under the dynamic conditions of the real world. Until the recent advances made by Professor Chamberlin, Mrs. Robinson, and others forced recognition of the limitations of theoretical analysis based upon assumptions of pure competition, most students of the kind of problems treated in this book had employed the apparatus of purely competitive theory (if not indeed of perfectly competitive theory!) to block out the questions or issues for factual study. The most serious shortcoming of this procedure was the conception of evils of monopoly as evils in the sense of much less desirable results than would accrue with pure or simple competition. The obvious requirements of comparison between the consequences of monopolistic forces and the results of some kind of competitive control were met by comparing the former with the outcome under hypothetical conditions of pure competition. Actually, however, the choice which public policy must usually make is between alternative mixtures of competitive and monopolistic elements, with or without different sorts of public control. The conditions of pure competition are neither a practicable nor a desirable alternative in many industries. For, although the great size attained by many industrial corporations is doubtless to be explained in substantial part by considerations of "power politics," advances in the technique of capital and administration have extended the most efficient scale for a firm in many industries to the point where it produces a sufficient portion of the total output to affect price or quality by its policy. Under these circumstances the appropriate comparison is between the results of simple oligopoly, oligopoly with agreements, single-firm monopoly, and each of these with different sorts of government control. Since the outcome with oligopoly may, as Professor Chamberlin has shown, vary all the way from that of pure competition to that of monopoly, the older measuring stick of "competitive" results turns out to be an elastic rod! It is necessary to study the particular sorts of market control which represent real alternatives to those actually in existence, to compare them with the latter, and to determine as well as can be which is best. It is plain, I think, that the technique for economic analysis of concrete problems of this kind needs improvement. It is my hope that the present volume may be of some stimulus to others who are interested in redesigning analytical tools for future study of these questions.

I had hoped to be able to deal more adequately with the issue of progressiveness. Unfortunately, difficulties with evaluation of standards and inability to obtain adequate material have made it impossible to devote to this problem the amount of space which its importance warrants. Treatment of this issue has been limited for the most part to consideration of progressiveness in the development of new alloys, fabricating processes, and finished goods.

In several instances, particularly in Part IV, the problems presented by concrete situations which were of relatively little importance have been subjected to detailed analysis because of the significance of the principles involved. Since government agencies often failed, in my judgment, to use correct principles in their analysis of competitive methods, it seemed important to devote considerable space to the development of the proper principles.

No plan for economic reform in the aluminum industry is advocated here. More adequate information is required to determine whether the undesirable consequences of the existing types of market control are of sufficient magnitude to make some sort of change worth while. Moreover, examination of the more purely governmental problems attending various kinds of public control — a task for which the economist is often not well fitted — is necessary before any final pronouncement on the relative merits of alternative schemes for improvement. In Chapters XV and XX, however, I have considered the economic problems presented by different devices for bettering the relations between investment, output, and demand in this industry, and have appraised the relative merits of the several alternatives according to economic considerations. In this analysis some schemes, particularly of government regulation in the narrow sense, have been discussed in detail which their seeming impracticability might not appear to warrant. But it is of the utmost importance to realize fully the breadth and complexity of the problems and the extent of detail in measurements and in policies which would be required for successful regulation. We must recognize that in an industry of this sort regulation according to simple rules is not likely to produce desirable results and that serious difficulties are to be encountered in the formulation and administration of complex rules.

PART I

HISTORICAL DEVELOPMENT

CHAPTER I

DEVELOPMENT OF MARKETS, 1888–1914

1. Silver from Clay

In the year 1854 there was called to the attention of Napoleon III a new metal which came to be popularly known as "silver from clay." The interest of the Emperor was caught by the possibility that aluminum, as the professors called it, might lighten as well as brighten the accoutrements of his army. He granted a liberal sum for the research necessary to perfect its manufacture. Although Napoleon's hope of its military serviceability was not realized until after his death, presumably the Prince Imperial enjoyed his aluminum rattle, and for a few years aluminum jewelry by Christofle was quite fashionable. Knives and forks of the new metal appeared at court banquets. The king of Siam wore an aluminum watch charm. Yet the new "silver" was nearly twice as abundant as iron, for aluminum constitutes almost eight per cent of the earth's crust and is exceeded in amount only by the elements oxygen and silicon. What paradox of supply and demand could elevate a substance existing in every clay bank to a position in the scale of values close behind the noble metals? The answer lay in the exceeding difficulty of reducing aluminum, which occurs nowhere in a free state, from its compounds. Indeed, so great is its tenacity for oxygen, with which it is usually combined in nature, that definite ascertainment of the existence of this element a century ago confirmed only the surmises of a few men of science. Ordinary smelting methods which had made iron, copper, tin, and other metals useful articles of man's world for several centuries were unavailing to obtain pure aluminum. Not only was it impossible to deprive aluminum of its oxygen by smelting with carbon or other common agents at the temperature attainable in the fuel furnace; even if this could have been accomplished the resulting aluminum would not have been pure enough for commercial uses. Silicon, iron, titanium, and other substances contained in the common aluminum ores, can be more easily reduced than aluminum itself. Moreover, carbon smelting would have yielded only useless carbides. It was a very expensive chemical process which finally brought forth a small-scale aluminum industry in the late fifties. For thirty years thereafter the metal was used only in rings, brooches, statuettes, and other ornaments, and sold for about twelve dollars a pound.

The reduction of aluminum from its association with the noble metals and its affinity for "conspicuous consumption" to a commonplace metal of mass consumption was accomplished by electrochemistry. And it was the search for a cheaper process of reducing aluminum which played the chief part in an outburst of experimentation and practical application in electrochemistry, beginning about 1880, the revolutionary consequences of which emerged in an extensive array of new products and processes. Industrial accomplishments in electrochemistry could not appear until the perfection of the dynamo, which occurred during the seventies; but scientific research and formulation of the general principles governing the relations of electric currents to chemical changes began with the science of electricity and proceeded throughout the nineteenth century. The electrochemical revolution consisted in the successful economic adaptation of these general principles to particular problems through further scientific investigation and practical industrial experiment.[1]

The introduction of an effective dynamo gave a great impetus to experiment with electrometallurgical and electrochemical methods of reducing highly refractory metallic compounds. During the eighties a host of inventors in America and Europe turned their attention to the "silver from clay" which still excited interest at expositions and commanded a price of twelve to fifteen dollars a pound. The possibility of discovering an electric key to unlock this abundant treasure must have been no less exciting to these inventors than the alluring dreams of synthetic gold which had earlier fired the alchemists. The more visionary saw the unfolding of an Aluminum Age in which the light metal would revolutionize transport, construction, and architecture.

In 1886 a successful process of electrolytic reduction was dis-

[1] A description of the early history of scientific discovery and invention as related to aluminum, and the repercussions of this work in other branches of electrochemistry, is given in Appendix A.

covered independently by Charles M. Hall in the United States and M. Paul L. T. Héroult in France. Two years later the Pittsburgh Reduction Company, which subsequently became the Aluminum Company of America, was formed by Hall and several Pittsburgh capitalists to produce aluminum commercially.[2] For some years aluminum-copper alloys had been produced in the electric furnace by the Cowles Electric Smelting and Aluminum Company of Cleveland, which had been responsible for important pioneer work in the development of the electric furnace. The Cowles brothers had experimented for several years upon the production of pure aluminum and had made an optional agreement with Hall in 1887 to buy his patent if he were able to demonstrate by work at their plant that his process could be satisfactorily adapted to commercial operation. They allowed the option to lapse, after which Hall found support in Pittsburgh. In 1891, perceiving that future profits lay in the production of pure aluminum rather than the none-too-serviceable alloys which they had been making, the Cowleses began to produce aluminum by an electrolytic process essentially the same as that of Hall. An infringement suit was decided against them in 1893. Determined to participate in the commercial success of the new metal, the Cowleses now countered with a suit alleging that the Hall process infringed a broad process patent issued to C. S. Bradley, possession of which they had secured by a combination of shrewdness, luck, and persistent litigation. This suit was finally decided in favor of the Cowles company by the Circuit Court of Appeals in 1903. It was held that the Bradley patent, application for which had preceded the filing of Hall's application, specified one essential element of the process which was not included in Hall's patent.[3] As the situation stood after this decision neither the Pittsburgh Reduction Company nor the Cowles firm could legally produce aluminum by the only successful method known without obtaining a license from the other.

Had the stalemate been resolved by a cross-licensing agreement, there might have been two established producers of aluminum in the United States when the basic patents expired. The Cowleses,

[2] Mr. A. W. Mellon and Mr. R. B. Mellon became stockholders of the company when its capital was enlarged two or three years after its formation.

[3] The patent litigation is described in Appendix B.

who were perhaps better fitted to enter this industry than any other group of men in the country, except Mr. Hall and his associates, decided, however, to withdraw with a lump-sum payment and annual royalties. The result of the patent litigation itself, as well as the decision of the Cowleses to refrain from entering the industry, helped to preserve monopoly in the United States. The Hall patents expired in 1906 during the boom of 1905–1907. The Bradley patent, which had not been issued until nearly ten years after the filing of the original application, endured, however, until 1909. The period of legal monopoly was thus carried through the boom, when competition might have developed, while the company gained an additional three years in which to fortify its position so strongly that fresh capital and enterprise did not enter the industry after expiration of the patent.

The Héroult process, which was in all essentials the same as that of Hall, was first operated commercially in 1889 at Neuhausen, Switzerland, by the Aluminium Industrie A. G., a company formed by Swiss industrial interests and the Deutsche Edison Gesellschaft (later the Allgemeine Elektrizitätsgesellschaft). Electrolytic production of aluminum was undertaken at the same time in France by the Société Électrométallurgique Française, which was founded by Héroult with the aid of the Neuhausen firm. The British Aluminium Company purchased the Héroult patent for Great Britain and began operations in Scotland in 1896. About the same time the French chemical firm which had produced aluminum for thirty years by the old chemical process engaged in electrolytic reduction under Hall's French patent. No other enterprises were established until after the expiration of the basic patents.

The efforts of inventors to discover a more economical method of obtaining aluminum have not yet borne fruit in commercial application. The Hall and Héroult process was essentially too simple to permit patentable modifications or variations upon the basis of which competing firms could be established during the life of those two patents. It was as if a law of nature had decreed that the intense competition of many would-be parents should yield a single type of child fitted for survival, while man-made law prohibited imitation during its youth. The competition of inventors issued in monopoly of production.

The chapters in Part I contain a brief history of the aluminum industry in America and Europe from 1890 to the present. The development of the American industry will be surveyed in some detail. While a somewhat larger degree of monopolistic market control has obtained here than abroad, the developments in technology, industrial structure, and uses have been quite similar on both sides of the Atlantic. Hence, treatment of the European industry may be limited to a brief description of the enterprises and to a consideration of the chief differences in the development abroad.

A description of the processes by which bauxite is converted into consumable aluminum goods will be presented before the fortunes of the Pittsburgh Reduction Company are taken up.[4] With a few exceptions, which will be noted in due course, no fundamental alterations of process or apparatus have occurred since the birth of the industry. Bauxite, the ore from which aluminum is produced, is a mixture of hydrated oxides of aluminum containing silica, ferric oxide, and other impurities. Since the contained metals are separated from oxygen with less difficulty than aluminum, they will, unless removed previously, appear in the aluminum after electrolytic reduction, rendering it unfit for use. Hence the bauxite must be submitted to a refining process which yields alumina (aluminum oxide) that is almost entirely free from such impurities. The Bayer alumina process, which was universally employed until quite recently and is still used in most works, involves a complex set of operations consisting of digestion of bauxite with caustic soda, filtration, precipitation, calcining, and cooling.

After being pulverized the alumina is sent to the reduction plant, which contains scores of electrolytic cells. Each cell or pot consists of a relatively small rectangular steel or iron box, lined with a layer of hard-baked carbon which forms the negative electrode, with connections made to the metal casing. Carbon anodes are suspended from a copper bus bar running above the pots. The anodes, of which there are several, are adjustable, so that the distance which they project into the bath may be regulated. Operations are started by placing in the cell certain fluoride salts

[4] The operations of the various stages in this conversion are described in more detail in Chap. VIII.

(chiefly cryolite and fluorspar), in which alumina dissolves readily, lowering the anodes until they touch the carbon lining, and turning the current on. As the bath material melts, more is added until the required volume is secured, and the anodes are withdrawn from direct connection with the cathode. Alumina is then dumped in, and electrolytic action begins as soon as the alumina is in solution. Reduction ensues according to the equation $Al_2 O_3 = 2Al + 3O$. Since aluminum is of lower specific gravity than the bath at the operating temperature, it collects at the bottom of the furnace, where it is tapped off. The oxygen combines at the anodes to form carbon monoxide and, upon escape into the air, carbon dioxide. As the anodes are gradually burned away they are pushed into the bath and finally replaced. A layer of powdered carbon is kept on the top of the bath as insulation to prevent heat losses and danger to workmen. The process has always been operated continuously, alumina being added as needed. Large currents in amperes at low voltage are sent through a line of reduction cells arranged in "series."[5] The metal obtained is remelted after analysis and cast into pigs or ingots. The succeeding operations of rolling, drawing, stamping, casting, and forging are familiar enough to need no further comment. Later sections will describe the various sorts of alloys and products which have been developed.

2. Early Marketing Difficulties

The patent litigation mentioned in the first chapter would not have been pushed so strenuously had the prospects of the Pittsburgh Reduction Company been less favorable. The early years of this firm's life exhibited a steady growth which was to prove characteristic of its later experience. This period was occupied with the typical problems of perfecting the technique of industrial operation and developing markets. Frequent breakdowns of the electrical apparatus, which was still in its infancy, impeded production. The tips and armatures of the dynamos often burned out. Caking and clogging of the electrolytic bath presented a two-

[5] Eight thousand and thirty thousand seem to be the minimum and maximum amperage limits. The current is sent through a line of cells (each taking 5 to 7 volts) because it is not economical to generate electricity at low voltages.

fold problem. Current efficiency was lessened, and the aluminum contained occluded gases or impurities reduced from the bath or the electrodes. The latter necessitated the development of a satisfactory process of making these carbons. This was one of the first steps in integration. Perfecting of technique resulted in an improvement in the quality of the aluminum and a reduction in cost. In 1892 it was said that tons of metal were being turned out containing over 99 per cent aluminum, but that only a few hundred pounds had ever exceeded 99.75 per cent.[6]

In its first decade the new company found its progress beset with serious marketing problems. When production was started aluminum was selling for about ten dollars a pound. Very small amounts produced by the old chemical process were used in the making of jewelry and a variety of novelty articles. This market was easily captured when electrolytic aluminum was offered at five dollars, but it provided no basis for a growing industry. Years before the Cowleses were declared victorious in the patent struggle they had been definitely vanquished in the commercial arena. Founders who had been purchasing the Cowles aluminum alloys discovered that the desired proportions for alloys could be more closely controlled by melting pure aluminum with the other constituents. Here again, however, there existed no developed market of sizable dimensions. The Cowles alloys had been sold for less than a decade. There was almost no body of experience regarding the uses of aluminum alloys, or the methods of their casting. Engineers, founders, and manufacturers were in almost complete ignorance in these respects.[7]

In 1890 a price of two dollars a pound in half-ton lots was quoted in an attempt to widen the market. Shortly thereafter Mr. Hall wrote to a friend:

> The mention of $2 in 1,000-pound lots didn't seem to interest anyone. I know a good many people look at it as a big guy, and they have reason to do so, as they know that the total consumption of aluminum in the U. S. has hardly been 1,000 pounds a year. People have said we didn't have 1,000

[6] E. P. Allen, in *Cassier's Magazine,* I, 301 (February 1892). The metallic content of aluminum is measured by what is called the "difference" method — the amount of impurities is determined by chemical analysis, and the difference is assumed to be the quantity of aluminum.

[7] Dr. Leonard Waldo, in *Transactions of the American Institute of Electrical Engineers,* VIII, 414 ff. (May 1891).

pounds. They were wrong, but they might have said, that so far as the users of aluminum were concerned, practically no one wanted 1,000 pounds.[8]

The chief problem was to develop new uses. The ready-made demand was small; aluminum had never heretofore competed with the older metals, copper, tin, brass, zinc, steel, which it must partially replace if a substantial market were to come into being. Almost every demand for aluminum had to be won in contest with well-established metals of which the qualities and suitabilities for specific purposes were well known. A campaign of demonstration and education could not bear immediate fruit. In the interim the steel industry provided a source of expanding consumption. A few years before the establishment of the electrolytic industry it had been discovered in Europe that the addition of minute quantities of aluminum to molten steel resulted in more complete deoxidation, thus improving quality by reducing blow holes and occluded gases. Within a few years the use of aluminum had become general practice in the making of steel, cast iron, and the "Mitis" castings of wrought iron.[9] For four or five years after 1889 a large part of the output of the Pittsburgh firm was taken by the iron and steel industry.

With the exception of this market, aluminum could hardly be sold in ingot form. At first attempts were made to interest foundries, rolling mills, and wire-drawing plants in the use of aluminum, but inertia and ignorance proved more powerful than curiosity. Moreover, as subsequent experience demonstrated, the working of aluminum required methods different from the empirical rules developed by long experience for the casting and rolling of brass, copper, and steel. A representative of the company tells us that

> lack of familiarity with the metallurgical characteristics of the light metal led to blisters, slivers, blow holes, and every ill to which metal fabrication is heir.

[8] J. D. Edwards, F. C. Frary, and Zay Jeffries, *The Aluminum Industry: Aluminum and Its Production* (New York: McGraw-Hill Book Company, 1930), p. 25. This volume will be designated hereafter as *Aluminum Industry,* I. A second volume, *The Aluminum Industry: Aluminum Products and Their Fabrication,* will be cited as *Aluminum Industry,* II.

[9] Bureau of Mines, *Mineral Resources of the United States* (1882), p. 220 (hereafter cited as MR). J. W. Richards, *Aluminium* (Philadelphia, 1896), pp. 575 ff. Also aluminum soon replaced magnesium in making nickel and was used in the production of brass.

Scrap losses and returned shipments were often greater than the metal that could be utilized.[10]

Modern methods of technical research were almost nonexistent at this time. Hence, although troubles were painfully apparent, remedies were uncertain. In some respects the members of the company were quite as ignorant as the users of metals. This was particularly true of alloys, with which experimentation was begun early because pure aluminum did not possess the strength or stiffness required for many purposes. Aluminum received a bad name in several industries during the nineties owing to its attempted application to uses for which it was not fitted or to application before research had developed proper methods of working and suitable alloys for specific uses.[11] A few years of this sort of experience were sufficient to convince the company that the campaign of demonstration and education must include fabrication. A foundry, a wire-drawing plant, and equipment for rolling sheets and fabricating shapes and tubing were soon installed at New Kensington. By the middle nineties this policy had begun to bear fruit. Several cooking-utensil companies were established about 1893. In heat conductivity aluminum is exceeded only by silver, copper, and gold. This advantage, combined with its untarnishability, lightness, and resistance to corrosion by organic acids, won a rapidly developing market, which soon absorbed more aluminum than the steel industry. It appears that between a third and a half of the total output of the middle nineties went into cooking utensils, and about half this amount to the steel industry.[12] While the cooking-utensil industry continued to be the largest market

[10] *Aluminum Industry*, II, 4.

[11] In addition to the ignorance and inertia described, the company was troubled by the circulation of many extravagant, erroneous, and misleading claims by inventors of remarkably cheap processes for aluminum production. See an article by Mr. Hunt, the president of the company, in *Engineering and Mining Journal*, LI, 280, March 1891 (hereafter EMJ). During its early life the *Aluminum World* rendered good service in pointing out both the misconceptions and the truth about the usefulness of the new metal. The members of the company itself, Mr. Hunt in particular, published many articles in the trade and scientific journals and presented speeches before trade conventions and scientific societies. The limitations of aluminum as well as its advantages were carefully explained in this campaign of education.

[12] Estimates of monthly consumption in these two markets were given in the *Aluminum World*, II, 147–148 (April 1896).

until the turn of the century, other new demands appeared every year. Aluminum was substituted for expensive lithographic stones with the advantage of greater durability and a saving of two-thirds of the original cost.[13] Characteristic of subsequent development was the broadening range of products which took smaller amounts. Among these were bicycle parts, reflectors in locomotive headlights, cameras, flashlight powder, semaphores, bathtubs. Aluminum leaf, beaten after rolling, had nearly superseded silver foil in decorative art work. The Smith pressure casting process had been successfully adapted for art work. Aluminum plates on the Herreschoff *Defender* of 1895 provoked a stormy controversy, in which the conservative tendencies of certain high navy officials became a standing joke. Difficulties with corrosion and soldering prevented the use of aluminum in countless employments, however, for many years. The fact that aluminum is electropositive to most metals resulted in corrosion and in the failure of soldered joints through galvanic action when the presence of moisture acted as an electrolyte between aluminum and the metals adjacent to it or alloyed with it. Many early trials of aluminum boats or boat fittings failed after a time for this reason.

The success of the introductory market campaign is evidenced by the expansion of the company. The output of the Pittsburgh plant in 1890 was 58,000 pounds. Four years later 550,000 pounds were produced in the works at New Kensington, to which operations had been transferred to take advantage of natural gas, which provided cheaper power. In 1895 the company moved to Niagara Falls and there became the first customer of the Niagara Power Company, which was just completing the first large hydroelectric development in the United States.[14] The production of ingot at New Kensington was abandoned in the following year, when the second of two reduction plants built at Niagara Falls

[13] This use exerted an important influence upon the improvement of early fabricating methods because large sheets with surfaces as nearly perfect as possible were demanded.

[14] Since 15–17 h.p. hours were required to produce one pound of metal, the company would probably have used cheap water power from the start had it been available. The Niagara Power Company agreed to furnish current transformed to suit the aluminum firm, a service which was evidently not accorded to others, for eighteen dollars per h.p. year, a rate which appears to have been no higher, at least, than the rate made to other industrial consumers (*Electrochemical Industry,* I, 1902, p. 49).

went into operation.[15] Output reached 1,300,000 pounds in 1896 and jumped to 4,000,000 pounds in 1897.

Cost reductions incident to the perfecting of technique and to cheaper energy were reflected to some extent in the course of prices. A short price war, precipitated by the temporary appearance of the Cowles firm as a seller of pure aluminum, had reduced the price to about 50 cents a pound in 1891.[16] For the next two years it fluctuated in the neighborhood of 75 cents. In the succeeding years the quotation was several times reduced, until it came to rest in 1899 at 33 cents, where it remained until the boom beginning in 1905.

The principal developments of the first eight or nine years of the electrolytic aluminum industry were these. Technical problems incident to the perfection of the Hall process in industrial application had been solved, or were well on the way toward solution. The market of the small-scale aluminum industry existing when the Pittsburgh Reduction Company began production had been captured and, of more importance, new demands for the metal had been created by an intensive campaign of familiarization. A steady and rather rapid growth in production was enabled by the widening markets. With the move to cheaper power at Niagara Falls the company embarked on a policy of greatly enlarging capacity and lowering prices to develop markets for its bounding production. Evidently this was the beginning of a campaign to invade the markets of the common metals, iron, copper, brass, steel, and zinc, which had been touched but superficially heretofore.

In the following ten or fifteen years it was shown that the new metal could compete successfully with many of its predecessors. Aluminum invaded the field of electric transmission lines, and captured a goodly share of the demand from the youthful automobile industry. But the potential market in the general engineer-

[15] The New Kensington plant was thereafter devoted to rolling, drawing, and stamping. A few years later the manufacture of cooking utensils was begun at that location.

[16] Prices given here are for No. 1 aluminum ingots in ton lots or over. No. 1 ingot was at first guaranteed over 98 per cent pure. In the late nineties this was changed to over 99 per cent and a little later to 99.75 per cent. Prices of No. 2 ingot and of fabricated products usually changed with that of No. 1 ingot. Records of prices are to be found in *The Mineral Industry, Mineral Resources,* and the trade journals.

ing trades was not appreciably developed until much later. Two other important developments were soon to appear. In the middle nineties the company initiated a program of vertical integration which continued through the first decade of the new century. At the same time horizontal expansion contributed to the entrenchment of its position as the single domestic producer of virgin ingot. This two-dimensional growth will be described in the next chapter after we review the further development of markets which made it possible.

3. Transmission Lines and the Automobile

Writing in *The Mineral Industry* for 1909, Professor J. W. Richards remarked that

> aluminium seems finally to have attained a position among commercial metals where it is treated entirely on its merits. In the early days of the industry the claims for aluminium with regard to its noncorrosive qualities, lightness, and other distinguishing characteristics, were so exaggerated that it failed to measure up to expectations thus created. It was tried in many uses to which it was not suited, and a reaction occurred, so that the real merits which the metal possesses have been somewhat discounted for a number of years. This condition no longer exists and today aluminium is ranked among metals according to its real value.

The new metal had finally gained a foothold in competition with the common metals of the present industrial civilization in those uses for which its true qualities were well adapted. Its adolescent growth was the result of the determined marketing crusade of the Pittsburgh firm, accompanied by an alteration in its "real value" in quite a different sense from that meant by Professor Richards. Marked increases in the prices of other metals in the late nineties and again after 1904 afforded an opportunity for the aluminum producer to obtain a hearing by keeping its price down. The attention of metal users was caught, and they began to learn that aluminum was quite suitable for many employments.

Aluminum wire, which was among the fabricated articles produced by the Pittsburgh Reduction Company, apparently commanded little attention until the late nineties, when a rise in the price of copper facilitated its adoption for electric transmission cables. Although No. 1 ingot aluminum was sold for about 33 cents per pound between 1899 and 1904, the company inaugurated in 1898 a special price for aluminum conductors of 29 cents a pound, delivered at the point of consumption.[17] Since copper was then selling at 14 cents, this put the two metals just about on a par with regard to cost, as will be explained in a moment. Sales of aluminum wire aggregated 1,300,000 pounds in 1898.[18] The first important contract for installation was completed early in that year with the erection of a forty-six-mile three-phase line for the Standard Electric Company of California. Several large installations were projected or completed by the end of 1899.[19]

The outstanding advantage of aluminum transmission lines is their lightness, which permits the use of lighter towers and supports and diminishes the costs of construction. The same volume of aluminum weighs only 30 per cent as much as copper. For an aluminum transmission line of the same size as one of copper only three-tenths as many pounds would be needed. The electrical conductivity of hard-drawn aluminum wire is, however, only about 63 per cent of that of commercial copper wire, so a conductor of approximately 59 per cent greater area in a unit of length is required to give equal conductivity.[20] This means, of course, that the weight of an aluminum line of equal conductivity is a little less than half (about 48 per cent) that of a copper conductor.[21] Disregarding other factors for the moment, it is obvious that when aluminum is twice the price of copper per pound a slight advantage exists for the former.

During the early years of the present century the aluminum producer maintained a marked price-differential in its favor, with the result that the white metal made substantial inroads upon the transmission field. Unfortunately few statistics are available to show the extent of this invasion. Perrine stated in 1902 that aluminum had already passed through the experimental stage and was

[17] *Aluminum World,* IV, 81 (February 1898).

[18] MI, VII, 24 (1898).

[19] A list of these is given by J. B. C. Kershaw, MI, VIII, 23 (1899).

[20] The figures are taken from *Aluminum Industry,* II, 701–702; and R. J. Anderson, *The Metallurgy of Aluminium and Aluminium Alloys* (New York, 1925), pp. 286–287. They relate to present conditions. F. A. C. Perrine, in his *Conductors for Electrical Distribution,* first published in 1902, gives figures differing but little which he derived from tests on commercial aluminum.

[21] $159 \times 0.3 = 47.7$.

thoroughly established as a conductor material.[22] Out of sales totaling 6,426,000 pounds of aluminum in all forms in 1903 wire accounted for 2,385,000 pounds or 37 per cent.[23] By 1908 fifty-six transmission lines had been equipped with aluminum cable in the United States, a fact which resulted in the consumption of several thousand tons of the light metal.[24]

Whenever great strength was required aluminum conductor was still at a disadvantage. In 1908 William Hoopes, electrical engineer of the Aluminum Company, removed this handicap by inventing a new type of cable which consisted of aluminum strands spiraled around a core of high-grade galvanized steel. The whole of the weight advantage was not sacrificed to greater strength.[25] A steel-cored, six-stranded cable equal in conductivity to a pure copper line weighed only 80 per cent as much as the latter and possessed 57 per cent greater strength.[26] Meeting with a favorable reception as soon as it appeared upon the market, the reinforced cable was used increasingly in subsequent years. It was said in 1912 that about 20 per cent of annual ingot production, or roughly 7 to 8 million pounds, was being sold in the form of wire.[27] In this field aluminum had become a recognized member of the family of common metals before the World War.

To a lesser extent it was beginning to be substituted for tin, brass, and bronze, and for copper in uses other than electric cable. The figures of Table I show that price ratios were altered in favor of aluminum during the upward thrust of the business cycle.[28]

[22] *Op. cit.*, p. 14.

[23] *Aluminum Industry,* II, 6.

[24] J. B. C. Kershaw, *Electro-metallurgy* (London, 1908), p. 34.

[25] Obviously the proportionate weight of aluminum to copper for equivalent electrical conductivity for a given distance was not affected. The price ratio of 2–1 was lessened by the cost of the steel.

[26] MI, XXIII, 22 (1914). Among other advantages are the following. The steel-cored conductor generally gives smaller inductance than develops in copper because the current does not flow readily in the core. Current leakage into the air, or "corona," as it is called, which has become important with high-tension lines in high altitudes, is lessened and sometimes eliminated by the use of aluminum conductors because of the larger diameter.

[27] EMJ, XCIV, 529 (September 1912).

[28] The Pittsburgh Reduction Company had initiated a campaign of competition with brass in 1897 by lowering its prices on sheet and calling attention to the fact that, on a volume basis, aluminum sheet was cheaper. See advertisements in the *Aluminum World* during 1897 and the following years, and editorial, *ibid.*, IV, 7 (1897). A few years earlier the company had introduced a special casting alloy to compete with brass, which was offered at a "development price."

TABLE 1

AVERAGE YEARLY PRICES OF ALUMINUM AND PRICE RATIOS OF ALUMINUM AND OTHER NONFERROUS METALS IN THE UNITED STATES, 1900–1907

Year	Average Price of 99 Per Cent Aluminum Ingot (*Cents per Pound*)	PRICE RATIOS Aluminum to Copper	Aluminum to Tin	Aluminum to Zinc
1900	33	2.04	1.10	7.51
1901	33	2.05	1.24	8.11
1902	33	2.84	1.23	6.82
1903	33	2.50	1.17	6.36
1904	33	2.58	1.18	6.69
1905	35	2.24	1.11	6.11
1906	36	1.88	0.98	5.95
1907	42	2.10	1.10	7.23
Indifference ratios		3.30	2.70	2.60

A comparison of the relative prices per unit of volume is more significant with respect to most employments of aluminum and the other metals than comparison of prices per pound. The table includes "indifference ratios," i.e., the ratios of the prices per pound of aluminum and the other metals at which equal volumes of aluminum and each of the other metals would cost the same sum. The term "indifference" here applies only to equivalence of prices per unit of volume. Naturally each metal possesses certain superior qualities for some uses.

The aluminum prices are quotations of the Aluminum Company as reported in the *Mineral Industry* and *Mineral Resources of the United States.* Average yearly prices for the other metals have been taken from the *Yearbook of the American Bureau of Metal Statistics.* They are prices of electrolytic copper at New York, tin at New York, zinc at New York for 1900–1902, thereafter at St. Louis.

Aided by more adequate knowledge and less misinformation about the white metal, users could adapt it more intelligently to their purposes; and undoubtedly they contributed much in return to the general store of knowledge. It is impossible to estimate the extent to which aluminum invaded these markets.[29] Evidently the established metals did not suffer much, but aluminum had caught a foothold. In general, the uses in which it was beginning to be substituted for brass, zinc, tin, and iron were in machinery, where

[29] *The Mineral Industry* and trade journals glow with vague and uninforming enthusiasm.

light weight is important, particularly in reciprocating parts; in electrical apparatus; in vats, tanks, and vessels employed in fruit preserving, wine making, brewing, the manufacture of linseed oil, varnish, stearic acid, sugar refining, explosives, rubber, and several chemical products.[30] In the latter uses aluminum has repaid principal and interest upon the chemical research which gave it birth. Aluminum foil began to be used for wrapping food products.[31] The manufacture of vats, tanks, and pans, and the production of castings were encouraged by the development of successful welding of joints with the aid of the new oxyacetylene burner. (Aluminum had heretofore been denied many uses because of the very great difficulties of soldering or welding it.) Aluminum paints were introduced about 1900. Calorizing, or the formation of a surface alloy of aluminum on ferrous or nonferrous metals and alloys, was first developed in 1911. The calorizing of steel pipe, engine pistons, and other apparatus gives greater protection from heat and air, because the aluminum surface becomes covered with a thin film of aluminum oxide when exposed to the air. A violent explosive called "ammonal," consisting of ammonium nitrate and aluminum powder, was discovered in 1901 and was used in mining until the World War bestowed upon it a wider field of endeavor. Extruded shapes and tubing began to have fairly large consumption after 1910.

Before introducing the automobile there remains to be mentioned one new use for aluminum which immediately developed a high importance in the metal trades, even though it never took large amounts of aluminum. It had long been known that aluminum, because of its exceedingly great affinity for oxygen, could reduce many other metals from their oxides. Such reduction could only occur at extremely high temperatures and proceeded with explosive violence. Shortly before 1900 Dr. Hans Goldschmidt of Germany discovered that it was only necessary to start the reaction at one point of the mixture of aluminum and the oxide of an-

[30] Its advantage in chemical employments is freedom from corrosive action of foods and many chemicals. In general, aluminum is resistant to organic acids and neutral organic solutions, but it is attacked by most inorganic acids, alkaline media, or solutions of the salts of heavy metals.

[31] It was said that aluminum could be rolled so thin that eight times as much foil could be made from aluminum as from an equivalent weight of tin. See Tariff Hearings, 1908–1909, House Doc. no. 141, 60 Cong., 2 Sess., p. 2268.

other metal, since the heat evolved by reduction at that point was sufficient to raise the adjacent portions to the temperature of reaction, and the reduction proceeded cumulatively. Dr. Goldschmidt used barium peroxide to ignite the mixture. A temperature of about 3000° centigrade was attained in this process. One of the principal applications of the "thermit" process, as it was called, was in the production of pure chromium and manganese, which were found to improve the qualities of steel alloys. Vanadium, tungsten, molybdenum, and others of the rarer metals were also produced comparatively cheaply by this process. More recently the most important use of thermit reduction has been in the production of vanadium alloys. Since the early years of the century thermit — *der Hochofen in der Westentasche* — has been employed extensively for the welding of rails, propeller shafts, beams, and so on.

During the period covered by this chapter the older uses of aluminum continued to expand, and, as has been shown, many new employments were added. However, none of these markets possessed the significance for the industry which attended the adoption of the metal by the early "horseless carriage." It was the growth of large-scale production of automobiles which enabled the output of aluminum to expand so rapidly before the World War, and motor-car production absorbed nearly half of the output for many years thereafter. Furthermore, the automobile gave a strong stimulus to the study of aluminum alloys. Spurred by the added influence of aviation and the war, the development of particular alloys for specific purposes has become the dominant trend in the more recent history of this industry.

It was apparent to motor-car manufacturers at the start that the light weight of aluminum would be advantageous in vehicle construction. Since the low relative strength and the softness of the pure metal constituted a bar to this use, the Pittsburgh Reduction Company directed more attention to the production of aluminum alloys. A few copper and zinc alloys of aluminum were developed, possessing greater strength, more stiffness, hardness, and more elasticity than the pure metal.[32] With a price advantage over some

[32] The tensile strength of annealed commercial pure aluminum is about 14,000 lbs. per square inch. The early aluminum-copper binary and ternary alloys showed a tensile strength of about 15,000 to 20,000 lbs. per square inch if properly cast.

of the older metals, castings from these alloys immediately met with a hearty reception from motor-car builders which more than compensated for the lack of any real enthusiasm in the metal trades in general. Mentioning the automobile in 1900 for the first time, the *Mineral Industry* reported that in France the use of aluminum alloys for motor-car construction had already gone beyond the experimental stage. The same year marked the beginning of the use of substantial quantities of aluminum for this purpose in the United States.[33] The second automobile show in New York (1901) exhibited much aluminum in parts such as casings or housings, where no great strength was required. The demand for aluminum alloys for casting such parts grew *pari passu* with the increase in motor-car production.

In general the same parts for which aluminum was originally adopted by car manufacturers continued to be made of this metal for more than a decade. Typical parts were the crankcase, gearset housing, rear-axle housing cover, fan cowl, oil pan, bonnet sides, and various housings and caps throughout the assembly.[34] By 1914 about 80 per cent of the cars made in this country contained aluminum crankcases and gear cases. Lower raw material costs of iron and bronze castings, the lack of knowledge concerning the properties of aluminum alloys, and the unfamiliarity with its casting qualities were offset by its weight advantage and the ease of machining aluminum castings. About 90 per cent of the consumption of aluminum in automobiles took the form of castings. Some cars also adopted aluminum radiators and dash fittings, while alloy sheet and extruded moldings were employed to some extent in body construction. In 1915 it was estimated that at least one-quarter of the annual production of aluminum was consumed in the form of light, stiff alloys, most of which went into motor cars.[35] Table 2 indicates in a general way the relative importance of the automobile.

These were the alloys used for casting in 1902, according to the *Metal Industry,* I, 5 (January 1903). See also *ibid.,* IV, 12 (January 1906).

[33] *Ibid.,* VII, 9 (January 1909). For a description of the early introductions, see *Aluminum World,* VI, 130 (April 1900), and VIII, 177 (June 1902).

[34] J. E. Schipper, "Aluminum — a Feather Weight," *Automobile,* XXX, 673 ff. (March 1914).

[35] *Ibid.*

TABLE 2

PER CENT OF ALUMINUM SOLD IN DIFFERENT FORMS IN THE UNITED STATES IN CERTAIN YEARS

Form	1893	1903	1912
Ingots	93.0	29	35
Wire	0.5	37	20
Sheet	4.5	34	
Other finished forms			45

Percentages for 1893 and 1903 are calculated from sales figures of the company given in *Aluminum Industry,* II, 6. Percentages for 1912 are the result of "expert estimates," EMJ, XCIV, 529 (1912). For 1893 and 1903 sheet should be considered to include goods fabricated from sheet.

In 1893 the company was just entering upon fabrication. The extent to which it went into this and the large demand for wire are indicated by the 1903 figures. Part of the growth in alloy castings is evidenced by the substantial increase in ingot sales in the next eight years. The classification "other finished forms" undoubtedly includes cast auto parts as well as sheet and its products. By 1915 some automobile companies were using over a million pounds a year, while the requirements of a number of companies ranged in the hundred thousands. The airplane and airship had also entered the market and were helping to heighten the interest in light alloys, although their full influence did not make itself felt until the war years. Aluminum did not, however, penetrate farther into the automobile engine until scientific study had developed alloys which were capable of replacing the cast-iron piston.

> When the manufacture of motor cars commenced on a large scale, aluminium alloys could not be employed for parts that might be stressed in service because of the general non-uniformity and unreliability of such alloys in the form of castings . . . specifications, even though quite lax, could not be met with certainty, and the light aluminium alloys were considered, therefore, to be unsafe except for use as ornaments, trimmings, oil pans, and as covers for some housings in automobile work.[36]

Before the war only a very few alloys had any wide use. Generally an 8 to 12 per cent copper alloy was employed in automobile

[36] Anderson, *The Metallurgy of Aluminium* (Henry Carey Baird and Company, New York), p. 300.

work and for many other uses. Or, if cheaper cost were a necessity, or the requirements for strength were greater than for lightness and elasticity, the maker shifted to a standard zinc alloy. There had been almost no development of particular alloys for specific purposes. However, experience had bettered foundry practice. More important, research upon aluminum alloy systems had begun, as the following statement testifies.

> The automobile and the aeroplane have forced the aluminum and iron alloys to make rapid strides. . . . The physical chemist has started along the way of a systematic coördination of certain properties of binary and, in a few cases, of tertiary alloys.[37]

The adoption of the microscope and pyrometer by some metal research workers gave promise of rapid maturity through determination of the constitution of alloys. Without microscopy it would have been impossible to deduce, from the knowledge of constitution, the properties of various alloys under different conditions of heat treatment, quenching, and cold working, even though the importance of these operations was empirically known. The results of research were emphatically demonstrated by the introduction, about 1909, of an aluminum alloy comparing favorably with mild steel on a strength-weight basis. "Duralumin," as this alloy was called, followed the discovery by Alfred Wilm in Germany that heat treatment could materially increase the strength of wrought alloys. While Wilm ascertained much about the various possibilities in heat treatment and their consequences, the subject was not thoroughly understood until after the war.

Another pre-war achievement in the field of alloys was the development of die-casting by H. H. Doehler and others. Pressure-casting through dies produced finished shapes to within .005 of an inch of requirements, thus eliminating much of the machining which was necessary with sand or permanent mold castings. It was estimated that die-castings were being used in the automobile industry to the extent of 900 tons a year in 1915.[38] Lastly, the

[37] W. R. Whitney, of the research laboratory of the General Electric Company, in a paper printed in *Metal Industry,* IX, 294 (July 1911). Cf. C. A. Edwards and J. H. Andrew before the British Institute of Metals, abstracted in *Electrochemical and Metallurgical Industry,* VII, 493 (1909), and J. E. Schipper, *op. cit.,* p. 676.

[38] MI, XXIV, 19 (1915), reporting Charles Pack in the American Institute of Metals, October 1, 1915.

aluminum alloy piston which was to carry this metal into the motive part of automobile and airplane engines had made its bow in racing cars.

The advent of the motor car directed attention to the possibilities of increasing the strength, hardness, stiffness, and elasticity of aluminum by alloying it with other metals. A few alloys were developed which enabled the utilization of substantial amounts of aluminum in those parts of the automobile engine for which great strength was not required. The necessity of employing alloys of some sort rather than pure aluminum, and the possibilities of further widening the field of employment, stimulated the beginnings of research upon alloys, which was later to step into the major role in the development of the industry.

Under the influences described in this chapter production expanded rapidly in both America and Europe after 1903. Estimated output in America grew from 3,200 metric tons in 1900 to 11,800 tons in 1907 and to 27,400 tons in 1913. In Europe production was estimated at 4,100 tons in 1900, 11,800 tons in 1907, and 36,400 tons in 1913. Estimates of production by countries for the years 1890–1935 are given in Table 38. Chart I portrays the growth of output between 1900 and 1935.

CHAPTER II

INDUSTRIAL STRUCTURE AND MARKET CONTROL

1. The Aluminum Company of America

The growth in demand which has just been described was accompanied by change in industrial structure and the continuance of monopoly. It has been explained that the Pittsburgh Reduction Company, finding no market for ingot aluminum, had early been forced to roll sheet and fabricate sundry articles in order to familiarize the metal trades and consumers with the various uses of the metal. The company has since continued to carry much of its product through the semi-finished stage — sheets, bars, rods, tubes, and so on — and to make an ever-growing number of finished products. About 1901, when the United States Aluminum Company was incorporated as the principal fabricating subsidiary, the production of stamped cooking utensils was undertaken. In the sale of this ware the Aluminum Cooking Utensil Company, a selling organization formed at the same time, carried integration all the way to the ultimate consumer by inaugurating a scheme of selling direct to housewives which it has employed ever since.[1]

The next move toward further integration was in the direction of raw material. In the first few years alumina had been imported from Germany. As it happened, deposits of bauxite were discovered in Georgia and Alabama at just about the time when Hall was producing his first aluminum. For thirty years or so after 1891 the alumina used in aluminum reduction in this country was derived chiefly from American bauxite. The Pittsburgh Reduction Company began to acquire bauxite deposits in 1894, when the Georgia Bauxite Company, which had been created for this purpose, secured several leases in the state for which it was named. For two

[1] Federal Trade Commission, *In the Matter of the Aluminum Company of America*, Docket 1335, Record, p. 5233. This docket comprises complaint, answer, testimony, exhibits, and briefs in a case before the Commission involving alleged unfair methods of competition in the sand-castings and cooking-utensil industries. It will hereafter be referred to as FTC Docket 1335.

or three years the bauxite was converted into alumina by the Pennsylvania Salt Company, an independent chemical concern. When the first deposits in Georgia and Alabama showed signs of approaching exhaustion the aluminum company transferred its interests to the hitherto undeveloped Arkansas field.[2] In the early years of the present century the company secured the large holdings in this field which supplied the major part of its ore until after the war.[3]

The gap between mining and aluminum reduction was closed in 1902 by the construction of a six-acre plant at East St. Louis for refining bauxite to pure alumina, and sometime thereafter a coal mine was added to the company's property. In succeeding years two short railroads were built by the company to afford trunk-line connections for the Arkansas mines and the alumina plant. A loading station on the Mississippi provided transshipment facilities for carriage of ore to East St. Louis by river boats. For some time the company had manufactured a portion of its carbon electrodes at Niagara Falls, employing an electric-furnace process patented by Mr. Hall. About 1905 the old carbon plant was torn down and rebuilt as one of the largest electrode factories in the world. Subsequently the company has made all its own carbons.

It has been explained that electric power is one of the chief items of expense in the manufacture of aluminum. About 1900 the company inaugurated a policy, which has continued to the present day, of acquiring water-power sites and developing its own energy. The first development took place at Shawinigan Falls on the St. Maurice River in Canada, where a powerhouse was built near that of the Shawinigan Falls Water and Power Company, and a reduction plant was erected to use 5,000 h.p. These facilities were operated by a wholly owned subsidiary, the Northern Aluminum Company of Canada. When a third reduction plant was constructed at

[2] Deposits had been discovered in Arkansas nearly ten years before, but had not been used to any extent. The failure of the railroads to make rates which reflected the market competition with the Georgia-Alabama field was partly responsible. See *Aluminum World*, VI, 46 (February 1900).

[3] During the few years which intervened between the slackening of bauxite production in the old field and the development of the Arkansas deposits French bauxite was used to meet increased demands. Fortunately, ocean freights were unusually cheap at this time. In fact, French bauxite could be laid down, duty ($1 per ton) paid, in cities of the East at less than the prices at which southern domestic ore could be delivered. See MI, x, 13 (1900).

Niagara Falls about 1906 the Pittsburgh company built a power plant located so that its generators were turned by the turbines of the Niagara Falls Hydraulic Power and Manufacturing Company.

In 1903 a ten-acre reduction works and electrode plant were completed at Massena, New York, and a wire mill was finished a few years later. Energy was taken originally under a long-time contract with the St. Lawrence River Power Company, which had created an artificial fall on the Grasse River by diverting water from the near-by St. Lawrence through a canal three and a half miles long. Three years later the power company, which was experiencing financial difficulties, sold its entire property, including power plant, canal, and water rights, to the aluminum enterprise.[4] Shortly thereafter the capacity of the power plant was doubled. Several other electric companies in this district were subsequently taken over to assure an unfluctuating supply of energy during the ice season.[5]

With the purchase of the St. Lawrence River Power Company the aluminum firm embarked upon an ambitious power program in this region. Extensive riparian rights were purchased along the Long Sault section of the St. Lawrence, which is adjacent to Massena, and application was made to the various interested governments for permission to dam the river to develop 800,000 h.p.[6] The New York state legislature granted its permission, but after some delay the Canadian government and the United States Congress refused permission. Balked in this direction, the Aluminum Company turned its attention to the Little Tennessee River.[7] With the purchase of the Knoxville Power Company and individually owned tracts, as well as the Union Development Company and the Tallassee Power Company, it became the owner of nearly all power rights and privileges along a forty-mile stretch of the river

[4] The power facilities were purchased by a subsidiary created to administer them, which exchanged $1,450,000 collateral trust bonds for the entire stock of the St. Lawrence River Power Company. See *Moody's Manual,* 1907–1911.

[5] Most of these companies were small, the principal one being the Hannawa Falls Water Power Company.

[6] See the testimony of an officer of the company before the Ways and Means Committee in Tariff Hearings, 1912–1913, 62 Cong., 2 sess., House Doc. no. 1447, II, 1499. The Long Sault Development Company had been formed to carry out that particular part of the St. Lawrence project.

[7] On January 1, 1907, the name of the corporation was changed to the Aluminum Company of America. No change in organization was involved.

in the Great Smoky Mountains.[8] Plans were announced for the building of a series of dams to develop 400,000 h.p.

The company also increased its interest in the castings branch of the industry by the ownership of 50 per cent of the stock of the Aluminum Castings Company, formed in 1909 as a consolidation of several foundry concerns, and the foundry department of the United States Aluminum Company. At about this time a plant was established at Dover, New Jersey, for the manufacture of aluminum bronze powder.[9]

Thus by 1910 the Aluminum Company had become a highly integrated concern. The ore produced at its mines was run through a crushing, grinding, and drying plant, and then sent to East St. Louis, where it was converted into aluminum oxide for the reduction furnaces at Niagara Falls, Massena, and Shawinigan Falls. Most of the electricity fed into the reduction cells was generated by the company, which also owned a substantial part of the rights to the water power which turned its dynamos. A campaign to develop water power which would, by comparison, dwarf the output of these holdings was being aggressively pushed. Carbon anodes and furnace linings were manufactured at Niagara and Massena. A substantial portion of the aluminum run into ingots at the reduction plants was later rolled into sheets or rods at Niagara and New Kensington; fabricated into tubes, shapes, cooking utensils, or other manufactured articles at the latter place; or emerged in the form of wire from the drawing plants at Niagara, Massena, and New Kensington. A large part of the ingot went to foundries of the Castings Company situated in the East and Middle West, and a small portion to Dover for the manufacture of aluminum bronze powder. Undoubtedly a substantial reduction in cost resulted from this vertical control, which brought cheaper power, a better quantitative and qualitative adjustment of materials between the various stages, and closer touch with markets.

About 1905 the Aluminum Company embarked upon an extensive program of horizontal expansion at several stages. Some parts of this extension have already been mentioned in connection with

[8] *Moody's Manual,* 1912, p. 2818.

[9] The manufacture of aluminum bronze powder was moved to New Kensington in 1913 with the erection of one of the most complete plants in the world for this product.

integration. The increase in capital investment from $6,000,000 in 1905 to $30,000,000 at the beginning of 1913, and to $43,000,-000 two years later, affords some indication of the extent of horizontal expansion, because the major part of the original extensions of vertical control over the ore, alumina, and power stages, and some branches of the semi-fabricating and finishing stages, had occurred before 1905.[10] One consequence of the rapid expansion was that entry into the industry was rendered more difficult and less attractive. Acquisition of a large part of the domestic deposits of bauxite suitable for aluminum reduction and restrictive agreements with firms receiving bauxite from the Aluminum Company increased the obstacles to entry; while rapid extension of operating capacity, combined with acquisition of enormous reserves of undeveloped power, seemed to leave little room for fresh capital and enterprise.[11] The company had acquired many ore properties in the southern field and in Arkansas prior to 1905. By 1909 purchase of the mining subsidiaries of two chemical firms brought the bauxite holdings of the Aluminum Company up to a large proportion of the domestic ore. A consent decree in 1912 did not disturb ownership of the deposits.

In 1906 and 1907 power and reduction facilities were hardly sufficient to satisfy the booming demand. By the beginning of 1908 the completion of plant which had been building for two or three years increased reduction capacity to nearly three times that of 1906! It was not until 1912 that the company was again selling as much metal as it could produce. Between 1906 and 1912 power rights at the Long Sault and on the Little Tennessee had been acquired. When the bounding sales of 1911 and 1912 seemed to forecast a sustained upward movement of demand, the company began its development in the South with the erection of a million-dollar reduction plant at a company town called Alcoa, built near Maryville, Tennessee. A large increase in the Massena works was in progress when the Alcoa cells went into operation in 1914 under purchased energy to be used during construction of the first unit

[10] Figures of capital investment for certain years were given by an officer of the company (Tariff Hearings, 1912–1913, pp. 1493 ff.). The figure for 1905 appeared in a balance sheet in *Moody's Manual,* 1906, p. 1891.

[11] Analysis of the factors contributing to continuance of monopoly is contained in Chap. V. Footnote references appearing there are omitted here.

of the Cheoah hydroelectric plant. Electrode works, rolling mills, and fabricating facilities were greatly extended. For most of the time between 1908 and the latter part of 1915, when the abnormal war demands began to be felt, the company had in operation, or under construction, capacity sufficient to satisfy the market at prices considerably lower than it cared to charge. And in the background loomed a potential power development of staggering proportions.

The Bradley patent expired early in 1909 in the middle of a business depression. By the time that new enterprise might have been expected to make its bid a vigorous policy of expansion (which included the promise of extensive development in Tennessee and perhaps at the Long Sault), combined with integration and ownership of a large part of the domestic ore suitable for aluminum, must have conferred an impressive formidability upon this corporation, information about the earnings and operations of which was exceedingly meager.

Apparently the only determined attempt to enter this industry in the United States before the war — and the only new venture which has ever reached the stage of building plant — was made by a group of experienced French aluminum producers who possessed their own bauxite. When the outbreak of the war prevented further financing in Europe to complete their partially constructed power plant and reduction works in North Carolina, no American bankers could be found to supply the necessary capital. The stockholders sold out to the Aluminum Company of America, which appeared to be the only potential buyer.

The rapid expansion of the Aluminum Company was financed almost altogether out of earnings. While actual income figures for this period were not published, calculations of earnings from the growth of capital by reinvestment indicate that the company had an unusually prosperous record from 1905 to 1912. Large profits were facilitated by tariff protection.[12] Commercial production was

[12] In 1890 a duty of 15 cents per pound was set on ingot aluminum. The rate became 10 cents in 1894, 8 cents in 1897, 7 cents in 1909, and 2 cents in 1913. The duty on aluminum manufactures was 35 per cent ad valorem in 1894. In 1897 it was raised to 45 per cent, where it remained until 1913, when rates of 25 per cent, 15 per cent, and 20 per cent were set for utensils, wire, and other manufactures respectively. Duties on plates, sheets, bars, etc., were 13 cents per pound in 1897, 11 cents in 1909, and 3½ cents in 1913.

started with a capital stock of $1,000,000, part of which represented cash investment, the rest having been exchanged for patents. An officer stated that at the end of 1912 the capital and surplus was just over $30,000,000.[13] Of this increase, only $1,000,000 was secured by sale of stock, making the total of capital contributed in this way but $2,000,000.[14] Evidently during the twenty-four years, 1889–1912, there was reinvested out of earnings $28,000,000.[15] In addition, small cash dividends totaling roughly $5,000,000 were paid out to stockholders, bringing the aggregate net earnings for the period to something like $33,000,000.

The company published a balance sheet in 1905, and no securities were sold during the rest of the period under consideration. Hence the constant rate of return which would, by reinvestment, bring the capital to certain figures at later dates can be computed. Total assets on August 31, 1905, were stated to be about $6,300,-000.[16] The assumption that investment at the beginning of 1905 was $6,000,000 is probably an overstatement, since this was a good year. By the beginning of 1909 capital had increased to $21,000,-000 at least, and at the end of 1912 it stood at $30,000,000.[17] The increase during the four years 1905–1908 would have been accomplished by earnings at the rate of 37.5 per cent each year, all reinvested. The growth of investment during 1905–1912 indicates a constant rate of 22.3 per cent. The assumption that capital grew at a constant rate during this eight-year period is, of course, not valid, as is obvious from the fact that the rate was higher during 1905–1908. An officer of the company stated that earnings for the three or four years ending with 1912, at least half of which were depression years, amounted to 15-17 per cent per annum.[18] If the company's figures are correct it would appear that the rate of earnings in the eight years ending 1912 averaged somewhere in the neighborhood of 26 per cent, exclusive of the small amounts paid

[13] The following figures for capital investment and earnings were given by an officer (Tariff Hearings, 1912–1913, p. 1493).

[14] $600,000 of preferred stock was sold for cash in 1899 but retired in 1909, presumably out of earnings. No bonds were issued until the post-war period.

[15] According to testimony of an officer all but $2,000,000 of the $30,000,000 of capital and surplus in 1912 represented reinvested earnings (Tariff Hearings, *loc. cit.*). Apparently there were no revaluations of assets upward during this period.

[16] *Moody's Manual,* 1907, p. 1858.

[17] Tariff Hearings, *loc. cit.*

[18] *Ibid.*

out in dividends. In 1913–1915 the rate of earnings apparently averaged about 20 per cent.[19] If the assets had grown only through reinvestment of earnings during the first seventeen years (1889–1905) of the life of the company, a constant rate of return of about 12 per cent would be indicated. As a matter of fact, $1,600,000 was invested by purchase of securities during this period, so that the earnings probably averaged somewhat less.

2. The Four European Producers

Introduction of the European producers affords a convenient place for a brief discussion of the conditions of economical aluminum production. The important materials are electric power, bauxite, coal, and caustic soda (used in the preparation of alumina), and carbon electrodes. One metric ton of aluminum requires three to four h.p. years of electric energy, four to five tons of bauxite, four to five tons of coal, a substantial amount of water, about one ton of caustic soda, and 0.5 to 0.6 of a ton of electrodes — altogether, over ten tons of materials, plus a large quantity of power.[20] In the absence of large differences in labor and capital costs between regions, it is obvious that aluminum could be produced at the lowest expense where cheap water power, good bauxite, and cheap coal exist in fairly close proximity. There is no place in the world, however, where these three are localized within a fairly small region.[21] The United States possesses all three, but they are separated by rather long distances. The same appears to be true of Russia. In the southeast of France cheap water power is available close to extensive deposits of the best bauxite in the world.[22] Coal, however, is dear in France. Before the war and perhaps later, the French producers were able to secure sufficient amounts of the cheaper lignite existing in that region to make some

[19] See Appendix C.

[20] These figures refer to present requirements, which are somewhat lower than those of former years. Four to 5 tons of bauxite give 2 tons of alumina, which in turn yield 1 ton of aluminum. If brown coal is used in the making of alumina, consumption is at the rate of 10 to 12 tons per ton of aluminum.

[21] A good discussion of the relative advantages of various countries is given by R. Pitaval, JFE, XXXII, 88 ff. (July 1923).

[22] It has been estimated that these deposits contain 60,000,000 tons of excellent ore. See Schoenebeck, *Das Aluminiumzollproblem* (Berlin, 1929), p. 24.

of their alumina without dear coal. Switzerland and Norway have cheap power only; while England and Germany have cheap coal, but little inexpensive water power, and almost no bauxite suitable for aluminum. Italy has some good bauxite, most of which belonged to Austria before the war. Austria has water power, while several central European and Balkan countries possess good ore. Canada has cheap water power. With the exception of France and Russia there is no country of Europe within which the whole process of aluminum extraction can be carried out as cheaply as is possible when the various operations are performed in different countries. Outside of France the cheapest water power which is well located with respect to the other materials and the markets of aluminum probably exists in Switzerland and Norway, while Germany and England have the cheapest and best-located coal. Reduction plants must, of course, be built close to the source of power. It appears to be most economical to locate alumina works nearer to coal than to bauxite. In the absence of market restrictions one would expect that France, Switzerland, and Norway would become the largest European producers of aluminum, using French bauxite prepared in Germany or England.

In describing the growth of the European companies it would be desirable to present capacity figures in tons per annum — that is, the capacity of the existing reduction works when operating with the typical energy load. In the absence of such data one must fall back upon estimates of installed horsepower, although these figures do not afford a very good indication of the amount of aluminum which can be produced. The main reason for this is not that some of the figures are estimates — the companies have not been so chary with figures of installed energy as with other data. It proceeds rather from two other facts. Owing largely to climatic conditions, the maximum power which can be generated varies, often greatly, from season to season. Secondly, most of the European companies use a changing proportion of their total energy for the production of goods other than aluminum. In spite of these difficulties, however, the figures given here and in subsequent chapters do indicate very roughly the trends of development, for all of the important companies normally use more than half their energy for aluminum. It should be understood, however, that unless otherwise specified the figures refer to total installed energy, rather than to the average amount of power which can be generated per unit of time, or the amount which is intended for aluminum reduction.

The rapid development of the aluminum industry in America, which has been described, was paralleled abroad, where four companies established to work the original patents continued to be the most important producers until after the war. As in America, these firms were occupied with the perfecting of the industrial application of the process, integration back to the ore, development of markets, and expansion of capacity. The markets for the new metal developed in rather similar fashion on both sides of the Atlantic.[23] It seems to have found a wider employment in soldiers' equipment in Europe and much less extensive use in the form of electric cable than in America. For a few years after 1895 the aluminum-copper price differential in Europe was favorable to the lighter metal. Again in 1907 aluminum was benefited by an unusually high price for copper, which resulted in the first substantial use of aluminum by the electric industry. Very low aluminum prices during the ensuing depression helped to broaden markets. The total output of the European firms was a little larger than that of the American company until the World War.[24]

The Aluminium Industrie A. G. of Neuhausen, Switzerland, founded by the Schweizerische Metallurgische Gesellschaft and the AEG, held the leading position among the European producers for most of the period surveyed in this chapter. The water-power plant at Rheinfall in Neuhausen was enlarged to about 4,000 h.p. in the nineties.[25] Just before the turn of the century two new power

[23] Cf. with Chap. I the description of market development in Europe given by Adolphe Minet, *The Production of Aluminium and Its Industrial Use* (New York, 1905), pp. 191–215; Alfred Gautschi, *Die Aluminiumindustrie* (Zürich, 1925), pp. 21–40; Wilfried Kossmann, *Über die Wirtschaftliche Entwicklung der Aluminiumindustrie* (Frankfurt, 1911), pp. 71–77; C. Dux, *Die Aluminium-Industrie-Aktiengesellschaft Neuhausen und Ihre Konkurrenz Gesellschaften* (Lucerne, 1912), pp. 9–11.

[24] Table 38 shows estimated production of aluminum in Europe and America. The table does not reflect relative consumption after 1908, because several thousand tons a year were exported from Europe to the United States during the years 1909 to 1913.

[25] Most of the information about this company presented here has been obtained from the works of Schulthess, Gautschi, Kossmann, Dux, the *Mineral Industry*, annual reports of the company, and an interview with an officer of the concern. The annual reports for 1901–1910 are reprinted in Dux, *Die Aluminium-Industrie-Aktiengesellschaft Neuhausen*.

developments were made at Rheinfelden in Baden and Lend in Austria. With the addition of a power plant at Rauris, near Lend, and completion of the developments at the other two sites, this firm had a total of about 24,000 h.p. in 1903.[26] In 1906 the company reported that, as a result of acquisition of bauxite deposits in southern France, it possessed ore reserves sufficient for many years.[27] Before 1900 an interest had been secured in the Goldschmieden alumina works in Silesia, Germany. Subsequently this firm was taken over entirely, and another plant, located at Trotha, Germany, was purchased. About 1907 a third alumina plant was built in Marseilles near the ore. The AIAG began early to make other products in addition to aluminum. To the production of calcium carbide, undertaken in the nineties, there was added electric steel (produced under Héroult patents) a few years later, and nitric acid and nitrogen products about 1910. The company was exceedingly profitable. While distributing average dividends of 10 per cent on paid-in share capital in the nineties and 17 per cent in the next decade, it reinvested substantial amounts in expanding its facilities.[28]

For a few years the Société Électrométallurgique Française, founded at Froges in 1888, remained the only producer of aluminum in France.[29] When successful commercial production was assured, this company erected a power plant at La Praz on the river Arc, which was enlarged to about 13,000 h.p. by 1900. Extensive ore deposits were purchased in the Department of Var. Alumina was produced at Gardannes (Bouches du Rhône), and carbon electrodes were made at Froges and La Praz. In the early nineties Minet had produced aluminum on a semicommercial scale in a plant at St. Michel-de-Maurienne belonging to Bernard Frères of Paris. About 1894 the Société Industrielle de l'Aluminium was formed to operate the Hall process in this plant, which was enlarged for commercial production. M. Péchiney of the Compagnie

[26] Annual report, 1903.

[27] Annual report, 1906.

[28] See Dux, *passim,* and the annual reports.

[29] Information concerning the development of the French industry has been obtained chiefly from the following sources: Minet, *op. cit.,* pp. 81, 119–122, 139–140; MI, *passim;* MR for 1885; *Aluminum Industry,* I, chap. III; Kossmann, *op. cit.;* Rudolph Debar, *Die Aluminiumindustrie* (Brunswick, 1925); and Jean Escard, *L'Aluminium dans l'industrie* (Paris, 1925).

de Produits Chimiques d'Alais et de la Camargue — the firm which had produced aluminum by Deville's chemical process for thirty years prior to the discoveries of Hall and Héroult — was finally convinced that electrolytic reduction of pure aluminum would be commercially successful. Accordingly, about 1896 the Compagnie Alais took over the St. Michel enterprise. As its name

CHART I

Estimated World Production of Primary Aluminum, 1900–1935.

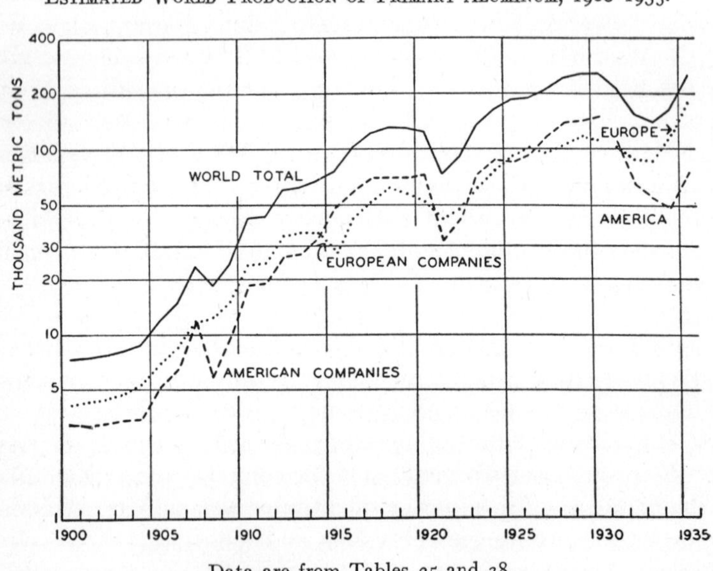

Data are from Tables 25 and 38.

implies, the Alais firm made various chemicals, among which were several electrochemical products, and was equipped to produce alumina at its plant at Salindres. It also acquired bauxite reserves in the departments of Var and Hérault, where the French deposits are largely localized.[30] These three leading aluminum producers of Europe all utilized part of their power for the production of commodities other than aluminum — Froges made calcium carbide, electric steel, and ferro-alloys — shifting the proportions of energy used for the various products in accordance with changes in

[30] This company also owned arable lands and vineyards from which it was said to enjoy a large revenue.

their market conditions. It appears that the two French concerns were quite profitable, although they did not equal the enviable record of the AIAG.[31]

In England the electrolytic process was not used industrially until after 1896, when the British Aluminium Company began operations in a 3,500 h.p. plant at Foyers, Scotland, under Héroult patents obtained from the Schweizerische Metallurgische Gesellschaft.[32] This corporation remained for several years much smaller and much less successful financially than its continental associates. It was, however, more highly integrated from the start. Irish bauxite was first acquired. When this proved to be of poor quality, the British Aluminium Company joined the owners of French ore lands. Low rates charged by English coal ships for back loads from Mediterranean ports helped to make the use of French bauxite economical. An alumina works was established at Larne, Ireland, and an electrode plant at Greenoch. Castings, sheet, rod, and tubing were made in plants of this company located in Milton and in Scotland. Unlike the companies on the other side of the Channel, this firm did not go extensively into the production of other commodities than aluminum. Five or six years after its birth, falling metal prices and increasing competition from France and Switzerland forced default on the debentures which had been used to finance a part of the heavy investment in Scotch water power and Irish bauxite. After reorganization the company made small profits under the helpful influence of an international cartel. It succumbed again during the depression of 1908–1909 and was reorganized a second time in 1910.

The first international aluminum cartel was formed by the four European producers during the business recession of 1901, just before the Héroult patents were to expire. Under the leadership of Neuhausen, whose annual output was double that of the combined total of the two French firms, an agreement was reached by which home markets were reserved to domestic producers, and the competitive market (chiefly Germany) was divided in stipulated proportions.[33] A minimum price for sales at home and abroad was

[31] See balance sheets and profit accounts given in Kossmann, *op. cit.*

[32] Details of the English aluminum industry have been taken from *The Mineral Industry, Aluminum Industry,* I, and the works of Schulthess, Gautschi, and Kossmann already cited.

[33] Gautschi, p. 46; Kossmann, pp. 111–112.

set from time to time. The Northern Aluminum Company, Canadian subsidiary of the Pittsburgh Reduction Company, became a member of the association.[34] The cartel at once increased the European price considerably and raised it higher during the marked business upswing of 1904–1907.

Coincident with the formation of the cartel the Froges and Alais concerns began the construction of new power plants (at La Saussaz and Calypso respectively) which brought their joint production up to that of Neuhausen by 1906.[35] As profits increased in 1905 and 1906 the cartel members embarked upon extensive programs of expansion. The Neuhausen firm began the development of Alpine water power near Chippis in Canton Wallis, which has since remained the chief center of this company's activities. In 1905 installed capacity in the plants at Neuhausen, Rheinfelden, and Lend amounted to about 24,000 h.p. Thirty-two thousand h.p. was added by a plant on the Navizance completed in 1908; and the near-by developments on the Rhône and the Borgne, finished in 1911 and 1913, enlarged capacity by 52,000 h.p. and 35,000 h.p. respectively.

In France the Compagnie Alais, which operated about 14,000 h.p. at Calypso and St. Félix, constructed a 20,000 h.p. plant at St. Jean-de-Maurienne, which delivered an initial 12,000 h.p. to the reduction cells in 1907. Development of 35,000 h.p. at L'Argentière was initiated in 1907 by the Froges concern, which then possessed 20,000-30,000 h.p. at La Praz and La Saussaz. The Argentière powerhouse began to operate three years later. About 1905 the British Aluminium Company, which was then using 5,000 h.p. at Foyers, undertook its first expansion. A 20,000 h.p. plant at

[34] Testimony of an officer of the Aluminum Company in the case of *Baush Machine Tool Company, Plaintiff-Appellant* v. *Aluminum Company of America, Defendant-Appellee,* U. S. Circuit Court of Appeals, Second Circuit, Record, fol. 680. The circuit court remanded this case to the district court for a second trial. When the second trial resulted in a verdict for the plaintiff the Aluminum Company appealed the case and gained a reversal in the circuit court. The record of the first trial will hereafter be cited as BMTC appellant v. ACOA and that of the second trial as BMTC v. ACOA appellant. Much, but not all, of the two records is nearly identical. Together they represent the most extensive history of this industry which has ever been gathered together in one place. The issues in this case are explained below, p. 481.

[35] The Alais concern also acquired about 1902 a small plant at St. Félix, which was apparently abandoned some years later. Evidently production of aluminum at the St. Michel works was discontinued when the Calypso plant began to operate.

Kinlochleven, Scotland, was finished in 1907.[36] In the preceding year the British company had purchased a partially developed water power at Stangfjord in Norway. A power plant of about 5,000 h.p. and reduction works were completed at this location in 1908. The cartel members possessed altogether about 65,000 to 70,000 installed h.p. in 1905. By 1908 this had been about doubled, and at the outbreak of the World War the total capacity of these concerns (including 14,000 h.p. obtained by the British Aluminium Company through purchase of an independent Norwegian firm) had risen to a little over 300,000 h.p. A substantial but varying proportion of this was, however, not used for aluminum. Total aluminum reduction capacity in Europe was between 40,000 and 50,000 tons per annum in 1914, or somewhat larger than capacity in America.

The development attained by the aluminum industry in 1914 is indicated in Table 3. In consulting the table it should be borne in mind that the continental companies used a considerable amount of power in the production of other electrochemical products.

The rapid expansion of demand for the new metal which began in 1904 and the profitable record of the four established companies attracted several new enterprises. Of the seven new firms which constructed facilities between 1906 and 1910, three were bought by the old companies. The four which continued independent remained quite small. While most of these new firms were relatively small, their aggregate capacity, completed or in prospect in 1908, represented about half of the total capacity of cartel members. Hence their impact upon the price structure during the depression, which began at the end of 1907, was heavy enough to jeopardize market control. With demand for aluminum greatly reduced, the cartel attempted to survive by a price reduction of about a third early in 1908. When the outsiders continued to undersell and the American market loomed more and more attractive, because the Aluminum Company apparently intended to stabilize at the 33 cent price of 1900–1904, all agreements were dissolved at the end

[36] When this company also purchased power rights in Switzerland at Orsières, the Aluminium Industrie A. G. objected that the contemplated production there would violate the contract of 1895, by which the British had agreed, in exchange for the patent and information upon industrial operation of the process at Neuhausen, to restrict the use of this to the United Kingdom and its colonies. An exception must have been made for Norway, however.

of September.[37] Price in Europe thereupon fell to less than half of the 1907 quotation, while foreign metal began to pour into New York in larger and larger quantities. One result of the increased investment in Europe and the breakdown of the cartel control of price was a tremendous expansion in the use of the new metal. The prices of 1902–1907 did not return with resumption of prosperity and reconstitution of market control. Continued exportation to America, until war broke out, of a substantial portion of the European production was another consequence of the large increase in investment in Europe.

With improvement in business conditions the cartel was again set up in 1912. A year earlier the French had formed a domestic sales syndicate called l'Aluminium Français, which included two of the new firms.[38] L'Aluminium Français has represented the French concerns in all marketing operations ever since. The new international cartel agreement was disrupted by the war, during which the AIAG supplied the Central Powers with metal. There is no evidence of an agreement between the Europeans and the Aluminum Company of America governing importation into the United States. The latter consented in 1912 to a decree enjoining it from entrance into any agreement curtailing importation of aluminum into the United States or affecting the domestic price.

During the years 1909–1912 the price in the United States was affected appreciably by imports from Europe. The apparent purpose of the Aluminum Company to stabilize price, after the crisis of 1907, at the 33-cent level of 1900–1904 was defeated by the low prices quoted on imported metal. Until the Underwood tariff reduced the duty from 7 cents to 2 cents, the foreign "control" of price in the United States was, of course, control only by suffer-

[37] At about the same time an agreement was made between the Canadian subsidiary of the Aluminum Company and the Neuhausen firm by the terms of which sales in the United States were reserved to the Aluminum Company of America, sales to the governments of Switzerland, Germany, and Austria-Hungary were reserved to the AIAG, and the total sales of the Canadian and Swiss companies in all other markets were divided according to stipulated terms. It has recently been testified that this agreement was canceled by the Northern Aluminum Company in the summer of 1911 (BMTC v. ACOA appellant, fol. 6411). The consent decree of 1912 contained a clause annulling certain portions of the agreement and enjoining similar activities in the future (see Appendix D). The agreement is reprinted as Exhibit 18, BMTC v. ACOA, appellant.

[38] The Metallgesellschaft of Frankfurt-am-Main, which became the German representative of l'Aluminium Français, took part in the organization of this syndicate.

TABLE 3

ESTIMATED CAPACITIES OF ALUMINUM PRODUCERS OF THE WORLD IN 1914 *

Company	Plants	Initial Year of Aluminum Production	Installed Horsepower in 1914 (*Thousands*)	Reduction Capacity (*Thousands of Metric Tons*)
AIAG				
	Neuhausen	1889	4.8	
	Rheinfelden	1897	6.0	
	Lend-Rauris	1898	15.0	
	Chippis			
	Navizance	1908	32.0	
	Rhône	1911	52.0	
	Borgne	1913	35.0	
Total			144.8	
Gebrüder Giulini				
	Martigny	1910	3.0	
Total for Swiss companies			147.8	15–20
Société Froges				
	La Praz	1894	13.0	
	La Saussaz	1903	17.0	
	L'Argentière	1910	35.0	
Total			65.0	

* The information upon plants and installed horsepower of the European companies has been taken chiefly from Schulthess and Gautschi. The latter prints somewhat the same type of table, pp. 18–20. Figures given by these two have been checked as far as possible in other sources and revised when the weight of testimony was against them. Estimates of capacities of European reduction works have been made from consideration of the estimates of others, particularly Bannert, Barut, Escard, and *Echo des Mines* (reported in the *Economist,* LXXXIV, 775, 1917), of maximum production during 1912–1914, and of other relevant information. Estimates of installed horsepower and reduction capacity of American works have been made from study of estimates published in trade journals and annual reviews and of probable power requirements in years when plants were worked to capacity.

TABLE 3 — *Continued*

Company	Plants	Initial Year of Aluminum Production	Installed Horsepower in 1914 (*Thousands*)		Reduction Capacity (*Thousands of Metric Tons*)
Compagnie Alais					
	St. Félix	1902	2.0		
	Calypso	1905	16.0		
	St. Jean de Maurienne	1907	20.0		
	Auzat	1908	12.0 †		
	Chedde	1906	13.0 †		
Total			63.0		
Société d'Électrochimie					
	Prémont	1906	10.0		
Total for French companies				138.0	15–18
British Aluminium Company					
	Foyers	1896	6.0		
	Kinlochleven	1907	20.0		
	Stangfjord	1908	5.0		
	Vigelands	1909	14.0		
Total			45.0		
Aluminium Corporation					
	Dolgarrog	1908	7.0		
Total for British companies				52.0	9–11
Societa Italiana per la Fabricazione dell' Alluminio					
	Bussi	1907	5.0	5.0	1
Total for Europe				342.8	40–50
Aluminum Company of America					
	Niagara Falls	1895	50.0		
	Massena	1903	55.0		
	Alcoa	1914	20.0		
	Shawinigan Falls	1901	40.0		
Total			165.0	165.0 ‡	35
Total for World				507.8	75–85

† The Compagnie Alais acquired the works at Auzat and Chedde from outsiders in 1914 and 1916 respectively.

‡ By 1917 American capacity had doubled as a result of expansion of facilities in the United States, while that of the leading European firms had increased but little.

ance of the Aluminum Company, which was making liberal depression profits.[39] Evidently the American company preferred to take the chance that the loss of some of its market would be temporary, and to use the foreign invasion as a plea for retention of tariff protection.[40] Reluctance to lower prices in 1907–1908 and to raise quotations late in 1912, when a sudden spurt exhausted its stocks, indicates a disposition towards price stabilization [41] which has continued, although the stability of 1912 might be explained by a desire to induce the domestic market to "go native" once more, or a wish to appear in a favorable light before the Congressional committees, which might raise unpleasant questions.

[39] See above, p. 30.

[40] The nature of the testimony in the 1913 tariff hearings suggests this conclusion. See brief of the Aluminum Company of America and testimony, Tariff Hearings, 1912–1913.

[41] Cf. remarks of an officer, Tariff Hearings, *op. cit.*, pp. 1485–1486.

CHAPTER III

RESEARCH AND MARKETS, 1915–1935

1. The War Stimulus

A detailed history of aluminum during the World War is without the compass of this study. The present section will merely note the significance of the war for the development of the industry. The appearance of a war demand for this metal was signaled by the jump in the New York open market price from 20 cents to 60 cents in the latter half of 1915.[1] Although imports had diminished to almost nothing, the increasing requirements of the belligerents could not be satisfied at home, so the European demand for American aluminum mounted steadily during the next year. With imports stopped and the output of Shawinigan commandeered by the Canadian government, the Aluminum Company's capacity, now doubled over that of 1913, was scarcely able to meet the booming home demand of 1916, occasioned by the heightened business activity. Until the United States entered the war, the company sold principally to the domestic market at contract prices equal to about half the open market figure. At these prices it apparently earned about 30 per cent on its investment in 1916. The exceptional prices for export metal turned attention for the first time to the recovery of scrap aluminum upon a substantial scale, with the result that many producers of "remelt" sprang up.

In France and England it was impossible under war conditions to expand plant for aluminum production. Det Norske Nitridak-

[1] The Aluminum Company of America has always made a large part of its sales on long-time contracts (i.e., contracts for six months or a year). Imported metal is sold to a large extent in the same way. The "open market" in aluminum is comprised of relatively small amounts of imported metal, "remelt," and portions of contract purchases available for resale. Until the substantial scrap recovery of recent years the open market was of little importance, except in a few years of distress imports. Open market dealings are "spot" sales; there has never been an organized future market in aluminum. Open market prices have, at times, diverged from the contract prices of the Aluminum Company because the latter did not adjust its quotations to the demand conditions felt in the open market. The most striking instance of this was in 1915 and 1916. In Europe sales are also made largely on long-time contracts, and there are no organized future markets.

tieselskab, a Norwegian corporation financed by French capital, which had started to produce a little aluminum just prior to the outbreak of war, was taken over by the Compagnie Alais. Plant was rapidly extended,[2] but total production during the war remained relatively small. The chief part of the burden of supplying the immense requirements of the Allied governments fell upon the Aluminum Company of America. The Neuhausen concern sold most of its metal to the German government, which also built aluminum works of its own as rapidly as possible. The governments of nearly all belligerent countries assumed control of the utilization and price of this new war material.

After the entry of the United States into the war in 1917 production was increasingly diverted from the markets of peace to military uses. Owing to the inelasticity of the power facilities for aluminum reduction, which was increased by abnormal labor conditions and the heavy claims upon the construction industries for more obvious war needs, this was the only way in which the unprecedented demands for the metal could be even partially satisfied. When the war program of the United States got into full swing the government was using 63 per cent of the domestic supply, and 27 per cent was going to the Allies or into indirect war uses.[3] Yet in spite of this diversion a considerable expansion of facilities was required to produce a substantial portion of the amounts called for by ambitious government plans. Purchases of power for the Maryville and Massena cells were increased, and work upon the Tennessee power development was pushed as fast as war conditions would allow. The dam and power plant upon the Yadkin River in North Carolina was completed in 1917, increasing the capacity at Badin to an average of about 70,000 h.p. While these additions could not satisfy completely the demands of the government, they did raise production in the United States from 40,000 tons in 1915 to nearly 60,000 tons in 1917 and 1918.

What sort of demands had called into being this remarkable growth?[4] Some of the uses, in such things as machine-gun radi-

[2] In 1916 one reduction plant was completed in less than a year by transferring from America to Norway the construction materials, electrical equipment, and other facilities which had been assembled for the contemplated plant of the Southern Aluminium Company. (See JFE, XL, 242 ff., 1931.)

[3] War Industries Board, Price Bulletin no. 34, p. 57.

[4] In assessing the significance of the war for the growth of the industry it must

ators, time fuses for shrapnel, aluminum powder for explosives, wire in bullets, helmets, and various sorts of accoutrements for troops, would largely disappear with the return of peace. But the uses which consumed great tonnages during the war were those destined to endure in peace or to stimulate, in one way or another, the emergence of new employments. Aluminum and its strong alloys were used in great amounts as sheet for automobile and airplane manufacture, and alloy castings were widely applied in the construction of trucks, motor buses, aircraft, and oil-burning engines for submarines and destroyers. The airplane had the greatest significance for aluminum both in its immediate demands and its indirect repercussions. War necessities quickly demonstrated the limitations of steel and wood for airplane construction, and the compelling importance of light weight turned attention to the light, strong alloys of aluminum.[5] The all-metal plane was really the outgrowth of the intensive development of structural design during the war. It is said that in 1918 the Allied governments put more than 90,000 tons of the light metal into aircraft.[6] Sheet went into fuselage, aileron frames, cowling, instruments, and so on. Even greater reduction of weight was achieved by the use of aluminum-alloy castings in the engine. About one-third of the weight of the Liberty motor was composed of aluminum in forty separate parts.[7] Aluminum-alloy pistons, which had just begun to receive the attention of a few motor-car builders before the war, were used extensively in airplane motors.[8]

The enthusiastic interest in aviation and the intensive development of all phases of aeronautics which grew out of the struggle created a new peace-time market for substantial quantities of aluminum. Although the production of pleasure cars had been restricted to almost nothing in 1918, the automobile industry had also received a big stimulus and understood aluminum castings better after the war than before. Furthermore, the war brought some familiarity with the metal in the general engineering trades,

---

be recognized that an increase in consumption of nearly 100 per cent had occurred between 1913 and 1915 owing to an exceptional expansion of peace-time uses, chief among which was the automobile. Perhaps it would not be far wrong to say that about half of the 275 per cent increase in production in this five-year period was caused by war requirements.

[5] R. J. Anderson, *op. cit.*, p. 315.

[6] *Ibid.*, p. 274.

[7] EMJ, CVII, 288 (February 1919).

[8] Anderson, *op. cit.*, p. 310.

which had previously displayed but scant interest in it.[9] Undoubtedly its prominent adoption for aircraft played a large part in awakening this interest, but aluminum was also used in countless applications where it had formerly enjoyed but little attention. This was especially true in Germany, which was cut off from copper; but everywhere the possibilities of substituting aluminum and its alloys for copper, tin, bronze, brass, and zinc were demonstrated. More intangible, but perhaps of equal importance, was the captured interest of millions of persons who had thought of aluminum, in the days before the advent of Zeppelins and Liberty motors, as a material for pots and pans. The war accomplished more advertising in two years for this industry than a decade of New York copy could have done. The figures of Table 4 indicate the significance of the war in raising aluminum to a prominent position among the nonferrous metals.

TABLE 4

RATIOS OF WORLD PRODUCTION OF ALUMINUM IN CERTAIN YEARS TO WORLD PRODUCTION OF OTHER NONFERROUS METALS BY VOLUME

Year	Aluminum to Zinc	Aluminum to Copper	Aluminum to Lead	Aluminum to Tin
1900	0.04	0.05	0.03	0.23
1910	.14	.16	.16	1.02
1920	.47	.44	.62	2.76
1930	0.49	0.55	0.69	3.97

The ratios express relative *volumes* rather than relative weights because the industrial importance of metals is more adequately measured by the former relation. Figures of aluminum production are given in Table 38. For computation of the ratios statistics of zinc production, mine production of copper, and smelter production of lead and tin have been taken from the *Mineral Industry* for 1900 and 1910 and from the American Bureau of Metal Statistics for 1920 and 1930.

Most important of all for the future course of development in the industry, the war airplane focused attention clearly upon the

[9] Practically every government department was a purchaser of aluminum. "Some of the amounts used were relatively unimportant from the standpoint of tonnage, but the diversified adoption of aluminum indicates a growing knowledge of methods of handling this metal" (EMJ, *loc. cit.*).

importance of research in the constitution, properties, and industrial applications of the strong, light alloys.

> Aluminium had rapidly extended its field of usefulness prior to the war, but that event was the cause of unprecedented interest in its properties, uses, and alloys, principally because of its applications in aircraft. The demands of the war had an exceedingly salutary effect upon the aluminium industry in general, and what is more important from the engineering point of view, metallographic research on aluminium and its light alloys was undertaken by a number of institutions and individuals, both in the United States and abroad, with a view to ascertaining what assurance of uniformity in quality might be made reasonably certain in the production of parts for aircraft in particular.[10]

The war demands both demonstrated the need for research, which had not been clearly understood before, and gave promise of a sufficient future market for its results to make it profitable. After the war the several aluminum firms greatly enlarged their research staffs and laboratory facilities, and broadened the scope of research work. Out of this came the development of special alloys for particular purposes.

A prominent part of the study of aluminum and its alloys was taken by government bureaus the world over as soon as it became manifest that this metal was to be regarded as an important war material. Since the war, governments have increasingly adopted measures to secure national self-sufficiency in the production of a metal which is indispensable for war. In countries which are poorly supplied with copper and other metals for which aluminum can be substituted, government encouragement has also been motivated by a desire to escape dependence upon foreign raw materials in peace as well as in war. Finally, the war broke up established commercial relationships in Europe and endowed the Aluminum Company of America with financial resources for expansion at the same time that the growth of the European companies, with the exception of the new German state corporation, was retarded.

2. New Alloys and New Markets.[11]

The post-war period has been characterized by a growing "extensive cultivation" of the possibilities of aluminum in so many directions that a summary treatment can do no more than note

[10] Anderson, *op. cit.*, p. 9.

[11] Much of the information used in this section has been drawn from *Aluminum*

the market trends and explain very briefly the connection between scientific research and those new uses which are fast making aluminum a metal of commanding importance. The dominant trend has been the development of special alloys, improved fabricating processes, and new products to render this metal more satisfactory in old employments and to widen the range of uses. As a result, an increasing diversification of markets has been attended by the diminishing importance of the automobile industry as the chief outlet for the light metal. The principal developments in alloys, processes, and products will be described before the changes in markets are surveyed.

The new strong alloys which place aluminum on a competing basis with steel, at least as far as moving structures are concerned, are the most spectacular accomplishments; but closely associated with them has occurred the development of an ever-increasing number of alloys, each suited in physical, fabricating, or casting properties for a particular use. Changes and improvements in methods of casting and fabricating have contributed to the development of better alloys and products.

Before the war empirical methods had introduced a few "general purpose" alloys. Until about ten or twelve years ago nearly all aluminum castings made in the United States were produced from an alloy containing about 92 per cent aluminum and 8 per cent copper (known to the trade as "Number 12"), and were poured in sand molds. In Germany the general casting alloy contained 4–10 per cent zinc and 2–4 per cent copper. While the majority of sand castings were still made from these general-purpose alloys six or seven years ago, the variety of alloys adapted for sand work has multiplied in the past decade. With the growing knowledge of aluminum alloys, a larger and larger proportion of castings has been made in permanent molds and pressure dies. The same sort of development has been characteristic of the wrought alloys — i.e., alloys used for rolling, stamping, and

---

Industry by members of the Aluminum Company of America's research staff; from R. J. Anderson, *The Metallurgy of Aluminium and Aluminium Alloys;* and from advertising literature of the Aluminum Company of America and foreign producers. These sources will not be cited except for special points. None of these sources provides a connected historical account of the developments of this period which is adequate. I have found no book or article which really attempts to present such an account.

forging. An aluminum-manganese composition was the only wrought alloy of importance in the United States prior to the appearance of the strong alloys. Since the discovery which was instrumental in launching alloy development was concerned with wrought alloys, the developments in the latter class will be considered first. It should be understood that each alloy referred to in the following brief survey constitutes a type, of which several variations have been developed to fit particular uses.

After several years of experiment directed toward the development of an aluminum alloy for Zeppelin construction, Alfred Wilm of the Centralstelle für wissenschaftliche und technische Forschungen in Germany announced to the world of metallurgists and engineers in 1909 that the mechanical strength of some alloys could be increased substantially by subjecting them to a process of "heat treatment." This consists of raising the alloy to a temperature of about 500° C., quenching in water or oil, and aging either by allowing it to stand for some days at room temperature or reheating it for a few hours at a slightly elevated temperature. The epochal character of Wilm's discovery may be appreciated when it is understood that the strong alloys of aluminum bear the same relation to their base metal as the various types of steel, that is iron alloys, bear to iron.[12] The uses of aluminum for structural purposes had formerly been quite restricted, because none of the known alloys attained a tensile strength of more than 40,000 pounds per square inch and their ductility was limited. By raising the tensile strength to 60,000 or 70,000 pounds and improving ductility,[13] Wilm's heat treatment introduced an alloy which possessed the mechanical properties of mild steel with only one-third the weight.

The alloy[14] which Wilm found to give the best results when

[12] See MI, xxxv, 37 (1926); and E. E. Free, in *Review of Reviews,* March 1931, p. 55. Prior to Wilm's discovery, steel was the only known material which could be hardened by heat treatment. Since the elucidation of the principles of heat treatment and age hardening there has appeared an ever-growing variety of alloys capable of heat treatment. Up to the present alloys of aluminum seem to have been the most numerous and most important (see Rosenhain, in *Engineering,* CXXXVI, 725, December 29, 1933).

[13] See J. D. Edwards, *Industrial and Engineering Chemistry,* XVIII, 932 (September 1926).

[14] Its ingredients were about 4 per cent copper, 0.5 per cent magnesium, 0.5 per cent manganese, and the silicon and iron present as impurities in the ingot aluminum.

subjected to heat treatment was produced before the war under the trade name of "Duralumin" by the Dürener Metallwerke A. G. Apparently it was also introduced at that time into France and England by Vickers Sons and Maxim, who acquired the patent rights for those countries.[15] During the war it was employed extensively in the construction of aircraft. In 1920 the Zeppelinwerke produced the first all-metal monoplane of duralumin.[16]

Under the stimulus of military importance as a material for aircraft much attention was devoted by government research staffs during and after the war to variations of duralumin and to the study of heat treatment, of which Wilm and his commercial associates had gained only an empirical knowledge. Understanding of the scientific principles of this process, so necessary for industrial perfection, seems to have come from the studies of P. D. Merica and his co-workers at the United States Bureau of Standards, of Rosenhain and others at the National Physical Laboratory in England, and of Guillet and his associates in France.[17] The results of their research were published in 1919 just as the aluminum firms were installing research staffs. Pre-war German experience, the war development of aviation, and research in government laboratories had made it clear that there were immense possibilities for the successful commercialization of better aluminum alloys, while the severe tests to which the existing alloys had been subjected by the war revealed the need for great improvement.

In the past fifteen years the scientific experts of the aluminum firms have invented many new alloys of the duralumin class (that is, alloys possessing mechanical qualities comparable, or nearly comparable, to the original duralumin), improved industrial application of the heat treatment process, and developed satisfactory fabricating machinery and methods. Experience during the war had shown that the duralumin alloys of Wilm's composition, or slight variations of that, were difficult to fabricate. In 1921 the Aluminum Company introduced a copper-silicon-manganese alloy

[15] *Bulletin bibliographique* of the *Revue de l'aluminium,* June 1924, p. 14.

[16] *Aluminium,* III, no. 15, p. 1 (April 11, 1921).

[17] *Aluminum Industry,* II, 135; *Engineering,* CXXXVI, 726 (December 29, 1933); Gaëtan Py, *Progrès de la métallurgie et leur influence sur l'aéronautique* (Paris, 1928), pp. 19 ff.

of aluminum (25S) which possesses many of the physical properties of the former and superior hot-working qualities. It has met with considerable use in forged connecting rods and aircraft propellers, and is beginning to be adopted for locomotive side rods. French research workers had been experimenting independently along these lines at the same time, with the result that a similar alloy was brought out shortly thereafter by l'Aluminium Français. The same type of alloy was subsequently made in Germany and sold by the Vereinigte Aluminiumwerke as Lautal.[18] As a cheaper substitute for duralumin it has achieved wide success in Germany in uses where the strength requirements are not as great as those furnished by its remarkable predecessor.

Of somewhat lesser physical qualities but greater facility in mechanical working is a class of aluminum-magnesium-silicon alloys. The foremost to appear were Aludur, introduced by the Giulini firm in Switzerland, and 51S of the Aluminum Company of America. L'Aluminium Français later developed Almasilium. This class of alloy has been used extensively in forgings, for which it is in general more economical than other alloys. Moreover, it is doubtful if certain designs of forgings could be produced from any other alloy or from steel. In the last five years or so, 51S has found another use in the manufacture of aluminum furniture. After heat treatment and age hardening, alloys of the 25S and 51S types attain physical qualities which are markedly greater than those of commercially pure aluminum, although they fall short of the qualities of Duralumin proper. In general they are more economical than the latter for uses to which they are suited.[19]

A few years ago a new type of alloy in this class was developed by Victor Hybinette and his son, and introduced by the Sheet Aluminum Corporation. Hyblum, as this series of alloys is called, contains nickel and chromium. Strength for strength, this ma-

[18] Lautal at first contained 2 per cent silicon, whereas 25S contained less than 1 per cent silicon. It is said that the composition of Lautal was shortly altered to correspond more nearly to that of 25S.

[19] In Europe several alloys of the 51S class have been developed in an attempt to replace steel-cored aluminum cable with a high-strength alloy conductor. Aldrey of the AIAG and Almélec and the J. L. alloy of l'Aluminium Français have been used for this purpose (*Revue de l'aluminium,* January–March 1928, p. 586). There appears to be some doubt whether they are as serviceable as the steel-cored aluminum cable (*Aluminum Industry,* II, 704).

terial in the heat-treated temper appears to possess somewhat greater ductility and to resist corrosion better than ordinary commercial grades of duralumin, with which it compares favorably in specific gravity.[20] Furthermore, unlike most other aluminum alloys made in this country, Hyblum is said to be suited for both castings and wrought work. Its chief products have been automotive accessories, aircraft equipment, kitchen equipment, and ornamental trim.

Since the war the Aluminum Company has produced several varieties of alloy closely akin to the original composition of Wilm's duralumin which have enhanced the adaptability of this type. In addition, its research staff has devised two compositions known as "super-duralumin" which are used wherever greater strength and hardness than that of duralumin proper is desired.[21] A tensile strength of 75,000 pounds per square inch puts these alloys in a class of structural materials never before invaded by such a light metal. Various alloys of the 17S line have been used extensively in aircraft construction in the form of sheet and other thin sections, for they possess the maximum combination of high strength and resistance to corrosion of all the strong alloys.

The corrosion resistance of the strong alloys is somewhat less than that of commercially pure aluminum. In some uses for which these alloys are otherwise particularly suited, such as aircraft, pontoons, and motorboats, corrosion is likely to be serious. To meet this problem the United States Bureau of Standards developed a very satisfactory process of spraying alloy sheet with a coating of the pure metal. Shortly thereafter, in 1927, members of the Aluminum Company's technical staff invented a method of grafting a layer of pure aluminum onto each side of alloy sheet. The process is such that the pure metal surface forms an intermediate layer of alloy with the material inside. Hence the surface is integral with the underlying base. The alloy is protected even at exposed edges, or holes in the coating, since the pure

[20]Archibald Black, "A New High-Strength Aluminum Alloy," *Airway Age,* June 1930.

[21] These are designated as C17S and 427. 17S is the Aluminum Company's base number for duralumin proper. A set of duralumin alloys is also produced here by the Baush Machine Tool Company. In the past few years the Aluminum Company has brought out two new wrought alloys (4S and 52S) containing magnesium which have been used extensively in the new streamlined trains. See *Mining and Metallurgy,* XVI, 51 (January 1935).

metal, having a higher solution potential, forms a galvanic couple with the alloy. Alloy sheet coated by this process is sold by the Aluminum Company under the name Alclad. At about the same time, or a little later, a somewhat similar process was developed by the Vereinigte Leichtmetallwerke, a subsidiary of the Vereinigte Aluminiumwerke and other firms.[22] A sheath of pure aluminum or some alloy of high corrosion resistance was grafted onto the alloy to be protected.

Another line of attack upon the problem of corrosion has resulted in the development of several methods of thickening and strengthening the protective surface film of aluminum oxide.[23] The oxide film does not, of course, possess the firmness or hardness of a layer of pure aluminum or aluminum alloy. Certain of these methods, however, render the film an excellent basis for a protective pigment or enamel, which also lends decorative utility. Various processes of enameling or lacquering have been used, particularly in Germany. Recently a method of anodic treating called the Alumilite process has been introduced commercially in this country by the Consolidated Stamping Company of Detroit.[24] It is said that the process is applicable to all types of sheets, stampings, and castings. The addition of aniline dyes to the electrolytic bath gives a high luster finish in a wide variety of colors. According to the claims of the inventors of this process, the plating forms actual fusion with the metal. Colored Alumilite products have been used in electric refrigeration and other nonautomotive fields. In 1934 the Aluminum Company acquired the plant and patents of Aluminum Colors, Inc., which was then operating the Alumilite process.[25]

Finally, certain alloys have been discovered which have better resistance to corrosion than pure aluminum or its other alloys.

[22] *Zeitschrift für Metallkunde,* XX, 294 (August 1928); *Hauszeitschrift der Vereinigte Aluminiumwerke,* January-February 1932; *Metallwirtschaft,* XIV, 1 (January 4, 1935).

[23] The more important methods seem to be the three following: (1) anodic oxidation, on which Bengough and Stuart of the Corrosion Research Committee of the Institute of Metals in England have done important work; (2) chemical processes developed chiefly by Bauer and Vogel and research workers of the Vereinigte Aluminiumwerke; (3) the Eloxal process of the same company. See *Engineering,* CXXXVI, 728 (December 29, 1933); *Hauszeitschrift der Vereinigte Aluminiumwerke,* III, 302 (September-November 1931); and *Aluminum Industry,* II, chap. XII.

[24] *Automotive Industries,* LXVIII, 298 (1933).

[25] *Poor's Manual of Industrials,* 1935, p. 1178.

Among these are the K. S. Seewasser alloys introduced in Germany, the Anticorodal line of the AIAG, and a new aluminum-magnesium-silicide alloy brought out in 1933 by the Aluminum Company of America. These alloys have resulted in marked improvement in resistance to salt-water corrosion in particular.

Another invention of importance for the alloy development should be mentioned. In 1919 William Hoopes of the Aluminum Company perfected an electrolytic process of refining the product of the Hall reduction cell, which rarely yields metal of higher than 99.7 per cent purity. By the Hoopes process any desired amount can be produced with an average purity of over 99.8 per cent. This enhanced purity enables the use of some alloys which would be less satisfactory with a higher iron and silicon content, for a difference of 0.1 per cent of impurities alters the qualities of aluminum appreciably.

The mere discovery of excellent alloys and effective heat treatments has not, of course, been sufficient to introduce them into industrial consumption. Scientific study has also had to develop a satisfactory technique of fabrication. The most interesting occurrence along this line has been the erection of new rolling mills and fabricating plants designed to handle the strong alloys. In 1928 the Aluminum Company of America constructed a blooming mill to roll billets and form structural shapes much akin in size and form to the products of a steel mill. Ingots for rolling in this mill weigh about a ton and a half, whereas the largest ingots rolled previously weighed only a few hundred pounds. Less obvious, but quite as important, has been the necessity to evaluate standards, specifications, and rules for use of the strong alloys as materials of "heavy" construction. Within the last few years the companies have published several pamphlets containing such information and have waged an intensive campaign to educate engineers to appreciate the utility of the strong alloys.

It would have been surprising if the success of heat treatment with wrought alloys had not suggested the possibility of increasing the physical qualities of casting alloys also. The pioneer work in this direction, which began just before the war, seems to have been done by Dr. Walter Rosenhain and his associates at the National Physical Laboratory in England.[26] From this study there issued

[26] "Aluminum Alloy Progress," *Iron Age,* CXXVI, 1455 (1930).

the well-known heat-treated "Y" alloy which has been used abroad, and to some extent here, in a variety of castings and forgings, principally in aircraft engine cylinders and cylinder heads, where high strength at elevated temperatures is imperative. In this country the research of Archer and Jeffries of the Aluminum Company resulted, about 1920, in the effective application of heat-treating methods to aluminum-copper alloy castings and the commercial introduction of two alloys especially adapted for such treatment (195 and 196). Proper heat treatment of castings made from these alloys confers greater strength and ductility than is possessed by the general casting alloys such as the No. 12 line. They stand in much the same relation to the latter as steel castings to cast iron. These alloys have been used particularly in crankcases and other motor parts in fire engines and buses. Recently a heat-treated aluminum-magnesium casting alloy has been introduced with success. Heat-treatment processes have also played an important part in the development of aluminum-alloy pistons to the point where they are standard equipment for a large number of motor cars and airplanes. The heat treatment lessens "permanent growth" of the piston, formerly a troublesome phenomenon, as well as conferring greater strength and hardness.

Previous to 1920 silicon had been generally considered as a worthless element for aluminum alloys, perhaps because the silicon impurities in commercially pure aluminum were known to lessen the usefulness of the metal for some adaptations. One of the most important steps in the development of aluminum alloys came in that year with the introduction of aluminum-silicon alloys by Aladar Pacz. Pacz not only discovered a useful type of alloy having high corrosion resistance, easy casting qualities, and fairly high strength; his work also led to an intensive study of the aluminum-silicon alloys and was mainly responsible, in indirect fashion, for all the developments along this line which have enabled these alloys to widen the range of uses of aluminum.[27] One of his chief discoveries was the fact that employment of a suitable reagent allowed a greater amount of silicon to go into solid solution with the aluminum,[28] resulting in considerable improvement

[27] See *Iron Age, loc. cit.,* and F. C. Frary, *Industrial and Engineering Chemistry,* XIX, 1094 (October 1927).

[28] This "modification" process, as it is called, has been shown to be applicable to other metals and alloys, with the result of widening the perspectives of metallurgists.

to some properties of this type of alloy. Subsequent studies by the staff of the Aluminum Company have demonstrated that much the same result can be gained by casting the alloys high in silicon in chill molds. In addition to the advantages mentioned above, the silicon alloys are lighter than most other alloys of aluminum and form dense, pressure-tight castings. As might be surmised, they are particularly adapted for die casting, which has made rapid strides since their introduction. Chief credit for the commercial development of the Pacz silicon alloys belongs to the Metallgesellschaft, which acquired patent rights for all of the world except France and America and introduced them in Germany under the name Silumin.[29] This alloy has enjoyed extensive use in Europe in the form of pistons, cylinders, and housings, containers for chemicals, and a variety of other castings. The Aluminum Company of America has also developed two alloys having less silicon than Alpax, as the Pacz alloys are called in America and France, from which castings satisfactory for many purposes may be made without "modification" and hence more cheaply. These have gone into architectural castings, marine fittings, chemical apparatus, auto-body castings, and the like.

Accompanying the development of special alloys since the war there has been a trend toward the increasing use of permanent molds and pressure casting in steel dies. Sand casting is the most flexible method of the three and must be used for the largest and most complicated work. It does not pay to set up fixed capital in molds or expensive pressure dies except for quantity production. Where the market is large enough to provide substantial utilization, permanent mold and die casting usually yield sounder and more economical products. The use of permanent metal molds, which was taken over from the brass industry, appears to have been first introduced into the making of aluminum castings in France before the war. It has since been employed intensively, particularly in the production of pistons.

Light weight and high thermal conductivity render aluminum peculiarly suitable for pistons. In 1917 the Aluminum Company of America developed a 10 per cent copper alloy (122) containing small amounts of iron and magnesium, for the purpose of making

[29] *Metal Industry,* XX, 183 (May 1922); *Revue de l'aluminium,* September 1925, p. 123, October 1928, p. 896.

pistons by the permanent mold method. Application of heat treatment subsequently improved the qualities of this alloy. The aluminum-alloy piston not only exhibits one of the most striking instances of enlarging the market by introduction of a new alloy; it illustrates also the importance of discovering proper product design. Aluminum pistons originally made in the solid barrel or trunk design copied from the cast-iron product were far from satisfactory, owing to the high expansion coefficient of aluminum, which resulted in seizing when hot and "piston snap" when cold. The "split skirt" type of piston, which made its appearance in 1920, embodied the first successful remedy for this difficulty. A combination of vertical and horizontal slots in the bearing faces permitted expansion without seizure even with a very small clearance between piston and cylinder. Three years later the "strut-type" piston, invented by an automotive engineer named Nelson, was put into production by the Bohn Aluminum and Brass Corporation, a large maker of castings. This design permitted closer fittings than its predecessors by controlling expansion of the aluminum with nickel-steel struts cast into the piston. It had a rapid growth in popularity and was consumed for several years in nearly equal proportions with the "split skirt" type. The Aluminum Company of America, which had sponsored the earlier design, has also produced the Nelson piston, and it has been used extensively in Europe as well. Recently the Aluminum Company has developed a new alloy (132) with high silicon content which has a coefficient of expansion materially lower than the alloys heretofore used for pistons. It is said that the new composition permits return to the barrel or trunk type of structure, which is actually the most satisfactory design from many standpoints.[30] Aluminum-alloy pistons of the simple trunk type have always been used in aircraft motors, where the advantages in weight and thermal conductivity have been much more decisive factors than in automobiles. Permanent mold casting has achieved such good results with pistons that it has been extended to other products as growth of demand has warranted.

The phenomenal growth in large-scale, rapid production of small castings, however, has occurred in die castings. This oper-

[30] *Aluminum Industry,* II, 631. The trunk type is better from the standpoint of efficient heat dissipation, lightness, low friction, large bearing area, and ruggedness.

ation, which has been borrowed from practice with other white metals, differs from casting in permanent molds chiefly in the use of pressure to force the molten metal into the mold or die. The product is practically finished in one operation. Machining and grinding are nearly eliminated, while closer tolerances are obtained in quantity production. The economy over the other casting processes is striking, particularly since the machinery can be made automatic to a large extent.[31] The use of this process has had a very rapid growth since the discovery of alloy steels suited for standing the strain on the dies and the development of the aluminum-silicon alloys, which are especially fitted for this type of casting. It has been adopted by many concerns for making small castings and intricate castings with extremely thin sections. A few typical products are fire boxes, automobile brake sholes, carburetors, and dash fittings, parts for vacuum cleaners, washing machines, and electrical apparatus. Sand castings have been supplanted in several uses, and it is said that some aluminum alloys, when die cast, are enabled to compete more effectively with many alloys of the other common metals which are not so well adapted for die casting.[32] While most die-cast products have weighed but a few pounds, we are informed that there are no inherent difficulties of design for machines to produce much larger castings.[33]

Extrusion, which is not to be confused with die casting, has been increasingly adopted since the war by fabricators of aluminum. Heated ingots are forced through a die — as tooth paste is squeezed out of a tube — by presses varying in capacity from 200 to 5,000 tons. In this fashion intricate shapes are produced which could not be made in any other way. Pure metal is extruded as molding for automobile bodies or as collapsible tubes. Some alloy extrusion products are I-beams, channels, other structural shapes, and sections for aircraft construction.

Throughout the post-war period experts of aluminum firms, of government bureaus, and of research institutions the world over

[31] "The daily output of a single casting die is frequently measured in thousands, where that of a permanent mold is in the hundreds, and of a sand-casting pattern in the dozens. Likewise, the dimensional variations of die castings are measured and held within thousandths of an inch, where those of permanent mold castings are in sixteenths" (Aluminum Company of America, *Alcoa Aluminum Die Castings,* p. 4).

[32] Hugh Farrell, *What Price Progress?* (New York, 1925), p. 100.

[33] *Aluminum Industry,* II, 328–329.

have devoted a great deal of attention to the study of theories of alloy construction and behavior. During the first part of the period the results of examination of the structure of alloys were limited to those obtainable with the microscope. In 1927 when the Aluminum Company of America, following the lead of some of the most progressive steel companies, installed X-ray equipment in the Cleveland foundry, metallurgists were beginning to appreciate the aid of this new tool of metallography. It has also been demonstrated that radiographic examination of castings of highly stressed parts, such as various airplane parts and brake shoes, is immediately economical.[34]

The brief survey of progress in the development of alloys and processes may now be summarized. The true potentialities of aluminum were not well understood until the war printed unmistakably in the sky what research in Europe had already intimated. During the war some progress was made in the development of strong alloys by government bureaus. With the resumption of peace the aluminum firms established well-equipped laboratories and research staffs, which began almost immediately to develop new alloys and improved processes of fabrication. Until the middle twenties the number of alloys in extensive commercial use appears to have grown but little as compared with ten years earlier. It was in the latter twenties that some part of the immense potentialities of research upon alloys and processes began to be realized. Industrial adoption of new alloys designed for particular purposes seems to have proceeded rapidly for four or five years until it was slowed down by depression. Among the scientific achievements which have broadened the market horizons three stand out — the development of heat treatment and age hardening, the discovery of methods which have given the silicon-aluminum alloys great utility, the invention of processes of coating alloy sheet with a protective layer of pure aluminum or some alloy with high corrosion resistance. In the case of the first two of these achievements the original discoveries and an important part of the subsequent work of development and improvement must be credited to persons outside the employ of aluminum firms.[35]

[34] See W. L. Fink and R. S. Archer, in *Transactions,* American Society for Steel Treating, XVI (1929), 551 ff.

[35] Cf. Léon Guillet, *L'Évolution de la métallurgie* (Paris, 1928), p. 148; Suhr, in

While scientific progress in the development of alloys, processes, and products has been very rapid since the war, it would not seem that a state of maturity has been reached. Indeed, the most important developments may come in the future, for it appears that the general laws of alloy construction and behavior still remain to be discovered.[36]

We now turn to the significance marketwise of the progress just described. Changes in the markets for aluminum in the United States will first be surveyed. Previous to the war the largest annual production in the United States was about 20,000 tons, nearly half of which went into the automobile industry. The greater portion of the automotive consumption was in the form of No. 12 alloy castings for parts subject to little stress; a much smaller amount was used as sheet for body construction. Utensils probably accounted for about a third of the annual production, while most of the remainder found its market as transmission cable and deoxidizing ingot for the iron and steel industry. A few years after the war the market situation was not strikingly different. A doubled output was distributed in larger percentages to the automobile and electric industries and in smaller proportions to the other old markets. In fact, if the estimates of the National Automobile Chamber of Commerce (shown in Table 5) are fairly correct, the rapidly expanding motor-car industry took much more than half of the annual increment of the metal during 1921–1923. Whether or not the actual figures here shown are accurate, there was undoubtedly a marked downward trend in the absolute tonnage used in automobiles for three or four years after 1923. Yet output of aluminum ingot, which by 1923 had again reached the war level, grew steadily in the following years.

Material for quantitative measurement of the shifting markets is disappointingly meager.[37] However, there are indications that

---

Revue de l'aluminium, September 30, 1925, p. 123; Mortimer, *ibid.,* October 1928, p. 896; and *Metal Industry,* XX, 183 (May 1922).

[36] Rosenhain writes: "It is not, perhaps, too much to hope that in the near future the general laws governing alloy construction and behavior will be discovered, and will bring order and understanding into the rather varied and troublesome collection of equilibrium diagrams with which the metallurgists have to deal at the present time" (see "Some Steps in Metallurgical Process, 1908–1933," *Engineering,* CXXXVI, 726, 1933). Cf. also Von der Porten, in *Die Versorgung der deutschen Wirtschaft mit Nicht-Eisen Metallen* (Berlin, 1931), p. 114.

[37] When it was found that the only continuous series of consumption figures was

net losses in the auto market were more than offset at first by an enhanced demand for electrical conductor and an ever-widening variety of casting and extruded products for household appliances, machinery, and aircraft and marine engines, while towards the end of the twenties the strong alloys began to enjoy a growing, if small, adoption as materials for construction and furniture.

TABLE 5

ESTIMATED ANNUAL CONSUMPTION OF ALUMINUM IN THE UNITED STATES BY THE AUTOMOBILE INDUSTRY, 1921–1934

Year	Estimated Annual Consumption (*Metric Tons*)	Percentage of Total Production of Primary and Secondary Aluminum
1921	19,540	60
1922	33,180	70
1923	41,810	54
1924	36,360	40
1925	26,360	25
1926	22,720	21
1927	19,090	17
1928	22,720	16
1929	33,630	23
1930	18,180	13
1931	12,720	12
1932	10,900	16
1933	9,090	13
1934	10,910	14

Consumption figures, 1921–1934, from Automobile Manufacturers' Association, *Automobile Facts and Figures*. Since there are no good figures of total consumption of aluminum in the United States, consumption by the automobile industry has been related to estimated output of virgin and secondary aluminum which is given in Appendix F.

---

that for the automobile industry, a request was made to the Aluminum Company of America for figures showing the changing trends in their markets. It was replied that such data were not kept by the company and that the expense of compiling them would be too great. No adequate consumption statistics for Europe have been published. The rough estimates which appear in the rest of this section have been made from scattered figures, often estimates themselves, in trade journals and in *Aluminum Industry*. Frequently they are somewhat conflicting. To cite all the sources and explain how my estimates were reached would lengthen the footnotes unduly.

The figures for automobile consumption obscure a decisive change in the character of the demand from this industry. In 1922 and 1923 about half the tonnage used in motor cars took the form of crankcases, and a substantial amount of sheet went into body construction. Within two or three years the sheet demand had almost completely evaporated, and cast-iron crankcases were rapidly replacing aluminum.[38] In this country steel sheet is still employed almost exclusively by body builders, while at the beginning of the depression the aluminum in crankcases was apparently less than a fifth of the total amount consumed in the automobile industry. It was the adoption of the aluminum-alloy piston and connecting rod and the alloy-sheet bus body which offset the loss of these other demands. From two and a half million pistons in 1924 consumption rose to six million in the following year. By 1928–1929 somewhere between twenty and thirty million were being sold as standard equipment for nearly every car except the Ford. After its introduction in 1922 the forged duralumin connecting rod came to have almost the same wide demand. The increasing call for these two items was mainly responsible for turning the automotive demand upward again.

Aluminum alloys have, of course, held a prominent position in aircraft structure and motors ever since the war, but this industry has remained too small to consume a very large proportion of the metal. Furthermore, it appears that no aluminum alloy has yet been developed wtih the strength per unit of weight equal to some of the special alloy steels which are used for the strength members of planes. After the perfection of processes for producing large heat-treated forgings from the strong alloys about 1923, the aluminum-alloy propeller came into general use. In the modern aircraft engine about one-half the weight and three-quarters of the volume consists of aluminum alloys. The trend toward the all-metal plane has enlarged the demand for the wrought alloys. Several dirigibles have been built with an entire "skin" of duralumin as well as the customary framework. In 1928 the consumption of aluminum alloys in the American aircraft industry trebled that of the previous year.[39] The very swift growth of die-cast parts for vacuum cleaners, washing machines, radio sets, elec-

[38] The reasons for these changes are explained below, p. 254.
[39] *New York Times*, May 9, 1929, p. 42.

trical appliances, textile machinery, and so forth, as well as for automobiles, is evidenced by the fact that the output of this type of casting doubled every five years between 1915 and 1930.[40] Nearly 10,000 tons, equivalent to about 10 per cent of the total output of virgin aluminum in the United States, went in one form or another into washing machines in 1928.[41] Accompanying the expansion of the electric light and power industry the mileage of aluminum conductor in service increased from 125,000 in 1924 to well over 300,000 five years later. Before the depression this use was taking over 10,000 tons a year. New forms of aluminum developed in the past decade which received a favorable reception were aluminum paint, foil, roofing material, screens, and a host of other products of scientific study which consume in the aggregate as much metal as was produced twenty years ago. The outboard motor would have been impossible but for the aluminum-alloy piston.

On the eve of depression the fate of the wrought alloys in their severe competition with steel remained undetermined, but it could be said that they had caught a foothold. By the end of 1928, 388 cars containing substantial amounts of aluminum alloys had been built for steam railroads, and 108 for electric railroads.[42] A much larger number ordered in 1929 and 1930 indicates that, although the tonnage was still relatively small, aluminum alloys were generally accepted as a satisfactory material for car construction.[43] Cast alloys were penetrating the architectural field for ornamental work, but the wrought alloys had been employed as structural members only in the top floors of some of the highest skyscrapers. Aluminum furniture, introduced about 1926 by the Aluminum Company, gave evidence of successful competition with steel.[44]

[40] *Alcoa Aluminum Die Castings,* p. 3.

[41] *Aluminum Industry,* II, 504.

[42] *Electric Railway Journal,* November 3, 1928.

[43] *Ibid.,* April 1930. New York City ordered 300 subway cars in 1930. All the sheet except the outside sheeting was of aluminum alloys.

[44] Herewith the interest of our resourceful navy in aluminum furniture:

"The naval limitation of arms conference resulted in a more careful study of the usefulness of aluminum on board ship. The heart of the limitations agreed upon is the fixed limit in displacement, and naturally that country which could build the best ship within the limits set, had the advantage. It was obvious that the best way would be to reduce to a minimum the weight of all desired essentials in the way of equipment without sacrificing strength. Every pound so saved can be utilized to increase the number of guns, the speed, fuel capacity, or any one or more of the

TABLE 6

ESTIMATED CONSUMPTION OF ALUMINUM IN VARIOUS MARKETS IN THE UNITED STATES IN 1930 EXPRESSED AS PERCENTAGES OF ANNUAL OUTPUT

Market	Percentage
Transportation (auto, steam and electric railways, air, and marine)	38
Cable, busbars, and electric equipment	16
Cooking utensils	16
Machinery	8
Iron and steel industry	8
Nonferrous foundries and metal manufacturers	4
Building	3
Chemical industry	2
Food industry	1
Miscellaneous	4
Total	100

These figures are given by C. L. Mantell, EMJ, CXXXI, 102 (February 1931). It is not indicated whether they refer to distribution of primary aluminum only or to the total annual production of primary and secondary.

Table 6 exhibits an estimate of the proportions in which consumption in the United States was divided between various uses in 1930. It is plain that substantial changes occurred between 1920 and 1930 in the markets for aluminum. Evidently consumption by the motor-car industry dropped from over one-half to less than one-third of the total output; and most of the aluminum used at the later date took the form of pistons and rods, which had been employed but slightly ten years earlier. Transmission lines were apparently absorbing an increased proportion, while the opposite was true of utensils. Consumption of the strong wrought alloys had grown to perhaps 10 per cent of the total. At least a quarter of the total output was probably used in products (other than pistons and connecting rods) which took hardly any aluminum in 1920. In general the market had become much more diversified

offensive and defensive qualities of the ship" ("Aluminum Furniture Proves Successful on Warships," by Commander F. G. Crisp, shop superintendent, Norfolk Navy Yard, reprinted from the *Marine Review,* June 1929).

under the influence of stronger competitive forces in the development of new variations or adaptations of the basic product.

It is quite impossible to estimate the effects upon markets of the diverse currents of the depression, but there do not appear to be any signs of striking changes. Advances have been made in a few new fields. The development of high-compression automobile motors seems to have resulted in some increase in the use of aluminum-alloy cylinder heads.[45] Resumption of brewing has opened an outlet for aluminum in vats, storage tanks, coils, and barrels.[46] It has also been adopted recently for dairy equipment. Aluminum steam-shovel buckets have begun to compete with steel. The most spectacular developments have occurred in transportation. The Brazilian Clipper, built in 1934, which with a weight of nineteen tons became the largest American air liner, was constructed chiefly of aluminum alloys. In the same year the Union Pacific put into operation two streamlined articulated trains built largely of strong aluminum alloys. Early in 1935 two additional aluminum trains were under construction for the Union Pacific, one for the New Haven, and one for the Baltimore and Ohio.[47] A bridge in Pittsburgh has lately been rebuilt with duralumin alloy. The Aluminum Company has recently had a steamship built from strong aluminum alloys.

Somewhat similar consumption trends have occurred in Europe. Before the war kitchen equipment absorbed the largest proportion of aluminum in Germany, where the greater part of the fabrication of ingot was carried on. The amounts consumed by the automobile industry were, of course, increasing. It is said that in 1921 the motor car took about 75 per cent of the output of the French, British, and Swiss producers.[48] About 1923 half of the annual production in Germany took the form of castings,[49] which seems to indicate large automobile consumption, although tanks, vats, and other apparatus for the chemical industry probably accounted for an appreciable part of the castings. Adoption of aluminum pistons and connecting rods seems to have broadened throughout the

[45] *Mining and Metallurgy,* XV, 39 (January 1934).
[46] *Iron Age,* CXXXIII, 77 (January 4, 1934).
[47] *Mining and Metallurgy,* XVI, 51 (January 1935).
[48] EMJ, CXXXI, 102 (February 9, 1931).
[49] Heinrich Buschlinger, *Entwicklung und Aufbau der Aluminiumwirtschaft* (Hamburg, 1924), p. 369.

twenties. A growing diversification of markets and the marked increase in consumption of aluminum by the electric industries pushed the kitchen utensil and the automobile out of their dominant position among consumption fields. By 1924 only 30 per cent of the German production was fabricated into utensils,[50] and this proportion sank to 15 per cent five years later. In France the weight of aluminum used per automobile in the chief makes is said to have tripled between 1920 and 1923.[51] Citroën and Renault apparently absorbed together an amount of aluminum equivalent to one-third of the virgin production of France about 1924.[52] At this time a considerable amount of the light metal was being used in the buses and street cars of Paris. Aluminum was also being introduced into railway coaches.[53] In 1928 it was said that in France aluminum was mainly consumed in the automobile, aeronautic, and electric industries.[54] Diversification of markets in the later twenties throughout Europe brought it about that automotive consumption was reduced to 21 per cent of the total output of cartel members in 1929.[55]

Until about ten years ago the use of aluminum for high-tension transmission lines was apparently farthest advanced in the United States, England, and France. Of 19,000 kilometers of transmission lines installed in the leading European countries between 1917 and 1926 it appears that nearly 40 per cent were aluminum lines.[56] During the twenties adoption of aluminum conductor increased rapidly, particularly in Germany, with the result that 65 per cent of the installations between 1920 and 1930 were aluminum.[57] More than 12,000 miles of steel-cored aluminum cable and earth wire, equivalent to about 8,500 tons of aluminum, were specified for the main lines of the "grid" system in Great Britain. It was expected that secondary low voltage lines would absorb an equal quantity of aluminum.[58] Other countries also took increas-

[50] *Ibid.*, p. 224.

[51] *Revue de l'aluminium*, April-May 1924, p. 5.

[52] *Ibid.*, September-October 1924, p. 36.

[53] *Ibid.*, March 1925, p. 88.

[54] JFE, XXXVII, 173 (June 1928).

[55] EMJ, CXXXI, 102 (1931).

[56] *Wirtschaftsdienst*, XIV, 446 (1929).

[57] Von der Porten, *op. cit.*, p. 109. The figure given presumably applies to Germany, but the text is not clear.

[58] *American Metal Market*, April 1, 1931, p. 4.

ing amounts of the metal in this form, and its employment in various sorts of electrical apparatus was appreciably extended. Table 7 shows an estimate of the length of high-tension aluminum lines in various European countries in 1933 compared to the totals.

TABLE 7

ALUMINUM CONDUCTORS IN HIGH-TENSION TRANSMISSION LINES IN CERTAIN COUNTRIES IN 1933 *

	(1) Total High-Tension Lines (*Kilometers*)	(2) Total of Aluminum, Steel-Aluminum, and Aluminum Alloy Lines (*Kilometers*)	Percentage (2) of (1)
France	32,624 †	9,882	30.3
Germany	27,034	10,185	37.7
England	9,114 ‡	7,727 ‡	84.8
Switzerland	7,309 †	2,625	36.9

* The figures for this table have been taken from *Alluminio,* II, 35–39 (1933), abstracted, *Journal of the Institute of Metals,* LIII, 590 (1933).
† Over 30 kv.
‡ Does not include new lines of grid.

TABLE 8

ESTIMATED CONSUMPTION OF ALUMINUM IN VARIOUS MARKETS OF GERMANY IN 1929 EXPRESSED AS PERCENTAGES OF TOTAL CONSUMPTION

Market	Percentage
Kitchen equipment	15
Chemical and brewing equipment	15
Foil	15
Electrical equipment	15
Silumin castings	12
Half-products for export	15
Miscellaneous	13
	100

The figures of this table are taken from Von der Porten, *op. cit.,* p. 109.

One of the most striking developments in the German market was the exceptionally swift replacement of tin foil by aluminum foil, which as a consequence accounted for 15 per cent of the total consumption of aluminum just prior to the depression. An estimate of the division of German consumption at the end of the post-war decade is given in Table 8. No estimates for other countries of Europe are at hand.

The progress of the strong alloys in Europe appears to have been no more rapid than in this country. They have been used extensively in bus and truck bodies. A beginning has been made in the construction of railway cars from aluminum.

CHAPTER IV

EXPANSION AND POLITICAL RELATIONS, 1915–1935

1. The Aluminum Company of America and Aluminium Limited

In 1915 the Aluminum Company of America had a capital investment of $50,000,000. By 1928 it had risen to a modest position in the ranks of mighty corporations by the growth of book capital to about $250,000,000. The expansion reflected in these figures has been characterized by aggressive acquisition of large reserves of foreign ore and power, purchase of foreign reduction plants, and an extension of facilities at home.

The first move in the acquisition of ore beds took place in South America. Extensive bauxite deposits in British Guiana had been reported in 1897 and again in 1910 through government documents and magazine articles, but they attracted little attention until 1912, when the Aluminum Company of America dispatched an engineer to investigate.[1] Acquisition of ore lands began almost immediately.[2] Apparently a fruitless endeavor was made about the same time by a civil engineer with a promoting turn of mind to interest the British Aluminium Company in the British Guiana deposits.[3] In 1916 the Demerara Bauxite Company, Ltd., was incorporated locally to hold and operate the ore lands leased or purchased by the Aluminum Company of America. Two thousand tons of ore were shipped to the United States in 1917. It appears that by this time the Demerara Company had, with vigorous enterprise, obtained control of most of the deposits formerly owned by private individuals and had leased a substantial portion of the ore lands belonging to the British Crown and the Colony of Christianburg.[4] Some of the leases of public deposits were granted upon

[1] BMTC v. ACOA appellant, fols. 5215–5216.

[2] It is said that a party was sent to Demerara, British Guiana, in 1914 to stay until all the workable deposits had been optioned or acquired for the Americans (Lloyd T. Emory, "Bauxite Deposits in British Guiana," EMJ, CXIX, 687, April 1925).

[3] *Ibid.*

[4] EMJ, CVIII, 243 (August 1919). In 1916, 1795 acres of Crown lands and 1718

condition that the Americans establish upon British soil a plant for the preparation of aluminum oxide.[5] Bauxite from public properties already leased was to be placed at the disposal of the British government.[6] Recognition of the military importance of aluminum also led at this time to a stand against further exploitation of public properties by foreign capital. For several years no more leases were granted. Nevertheless, in spite of belated efforts of others to obtain deposits, it appears that the Aluminum Company of America, through persistent negotiation, litigation, and compromise, had acquired a very large proportion of the suitable bauxite of British Guiana by 1925.[7] Additional ore was acquired in subsequent years. Over $5,000,000 was spent in development and equipment consisting of mining facilities, a crushing and drying plant, a powerhouse, and a twelve-mile railroad to carry ore from the mines to the plant, which was situated at the head of navigation on the Demerara River, whence the Aluminum Company's ocean-going steamships transported it to Gulf ports for shipment to East St. Louis. Shipping terminals were built by the company at Baton Rouge, Louisiana. The British Aluminium Company later leased some public deposits, but the major portion of the ore was evidently still controlled a short time ago by Aluminium Limited,[8] a Canadian corporation to which most of the foreign properties of the Aluminum Company were transferred in 1928.[9]

The history of deposits in Dutch Guiana appears to be similar. Within a few years after 1912 the Aluminum Company had evi-

---

acres of colony lands had been leased to the Demerara Bauxite Company (British Guiana, Combined Court, *Report on the Condition of the Colony of British Guiana during the Great European War,* 1919, p. 84).

[5] *Haskell* v. *Perkins,* Record, pp. 458, 1734. (This reference is explained below, p. 132, footnote 6.) Evidently the Americans agreed to build on British soil by 1923 an oxide plant with a daily capacity of 50,000 pounds (*ibid.,* p. 1899). For some reason, however, an alumina plant was not established in British territory until 1928, when one was built at Arvida in Canada.

[6] EMJ, CIV, 999 (December 1917).

[7] Emory, *loc. cit.* The British Guiana Handbook for 1922 reported that the Demerara Bauxite Company controlled by far the major portion of the Christianburg-Akyma deposits, which were the most extensive known deposits in British Guiana (pp. 116–117). At that time no leases had been issued to others by the government.

[8] A description of Aluminium Limited is given below, p. 74.

[9] Report of the Land and Mines Department of British Guiana for 1930, p. 9; *Bulletin of the Imperial Institute,* XXXI, 396 (1933).

dently secured control of a large proportion of the ore and established the Surinaamsche Bauxite Maatschappij to hold and operate these deposits.[10] Until recently, at least, no mines in either of these colonies have been operated by any persons but the subsidiaries of the Aluminum Company and of Aluminium Limited.[11] The Guiana deposits are among the most extensive of the world's bauxite directly accessible to ocean shipping.[12] In quality they are somewhat superior to the Arkansas ore because of a slightly higher alumina content and less silica. With the necessity of resort to underground mining in Arkansas early in the last decade the Aluminum Company began to import substantial quantities of Guiana ore for its domestic reduction plants. In the years 1925–1930 more than 500,000 tons per annum were brought in.

Very likely the vigorous campaign of the Aluminum Company in South America hastened the worldwide scramble for the better bauxite properties which occurred after the war stimulus to the aluminum industry. In this international competition the American company took a prominent part, with the result that it came to possess fairly large ore reserves in various portions of Europe.[13] An ore tonnage in France nearly as great as that now remaining in Arkansas was acquired by the Bauxites du Midi, a 100 per cent subsidiary. Istrian deposits of about the same extent were held by another completely owned subsidiary, the Societa Anonyma Mineraria Triestina. Jugo-Slavian ore was obtained through acquisition of majority stock holdings in the Jadranski Bauxite Dioni'co Drus'tvo and the Primorske Bauxite Dioni'co Drus'tvo, which have recently been merged.[14] It has been reported that the Aluminum Company made an unsuccessful attempt to buy into the Bauxit Trust, a subsidiary of the German aluminum monopoly.[15]

Interests in several companies producing aluminum and owning

[10] MI, XXXIII, 40 (1924).

[11] FTC Docket 1335, Record, p. 675; Yearbook of the Bermudas, the Bahamas, British Guiana, etc., 1931, p. 361.

[12] It is reported that the reserves of bauxite in British Guiana exceed 10,000,000 tons (*ibid.*).

[13] Its European acquisitions started in 1912, when some Dalmatian and French lands were bought, but most of its purchases have occurred since the war.

[14] Information on European acquisitions is contained in the testimony of an officer of the Aluminum Company, FTC Docket 1335, Record, pp. 18 ff.; and in BMTC appellant v. ACOA, fols. 772 ff.

[15] Czimatis, *op. cit.*, p. 102.

desirable water-power rights were also obtained by the American company. In 1921 an arrangement was made with the nearly bankrupt Aktieselskab Hoyangfaldene Norsk Aluminium Company, whose plant had been built at high cost during the war.[16] The Aluminum Company purchased a half interest in a new Norsk Aluminium Company which acquired all the assets of its predecessor except water rights and power plant. (Norwegian law provided that the latter must remain in the hands of citizens of Norway.) The Americans were to have a majority of the directors of the new Norsk Company. This firm possessed bauxite in France and in Dutch Guiana, an alumina plant in France, 80,000 potential horsepower, of which 30,000 h.p. had been developed for reduction works with a capacity of 7,000 tons per year, and a fabricating plant. It produced at least 6,500 tons of aluminum annually, much of which was at first imported into the United States. In 1923 the Aluminum Company bought a one-third interest in Det Norske Nitridaktieselskab, which had begun aluminum manufacture in 1912.[17] The British Aluminium Company and the Compagnie Alais, Froges et Camargue, each own one-third of this corporation, whose capacity is about 15,000 tons. The Norsk company indirectly owns the Kinservik water-power site capable of developing about 125,000 h.p.; and Det Norske Nitrid holds a part interest in the undeveloped 175,000 h.p. Tysse site. The Americans also took a majority control in Det Norske Aktieselskab for Elektrokemisk Industri, which, in addition to possessing valuable electrode patents, had a substantial interest in the Tysse water power. With characteristic foresight and vigor, the American aluminum interests acquired control of a considerable portion of water power in the country which is regarded by many as the logical resort for electrochemical industries in the future.

[16] *Industrial and Engineering Chemistry,* XVI, 979 (September 1924). Other information about the Norwegian subsidiaries has been obtained from *Aluminum Industry,* I, 54–57; testimony of officers of the Aluminum Company in FTC Docket 1335, Record, pp. 39 ff., and 669–671; testimony and exhibits in the cases of *Haskell* v. *Perkins* in the District Court of New Jersey, and BMTC appellant v. ACOA.

[17] Before completing these purchases the Aluminum Company made formal application to the Attorney General for a modification of the consent decree of 1912. Upon consent of the Attorney General the United States District Court for the Western District of Pennsylvania, on October 25, 1922, issued a supplementary decree modifying the original decree by permission to make the desired purchases. No other modification was made at that time.

In 1925 the Aluminum Company turned to France, where it purchased the Société Anonyme des Forces Motrices du Béarn, thereby securing 20,000–30,000 h.p. on the Aspe River in the Pyrenees. This is at present all sold to the P. L. M. railway. The American enterprise also acquired from the French aluminum interests the Societa dell'Alluminio Italiano, which produces about 1,400 tons of aluminum a year. Semifabricating and finishing plants were located at strategic points in various foreign countries, and American sales offices maintained in every country of importance in the world.

While the growth of American investment in Europe influenced international relations in this industry appreciably, the most important plant development of international significance since the war occurred in Canada. Some years ago Mr. J. B. Duke, who was famous for his interest in water power as well as tobacco, purchased with Sir William Price certain riparian properties on the Saguenay River where more than a million horsepower might be developed in the heart of a wilderness. In 1924–1925 Mr. Duke and Mr. George Haskell, an American manufacturer of duralumin alloys, were apparently considering the production of aluminum with part of the 360,000 h.p. to be generated in a nearly completed plant of the Duke-Price Power Company at Isle Maligne, the so-called Upper Development.[18] However, during 1925 Mr. Duke and the Aluminum Company of America reached an agreement whereby a new Aluminum Company of America was incorporated to take over the assets of its predecessor in name and those of the Canadian Manufacturing and Development Company.[19] The whole property of the latter had consisted of the entire stock of the Chute-à-Caron Power Company, whose name was soon changed to Alcoa Power Company, which in turn owned the riparian lands and water rights for the Lower Development, some miles below Isle Maligne, where it is estimated there is nearly a million potential horsepower. Less than a year after the merger the Aluminum Company acquired about 53 per cent of the stock

[18] See below, pp. 132 ff.

[19] Prior to the merger the Aluminum Company had common stock outstanding of $18,729,600 par value and a surplus of about $100,000,000. As a result of the financial arrangements incident to the merger, by which most of the surplus was capitalized, the capital structure came to include preferred stock of $147,262,500 par value and 1,472,625 shares of no par common carried at $5 per share.

of the Duke-Price Power Company, which owned the Isle Maligne hydroelectric station.[20] This plant was then generating 360,000 h.p. and has since been enlarged to an installed capacity of 400,000 h.p., which yields about 350,000 constant h.p.[21] Coincident with this transaction the Aluminium Company of Canada (the Northern Aluminum Company, in more appropriate appellation) contracted to take from the Duke-Price Company (the name of which is now the Saguenay Power Company, Limited) for fifty years 100,000 h.p. per annum at its new reduction plant, which was then being constructed within a few miles of Chute-à-Caron. In July 1926 energy began to flow into the cells of this plant at Arvida, as the new city built by the Aluminum Company is called. The works at this location include an electrode factory and a "dry process" oxide plant, employing electric energy instead of coal, which went into operation in 1928. Bauxite is shipped from British Guiana in ocean steamers to Port Alfred, which is only twenty miles from Arvida. The total reduction capacity in Canada is now about 40,000 tons; but it appears that this can be rapidly increased by about 50 per cent whenever demand warrants.

The first stage of the Lower Development, upon which work was begun in 1927, was completed in 1931 with a dam and powerhouse with 260,000 installed horsepower. A projected dam on a second branch of the Saguenay would bring the Lower Development to 800,000 h.p.

Its new position as an owner of extensive properties in various parts of the world was recognized by the Aluminum Company in June 1928 by the creation of a Canadian corporation, Aluminium Limited, which took legal title to nearly all the foreign holdings of the parent with the exception of the Surinaamsche Bauxite Maatschappij and the Alcoa Power Company, which owns the first power plant and the riparian rights at the Lower Development on the Saguenay.[22] Legal relations between the Aluminum Company of America and Aluminium Limited were terminated when the former

[20] Twenty per cent of the Duke-Price stock was taken at this time by the Shawinigan Water and Power Company. The Price interests and the Duke estate (Mr. Duke died shortly after the merger) retained about 27 per cent of the stock. The financial transaction was reported in the *New York Times*, April 29, 1926, p. 31.

[21] BMTC appellant v. ACOA, fols. 1000 ff.

[22] FTC Docket 1335, Record, pp. 5533 ff.; and BMTC appellant v. ACOA, fols. 992 ff.

distributed to its stockholders the shares of the new Canadian company. Mr. E. K. Davis, for many years vice-president of the Aluminum Company, became the president of Aluminium Limited, and several experienced members of the older company went with him.

The two corporations have had no directors or officers in common. On June 4, 1928, 490,895 common (voting) shares of Aluminium Limited were issued, of which twenty went to the applicants for the charter of incorporation. Of the 490,875 common shares of Aluminium Limited distributed on that day pro rata to the stockholders of the Aluminum Company of America, 270,431 shares, or approximately 55 per cent of the total voting shares, were received by four shareholders of the Aluminum Company of America. These four persons must then have owned about 55 per cent of the common voting stock of the Aluminum Company, since the distribution was in direct proportion (1 to 3) to holdings of stock in the latter. On December 31, 1931, there were outstanding 592,299 common voting shares of Aluminium Limited,[23] of which 307,696, or about 52 per cent, were held by two of the four persons referred to above, a corporation wholly owned by the third, and a corporation owned jointly by the fourth and his son and daughter. The four persons in question held on December 31, 1931, about 41 per cent of the common stock of the Aluminum Company of America. The total of their holdings and those of members of their immediate families and those of the president of Aluminium Limited amounted to slightly more than 50 per cent of the common stock of the Aluminum Company. A part of the rest of the stock of this company was owned by directors and officers and the heirs of deceased directors.[24] For a short time the secretary, treasurer, and auditor of Aluminium Limited had their offices in Pittsburgh, where the accounts were kept. Default on preferred stock dividends by Aluminum Limited in 1932 and 1933 gave the preferred stock the right to elect two directors. This may have accounted for two changes in the directorate of Aluminium Limited in 1933, when two former members were replaced by the treasurer of the company and a vice-president of a

[23] Annual report.

[24] The facts presented above have been taken from data in stipulations of the attorneys for the Aluminum Company of America (BMTC appellant v. ACOA, fols. 6019 ff.) and testimony (*ibid.*, fols. 1105–1106), and from Poor's Industrials.

Pittsburgh bank which acted as transfer agent for the company. In 1933 the preferred stock of the Aluminum Company became entitled to vote as a result of default on preferred dividends. According to the annual reports no changes occurred in the make-up of the board of the Aluminum Company between March 1932 and March 1936 except the reduction of its size occasioned by the deaths of two members, who were not replaced. One of the men who died was the owner of one of the personal corporations mentioned above. At the end of 1931 there were many stockholders not common to the two corporations, and it is said that the number has since increased. It has been impossible to obtain information upon changes in stockholdings since 1931.

It has recently been testified by representatives of both companies that no common control of the two corporations has been exercised.[25] The two corporations seem to have been distinct operating units.[26] Since 1928 the Aluminum Company has occupied itself principally with production for the United States market, while Aluminium Limited has been concerned with the problem of developing foreign markets. Owing to the tariff policies of the principal consuming countries of Europe, the chief outlets for its product have been in Asia and various parts of the British Empire. In corporate domicile, location of mines and plants, and in nationality of labor Aluminium Limited is a British firm enjoying intra-Empire trade privileges.

In the post-war decade the Aluminum Company of America rose to a position which implied some threat to the markets of the less powerful European producers. Cheap power and suitable bauxite had been acquired in places which were reciprocally well located from the point of view of cheap transportation. The heavy investment in reserves of ore and power suggested a strong incentive to cultivate world markets in vigorous fashion, while a very low cost of production at Arvida and the United States tariff may have made the Americans less apprehensive than the foreign producers about a real trial of strength. The formation of a new aluminum cartel by the leading European producers in 1926 doubtless reflected fear of the Americans as well as a desire to regulate competition among themselves. In the closing years of the twenties

[25] BMTC v. ACOA appellant, fols. 1080–1083, 5681, and 5687.
[26] *Ibid.*, fols. 980–996, 1025–1028, 5920–5921.

increasing exports from Canada were sold in markets formerly served by the Europeans. In 1931 Aluminium Limited formally entered the international cartel.

Up to the present Aluminium Limited has filled its power requirements almost entirely from its subsidiary, the Saguenay Power Company, which operates the powerhouse at Isle Maligne, the Upper Development. It appears that energy from the plant of the Alcoa Power Company at the Lower Development, ownership of which was retained by the Aluminum Company, has found little market as yet. The riparian rights for the second stage of the Lower Development are also owned by the Alcoa Power Company.

While the more spectacular expansion outside the borders of the United States introduced new elements into the home situation, the plants in this country have also been considerably extended. The Aluminum Company emerged from the war with a reduction capacity in the United States of approximately 315,000 h.p., probably equivalent to a maximum of 70,000 tons under average conditions of water flow, which would doubtless have taken care of the demand in 1919 and 1920 had it not been for abnormal labor and transportation conditions.[27] No substantial increases in power and reduction facilities in the United States came into operation until 1927. In the intervening years the company was not at all times able to satisfy the calls for metal, and, as foreign producing firms were acquired, their output was largely diverted to this country.[28] After the installation of the Cheoah dam and powerhouse in 1919 the projected developments on the Little Tennessee were halted for a time, partly owing to the higher costs of construction work and the uncertainties of business depression. In 1925 they were again resumed when a dam was started at Santeetlah. Before this was completed in 1928 a small power development had been built at High Rock, North Carolina, to serve the Badin reduction plant, and the third dam in the Little Tennessee series was under construction. The powerhouse at Santeetlah began to generate its maximum of 66,000 h.p. in 1929, while during the following year Calderwood, the third development, came into operation with two 56,000 h.p. units and room for a third unit of

[27] The capacity of reduction plants under normal conditions for the years 1919–1925 was given by an official of the company (FTC Docket 1335, Record, 3171 ff.).

[28] There were extenuating factors involved. See Chap. XVIII for discussion.

the same capacity.[29] Since the power capacity at Massena had also been extended, average reduction capacity in the United States had been raised to 550,000 h.p. or thereabouts, which is equivalent, perhaps, to 125,000 tons of virgin metal.

The plans of the Aluminum Company call for a total of eight dams along a 104-mile stretch of the Little Tennessee River which will completely utilize the entire flow of the stream.[30] The company owns between 60 and 85 per cent of the land required for the five remaining developments. Some work has been done on the fourth basin at Chilhowee, Tennessee, where about 35,000 h.p. will be developed.[31] The Fontana dam, which is the most important of the other four projected developments, is expected to have a capacity of about 300,000 h.p.[32] The whole Little Tennessee development will probably yield at least 800,000 h.p.

The government program for development and regulation of the Tennessee River system with respect to navigation and flood control may result in some curtailment of the freedom of the Aluminum Company to decide the amount and character of its future investment in power facilities on the Little Tennessee River and the conditions of their operation. Section 26a of the Tennessee Valley Authority Act as amended in 1935 provides that no dam or appurtenant works may be constructed and thereafter operated on the Tennessee or any of its tributaries until plans for construction, operation, and maintenance have been submitted to the board of the TVA and approved by it. However, if the board fails to approve the plans within sixty days after their submission, the requirements just stated are to be deemed satisfied if the plans are approved by the Secretary of War as reasonably adequate and effective for the unified development and regulation of the Tennessee River system.[33] Evidently the TVA or the Secretary

[29] *Power,* LXX, 341 (August 1929); LXXI, 801 (May 1930).

[30] Hearings on the Tennessee Valley Authority, House Committee on Military Affairs, 74 Cong., 1 Sess., II, 662 (hereafter cited as TVA Hearings).

[31] *New York Times,* August 9, 1929, p. 24; *Electrical World,* XCIV, 440 (August 1929). Actual construction of this dam at Chilhowee, Tennessee, was held up for some time by a dispute with the Southern Railway, which was intending to build an extension through territory that the Aluminum Company planned to flood as a storage basin.

[32] TVA Hearings, p. 672.

[33] Upon application to the Secretary of War due notice must be given the TVA and hearings held.

of War may require as a condition of approval whatever terms concerning construction, operation, and maintenance seem appropriate for realization of the program for flood control and navigation on the Tennessee River system. The government plans call for six dams on the Tennessee below the point where the Little Tennessee empties into it. At times of low water, stream flow on the Tennessee might be appreciably affected by the rate at which water was allowed to flow over the dams of the Aluminum Company on the tributary.[34]

The post-war decade has also witnessed a very great extension of this company's facilities at other stages of the industry. In 1924 the Aluminum Ore Company purchased an important group of fluorspar mines in Illinois and Kentucky which, added to small holdings previously obtained, made it the dominant producing unit in this field.[35] Fluorspar is used in making artificial cryolite to substitute for the natural product.

The plant devoted to semifabricating and finishing activities has probably been augmented in greater proportion than the reduction facilities. Towards the end of the twenties the company was selling a smaller proportion of metal in the form of ingot than had been the case earlier.[36] About 1920 two large mills were built at Edgewater, New Jersey, and Alcoa, Tennessee, to care for the enlarged sheet demand. More recently the promising development of the strong alloys has resulted in the erection at Massena of the first blooming mill to produce aluminum billets and structural shapes, and the building of a mill at Alcoa, designed to roll duralumin and "Alclad" sheet. In 1922 the Aluminum Company leased for twenty-five years all the plants of Aluminum Manufactures, Inc.[37] This concern was the old Aluminum Castings

[34] Before Section 26a was made a part of the law the TVA purchased two small plots of land in the projected reservoir for the Fontana development of the Aluminum Company, apparently with the object of being in a position to force the company to agree to whatever conditions the TVA considered appropriate for the realization of its program (TVA Hearings, pp. 669 ff., 705–706). It seems questionable whether such practices on the part of government agencies are desirable.

[35] *New York Times*, November 27, 1924, p. 31; MI, XXXIII, 18 (1924). The name of this alumina subsidiary has recently been changed to Alcoa Ore Company.

[36] See below, p. 251.

[37] At this time the Aluminum Company acquired stock holdings in two small subsidiaries of Aluminum Manufactures, the Aluminum Diecasting Corporation and the Aluminum Screw Machine Products Company.

Company, which had expanded under a new name just in time to suffer embarrassment in the automobile slump of 1920–1921. Thenceforth these plants were run by the United States Aluminum Company, thus substituting more direct operation for the majority stock control possessed at the time of the lease. The Aluminum Company later increased its ownership of common stock in Aluminum Manufactures to more than 70 per cent.[38] As soon as the lease was signed, there was launched a vigorous campaign which brought this castings unit out of the financial doldrums by regaining for it the prominent position in this branch of the industry which had formerly been held by the Aluminum Castings Company. About 1929 the Aluminum Company erected a magnificently equipped research laboratory at New Kensington which exhibits aluminum attractively in its design, fittings, and furnishings.

Finally, a word should be added about those activities of the Aluminum Company which are only indirectly related to the aluminum business. It will be recalled that several small electric companies in the neighborhood of Massena were acquired soon after the purchase of the St. Lawrence River Power Company.[39] These companies have distributed electricity for light, heat, and power to the general public, in addition to serving as auxiliary supply for the Massena reduction plant. In 1924 they were merged into the St. Lawrence Valley Power Corporation, which was created for that purpose.[40] In the South a subsidiary called the Nantahala Power and Light Company engages in an ordinary public utility business. In 1921 the Frontier Corporation was formed to consolidate the undeveloped riparian rights along the St. Lawrence owned by the Aluminum Company, the General Electric Company, and the Duponts. Since the Canadian and United States governments took the position that the St. Lawrence

[38] Report of Special Assistant to the Attorney General, William R. Benham, concerning alleged violations by the Aluminum Company of America of the decree entered against it in the United States District Court for the Western District of Pennsylvania on June 7, 1912, being Senate Doc. no. 67, 69 Cong., 1 Sess., p. 36 (hereafter cited as BR).

[39] In 1921 the Aluminum Company participated with General Electric and others in buying a controlling interest in the Niagara, Lockport and Ontario Power Company.

[40] This company should not be confused with the St. Lawrence River Power Company, which generates power chiefly for use in the reduction plant at Massena.

power should be developed as a whole, it was thought best to join interests. In exchange for its Long Sault lands the Aluminum Company received one-third of the common and 88 per cent of the preferred stock of this company. In September 1931 both the Frontier Corporation and the St. Lawrence Valley Power Corporation were sold to the Morgan interests for inclusion in the Niagara-Hudson Power Company. Two million five hundred thousand shares in the latter and one directorship came to the Aluminum Company. The Aluminum Company owns four subsidiaries which operate short railroads as common carriers in the vicinity of its plants. The Republic Mining and Manufacturing Company mines and sells large quantities of bauxite to the chemical trade. The Aluminum Company has also built and "operated" towns at most of its mines and power sites. Its own engineers build its dams and power plants. Another sort of relationship to aluminum is exemplified in the ownership of the American Magnesium Corporation, which has only one domestic competitor in the manufacture of this potential rival of the Aluminum Company's product.

The Aluminum Company of 1935 is a large, thoroughly integrated business unit with extensive reserves of ore and power, a prominent position in several fabricating branches, and no domestic competitors in the production of virgin ingot. Its impressive financial strength is enhanced by membership in a strong group of financial and industrial enterprises. Until the close of the war the Aluminum Company had financed nearly the whole of its expansion from earnings. In the post-war period about 40 per cent of the net addition to assets seems to have come from the sale of notes, bonds, and preferred stock.[41] Aluminium Limited, which became the owner of most of the foreign properties of the Aluminum Company in 1928, has attained a position of great importance in the world aluminum industry.

In the post-war period annual imports of aluminum ingot into the United States from European producers seem to have averaged 12–15 per cent of total domestic sales of primary aluminum in all forms.[42] Only in the depression years 1921 and 1922 did the estimated participation of European firms in the United States

[41] Below, p. 262.

[42] Annual percentages have been estimated as explained in the note to Table 27 on page 325. Data are from Tables 15, 17, and 27, and from BR, pp. 117–118.

market exceed 20 per cent. In the years 1923–1927 their proportion apparently averaged about 15 per cent, while during the period 1928–1933 it was evidently less than 10 per cent. Imports of fabricated aluminum have ordinarily been negligible. The period 1913–1922, during which the import duties were 2 cents per pound on ingot and 3½ cents per pound on sheet, was too abnormal to afford any indication of the adjustment which might develop with low duties. In 1922 the rates were raised to 5 cents and 9 cents respectively. The Smoot-Hawley tariff lowered them to 4 cents and 7 cents in 1930. Partly because of the tariff and partly for other reasons the European firms have apparently not established any substantial increment of capacity to serve the American market.

2. New and Old National Monopolies in Europe

During the war the production of aluminum, as well as of many other basic products, was expanded very much farther in America than in Europe.[43] But under consumption influences much the same as those in America the output of the European companies was nearly doubled between 1918 and 1929, while that of the Americans was increased by a little more than 100 per cent.

The only important additions to the family of European producers have been the German government corporation, which now ranks ahead of the older European firms in output; the Soviet government plants, of which two at least have gone into operation in recent years; and two Italian companies, fostered by the Italian government, which use Istrian bauxite acquired by the peace settlement. The German and Swiss aluminum companies participated in the financing and direction of the Italian enterprises. Aluminum capacity in Norway, which was small twenty years ago, has been extended substantially by the French, British, and American companies. The other two outstanding developments related to organization of the industry in Europe have been the increasing American investment in Europe and the growth of closer and closer coöperation between European firms for market

[43] Estimates of production are given in Table 25, p. 308, and Table 38, p. 569. See also Chart I, p. 35.

control, a movement to which Aluminium Limited formally attached itself in 1931.

The German corporation, whose striking growth is partially due to government protection from competition in a market which was already well developed,[44] had its inception in war necessity. In addition to the use of aluminum for the purposes already described, it was imperative for the Germans to substitute this metal for copper, from which they were largely cut off. Accordingly three small aluminum plants were built during 1915, 1916, and 1917 at points where sufficient electric energy was already at hand.[45] Two of these were very small, high-cost plants which were abandoned after the war.[46] In December 1917 the Erftwerk, with a capacity of 12,000 tons a year, was completed, and just prior to the armistice the Lautawerk began to operate with about the same capacity. Both of these plants were located in lignite regions, since power generated with brown coal is the cheapest kind of steam-produced electricity. The Erftwerk in Grevenbroich (Niederrhein) was served by the Rheinisch-Westfälisches Elektrizitätswerk A. G. A new power plant was erected by the government at Lauta in the lignite district of Saxony. At the same time plans were made to develop a large water power on the Inn River in Bavaria.

The original financing of these developments had been divided between the government and prominent private companies engaged in the metal, electric, and chemical industries.[47] The private companies owned minority stock interests in the Vereinigte Aluminiumwerke A. G., the corporation formed in 1917 to operate the new plants, and in its subsidiaries, Erftwerk A. G. and Innwerk, Bayerische Aluminium A. G. The private interests had, however, a majority of the board of directors of each. When the government decided after the war to continue its direct interest in the

[44] The German market had been supplied chiefly by the Neuhausen concern before the war. See below, p. 127.

[45] Details of the development of the German industry are to be found in Georg Günther, *Die deutsche Rohaluminium Industrie* (Leipzig, 1931); Hans Bannert, *Der Rohaluminiumweltmarkt* (Halle, 1927); Schoenebeck, *op. cit.;* Buschlinger, *Entwicklung und Aufbau der Aluminiumwirtschaft,* and periodical literature.

[46] The third, at Bitterfeld, was later sold to the I. G. Farbenindustrie and the Metallgesellschaft, which operate it jointly.

[47] Chief among these were the Metallbank group, Griesheim Elektron, Gebrüder Giulini, A. E. G., Siemens-Schuckert.

development of this industry, the product of which was so important to Germany in war and peace, the private concerns sold their holdings to the Reich.[48] Since 1925 aluminum-ingot production has been controlled exclusively by the Reich, except for the small plant at Rheinfelden operated by the Neuhausen firm and the Bitterfeld works now owned jointly by I. G. Farbenindustrie and the Metallgesellschaft. The Vereinigte Aluminiumwerke owns directly the Lautawerk and the Innwerk, which began to operate in January 1925 with a capacity of 10,000 tons, and all of the stock of Erftwerk A. G., which operates the plant of that name. The Vereinigte Industrieunternehmungen A. G. (VIAG), the Dachgesellschaft which holds the stock of industrial undertakings of the government, owns in addition to the Vereinigte Aluminiumwerke a corporation which operates the power plant at Lauta, and the company which owns the 100,000 h.p. plant on the Inn River. The Vereinigte Aluminiumwerke soon became a well-integrated concern. An electrode factory was attached to the Erftwerk at the start, and an alumina plant was built at Lauta. Bauxite deposits were acquired in Istria shortly after the peace settlement. In 1923 the German company, in conjunction with the Otavi Minen-und-Eisenbahngesellschaft, formed the Bauxit Trust A. G., which embarked upon a determined campaign of bauxite acquisition, particularly after the Italian government restricted exports. Extensive deposits in Hungary were purchased in 1925 and 1926, in addition to smaller ore holdings in other countries.[49] The power plants serving the Lautawerk and the Innwerk are owned by the VIAG, which also possesses a minority interest in the Rheinisch Westfälisches Elektrizitätswerk A. G., which supplies energy to the Erftwerk. The rolling, fabricating, and casting branches of the aluminum industry were well developed in Germany before the war. For the most part the Vereinigte Aluminiumwerke has contented itself with creating a well-integrated organization which stops short of extensive partic-

[48] It has been said that the government and the private interests were more or less at odds until the Reich took over complete ownership. See Albrecht Czimatis, *Rohstoffprobleme der deutschen Aluminiumindustrie in Rahmen ihrer wirtschaftlichen Entwicklung* (Dresden, 1930), pp. 92–93.

[49] The chief subsidiaries of the Bauxit Trust are the Aluminium Erz Bergbau und Industrie A. G., Budapest; the Tapolcza Mining Company, Budapest; and the Continentalen Bauxit Bergbau und Industrie A. G., Agram.

ipation in these branches. But it joined with the Metallgesellschaft in the production of the casting-alloy Silumin, and owns, in conjunction with the same company and several other firms, the Vereinigte Leichtmetallwerke, which is devoted to the development and production of strong, light alloys and their products.

It is probable that the German company operated under somewhat higher costs, at least until the latter twenties, than the older European firms.[50] War-time necessities bequeathed two plants at which steam power is used. At one of these electricity must be purchased from a private company. Reduction in the costs of generating power from brown coal has reached the point where such energy is no more expensive than the probable cost of electricity at European water-power sites which have not yet been used, with the exception of some in Norway.[51] Yet it is not certain that the steam power at the Lautawerk is as cheap as the Inn water power or the marginal water power which has recently been developed by aluminum companies in other countries. The war plants resulted in some overcapacity — at prices which the German company wished to charge — which was continued, in spite of the rapid growth in demand, by the building of the plant on the Inn River. The German company has always used a small part of its power for the production of *Nebenprodukte*. The bauxite costs are certainly higher for the German company, because most of its ore is somewhat inferior in quality to the French bauxite and the transport expenses are greater. Coal, on the other hand, is much cheaper in Germany. On the whole, it is doubtful if the German aluminum output in the latter twenties could have been produced much more cheaply anywhere else on the Continent south of Norway. The recent policy of self-sufficiency and rearmament in Germany has resulted in enlarging aluminum-ingot

[50] The German disadvantage has probably been overstated by Günther (*op cit.*, pp. 52–54) and several other authors of newspaper and periodical articles, who incorrectly reckon in the large reserves accumulated by the old companies. Unless the latter do not expect to earn returns on such part of their investments, their costs are not lowered by the fact that their capital is written down on their books. They are financially stronger only in the sense that at lower prices they could show a profit on their capitalization.

[51] Personal interviews. Cf. also Bannert (*op. cit.*, p. 51), who concludes that the cost per kilowatt-hour of hydroelectric energy at the great plant on the Inn River would have been higher than brown coal power if the water plant had not been built during the depreciation of the mark.

capacity from 45,000 tons to about 75,000 tons. What effect this has had on cost does not appear.

The lack of satisfactory bauxite in Germany has stimulated research upon methods of using various clays. Processes have now been developed which are technically satisfactory, so that Germany could provide herself with aluminum should war cut off her bauxite supplies. It is said that one of these methods yields alumina just as cheaply as the Bayer process, which continues to be used because of the heavy investment in Bayer plants and bauxite reserves.

The Neuhausen concern, whose position as leading producer of Europe had been challenged by the French in the pre-war decade, forged ahead again under the influence of great prosperity during and immediately after the war.[52] But during the twenties, excluded for the most part from the German market, which had formerly been the chief outlet for its metal, the production of the AIAG has grown much less than that of the other European concerns. Its earlier position as the largest producer in Europe was first lost to the Vereinigte Aluminiumwerke, which was in turn outstripped by the French toward the end of the decade. The Swiss home market has, of course, always been much smaller than that of any other important producer. Faced with the problem of finding new markets for its metal in international competition — a problem made more difficult by depreciation of the pound, mark, and franc — the Neuhausen firm extended its control into the semifabricating and finishing branches.

The AIAG has evaded in some degree the force of nationalistic measures restricting importation of aluminum into various countries by carrying out part of its expansion in facilities for producing the basic product in such countries, notably in Italy, Austria, Spain, and Germany, where it is now equipped to produce perhaps 19,000 tons in the aggregate.[53] At home capacity was expanded somewhat by the construction of two new power plants

[52] A large part of this was evidently the result of a very favorable war-time contract with the German government, under the terms of which the latter was forced to receive substantial amounts of metal for a year or two after the armistice and to pay in Swiss francs. See *Aluminium,* II, Heft 7, p. 11 (February 16, 1920).

[53] Speeches in general meetings of the corporation, 1925 and later years. Reported in *Aluminium,* VII, Heft 5, p. 10 (March 16, 1925); VIII, Heft 8, p. 9 (April 30, 1926); IX, Heft 8, p. 5 (April 30, 1927).

near Chippis, the Illseewerk and Turtmannwerk of about 11,000 h.p. and 21,000 h.p. respectively, which were completed in 1925 and 1926.[54] Since part of this power was devoted to other products, aluminum capacity was not increased markedly. The capacity of the reduction works in Switzerland at present is probably between 25,000 and 30,000 tons per annum.

At the outbreak of the war there were four aluminum companies in France united in a domestic cartel, l'Aluminium Français. In 1916 the more aggressive of the two older concerns, the Compagnie Alais, absorbed one of the two remaining outsiders, the Société des Forces Motrices de l'Arve. During the war the capacity for aluminum production was not enlarged in France, but the Compagnie Alais took over Det Norske Nitridaktieselskab, which had begun to manufacture aluminum in 1914 at Eydehavn, Norway, where 25,000 h.p. was used. In 1916 another reduction works was put into operation at Tyssedal, using 30,000 h.p., in order to meet the growing demand of the Allied powers for aluminum. Evidently French penetration of Norway was considered a war measure only.[55] The Compagnie Alais was at the time considering ambitious projects for expansion in Italy, the Balkans, and India. In any event two-thirds of the capital of DNN were sold in equal parts to the British Aluminium Company and the Aluminum Company of America in 1923.

A year or so after the close of the war the old Froges concern was absorbed by the Compagnie Alais, which thenceforth was known as the Compagnie de Produits Chimiques et Électrométallurgiques Alais, Froges et Camargue, or, in brief, the Compagnie AFC.[56] This merger, which brought seven of the eight reduction plants in France into one organization, left the Société d'Électrochimie, d'Électrométallurgie et des Aciéries Électriques d'Ugine as the only remaining company. In so far as it concerns itself with aluminum, which has never been its chief product, the interests of

[54] Schulthess, "Die Entwicklung der Aluminiumindustrie in der Schweiz," *Schweizerische Technische Zeitschrift*, Jahrgang 1926, nos. 34–35, p. 6.

[55] See Victor Barut, *L'Industrie de l'électro-chimie et de l'électro-métallurgie en France* (Paris, 1924), p. 160, who quotes M. Badin as follows: "Bah, nous en sortirons. . . . Ne faut-il pas être marin pour faire de l'aluminium en Norvège? Mais oui, parceque l'industrie y est surtout une question de transport maritime."

[56] Most of the following information about this company has been secured from its annual reports, JFE, and the *Revue de l'aluminium*.

this firm have been closely allied with those of the Compagnie AFC through l'Aluminium Français and one or two joint ventures. In 1928 some shares of the Société d'Électrochimie were bought by the Compagnie AFC. The latter, whose principal product is aluminum, has regularly produced 90 per cent or more of the total output.

After the addition of a small plant at Les Clavaux, belonging to the Société d'Électrochimie, about 1920, the aluminum capacity in France remained substantially the same until the middle of 1924, when a plant of the Compagnie AFC came into operation at Beyrède, where water power had been developed before the war. With a great growth in demand in the early twenties the company carried through a program of expansion which brought into operation between 1925 and 1929 three new water-power developments and adjacent reduction plants at Rioupéroux, St. Auban, and Sabart. A small plant at Venthon was opened by the other French producer in 1928. The total capacity for aluminum production in France, which had been, perhaps, 20,000 tons in the early twenties, appears to have been increased to more than 30,000 tons at the end of the decade. Three other water-power developments of the Compagnie AFC are just about completed. Whether reduction capacity has also been extended in the last few years does not appear. Perhaps one may infer from the growth of closer relations with a large power company in the same region that it is intended to sell a substantial part of the additional energy. In 1925 the plants of Det Norske Nitrid were expanded somewhat. Fabricating mills, foundries, and other facilities of the Compagnie AFC have been greatly extended, and this company, which is said to own about 40 per cent of the extensive French bauxite deposits, has recently been purchasing more ore lands abroad.

The capacity of the English aluminum industry remained unchanged for more than a decade after 1914, although the amount of metal produced under British control was increased by participation of the British Aluminium Company in Det Norske Nitrid of Norway. In accordance with an agreement between the English government and the British Aluminium Company plans were laid during the war for an extensive hydroelectric development near Lochaber in Scotland.[57] Parliament did not, however, pass

[57] Information upon this development is found in the following sources: *Econ-*

the requisite legislation until 1921. After further delay of four years the North British Aluminium Company was incorporated to undertake the development. A loan of £2,500,000 was guaranteed by the government to reduce interest charges. With the addition of about 45,000 h.p. upon the completion of the first stage of this development in 1929, capacity for aluminum production in the British Isles was more than doubled. At that time it was contemplated that the second stage, which would add about 75,000 h.p., making this the largest hydroelectric development in Great Britain, would be finished by 1938. In spite of low interest charges the energy cost at the new plant is probably quite high relative to that which would be incurred in Norway, where the British Aluminium Company was already firmly established. Since the war this company has climbed from its inferior pre-war position to a point where its output more nearly approaches that of the other leading European concerns. The small Aluminium Corporation undertook to secure for itself a substantial slice of the expected growth in business by development of 50,000 h.p. on the Glomfjord in Norway. This plant had scarcely been completed, however, when the depression forced the company to sell its entire property to the international cartel.

The largest increases in aluminum capacity in the thirties will take place in Russia if even a fractional part of the ambitious plans of the Soviet government is realized.[58] The known deposits of bauxite in Russia are estimated to contain at least 5,000,000 tons of ore. It is said that the Russians intend also to produce aluminum from alunite, supplies of which are estimated at 100,000,000 tons of metal. The first reduction plant produced 4,000 tons of aluminum in 1932 at Swanka, using power from the Wolchow River. In the following year the great Dnieper power plant began to deliver energy to near-by reduction works, which were brought to an annual capacity of 20,000 tons in 1935.[59] Aluminum plants at Rion and Kamensk are now under construction. The second five-year plan calls for a total capacity in these

omist, LXXXVI, 550 (1918), XCIV, 642 (1922), XCVIII, 704 (1924), C, 621 (1925), CXII, 747 (1931); *Canadian Mining Journal,* LI, 1204 (1930); *Engineer,* CLV, 559 (1933); *Engineering,* CXXXV, 636 (1933).

[58] *Industrial and Engineering Chemistry,* News Edition, XII, 247 (July 10, 1934).

[59] MW, XXIV, 154 (1935); *Minerals Yearbook,* 1936, p. 411.

four plants of nearly 100,000 tons by 1937. It provides also for five other aluminum plants which, if completed according to plans, will enlarge total capacity to 230,000 tons per year, more than the estimated present capacity of either Europe (exclusive of Russia) or America.

A material growth in international competition during the post-war decade is evidenced by the larger foreign trade figures and by the increased investments in foreign countries. The leading role in foreign expansion has been taken by American capital. In addition to the acquisitions already described, it was said in the European press that the Americans attempted to purchase substantial interests in the Vereinigte Aluminiumwerke and the Bauxit Trust, and negotiated with the Hungarian and Italian governments about concessions relative to the establishment of organizations for aluminum production in those countries. Whatever the truth concerning these matters, both the French and Swiss companies took steps, ostensibly directed against the Americans, to prevent purchase of control by outsiders. In 1927 the Compagnie AFC issued to some of its stockholders a small lot of class B shares, each of which carried twenty times the voting power of one old class A share.[60] This example was followed in the next year by the AIAG, which issued a new class of preference shares possessing a total voting power equal to that of the old shares.[61] Transfer of the voting rights by sale of such shares was made subject to approval of the board of directors.

In 1934 the B shares of the Compagnie AFC were withdrawn to meet a change in the law. Their owners received two new A shares for three B shares, with the result that the capital stock now consists entirely of one class of shares, each of which possesses one vote. Some measure of protection against foreign purchase still remains, however, inasmuch as French owners of registered personal shares are entitled to a double vote when their shares have been registered in their names for five years.[62] The

[60] This gave the class B shares 16,000 votes in the aggregate as against 20,000 for class A. When capitalization was increased in 1929, the number of class B shares was increased in such proportion that both classes possessed an aggregate of 25,000 votes. The class B shares had one vote each, while class A shares had one vote for each block of 20 (JFE, xxxv, 179, 1926, and annual reports of the company).

[61] See annual report of the company for the year 1927, and press comments.

[62] This applies both to the new shares and all old shares registered in the name of

AIAG has evidently decided that the threat of foreign control has passed. In the spring of 1935 it recalled the preference shares issued seven years earlier.[63]

Expansion across national borders has not, of course, been confined to the Americans. Under the influence of competition for limited resources of ore and power (a topic which will be discussed in some detail in a later chapter) the European companies have also made extensive investments abroad. This tendency has been heightened by nationalistic policies restricting export of ore or import of metal. Next to the growing exploitation of the opportunities in Norway the developments in Italy have been the most significant. Until a few years ago the production of aluminum in this country was of negligible importance.[64]

The peace settlement had given Italy large reserves of good bauxite in Istria. Many of these deposits had been purchased by foreign capital, in particular the German state aluminum enterprise, before the Fascist regime became well established. In the middle twenties the Superior Council of National Economy undertook an intensive study of aluminum, and power to regulate the production and export of ore was conferred upon the minister of national economy. Apparently no official decree was promulgated, but exports from Istria were in fact limited to 100,000 tons per year beginning in 1926, and it was made clear that the government intended to foster home production of aluminum.[65] It appears that the Italian government, in conjunction with private interests, then began to negotiate with the various existing producers of the world relative to the establishment of a domestic industry which might avail itself of their experience, capital, and ore resources. Evidently the Germans and the Swiss were prepared to make the best offer. The Societa Italiano dell'Alluminio was organized by the Vereinigte Aluminiumwerke and the Montecatini group of

---

the same owner or his legal heirs for five years (*Assemblées générales,* XXX, 689, June 23, 1934; hereafter AG).

[63] MW, XIV, 334 (April 26, 1935).

[64] One small plant was operated at Bussi by Italian enterprise from 1910 on until its recent sale to the Neuhausen firm. The Compagnie Alais operated another small plant in Italy for a few years during and after the war. This was also sold to Neuhausen recently.

[65] The Istrian mines were organized for a maximum output of 350,000 tons per year at that time. In 1928 the export contingent was raised to 130,000 tons.

Milan.[66] It has built an aluminum plant at Mori in the southern Tyrol, where it could develop its own water power. An alumina firm, owned in common, has a factory in Marghera, near Venice, for digestion of Istrian ore. The reduction works, which had a capacity of about 6,000 tons, began to operate at the end of 1928. In 1935 the whole of the German interest in the aluminum enterprise was sold to Montecatini.[67] The S. A. Veneta dell'Alluminio, formed by Neuhausen and an Italian group, also began to produce metal in 1928 in a plant at Marghera which had about the same capacity as that in Mori. Energy is generated at Cismon in the near-by mountains. Aluminium Limited purchased a small plant at Borgofranco which formerly belonged to the French. Until quite recently the capacity of plants in Italy was estimated at about 13,500 tons per year.[68] It appears that this has been enlarged somewhat in the last year or so. The Italians have also expended much effort in an endeavor to develop a satisfactory process for obtaining aluminum and potash from leucite, a type of rock which abounds in Italy. Although it is said that no process yet discovered can compete economically with the existing methods of reducing the metal from bauxite, it has been demonstrated that it is possible technically to secure aluminum from leucite.

In 1927 a small plant was opened in Spain by Aluminio Español, owned jointly by Neuhausen, the Compagnie AFC, and the Aluminum Company. The latter interest was later sold to the other two. The Spanish government is said to have encouraged the creation of this company. Whatever other favors may have been bestowed, the tariff was raised from 8 gold pesetas per 100 kilograms to a protective figure of 82.50 gold pesetas per 100 kilograms (equivalent to about 30 per cent ad valorem at the time). Several of the new countries of Central Europe strove for several years after the war to foster domestic aluminum industries. Jugo-Slavia, Hungary, and Rumania all placed restrictions on the export of bauxite and attempted to interest foreign or domestic capi-

[66] See annual reports of Vereinigte Aluminiumwerke. The Germans had an interest of 40 per cent (*Handbuch der deutschen Aktiengesellschaften,* Ausgabe 1934, I, 293).

[67] MW, xv, 313 (1936). The Germans retained ownership of all the shares of the Societa Allumina, but the alumina plant in Marghera was sold to Montecatini and a toll arrangement made.

[68] JFE, xl, 361 (1931), abstract from annual report of Montecatini. This estimate has been verified in other sources.

tal in establishing reduction works within their borders. When these endeavors proved fruitless some of the restrictions were abandoned. Hungary agreed in 1926 to allow the export of 350,000 tons per annum to Germany for twenty-five years. According to a recent report a reduction plant with a capacity of 6,000 tons was finally put into operation in Hungary during the early part of 1935 by the firm of Manfred Weiss. A protective tariff was imposed in April.[69] The government of Japan also made fruitless efforts for many years to foster the establishment of a domestic industry. The lack of bauxite in the Japanese islands directed attention to the use of native clay. A government commission reported in 1926 that aluminum could be made from this material at the cost of $750 per ton, which was $200 greater than the import price at the highest level of price during the last ten years. Within the past two years a small aluminum producing plant has finally been established in Japan by the Japan Aluminum Reduction Company. The growth of economic nationalism since the war is also exemplified by the general increase in tariffs on aluminum in the leading European producing countries which began about 1926.

It is not surprising that after the restoration of normal peacetime relationships the increasing competitive influences in Europe, supplemented, perhaps, by some fear of the growing American investment in Europe, resulted in the formation of a formal cartel organization in 1926 to replace the temporary gentlemen's agreements which had existed between some of the firms in the early twenties. The cartel was composed of l'Aluminium Français, representing the two French producers, the AIAG of Neuhausen, the Vereinigte Aluminiumwerke, and the British Aluminium Company. The Americans did not enter the organization. This cartel, which was renewed for three years in 1928, achieved no strong control over output, although it regulated price effectively. In 1926 substantial augmentations of capacity had just been made or were in process of completion, and costs had already been reduced somewhat. During the next few years the cartel lowered price by nearly a third of the 1926 high. Since costs were apparently reduced by 20–25 per cent between 1925 and 1930 the profits of the cartel members did not suffer greatly. Perhaps the cartel prevented quotations from falling as far as they otherwise would

[69] MW, XIV, 356 (May 3, 1935).

have, especially after the depression began. The success in the control of the ingot price was not, however, reflected in the world markets for half-products and finished goods, where price reductions were greater. It was clear by 1931 that a stronger organization would be necessary if stringent control of output were to be achieved. An additional incentive for a more closely-knit organization lay in a desire to pool the large stocks which had accumulated. Furthermore, Aluminium Limited was at last willing to join a strong cartel. As a result, there was incorporated in Switzerland late in 1931 the Alliance Aluminium Compagnie, which is a highly centralized cartel designed to regulate production, sales, and price. It also finances the pool of accumulated stocks. Although organized as a corporation, it is not, of course, a holding corporation which unifies control of investment and manufacturing policies.

The Alliance cartel includes all leading producers except the Aluminum Company of America, the Russians, and the Italian companies. It appears that the latter remained outside at the instigation of the Italian government, which desired to have the domestic industry developed free from restrictions. The Italian import duty was nearly doubled in 1931. Owing partly to government patronage the Italian firms have produced almost to capacity during the depression. In 1934 the government prohibited imports of aluminum ingot and scrap and restricted the import of copper in order to save exchange and to facilitate the maximum use of aluminum where it could be substituted for its dusky predecessor.[70] Apparently the German government has insisted that output for domestic consumption be free from cartel restrictions. Under the influence of Nazi rearmament and government encouragement of substitution for copper the production of aluminum in Germany in 1934 was twice that of the foregoing year, and nearly doubled again in 1935. During the summer of 1934 the use of copper for transmission lines was prohibited, and the government made serious attempts to encourage the adoption of aluminum for many other purposes formerly served by imported copper. The capacity of the government plants was increased during 1934 and 1935 from about 32,000 tons to perhaps 55,000

[70] *Metals,* V, 18, 31 (December 1934).

tons, while the other two plants in Germany were also enlarged. Consumption, which had been less than 35,000 tons in 1933, rose to nearly 60,000 tons in the following year, and exceeded 90,000 tons in 1935.[71] In France the government acted to conserve future supplies of aluminum by a decree issued in April 1935 which forbade exportation of bauxite, alumina, or aluminum.[72]

In 1932 the total capacity of European aluminum works, excluding the first small Russian plant, seems to have been about 180,000 tons per year, and that in America about 165,000 tons. Since the European capacity owned by Aluminium Limited was almost exactly equal to the difference between these figures, American companies possessed about 180,000 tons estimated capacity and European firms about 165,000 tons. In the past few years the situation has been altered. While American capacity seems to have remained the same, enlargement of the German works, supplemented by small additions in Hungary, Sweden, and Italy, has carried total European capacity, exclusive of Russia, to about 220,000 tons, less than 20,000 tons of which are owned by Aluminium Limited. In 1932 the German, Swiss, and French firms were nearly equal in size, each commanding about 40,000 tons' capacity. Now the German government corporation has reached a capacity almost equal to that of Aluminium Limited, while the others do not appear to have grown substantially.

In the three years 1928–1930 the output of the United States and Canada exceeded that of European plants by a small margin.[73] During the depression American production fell much farther than European volume. In 1935 European output, exclusive of Russia, substantially surpassed pre-depression figures and amounted to a little more than double the American production, which remained below its 1930 high. Output in plants of Europe outside Germany and Russia, however, still remained in 1935 about 15,000 tons below their total for 1929. Estimated capacities of the various aluminum companies and the distribution of capacity between countries are shown in Tables 9 and 10.

[71] MW, xv, 311 (1936).
[72] *Minerals Yearbook*, 1936, p. 409.
[73] Output figures are given in Table 38, p. 569.

TABLE 9

ESTIMATED CAPACITIES OF ALUMINUM COMPANIES OF THE WORLD IN 1936 *

Company and Plants	Estimated Capacities (*Metric Tons per Year*)	
Aluminum Company of America		
Alcoa, Badin, Niagara Falls, Massena, U. S. A.		125,000
Aluminium Limited		
Aluminium Company of Canada, Arvida and Shawinigan Falls, Canada	40,000	
Norsk Aluminium Company, Höyanger, Norway, and Mansbo, Sweden	13,000	
One-third interest in DNN, Eydhavn and Tysse, Norway	5,000	
Societa dell'Alluminio Italiano, Borgofranco, Italy ..	1,500	
Total	59,500	59,500
Vereinigte Aluminiumwerke A. G.		
Lautawerk, Erftwerk, Innwerk, Germany		55,000
Compagnie Alais, Froges et Camargue		
L'Argentière, Calypso, St. Jean, La Praz, La Saussaz, Auzat, Chedde, Beyrède, Rioupéroux, St. Auban, Sabart, France	30,000	
One-third interest in DNN	5,000	
One-half interest in Aluminio Español, S. A., Sabinanigo, Spain	1,000	
Total	36,000	36,000

* The figures of this table represent rough estimates of the yearly capacities of *aluminum reduction works* under typical conditions of power supply. They have been derived from study of estimates reported in the literature and production figures at times when it was known plants were working at capacity. The error in the figure for each company, with the possible exception of the Aluminum Company of America and the Soviet government enterprise, is probably less than 5,000 tons.

TABLE 9 — *Continued*

Company and Plants	Estimated Capacities (*Metric Tons per Year*)	
Aluminium Industrie A. G.		
Chippis and Neuhausen, Switzerland	27,000	
Lend, Austria	4,000	
Rheinfelden, Germany	8,000	
S. A. Veneta dell'Alluminio,† Marghera, Italy	6,000	
One-half interest in Aluminio Español, S. A.	1,000	
Total	46,000	46,000
British Aluminium Company		
Foyers, Kinlochleven, Lochaber, Scotland	20,000	
Vigelands and Stangfjord, Norway	4,000	
One-third interest in DNN	5,000	
Total	29,000	29,000
Aluminiumwerke, G.m.b.H.		
Bitterfeld, Germany		12,000
Societa Italiano dell'Alluminio		
Mori, Italy		7,000
Aluminiumwerke Manfred Weiss A. G.		
Csepel, Hungary		6,000
Alliance Aluminium Compagnie		
Dolgarrog, Wales	1,500	
Glomfjord, Norway	4,000	
Total	5,500	5,500
Société d'Électrochimie		
Prémont, Les Clavaux, Venthon, France		3,500
Gebrüder Giulini		
Martigny, Switzerland		1,500
Aluminiumwerke Steeg		
Steeg, Austria		1,000
Soviet Russia		30,000
Japan Aluminum Reduction Company		5,000
Total		422,000

† Owned jointly with an Italian group.

TABLE 10

ESTIMATED CAPACITIES FOR ALUMINUM PRODUCTION IN 1936 BY COUNTRIES

Country	Estimated Capacities (*Metric Tons per Year*)
United States	125,000
Germany	75,000
Canada	40,000
Norway	34,000
France	33,500
Russia	30,000
Switzerland	28,500
England	21,500
Italy	14,500
Hungary	6,000
Austria	5,000
Japan	5,000
Spain	2,000
Sweden	2,000
Total	422,000

PART II

NATURE OF MARKET CONTROL — THE BASIC PRODUCT

CHAPTER V

EARLY MARKET CONTROL

1. Monopoly in America

During the prosperous years 1905–1907 several new aluminum producers sprang up in Europe, where the basic patents had already expired. While they remained rather small relative to the old firms, some of them threatened to become capable competitors. Although the life of the Hall patent ended in 1906, the American company was protected throughout these boom years by the Bradley patent, which lasted until 1909. Conjecture as to what might have happened if no patent had been issued to Bradley, or if the delay in its issuance had not occurred, is not altogether idle. Had patent protection ended in 1906 it is highly improbable that the Aluminum Company would have attained by that time the degre of size, integration, and power which, after the intervening boom years, faced potential competitors in 1909. Furthermore, at the time when the Bradley patent lapsed, newcomers were confronted with the additional deterrents of industrial depression and low-priced imports from Europe. Extension of patent protection to twenty years destroyed the opportunity for competitors to enter at a period when conditions were perhaps more favorable than they have ever been since.

Part II of this study will consider the reasons for the continuance of the monopoly of primary or virgin ingot production in the United States for the quarter-century since the expiration of the basic reduction patents; the extent of competitive and monopolistic elements in Europe; and the nature of international relations in this industry.

In the year in which the Bradley patent expired *The Mineral Industry* made the following prediction concerning competition:

> Even with the large advantages of a long-established and well-organized business, ample facilities in the way of hydroelectric power and supplies of raw material, plenty of capital, and the fact that it is at present able to manufacture more than sufficient metal to take care of domestic consumption;

this company cannot hope to deter other metallurgical interests from engaging in the production of aluminum if it continues in the "standpat" policy of high prices, which it has followed to date.[1]

While misleading in its emphasis upon price policy,[2] this statement points the clues to the continuance of monopoly in the aluminum industry. During the last few years before expiration of the Bradley patent the Aluminum Company of America apparently received average earnings of somewhere between 30 and 40 per cent upon its total assets.[3] The average annual rate of return seems to have been over 15 per cent in the years 1909–1911 and nearly 20 per cent in the next three years.[4] A tariff of 9 cents per pound until 1909 and 7 cents until 1913 contributed in some measure to this record.[5] It must have been apparent to anyone familiar with the metal industries during the decade 1905–1914 that the increasing use of this new metal was proceeding by great jumps. Here, one would think, was an inviting prospect for new capital and enterprise. During this period the press contained announcements of projected companies about to enter the field, but no reports of actual productive operations. Why was this tempting invitation so completely declined? The answer seems to be found chiefly in the expansion policy, facilitated by tariff protection, upon which the Aluminum Company embarked four years before the patent expired and at the time when the upward swing of demand for the young metal began — an expansion which consisted in the purchase of a large part of the deposits of domestic bauxite economically suitable for aluminum reduction and a tremendous increase in power resources and power plant, reduction cell, and semifabricating capacity. Contributory elements were the scarcity of aluminum metallurgists and ignorance concerning the financial success of the company and the probable movement of demand.

Commercial bauxite consists of hydrated oxides of aluminum mixed with oxides of iron, silicon, titanium, and other elements

[1] MI, XVIII, 20 (1909).

[2] See below, pp. 111 ff.

[3] Above, p. 30.

[4] Sources of information upon profits are given in Chap. XI and Appendix C. Owing to inadequate data the figures may not be considered precisely accurate, but it is believed that they reflect roughly the changing trends.

[5] In 1913 the duty was lowered to 2 cents per pound.

in varying quantities. It ordinarily contains between about 50 per cent and 65 per cent of aluminum oxide. Although most bauxites can be used interchangeably in the manufacture of aluminum, alum and aluminum salts, artificial abrasives and refractories, to some extent it appears to be more economical to use bauxites of particular compositions for particular products.[6] Thus the purest bauxite seems to be especially suited to the requirements of the chemical trade.

We have seen that the Pittsburgh Reduction Company had purchased ore in the Georgia-Alabama field during the nineties. Just before the turn of the century it acquired several hundred acres of bauxite lands in Arkansas.[7] During the next ten years or so ownership of the deposits in both fields became concentrated in the hands of a few firms. By 1909 there were only four operators mining bauxite in the country.[8] Some of the deposits in the southern field were economically satisfactory for the production of aluminum, but it appears that two counties in Arkansas contained reserves of bauxite suitable for aluminum which were equal to many times the tonnage existing in the other field.[9] In 1905 the Pittsburgh firm purchased from the General Chemical Company all, or nearly all, of the stock of the General Bauxite Company, which owned a substantial portion of the Arkansas ore, as well as holdings in the southern states.[10] At the same time a long-term contract was made between the General Chemical Company and its erstwhile bauxite subsidiary, whereby the former was to be furnished with its ore requirements for the chemical business

[6] In 1916 about 70 per cent of the domestic bauxite mined was used for aluminum (MR, 1916, p. 168). In the twenties about 65 per cent was used for alumina (most of which probably went into aluminum), 16 per cent for chemicals, 14 per cent for abrasives, and 4 per cent for aluminous cement (BMTC v. ACOA appellant, fol. 5262).

[7] United States Geological Survey, 21st Annual Report (1899–1900), III, 467.

[8] Tariff Hearings 1908–1909, House Committee on Ways and Means, p. 775; MI, XVII, 79 (1908). The four firms included those whose product was used chiefly for the manufacture of chemicals and abrasives, as well as those mining bauxite for aluminum. While these four did not own all the deposits capable of economical exploitation they had been increasing their holdings.

[9] MI, XVI, 98 (1907), and other years, *passim;* United States Geological Survey, 21st Annual Report, *loc. cit.;* MR, *passim.*

[10] MI, XIV, 46 (1905); petition and decree, *U. S.* v. *Aluminum Company of America* (1912), reprinted in Tariff Hearings, 62 Cong., 3 Sess., House Doc. no. 1447, II, 1519–1537. The decree appears below, Appendix D.

and agreed in return not to use or sell to others for the purpose of aluminum reduction any bauxite or products thereof.[11] The General Bauxite Company agreed not to use or sell to others, without the consent of the General Chemical Company, any bauxite for manufacture in the United States of certain specified chemicals. Performance of the provisions of the contract was guaranteed by the Pittsburgh Reduction Company. This acquisition gave the aluminum firm ownership of a substantial part of the known bauxite deposits in this country.[12] In 1909 another long-term contract was made with the Norton Company of Worcester, a manufacturer of abrasives, by which this firm sold to the Aluminum Company the entire capital stock of the Republic Mining and Manufacturing Company.[13] The latter corporation owned large tonnages of ore suitable for aluminum reduction, as well as large reserves of the sorts of bauxite used in the preparation of chemicals and abrasives.[14] A forty-acre tract of bauxite land in Arkansas became the property of the Norton Company by the terms of the contract, which also provided that the Norton Company would not use or sell bauxite from that tract for the purpose of conversion into aluminum. The Norton Company also agreed not to sell or lease the forty-acre tract except subject to the same restriction. It further agreed not to use or sell any other bauxite or products thereof, later acquired in the United States or Canada, for the purpose of making aluminum. The Aluminum Company agreed to furnish the Norton Company with an annual maximum tonnage of ore suitable for its operations, and to refrain from competition with the Norton Company and from assisting others to compete with it in the manufacture and sale of its abrasive alundum. Certain patents were also interchanged.

In 1907 the Aluminum Company contracted to buy 37,500 tons of alumina from the Pennsylvania Salt Manufacturing Company over a period of five years, with an option to renew the contract for five years more in 1912. The latter firm agreed not to enter the business of aluminum reduction during the life of the contract. In a covering letter it assured the Aluminum Company that

[11] The contract appears in BMTC v. ACOA appellant, fols. 7108 ff.

[12] MI, xv, 15 (1906); Tariff Hearings, 1908–1909, I, 769.

[13] The contract is reprinted in BMTC v. ACOA appellant, fols. 7156 ff.

[14] Statement of its president (Tariff Hearings, 1908–1909, I, 775).

it would not assist anyone else to enter the aluminum business or sell alumina to anyone for the purpose of making aluminum.[15] The restrictive clauses concerning bauxite and alumina in the contracts with these three companies were annulled by court decree in 1912.

In denying that ore lands had been bought for purposes of control, an officer of the Aluminum Company said that "the Aluminum Company of America has purchased bauxite only for its own needs, and its reserve supply of bauxite is in fact too small a reserve for its own business." [16] Recently a representative of the company, in testifying upon the purchases of the General Bauxite Company and the Republic Mining and Manufacturing Company and the restrictive agreements incident thereto, has explained that the supplies of ore in this country commercially suitable for the manufacture of aluminum, chemical products, and abrasives respectively were so limited that the Aluminum Company, the General Chemical Company, and the Norton Company were all concerned about keeping or procuring adequate reserves for the future.[17] It was testified that the Aluminum Company considered that it possessed about one million tons of bauxite before the purchase of the General Bauxite Company, which brought the total to something like three million tons. It was estimated that acquisition of the Republic company added about two millions more.[18] In 1913 control of another bauxite property estimated to contain between two and three million tons was obtained.[19] No important domestic deposits were acquired between 1909 and 1913 or subsequent to the latter date.[20] Large amounts of the ore mined by the Aluminum Company have been sold to other industries consuming bauxite.

It is impossible to ascertain exactly what proportion the reserves of the Aluminum Company made of all known deposits in

[15] The contract and letter are reprinted in BMTC v. ACOA appellant, fols. 7228 ff.

[16] Tariff Hearings, 1912–1913, House Doc. no. 1447, II, 1512.

[17] BMTC v ACOA appellant, fols. 5525–5532, 5859–5869, 6101–6103.

[18] *Ibid.*, fols. 6104–6106. These estimates, made at the time, have apparently turned out to be too low.

[19] *Ibid.*, fols. 5204, 6107; BR, pp. 116–117. This transaction was completed after the assurance of the Department of Justice had been received that it would not appear to violate the consent decree which had just been entered against the company.

[20] BR, p. 98.

this country of bauxite economically suitable for aluminum reduction at the time of the expiration of the Bradley patent and in ensuing years. In its petition for an antitrust decree in 1912 the government alleged that the company had acquired 90 per cent of all the known deposits in the United States and Canada which were economically suited for production of aluminum.[21] The Department of Justice did not ask that the company be required to dispose of any of its deposits, and the consent decree which issued contained no finding about the proportionate ownership of bauxite.[22] A bauxite official of the Aluminum Company has recently stated that prior to 1913 the company had acquired less than half of the domestic deposits, and that the situation with respect to ownership of domestic ore is unchanged at present.[23] He added, "I honestly would not know whether to select that [owned by others] or our own if I had a chance to pick one or the other. I mean, because of location and quantity." [24] The same officer also stated several years ago that there was plenty of bauxite in the United States available to anyone wishing to enter the aluminum business.[25] This testimony is, however, quite as unsatisfactory as the simple allegation of the government, for it does not state what proportions of the domestic deposits economically suited, considering location, quality, depth, and so on, for immediate production of aluminum were owned or controlled by the Aluminum Company in various years from 1909 on. In 1912 it was said that the American Bauxite Company, which had been created to own and operate the Arkansas holdings of the Aluminum Company, owned 90 per cent of the bauxite lands in Saline County, where the larger part of the Arkansas ore was located.[26]

[21] In the Kitchen Furnishings Report published in 1924 the Federal Trade Commission, apparently relying upon this petition of the Department of Justice, stated that "while operating under the patent monopoly the company succeeded in acquiring a substantial monopoly of the commercial bauxite properties in the United States suitable for the manufacture of aluminum" (pp. 90–91). This allegation of the economic division of the Commission was not adjudicated in the proceedings initiated by the Commission in 1925 under a complaint which concerned competitive methods in the fabricating branches, although some testimony on ownership of bauxite was taken.

[22] See below, p. 108.

[23] BMTC v. ACOA appellant, fols. 5226, 5197.

[24] *Ibid.*, fol. 5199.

[25] FTC Docket 1335, Record, pp. 5300 ff. Cf. also *Aluminum Industry,* II, 19.

[26] *Second Biennial Report of the Arkansas Tax Commission* (1912), p. 27.

In 1926 the Department of Justice referred to the "fact that that company [the Aluminum Company] owns practically all of the bauxite lands in the United States."[27] How much exaggeration is contained in this statement is not known. In 1929 the president of the Dixie Bauxite Company, in a letter urging an increase in the import duty on bauxite, stated that there were millions of tons of bauxite in Arkansas not yet under lease.[28] Again it is not clear whether the ore referred to was commercially suitable for aluminum.[29]

Certain other considerations suggest that after 1909 there were not available in this country many large deposits of bauxite suitable in amount, quality, location, and reservation price to provide a basis for substantial ventures in the aluminum business. If that had been so there would have been little reason in the restrictive agreements made with the General Chemical Company, the Norton Company, and the Pennsylvania Salt Manufacturing Company. Officers of a French company which intended to produce aluminum in North Carolina have said that they could have secured bauxite in the United States in 1912; that it was decided, however, to bring over their own French ore because the use of the obtainable American bauxite would have been more costly, all things considered.[30] Presumably they would never have embarked upon the project at all unless they had believed that they could produce aluminum here as cheaply, or nearly as cheaply, as the Aluminum Company of America. Preference for their own ore implies that production with the remaining available American bauxite would have resulted in a higher cost of production than that of the Aluminum Company; unless use of their French ore would have enabled a much lower cost than the expense incident to use of the ore acquired by the Aluminum Company. But if that had been so, it is reasonable to suppose that the Aluminum

[27] BR, p. 84. This statement occurs in a sentence stating that the dominant position of the company in the aluminum industry at the time of the report was not related to its possession of bauxite in the United States.

[28] *Manufacturers Record,* XCV, 70 (May 9, 1929).

[29] The Dixie Company itself was mining ore from a depth of 150 feet and was described as "the only bauxite company in the world today using this system." If the bauxite not yet leased is all deep down and if the cost of deep mining is greater than that of shallow mining, the remaining ore would not, of course, be commercially on a par with the reserves of ore near the surface.

[30] BR, pp. 113–116.

Company would have acquired large deposits of low-cost French ore, whether or not it accumulated reserves of high-cost ore here. Although the bauxite situation in Europe before the war is not altogether clear, it would appear that deposits of the better French bauxite could have been secured at least up to the middle of the first decade of the century.

The restrictive clauses of the three contracts relating to bauxite and alumina were annulled by a consent decree in 1912,[31] but the ownership of the deposits was left with the Aluminum Company. The Department of Justice stated in its petition that it was

> advised that there are practically inexhaustible quantities of bauxite abroad, which may be mined and shipped into the United States at such prices as would enable independent companies to successfully compete with defendant were all other restraints removed from the aluminum industry. Hence, petitioner does not attack defendant's ownership of various deposits of bauxite to which it now has title.[32]

Doubtless the Department also felt loath to ask the court to disturb acquisitions which, viewed from one angle, merely constituted reserves for the future. The acquisitions of bauxite by the Aluminum Company obviously reduced the opportunity for others to enter the industry. Aluminum producers cannot afford to jeopardize their investments in power and reduction facilities by relying for long on independent sources of raw materials. All the substantial producers acquired reserves of bauxite in the first decade of the century. Ownership of ore may bring some technical advantage as well as the more obvious tactical ones. It seems significant that the only potential entrant which actually reached the stage of plant construction in this country possesed ore lands in Europe.[33] It may be concluded that cheap imports from European mining firms, as distinguished from ownership of foreign

[31] A representative of the Aluminum Company has recently testified that the clause preventing the Norton Company from using or selling bauxite for purposes of aluminum reduction was abrogated by mutual consent before the suit of the government was brought (BMTC appellant v. ACOA, fols. 703 ff.).

[32] In the tariff hearings of 1912–1913 the company maintained that bauxite could be purchased at cheap prices from European mining companies and that satisfactory deposits of ore in Europe were available for purchase. Undoubtedly these points were urged upon the Department during negotiations prior to the filing of the government's petition (Tariff Hearings, p. 1512).

[33] Above, p. 107, and below, pp. 115 ff.

bauxite, would hardly have provided a secure foundation for continuing existence. It is probable that deposits economically comparable with those being used could have been purchased in Europe in the last pre-war decade. Manifestly this would have involved somewhat greater difficulties and uncertainties than would have attended the utilization of native mines, even though operating cost including the import duty of one dollar per ton might not have been markedly different. And in the last few years before the war the best European bauxite was being expeditiously gathered in by the established producers of aluminum and alumina.[34] With the rapid growth of the aluminum industry, and the discovery and development of the commercial possibilities of deposits in South America and eastern Europe, foreign ore deposits later became both necessary and advantageous; until 1912 or so the Aluminum Company evidently considered a search for foreign bauxite neither exigent nor favorable. A potential rival might have found it undesirable.

In this instance of limited natural resources, as in others, the Sherman Act as interpreted did not prevent the diminution of opportunity for potential competitors to enter. It seems very unlikely that the court would have ordered divestiture of a substantial part of the bauxite lands if the Department of Justice had asked for that. No evidence has appeared that the Aluminum Company had repressed would-be purchasers, or coerced sellers, or paid artificially high prices. It may have been obvious in 1912 that the Aluminum Company would need all the bauxite it had acquired if it were to grow *pari passu* with the increase in demand in the next two or three decades.[35] It should have been quite as plain that if new firms were to emerge they would also require reserves, and that their existence would prevent the Aluminum Company from growing at the same rate as demand. While it is doubtless true that the company was quite sincere in the contention that all the reserves which it had purchased were needed for its future growth, it should also be recognized that possession of these re-

[34] Below, p. 119.

[35] Of the 7 or 8 million tons of domestic ore estimated to belong to the Aluminum Company in 1913, apparently about 5 million tons have been used to date. Underground mining has revealed larger reserves than the original estimates indicated, with the result that the Aluminum Company now appears to possess about 7 million tons of domestic ore (BMTC v. ACOA appellant, fols. 6106–6108).

serves tended to promote the continuance of more monopoly power than might otherwise have existed. Although possession of a large portion of the domestic ore suitable for aluminum might not alone have prevented the entry of new enterprise, it was a deterrent element which could, perhaps, have been removed more easily than the factor next to be considered.[36]

In addition to accumulating large reserves of bauxite the Aluminum Company expanded its operating facilities at all stages very rapidly after 1904 and purchased enormous reserves of potential power. In 1904 the installed energy of the company in the United States was about 15,000 h.p. and output was about 3,500 metric tons.[37] In the ensuing two years both capacity and output were nearly doubled.[38] By 1908 installed power had reached something like 70,000 h.p., a capacity which was not well utilized until 1910, when output amounted to 15,500 metric tons. During the four years 1908–1911 the company was producing more metal than it sold, partly because of large imports from European producers suffering from depression at home.[39] In 1912 and 1913, however, a spurt in demand exhausted accumulated stocks and strained the firm's capacity, which had grown by 20,000 h.p. since 1908. With the scarcity, which was accentuated by a railroad-car shortage, imports increased further. Expansion of operating equipment begun at this time raised capacity in the United States to perhaps 180,000 h.p. in 1915, or about double the capacity of 1912.

To conclude that prospects during the years 1911–1913 were inviting to promoters would overlook the force of facilities under construction and the potential capacity represented in riparian rights. Since 1906 the Aluminum Company had been industriously acquiring the necessary rights and endeavoring to obtain

[36] I imply here no judgment upon the relative desirability of predominance of competitive or monopolistic elements in the aluminum industry, for that cannot be made until after an examination of the consequences of monopoly. I simply point out that the antitrust law as applied permitted some diminution in the opportunity for others to enter the industry.

[37] Statistics of production are given in Table 38. Estimates of capacity have been made from study of unofficial estimates published in trade journals and annual reviews and of probable maximum and minimum requirements for outputs in years when capacity was said to be well utilized.

[38] During the same period capacity at Shawinigan increased from 5,000 to 15,000 h.p.

[39] Tariff Hearings, 1912–1913, pp. 1486 and 1507. See also above, p. 38.

the requisite legislation for the Long Sault project. Just before the war this was still hanging fire. In the four years 1909–1912 more than $10,000,000 of earnings were evidently reinvested.[40] A part of this was used to increase plant at other stages than ingot production. Much of the rest probably went into the acquisition of riparian rights for the Tennessee project. During 1913 construction of a million-dollar reduction plant was started at Alcoa, Tennessee, while Massena and Shawinigan were both undergoing enlargement. The year 1914 saw Alcoa taking an initial 20,000 h.p. of purchased energy to begin ingot production before the hydroelectric development was completed. Capacity at Shawinigan had now reached 40,000 h.p.,[41] and in the following year the Cedar Rapids Power Station of the Montreal Light, Heat, and Power Company began to feed energy into the cells of a greatly enlarged Massena works under a contract calling eventually for 85,000 h.p.[42] The quickened war demand pushed all the new facilities to full utilization as soon as they could be completed, but during 1914 and the first part of 1915 the company was storing metal once more. Thus from 1908 until the latter part of 1911 the Aluminum Company had operating capacity sufficient to meet demand at lower prices than it was charging. By the time that increasing demand strained the facilities at this level of price, the company had acquired extensive resources of power and ore and was announcing its intention to make enormous additions to capacity.

However it may have appeared at the time, hindsight does not demonstrate that the investment and price policy of the American company after 1908 was such as to leave no room for new firms. Between 1908 and 1915 the Aluminum Company itself more than doubled its installed horsepower. Construction of this additional capacity was begun before the outbreak of the war in Europe intimated an increase in demand from armament industries.[43] In 1912 the Southern Aluminium Company, formed by French

[40] Appendix C.

[41] *Economic Minerals and Mining Industries of Canada* (1913), p. 13.

[42] MI, XXIII, 15 (1914). Throughout this period rolling mills, wire mills, and other equipment at later stages were greatly extended.

[43] The company's capacity was more than doubled without counting the 25,000 h.p. secured by the purchase of the Southern Aluminium Company after the war demand had appeared.

aluminum producers, undertook the development of water power in North Carolina to be used for production of this metal. The profit record of the Aluminum Company of America appears to indicate that its added operating investment earned very good returns as soon as it began to produce. Hence it may be inferred that, in the absence of economic warfare, new firms of effective structure which possessed cheap enough ore and power could have profitably introduced at least a part of the new operating investment that was actually brought in by the Aluminum Company. This conclusion certainly seems true of the investment in operation by 1914. Whether that part which began to produce in the following year would have reaped satisfactory earnings immediately in the absence of the phenomenal war demand is a matter of conjecture. Probably the rapid growth of the automobile industry at this time seemed to forecast substantial profits within a few years.

The profits of the Aluminum Company after 1908 represented a much lower rate of return than the 30 to 40 per cent which seems to have been gained in the four years ending with 1908, during which prices ranged between 33 and 42 cents per pound. In the next four years price was below 20 cents much of the time, while sales did not absorb the full output of existing capacity until 1912; yet the average return on investment did not fall below 15 per cent. In 1913 and 1914 earnings evidently averaged at least 20 per cent with price fluctuating around 20 cents. Several factors accounted for the lower price after 1908. Desperation imports during depression were followed, after an interval of two years of business recovery, by a reduction in the duty on ingot from 7 cents to 2 cents per pound. Costs were evidently lowered substantially by the development of cheaper power, the extension of integrated control, and perhaps some horizontal economies. Doubtless demand in its new position was much more elastic within the price range between 20 and 35 cents than it had been earlier. Under these circumstances a marked fall in price was to be expected. It seems quite clear, however, that the Aluminum Company did not expand its operating investment far enough and reduce price sufficiently so that, in the absence of other deterrent elements, newcomers would have refrained from entering the industry simply because good returns to additional investment seemed unlikely to materialize for some years to come. This becomes even more evident when it is recalled that some part of the investment of the Aluminum Company represented inoperative reserves of ore and power which would not in all probability be used for many years. Price was still high in the sense that it permitted substantial elements of monopoly profit.

Why, then, did no firm other than the Southern Aluminium Company seriously attempt to enter the industry? Let us endeavor to place ourselves for a moment in the position of a promoter considering the prospects of a new firm. And let the unknowns be enumerated first. No financial reports were published by the Aluminum Company, the shares of which were closely held. Trade gossip would have indicated that the business was quite profitable, but until some meager data were divulged in the tariff hearings of 1912–1913, the promoter would have been forced to rely upon attempts to reckon costs of production in order to derive even an approximate guess as to how profitable. It might have been difficult to discover persons technically competent to perform this task. An editorial in the *Engineering and Mining Journal* implies that technical men familiar with aluminum constituted a "noncompeting group." "Copper metallurgists, steel metallurgists, lead metallurgists are common . . . but who ever heard of an aluminum metallurgist outside of those who hide their lights under Mr. Davis' bushel?"[44] Furthermore, although it must have been clear that the new metal was to enjoy a large and speedy growth, it was doubtless not easy to discern with even approximate sureness how great or how rapid it would turn out to be. And since it requires several years to bring a new power plant and reduction works into operation, any degree of inability to forecast demand is a serious element.

Contrast with these doubtful factors the elements which, in immediate appearance at least, seemed quite definitely known. A large part of the bauxite in the United States which was commercially suitable for aluminum had been acquired by the Aluminum Company. The striking expansion of the company *appeared* to keep in existence most of the time a greater operating capacity than was required to satisfy the existing demand, and at the same time provided large reserves of ore and power which suggested

[44] EMJ, CIV, 144 (July 1917).

that the Aluminum Company regarded it as natural business policy to equip itself to meet large increases in domestic demand for many years to come. Large imports from Europe during depression provided support for the illusion of marked overinvestment in this industry.[45] Continued imports after renewal of prosperity must have suggested that aluminum reduction could be carried out much more cheaply abroad than in this country. In any event, the decided drop in the tariff in 1913 acted in the direction of discouraging new ventures thereafter.

While the establishment of an efficient aluminum firm would not necessitate a huge initial investment, it was plain that a well-integrated structure which included economical ore and power would be essential almost from the start. This would probably require an investment of several millions. It appears that the chemical process of extracting alumina from bauxite is more complicated than most metallurgical processes. Although each of the other steps in the production of ingot is simple enough in itself, it was obvious that the complexity of the process taken as a whole would require an organization of diversified experts. Apart from uncertainty about the supply of experts, it must have appeared that, owing to the nature of the process and necessary organization, the gestation period for a new aluminum enterprise would be longer than that in many other industries. The efficiency of the existing firm was scarcely in doubt. Although it could not be foretold with assurance precisely how the Aluminum Company would react to the entrance of a newcomer, the aggressive acquisition of bauxite, the restrictive agreements — which also included contracts with two power companies whereby they agreed not to furnish anyone else with energy for aluminum reduction —,[46] and the expansion policy in general were not likely to encourage belief that the new enterprise would be tolerated without trial by combat. However vague the knowledge of the financial strength

[45] In January 1909 "Aluminum Man" remarked (*Metal Industry*, VII, 10) that the world's capacity for producing aluminum was "several times the ability of the market to absorb it at present and it now seems probable that no further extensions in capacity will be required for many years, and that none are likely to be made on account of the fact that the large overproduction . . . had rendered the business unattractive from a financial standpoint."

[46] Tariff Hearings, 1912–1913, p. 1496; and BMTC v. ACOA appellant, fols. 5853 ff.

of the Aluminum Company, it was well known, of course, that this corporation belonged to the powerful group of financial and industrial interests backed by the Mellons and their associates.

Some of the deterrent factors may be illustrated by the case of the Southern Aluminium Company, which was promoted by M. Adrien Badin and a group of associates in the French aluminum industry, who partially foresaw the phenomenal development in the production of this metal in the highly industrialized United States, where cheap coal, cheap power, and good bauxite were all to be found.[47] In August 1912 l'Aluminium Français, in conjunction with some large French and Swiss banks, incorporated the Southern Aluminium Company, all of whose stock was owned by Europeans. The new company took over property earlier obtained by an American group through purchase of control of a hydroelectric concern which had been suffering financial difficulties during its attempts to carry through a large power development on the Yadkin River in central North Carolina.[48] Under the supervision of M. Paul L. T. Héroult, construction was immediately begun upon a power plant and reduction works planned for a capacity of about 25,000 tons per year.[49] The new company was assured of an adequate supply of bauxite from the French mines of its owners. Plans to start production in 1913 or 1914 did not materialize when it was found that the partly constructed dam of the predecessor company was poorly located. Just prior to the outbreak of the war it was announced that production would begin in June 1915 at an initial rate of about 5,000 tons a year.[50] In October 1914, however, construction operations had to be suspended, owing to the impossibility of further financing in France under war conditions.[51] Fruitless endeavors were made to secure

[47] R. Pitaval, JFE, XXXII, 28 (March 1–15, 1923). M. Pitaval evidently had some connection with the aluminum industry in France. He does not say whether or not the French hoped to acquire bauxite in the southeastern states. It appears, however, that they intended to use their own bauxite at first, anyway. See above, p. 107.

[48] *Commercial and Financial Chronicle*, XCI, 1636 (1910), and XCV, 301 (1912); *Engineering News*, LXXI, 1279 (1914).

[49] *Chronicle*, XCVI, 1560 (1913); EMJ, XCIII, 1212 (1912). It was estimated that 100,000–120,000 h.p. could be developed at this site. There is a detailed account of the plant and plans of this company in an illustrated article by D. M. Liddell, "The Southern Aluminum Company," EMJ, XCVII, 1179 ff. (1914).

[50] *Chronicle*, XCVIII, 1923 (1914).

[51] Up to that time about $5,500,000 had been spent in acquiring properties and in

the necessary financial assistance in England as well as in France. When application to powerful financial interests in the United States also met with no success, the stockholders, faced with entire loss of the large investment already sunk, negotiated with the Aluminum Company of America, which agreed to purchase provided the Department of Justice did not regard the transaction as violative of the consent decree.[52] Upon receiving the desired assurance from the Attorney General that the Department saw nothing in the facts as presented which would call for action under the decree, the Aluminum Company purchased the plant on August 15, 1915.[53] It has been maintained by the Aluminum Company of America that the purpose in taking over the abandoned plant was to meet the great war demands of the Allied governments.[54]

The significance of the episode seems to be this. The only thoroughgoing and serious attempt to enter the field after the expiration of the Bradley patent was made by a group of aluminum producers with long experience, possessing its own bauxite. It was planned to employ an investment of about $10,000,000 in setting up a thoroughly integrated, large-scale concern. Horsepower equal to a third or a quarter of the American company's capacity was to be developed in the first instance and later doubled if circumstances warranted. When extraordinary conditions prevented the completion of financing abroad no American bankers could be induced in 1915 to supply the backing needed to bring operations into being — even with the beckoning force of a rising war demand for aluminum. In the large view, looking through the mists of war uncertainty, and the French nationality of the undertakers, this seems nearly equivalent to an unwillingness upon the part of our bankers to back new American enterprise for entry into the aluminum industry. The plant of the Southern Aluminium

construction work, and it was estimated that approximately $7,500,000 was needed to complete the plants (BR, pp. 5, 113, 116).

[52] BR, p. 5.

[53] It is stated in BR that the consideration was $5,030,000.

[54] FTC Docket 1335, Record, p. 706. An officer of the company stated that the French government asked them to take over and operate the plant. M. Pitaval remarks (*loc. cit.*) that the Aluminum Company of America "qui avait vu s'établir cette concurrence avec une certaine crainte, s'empressa d'acheter la nouvelle usine pour éviter qu'elle ne tombât en d'autres mains. Déjà les Américains profitaient de la guerre."

Company could not be removed to Europe, and it was as safe as any other domestic establishment from foreign depredations. In the event that the trained French management was drawn into military service, control of operations would presumably fall more and more into the hands of any Americans who had furnished financial backing.[55]

In the possession of the Aluminum Company of America the plant was pushed to rapid completion in order to meet the now bounding demands of the belligerents. The war called into being immense extensions of plant and yielded large returns with which to finance them. During 1915 and 1916 the company spent at least twenty million dollars in expansion and practically doubled its capacity. Investment jumped from a little under fifty millions in 1915 to a little over ninety millions in 1918.[56] Apparently no other attempts to enter the field were made until after the war, by which time the size and strength of the Aluminum Company had been appreciably enhanced.

[55] Since this account was written there has come to light a contract made in 1913 between the Société Générale des Nitrures (a subsidiary of the Compagnie Alais), the Southern Aluminium Company, the Northern Aluminum Company, and the Aluminum Company of America. The agreement related to the use of the American patents of the Société Générale des Nitrures for a process of making aluminum nitride from which alumina and ammonia could be derived. It provided for the formation of the American Nitrogen Corporation, which was to be owned jointly by the four parties to the contract, to take over the patents and operate them if experimental work being carried on in France demonstrated that commercial operation was likely to be attended with success. It appears that the process has never been operated commercially here or abroad (*Aluminum Industry*, I, 250). The terms of this contract do not deal in any way with the production of aluminum, but provide only for a joint venture in the manufacture of alumina and ammoniacal products. As far as this contract is concerned there seems to be no reason to change the interpretation given in the text. It does not seem that association of the Americans and the French in the American Nitrogen Corporation would have resulted in different policies with respect to aluminum than would have occurred in its absence. Membership of the French and the Northern Aluminum Company, a completely owned subsidiary of the Aluminum Company of America, in the cartels of 1901 and 1912 regulating European sales had already demonstrated that the parties to the cartel believed in coöperation rather than competition. Duopolistic rather than competitive policies were to be expected in any case. But it cannot be inferred from this that the French were dominated by the Aluminum Company and consequently did not seek energetically to dispose of their property elsewhere. The available evidence indicates that they did. Hence the important point seems to be found in the lack of interest in aluminum on the part of American promoters and bankers. The contract referred to in this note appears as Exhibit 553, BMTC v. ACOA, appellant.

[56] No indication has appeared of a revaluation of assets in these years.

The avowed object of the antitrust laws under the court interpretation permitting the possession of a considerable degree of monopoly power in the absence of repressive tactics, restrictive agreements, or effects patently injurious to consumers, was preservation of the freedom to compete or the maintenance of free enterprise. It was believed that if freedom to compete were maintained, "competition" would actually exist. In the case of production of aluminum ingot in this country it does not appear that there was any violation of the law as interpreted (with the possible exception of the restrictive agreements concerning bauxite, which were not adjudicated),[57] but no rivals entered the field.

By enabling the company to reap greater profits the tariff aided in the attainment of impressive size and financial strength. Finally, failure to require publication of data upon capacity, production, and consumption, and financial condition permitted the company to keep secret this basic information which promoters and bankers must possess if they are to exercise their social function of directing capital and enterprise into the channels where they will best meet consumer demands.

2. Competition and Monopoly in Europe

The existence of four important companies[58] across the Atlantic from the single American producer is attributable to several factors. We have seen that no one country enjoys sufficient differential advantages for aluminum production to grant it a monopoly. Although the promoters of the Schweizerische Metallurgische Gesellschaft held the Héroult patents for most if not all of the important European countries, they evidently did not wish to set up a monopolistic unit with plants in various countries, or else they were unable, on account of inadequate capital or connections, or for reasons inherent in the existence of different nations, to accomplish this. Apparently Neuhausen exerted some influence over both the Froges concern and the British Aluminium Company for a few years after their origin, but it is doubtful if this continued much beyond 1900. The granting of French patents to Minet and Hall (the Hall patent was the more significant) permitted the

[57] Above, pp. 103 ff.

[58] The early development of these companies has been described in Chap. II.

Compagnie Alais to enter the industry before the expiration of the Héroult patent. It was not strange that this concern, which had been engaged in the aluminum business before Héroult's discovery, and had had several years' experience in other branches of industrial electrochemistry, was able to become a formidable rival of the Société Froges within a short space of time.

The bauxite reserves in France were not early acquired by aluminum producers.[59] It appears that for over a decade after the birth of the electrolytic aluminum industry producers purchased their annual ore requirements at first from many small bauxite enterprises, and after the middle nineties from the few large concerns which emerged from a concentration movement in this mining industry. The cartel of 1901 arranged that the requirements of its members should be cared for by the two largest bauxite companies, the Union des Bauxites de France and the Société des Bauxites de France. Evidently it was not until competing aluminum companies were founded that the old firms began aggressively to acquire ore deposits. During the ensuing rivalry all of the good French bauxite came into the ownership of aluminum producers and a few manufacturers of alumina before the outbreak of the war.[60] By that time the British Aluminium Company had acquired control of the Union des Bauxites, and Neuhausen had purchased nearly all the shares of the Société des Bauxites.[61] Some of the new aluminum enterprises, shortly to be described, secured bauxite deposits of their own; others had close connections with established alumina firms whose market suffered as the older aluminum producers began to mine their own ore. In the earlier years entrance into the aluminum industry was also facilitated by the number of small but relatively cheap power sites in the mountains of southern France. Water power capable of economic development for aluminum reduction was also available in Norway and Switzerland.

It is likely that the cartel formed in 1901 by the four well-established European producers and the Northern Aluminum Company was motivated partly by a desire to present a united

[59] I have discovered little information about ownership of bauxite before the war. The account given here is based mainly on Kossmann, *op. cit.*, pp. 38 ff., and Escard, *op. cit.*, pp. 7 ff.

[60] Czimatis, *op cit.*, p. 78.

[61] Escard, *op. cit.*, p. 13.

front to potential competitors after the life of the basic patents ran out.[62] Nevertheless, in the five years following 1902 and 1903, when this occurred, none of the factors which combined to deter entrance into the field in America after 1909 existed in the same degree of force in Europe. In spite of their large additions to investment,[63] the existing incumbents were not formidable enough, even when acting in concert, to discourage all potential competitors. As we have just seen, bauxite and power were obtainable. The knowledge and experience concerning industrial electrochemistry in France, which had early taken a leading role in this field and in electrometallurgy, must have facilitated the establishment of new aluminum enterprises in that country.

A marked improvement in general business beginning about 1904 was magnified in the rapidly expanding demand for aluminum. Under these conditions the invitation extended by the cartel's policy of high prices was hardly likely to be refused. At least seven new enterprises constructed facilities for the production of aluminum between 1906 and 1910.[64] In 1907 a small plant was put into operation at Bussi on the Pescara River in Italy by a firm which owned bauxite in the province of Aquila. The Aluminium Corporation, founded in the same year, built a hydroelectric plant of 7,000 h.p. and reduction works at Dolgarrog in Wales. It immediately purchased ore deposits in Var and an alumina plant in England. A 3,000 h.p. plant at Martigny, Switzerland, owned by the Gebrüder Giulini, alumina producers of Ludwigshafen, Germany, began operations in 1910. The A. S. Vigelands Brug, incorporated in 1906 in Norway, developed 12,000–14,000 h.p. at Otterdal near Kristianssand.[65] Raw material supplies were obtained by a long-term contract with a Belgian alumina concern. In France the new aluminum undertakings were somewhat more substantial. The Société d'Électrochimie, which had produced chlorates by electrolysis since 1890, engaged in the reduction of aluminum about 1906 at Prémont,

[62] The details of this cartel agreement are not definitely known. Apparently it reserved home markets, apportioned sales quotas for the competitive market, and fixed minimum prices. See above, p. 36.

[63] See above, Chap. II.

[64] Details of these ventures have been taken from the sources of information on pre-war European developments cited above, pp. 33–36.

[65] This company was owned by a parent corporation known as the Anglo-Norwegian Aluminium Company.

where it soon developed 10,000 h.p. Bauxite lands were acquired near La Barasse, where the company erected an alumina works. Production of the light metal was also taken up in 1906 by another manufacturer of chlorates, the Société des Forces Motrices et Usines de l'Arve (founded in 1895), which possessed 13,000 h.p. at Chedde. In the same year this company set up a subsidiary, the Société des Produits Électrochimiques et Métallurgiques des Pyrénées, which installed 4,000 h.p., part of which was intended for aluminum reduction, at Auzat two years later. The Auzat plant was shortly enlarged to 12,000 h.p. It was said that the large alumina firm of Giulini took a financial interest in both of these concerns when they began to produce aluminum.[66] The Société des Pyrénées purchased bauxite deposits in Hérault. L'Aluminium du Sud-Ouest, formed in 1906, built a plant at Beyrède; while its subsidiary, Électrométallurgique du Sud-Est, constructed works at Venthon. It was reported that each of these plants had a capacity of about 10,000 h.p.,[67] but other indications point to a smaller initial installation. The latter, at least, had long-term contracts for purchase of supplies of alumina.

By 1908 the outsiders possessed a total capacity, completed or in prospect, of perhaps 70,000 h.p., which was just about equal to the additions made by cartel members in the three preceding years, and represented about half of the total cartel capacity. (In the absence of specific information it is, of course, impossible to know the proportion of capacity which either group intended to use in the production of aluminum as compared to that to be employed for other products.) Whether or not the dissolution of the cartel in that year was intended partly to give the members a freer hand for the subjugation of interlopers in their respective bailiwicks, the newcomers did not achieve any very substantial competitive strength even after the return of prosperity.

The Aluminium Corporation failed shortly after its appearance in the lists. Declining an offer to purchase by the British Aluminium Company, it reorganized and continued as a small producer.[68] The firms at Bussi and Martigny remained of negligible importance.[69] The Vigelands concern was sold to the British

[66] Kossmann, *op. cit.*, p. 37.

[67] Debar, *op. cit.*

[68] *Chemical and Metallurgical Engineering,* VII, 165 (1909).

[69] The Aluminium Corporation and the Bussi plant later came into the hands of the original companies.

Aluminium Company in 1912. While these three independents presented no obstacles to reconstruction of market control when business began to improve, the three outsiders in France were somewhat larger and possessed stronger support. Although they apparently did not use much of their energy for production of aluminum in 1909–1910 when output of the four old companies increased markedly, they presented some threat for both the domestic and the international markets.[70] The formation of a domestic sales syndicate called l'Aluminium Français in 1911 not only established control in the home market, but set up an organized unit for the subsequent dealings with Neuhausen with regard to a new international cartel. The Société d'Électrochimie and the Société des Forces Motrices de l'Arve, with its subsidiary, were admitted as active members of l'Aluminium Français, but their capacity remained small compared to that of the older companies. L'Aluminium Français bought out l'Aluminium du Sud-Ouest and its subsidiary. The plants at Beyrède and Venthon were not used for aluminum production until after the war. In 1914 the Compagnie Alais absorbed the Société des Pyrénées, and two years later the erstwhile parent of the latter, the Société des Forces Motrices de l'Arve, was brought into the same organization. Of the seven new enterprises which entered the field after 1905 three were purchased by the original companies in the following decade. Three of the four which continued independent were very small, while the Société d'Électrochimie possessed only a small part of the French market. Not only was the aggregate capacity of the invaders reduced by purchase to 50,000 h.p. in 1913, to 38,000 h.p. in the following year, and to 25,000 h.p. two years later; but in the period 1909–1914, during which the original cartel members more than doubled their facilities of 1908, the plants of outsiders were enlarged by only 14,000 h.p. (Table 11 shows the relative development of capacities of the old companies and the newcomers during the decade 1905–1914).

The invasion of new capital and enterprise did not undermine the established position of the four old companies. It hardly

[70] Kossmann (*op. cit.*, pp. 31–32) gives production of the French companies in 1910 as follows: Froges — 6,000 tons; Alais — 2,500 tons; total output of the four outsiders — 1,700 tons. Presumably these were provisional figures of the Metallgesellschaft, later revised, but they may indicate correctly the relative proportions in which output came from the old and the new concerns.

seems probable that the continued inferior position of all the new enterprises can be ascribed altogether to the depression which greeted their entry into the industry. While many suffered losses during their first few years,[71] none abandoned the field except to sell out to the old companies. The continued existence of the four

TABLE 11

ESTIMATED CAPACITIES OF OLD AND NEW EUROPEAN COMPANIES IN CERTAIN YEARS

(Thousands of Horsepower)

Company or Plant	1905	Additions 1905–1908	Total 1908	Additions 1908–1914	Total 1914
AIAG	24	32	55	90	145
Froges	20–30	0	25	40	65
Alais	14	12	25	25	50
British Aluminium Company	6	25	30	15	45
Total Old Companies	64–74	69	135	170	305
Martigny			3	0	3
Bussi			5	0	5
Dolgarrog			7	0	7
Otterdal			14	0	(14)
Prémont			4	6	10
Chedde			13	0	13
Auzat			4	8	(12)
Beyrède, Venthon			5–20	–	–
Total New Companies			55–70	14	38

The data for this table have been selected from estimates given in several of the sources cited in Chap. II, particularly Gautschi and Schulthess. Perhaps a third of the maximum capacity shown for the newcomers in 1908 was not completed. Parentheses indicate sale to the old companies, whose 1914 capacities include such purchases. The capacity of the Beyrède and Venthon plants is, however, omitted from the 1914 figures because it was not being used at all for aluminum production. It is doubtful if these two plants possessed in 1908 the 20,000 h.p. which some writers ascribed to them.

[71] See Kossmann, *op. cit.*, pp. 59, 111.

small firms suggests that the others would not have been eliminated by the operation of ordinary economic forces had they remained independent.[72]

The strength of the four old companies rested in factors connected with their own growth — their head start, connections, wealth, efficiency, and so on — rather than in concerted action. The first cartel was not a strongly centralized instrument for market control. It included no central sales agency or central administration with power to control rigorously the activities of its members.[73] It appears that from the beginning the French distrusted Neuhausen, which assumed the position of leadership.[74] Competition between the cartel members existed in the form of expanding investments, with the result that the French firms increased their capacity relative to that of the AIAG during the life of the cartel. The total French output, which had been but one-half of the Swiss production in 1901, equaled the latter in 1905 and ran ahead of it in the years 1906–1908.[75] Although French exports of ingot aluminum to Germany remained small during the life of the cartel, they grew from about one-tenth of the sum of French and Swiss exports in 1902–1903 to one-quarter of the common total in 1906–1907.[76] After the dissolution of the cartel the French sent to Germany in 1909 a tonnage nearly equal to that received from Switzerland. In the following four years they delivered about one-third of the common total. As the French companies began to overtake the Swiss, the friction between the two groups increased. It seems doubtful whether the cartel could have survived without breakup or reorganization in the absence of newcomers, even if its difficulties had not been increased by depression.

[72] The reasons for sale to the established firms do not appear clearly. It is possible that pressure was used. Or perhaps the newcomers were unable to secure reserves of economical bauxite and sufficient cheap power to enable them to become effective competitors. I have discovered no direct evidence upon this. The facts presented in this chapter do not support this hypothesis, but they do not contradict it entirely.

[73] As Bannert points out, the title "Internationales Aluminiumsyndikat" was a misnomer (*op. cit.*, p. 32).

[74] Gautschi asserts that this distrust of Neuhausen prevented strong centralization, which had been urged by the Swiss at the beginning.

[75] After 1906 a small part of the increase in French output was contributed by the two outsiders.

[76] Computed from official figures in the *Statistisches Jahrbuch für das deutsche Reich.*

Agreement between the French producers, which resulted in organization of l'Aluminium Français, began to take form late in 1910. About the same time serious negotiations were undertaken toward reconstitution of the international cartel. Temporary price agreements were once or twice terminated in the course of a determined struggle between the Swiss and the newly strengthened French over the allotment of the largest quota. In June 1912 an agreement was reached which created the Aluminium Association on the first of January 1913.[77] This cartel included the Aluminium Corporation and the Italian firm, as well as the members of the first cartel.[78] Its duration was fixed at ten years unless terminated sooner upon request of three of the four leading members. This cartel appears to have attempted more extensive control of the market than its predecessor, although the appearance may be deceptive, owing to lack of precise information upon the nature of the earlier agreement. The 1912 contract provided for regulation of all sales by members of aluminum, its alloys, half products, and manufactured articles in all markets except the United States. Output was not limited directly, but sales quotas were allotted to each company according to agreed proportions.[79] A firm which exceeded its quota at any time was to pass some of its orders to those who were still below their quotas or to buy from the latter at the standard price the amount of metal by which it had surpassed its allotment. A member in arrears who was unable to fill the orders passed to him within a month must surrender them for division among the others. Each member was free to sell his quota where-

[77] The agreement is reprinted as Exhibit 23A, BMTC v. ACOA appellant, fols. 7057 ff. The association actually began to function in the summer of 1912. Material contained in Exhibit 121A in the same case includes regulations for the central bureau of the association, regulations for the committee, and minutes of several meetings of the committee and the general assembly.

[78] The Anglo-Norwegian Company was a member for a few months, until its operating subsidiary, Vigelands Brug, was taken over by the British Aluminium Company. It is reported that the president of the AIAG said that the new cartel would suppress the influence of outsiders (AG, 1912, Partie sup., 180).

[79] The proportionate percentages fixed were as follows:

L'Aluminium Français	38.9
AIAG	21.4
British Aluminium Company	19.9
Northern Aluminum Company	16.0
Societa Italiana	1.9
Aluminium Corporation	1.9

ever he chose. The standard price fixed from time to time was a minimum price; higher prices might be charged by individual companies as they saw fit. It was agreed that no member would acquire an interest in any aluminum-producing firm which was not a party to the contract, or sell to or buy from any aluminum producer except through the association. There is no evidence of any agreement about the United States market with the Aluminum Company of America, which had just consented to a decree enjoining anything of that sort. Its Canadian subsidiary, the Northern Aluminum Company, which exported metal to Europe, was a member of this cartel as well as the earlier association.[80]

The provisions of the second cartel agreement, which was terminated by the war, have been described in detail because some of the essential features were adopted in post-war agreements. It would have been interesting to see if an agreement of this sort could have lasted for ten years. There was no definite control of output and no central sales agency, and the administrative instrumentality appears to have been none too well implemented for prevention of evasion.[81] During the two years of its existence internal disputes occurred over the purchase of Vigelands Brug by the British and the acquisition of the Société Générale des Nitrures by the Compagnie Alais. In the spring of 1914 vehement allegations of secret price-cutting were made, and the AIAG protested that its quota was too small, particularly relative to that of l'Aluminium Français.

In closing this section a word should be said about the supplementary relationships between branches of the pre-war aluminum industry in France, Switzerland, and Germany. These three countries formed a unified economic unit as far as aluminum was concerned.[82] A part of the French bauxite used by Swiss producers was converted into alumina in Germany, where coal was cheap. Some bauxite for the French producers was also sent to Germany

[80] At the request of the Aluminum Company the consent decree of 1912 was framed in such a way as to leave the Northern Aluminum Company free to enter into agreements of that sort which contained no provisions for control of the United States market. See letter of the Aluminum Company's attorney, BMTC v. ACOA appellant, Exhibit 122.

[81] See minutes of meetings, 1912–1914, reprinted as Exhibit 121A, BMTC v. ACOA appellant. Outside auditors were apparently not employed.

[82] See Czimatis, *op. cit.*, pp. 18 ff.; Schoenebeck, *op. cit.*, pp. 24, 40–41; Gautschi, *op. cit.*, pp. 30–32.

for preparation. Apparently part of the advantage of carrying French ore to German coal was supplied by low railroad rates on bauxite.[83] With the exception of the output of the small plant at Rheinfelden no aluminum was produced in Germany, but the Germans made much the largest part of the *Halbzeuge* and finished goods in Europe. According to Günther, the aluminum used in German rolling mills, finishing plants, and foundries, and in German steel mills and other works, amounted to about 10 per cent of world consumption of ingot aluminum in the years 1899–1903, 19 per cent in the next four years, and 25 per cent in 1909–1913. The relative consumption of aluminum by the utensil industries of various countries about 1910 was given by a utensil manufacturer as follows:[84] Germany, 2,000 tons; France, England, Switzerland, and Italy, each 250 tons.[85] After the Chippis works went into operation in 1905 most of the metal produced by the AIAG went to Germany. Only a small amount of aluminum products was made in Switzerland.[86] Although l'Aluminium Français engaged to some extent in rolling and fabricating, France also exported some aluminum to Germany for fabrication. While national boundaries are no necessary barrier to ownership, integration had not, in fact, been carried as far by some of the European producers as by the Aluminum Company of America. Neuhausen in particular was linked to the independent German rolling and fabricating industry.

The question has often been raised in German literature why no producing industry rose in that country until fostered by the necessity of war. We have already seen that Germany possesses almost no bauxite and little cheap water power.[87] But the same is true of Great Britain.[88] The most likely explanation seems to

[83] See Gautschi, *op. cit.*, p. 84.

[84] See Kossmann, *op. cit.*, p. 75.

[85] Upon the formation of the second cartel the German consumers of ingot banded together into a purchasing association.

[86] Gautschi, *loc. cit.* In 1905 there were only three small plants engaged in rolling or manufacturing aluminum products (*Ergebnisse der eidgenössischen Betriebszählung vom 9 August 1905,* I, Heft 8).

[87] In addition to these answers it is pointed out that before the war Germany had little experience with, and hence, it may be inferred, little interest in, the development of what water power she had. Czimatis alleges that economical bauxite would have been unobtainable, but it is to be doubted if that was true in 1904 or 1905.

[88] As far as the period before patent expiration is concerned, it is doubtless true that the Swiss did not care to license a producer in Germany, which had greater potentialities as an outlet for their metal.

be found partly in those undiscernible sets of relationships which we call chance; and partly in the fact that in pre-war Germany an immature daughter of science who was only beginning to resist relegation to the *Küche* was quite likely to be overlooked in the rush to take advantage of the many alluring opportunities offered by rapid industrialization of very promising resources.

CHAPTER VI

POTENTIAL COMPETITION — CONTROL OF ORE AND POWER

1. Potential Competition in America — The Uihlein and Duke-Haskell Episodes

Since the war the military and industrial importance of aluminum and its strong alloys has been definitely recognized, although the far-reaching significance of the alloy development has not yet been realized. Yet in spite of the extensive growth in demand which has brought aluminum to an established position among the chief metals, no new enterprises of any consequence have been set up in this industry, with the exception of those fostered or protected by national governments in countries hitherto non-producers, or companies promoted wholly or in part by the old firms. Elements which have discouraged entrance into this profitable field must have become stronger, for potential competition appears to have been more serious, at least in America, than it was earlier. The remarkable increase in consumption has brought about a resort to ore lands and power sites farther removed from consumption areas. We have seen that the Aluminum Company of America acquired some European bauxite and a large portion of the deposits in the Dutch and British Guianas, while European producers were buying ore lands in various spots. Under the circumstances, possession of a large part of the bauxite in the United States economically suited for aluminum lost much of its earlier significance, as was shown by the fact that for a time the Uihlein group of Milwaukee competed for South American ore with the apparent object of engaging in the production of aluminum in North America.

During the war the Uihlein family, owners of the Schlitz Beverage Company, had built a large carbon electrode plant near Niagara Falls with the immediate purpose of meeting war requirements. Whether or not they were then considering the production of aluminum when the struggle should cease does not appear.[1] A month after the signing of the armistice they turned their attention to aluminum as a means of utilizing a plant for the products of which peace held little prospect of demand. It will be recalled that carbon electrodes play an important part in aluminum reduction; until quite recently their tonnage consumption was nearly equal to the metal output.

In December 1918 Mr. Lloyd T. Emory, an expert bauxite engineer who had been general manager of the Demerara Bauxite Company before entering the United States Army, was engaged by the Uihleins to investigate deposits of this ore which were commercially available for aluminum reduction. Emory first visited Venezuela, but upon ascertaining that deposits there were not suitable, he went to British Guiana. Early in 1919 he apparently secured from a local solicitor an option upon the only two properties for sale there which he considered valuable. This option had been obtained from a client by the solicitor in his own name. For some time previous the Demerara Bauxite Company had been trying to obtain title to these properties.[2] An agreement had been signed with the owner, a Mr. Hubbard, who was adjudged insane before the transaction was completed. A contract made with two curators appointed after commitment of Mr. Hubbard was vitiated by the death of one of them. At this juncture the solicitor, who was or had been connected with the Demerara Company, sold the lands to this corporation instead of to the Uihleins. Emory then secured another option direct from the widow of the former owner of the lands and exercised it about the beginning of 1920. Suit was immediately brought by the Demerara Company.[3] While this case was making its way up to the Privy Council the Uihleins were busy endeavoring to ac-

[1] Part of the information about this episode is found in the testimony of L. T. Emory in FTC Docket 1335, Record, pp. 2152 ff. During the hearings, counsel for the Aluminum Company sought to prevent the introduction of Emory's testimony in whole or in part. Opposition was made upon the ground that it had no relation to the charges of the complaint. Additional information is contained in the testimony of officials of the Aluminum Company in the same proceeding and in BMTC v. ACOA appellant, fols. 668 ff. and 5215 ff.

[2] Testimony of an official of the Aluminum Company (BMTC v. ACOA appellant, fols. 5237–5240).

[3] *Demerara Bauxite Company* v. *Hubbard, Humphreys, and Emory.* Mrs. Hubbard was the widow and Mr. Humphreys the shrewd solicitor. The negotiations for this property had been complicated by a legal fight for title between Mrs. Hubbard and a relative of her insane husband.

quire additional bauxite. An option upon a property in Dutch Guiana was obtained, and Emory was negotiating for the purchase or lease of that part of the government-owned Christianburg deposits in British Guiana which was still available. With mining rights to this tract and a favorable decision in the above-mentioned litigation, the Uihleins "would have had sufficient ore in order to embark in quite a comprehensive program for the manufacture of aluminum."[4] The disputed deposits were decided by the court of first instance to belong legally to Emory. Upon appeal to the West Indian Court of Appeals this ruling was upheld. Thereupon the aluminum interests carried the battle to the Privy Council in London, which in May 1923 also decided against them.[5]

The Uihleins had contemplated the use of Niagara Falls power, but it appeared that sufficient energy was not available there at a satisfactory price. Other sources of power were investigated. In December 1924, while Emory was still engaged in active negotiation for part of the Christianburg deposit, he was informed by the Uihleins that they were disposing of their South American bauxite. New Year's Day of 1925 witnessed the sale of the Republic Carbon Company, which owned the ore deposits and options as well as the electrode factory, to the Aluminum Company of America, the Carborundum Company, and the Acheson Graphite Company, each of which took a one-third interest. Mr. Robert Uihlein, testifying at a hearing of the Federal Trade Commission, explained that the project was abandoned by his family and their associates not because of any lack of experience or difficulties in obtaining water power, but merely because it would occasion "too much work."

[4] Emory's opinion (FTC Docket 1335, Record, p. 2204).

[5] The point upon which the decision turned seems to have been that a solicitor may not obtain an option from a client and sell the property secured by its exercise at a profit unless the client is represented by independent advice (Emory, *op. cit.*, and BMTC v. ACOA appellant, fol. 5242). The bare facts of this incident seem to be as given here. Interpretation is avoided because testimony indicated hard feelings between Emory and officials of the Aluminum Company. After his discharge from the army Emory was not reëmployed by the latter. An officer of the Republic Mining and Manufacturing Company said that the property was practically stolen from them, and implied that Emory had been lax in following instructions to acquire it while he was with the Demerara Company. Emory felt that the solicitor had "double-crossed" him. It is possible, of course, that the Uihleins acquired bauxite only to resell it.

When Emory's services were no longer desired by the Uihleins he consulted with the Aluminum Company about a position, but no arrangement satisfactory to both parties was reached. Emory then found employment in the American Cyanamid Company, a J. B. Duke corporation. His transition from the Uihleins to the latter via negotiations with the Aluminum Company indicates nearly the whole extent of interest in aluminum production, actual and potential, in North America in 1925. Emory passes out of our story at this point after introducing us to Mr. Duke, whose attention was occupied, in much more profound fashion than that of the Uihleins, with aluminum as a means of utilizing a part of the tremendous amount of electrical energy which he hoped to develop on the Saguenay River in Canada.[6] Early in the century Duke had started to convert the water power of North Carolina into electric energy, and long before this was completed he turned to the Saguenay River with the idea of developing power for nitrogen fixation. With Sir William Price he purchased the necessary property, incorporated the Quebec Development Company, and in 1923 began the construction of a dam at Isle Maligne where ultimately over 500,000 h.p. was to be generated. By 1924 Duke had apparently given up the idea of nitrogen fixation and was endeavoring to discover other industries which would use his power. During the preceding year his engineers had made an exhaustive but fruitless survey to find, in that part of Canada, some basic product capable of consuming large quantities of electric energy in its manufacture.

In the meantime George D. Haskell, president of the Baush Machine Tool Company, had become interested in the production of aluminum. Mr. Haskell was anxious to avoid dependence upon the Aluminum Company for material with which to manufacture a

[6] Information about the Duke-Haskell episode has been obtained chiefly from the record and court opinion in *Haskell* v. *Perkins et al.*, 31 Fed. Rep. (2) 54. (The decision in this case is given below, p. 136.) The record included the whole of the voluminous testimony and exhibits submitted in the lower court. The testimony of Mr. Haskell at the Federal Trade Commission hearings (Docket 1335, Record, pp. 2394 ff.) has also been used. Haskell's testimony at these hearings was admitted over objection of respondent's counsel that it was irrelevant to the charges of the complaint, which concerned methods of competition in the fabricating branches of the industry. (See below, Chaps. XVII–XIX.) Officers of the Aluminum Company refused, upon advice of counsel, to answer questions concerning this affair. The leading official of the company testified upon this matter in the case of *Haskell* v. *Perkins* and in BMTC v. ACOA appellant (fols. 507 ff., 5661 ff. and 6036).

duralumin alloy, and also wished to participate in the remarkable future which he believed aluminum was to have. During 1921 Haskell and some Boston associates negotiated with the owners of the Norsk Aluminium Company, who, finding themselves in financial difficulties, were anxious to dispose of a half interest in their property. The latter evidently carried on negotiations with the Aluminum Company of America at the same time.[7] Several schemes were considered by the Haskell group, one of which involved the participation of Henry Ford, who was then interested in a supply of cheap aluminum. Owing to the withdrawal of Ford, lack of interest on the part of American bankers, and uncertainty regarding the tariff situation as well as general business conditions, the Haskell group was unable during the first half of 1921 to take advantage of a very favorable offer. In July the Aluminum Company made a conditional agreement to purchase a half interest in a new Norsk Aluminium Company. The agreement was carried out after modification of the decree of 1912 had been obtained.

With the revival of business Mr. Haskell began to investigate the possibility of establishing an independent aluminum enterprise on this continent. He studied methods of production, requirements, and markets through such sources as were available here and visited foreign producers. Certain bankers were apparently interested in his project. In 1924 he came into contact with Duke in the course of a search for cheap water power. After a few conferences between the two, Duke wrote a letter agreeing to reserve for four months 50,000 h.p. for Haskell's proposed company, which was to produce an annual output of about 10,000 tons of aluminum. Haskell then went to Europe, accompanied by two engineers and a metallurgist, where he acquired information concerning production methods and costs and the assurance of a supply of alumina from Germany. Upon his return he submitted to the officials of the Quebec Development Company such data concerning size of his proposed company, stock issue, markets, raw materials, and the like, as a power company would wish to have before making a long-time contract with a promoter for a large block of energy.

Haskell may have hoped from the beginning to interest Duke

[7] BR, pp. 6, 7.

in participation in the projected enterprise. Up to this point, however, the negotiations appear to have concerned only the prospective sale of power to the promoter of a new company. Now Duke began to evince an interest in participation. In the course of a trip to the Saguenay in Mr. Duke's private car in June 1924 conferences were held between Duke, Haskell, and their metallurgists and engineers. Haskell presented information obtained abroad. Other meetings were held in New York which were attended by Haskell's bankers. The outcome of these discussions was a verbal agreement between Duke and Haskell, the exact nature of which has never emerged clearly.[8] The inference from testimony and correspondence appears to be that Duke at least agreed to join with Haskell in promoting an aluminum enterprise to use Saguenay power, provided they could acquire sufficient bauxite of suitable quality and location for twenty years' production and provided he could be shown that they could produce aluminum of good quality more cheaply than anyone else. Dr. Landis, an engineer of the American Cyanamid Company, a Duke corporation, was instructed to search for ore and to investigate the technology and costs of the whole process of producing aluminum. Fifty thousand dollars was placed at his disposal. Haskell, who had already spent many thousand dollars on investigation, was to be relieved of further expenditure of this sort, but was asked to coöperate with Landis. At this time Haskell informed his bankers that their aid would not be required.[9] In the course of the next ten or eleven months an intensive investigation of ore, production methods, and costs was carried out by Dr. Landis with the assistance of Mr. Haskell and of technical experts, among whom were some who had at an earlier date been in the employ of the Aluminum Company of America. A total of $180,000 was spent by Duke on this investigation. As the work proceeded it appeared likely that Mr. Duke's condition would be met. It was estimated that cost of production would be relatively low at the

[8] The precise nature of this agreement of July 18, 1924, was not made clear in testimony taken in the case of *Haskell* v. *Perkins et al.* (executors of the Duke estate). Duke died before the suit began. A New Jersey statute prevented the plaintiff from testifying to any conversation with or in the presence of Mr. Duke. Few witnesses testified directly on the character of the agreement. Not all of those who might have been expected to know were called. One witness said that Duke had told him that "he had made arrangements with Mr. Haskell to go into the aluminum business" (Record, p. 346).

[9] *Ibid.*, pp. 1079 ff.

Saguenay, provided suitable ore close to tidewater could be found. From behind its screen of Canadian directors, the Quebec Aluminium Company, which seems to have been set up chiefly for this purpose, prosecuted a vigorous campaign in England — in which it enlisted the political influence of Lord Beaverbrook and other prominent English and Canadian gentlemen — which held some promise of resulting in leases of satisfactory bauxite deposits on Crown and colony properties in British Guiana.

In the fall of 1924 formal organization at an early date of a company to take over the investigational and promotional activities then being carried on by the Cyanamid Company seems to have been contemplated.[10] When the Quebec Aluminium Company was incorporated in December 1924, however, Haskell was not included among its directors, who were all Duke men until their replacement by Canadians for purposes of strategy in London. Mr. Haskell testified that late in 1924 and in the first half of 1925 he heard disquieting rumors of negotiations between Duke and the officers of the Aluminum Company of America. These appear to have been denied by the Duke officials whom he interrogated. In June 1925 Dr. Landis was instructed to discontinue his work on the aluminum project and withdraw applications for the public lands of British Guiana; and Haskell was told that the deal was off because an arrangement had been made with the Aluminum Company.[11]

Mr. A. V. Davis of the Aluminum Company testified that at the end of 1922 he and Mr. Aldred, president of the Shawinigan Water Power Company, who were both interested in obtaining additional power, arranged that the latter should negotiate with Mr. Duke for the acquisition of an interest in the Duke-Price Power Company.[12] Mr. Davis did not see Duke during the subsequent negotiations, which lasted about a year and came to nothing. In October 1924 Davis was approached by a representative of Duke, who offered to sell him a substantial block of power to be used for aluminum.[13] Davis replied that while he was in-

[10] *Ibid.*, p. 1638.

[11] One of Duke's officials evidently tried unsuccessfully to persuade the Aluminum Company to grant Haskell a long-time contract at a price below the market.

[12] The testimony of Mr. Davis is on pp. 955 ff. of the Record.

[13] According to testimony in recent proceedings, the offer referred to Saguenay power (BMTC v. ACOA appellant, fols. 5658–5659).

terested in the purchase of power he would prefer a participation in the power development on the Saguenay. As a result of several meetings in the next two months a merger agreement was reached in January 1925, the terms of which were worked out in detail during the spring.

The Duke interests received about $16,000,000 par value of preferred stock of the new Aluminum Company of America, equivalent to one-ninth of the whole issue, and 15 per cent of the no-par common. Doubtless Mr. Duke was chiefly gratified by the prospective realization of a dream in which hundreds of thousands of kilowatts were transformed into aluminum.

After an unsuccessful attempt to carry on with his scheme, in which he appears to have been aided by the Curtis and Wright airplane interests, Mr. Haskell resorted to litigation against the Aluminum Company and the Duke estate.[14] Two suits were initiated under the Sherman Act, but only one, the suit against the Duke estate, came to trial. In 1928 Haskell was awarded $8,000,000 damages against the Duke estate by a jury in the District Court of New Jersey.[15] Upon appeal the Third Circuit Court of Appeals, by a unanimous opinion of three judges, decided that the question of damages should never have been submitted to a jury, inasmuch as there was insufficient evidence to prove a contract between Duke and Haskell.[16] Certiorari was denied by the Supreme Court.[17]

The circumstances might suggest that the active interest of the Aluminum Company in Saguenay water power was enlivened, in part at least, by the knowledge that someone else contemplated its use for the production of aluminum. It is not clear whether provision for future growth of home demand and its export business would have led the company to acquire at that time rights to develop nearly a million horsepower. It already possessed undeveloped capacity in Tennessee which was probably sufficient to care for many years of growth at home, and undeveloped

[14] Mr. Duke died three months after the merger.

[15] A motion for triple damages was denied, since the guilty person was deceased (*Haskell* v. *Perkins et al,* 28 Fed. Rep. (2) 222).

[16] 31 Fed. Rep. (2) 54 (1929). Judge Buffington said: "Indeed the only definite, concise thing the situation developed was that Duke volunteered to make and pay for a research, and Haskell agreed to coöperate with him in doing so; and beyond that the future was not provided for by a contract then fixed and entered upon."

[17] 279 U. S. 872 (1930).

power in Norway for expansion abroad. However, the leading official of the Aluminum Company, who conducted the negotiations with Duke which culminated in the merger, has testified explicitly that during these negotiations he had no information that Duke or anyone else was considering the use of the Saguenay power for a new aluminum enterprise, and that the only purpose actuating the Aluminum Company was to obtain additional power for the enlargement of its business.[18]

2. Control of Bauxite and Power

It must be inquired whether the failure of new competition to arise since the war has been due to inability to obtain ore beds or power sites which would enable effective competition. The experience of the Uihleins seems to show that they were able to obtain satisfactory bauxite in South America. In the spring of 1925 Dr. Landis was fairly confident that good ore could be obtained in British Guiana in spite of the attempts of the British Colonial Office to restrict grants to British capital and enterprise. At that time several million tons of good bauxite belonging to the Crown or the Colony of Christianburg had not yet been leased. However, a very large part of the better ore in the Guianas was probably in the hands of the Aluminum Company or other aluminum firms by the latter part of the post-war decade. The situation in Europe seems to have been quite similar.[19]

Prior to the war almost the whole bauxite production of the world was concentrated in the United States and France. It will be recalled that French bauxite, which was nearly all owned by the established producers of aluminum and alumina, was converted into aluminum oxide in Germany by several producers of that product and by the Neuhausen aluminum firm. Within the confines of the old Austro-Hungarian empire there existed extensive deposits of bauxite which there had been no occasion to in-

[18] BMTC v. ACOA appellant, fols. 5661–5662.

[19] The chief sources from which material relating to the European bauxites has been taken are: Bannert, *op. cit.*, pp. 11–14; Czimatis, *op. cit.*, *passim;* C. S. Fox, *Bauxite and Aluminous Laterite* (London, 1932), *passim;* Günther, *op. cit.*, pp. 35–43; O. Hausbrand, "Bauxit und Aluminium," *Weltmontanstatistik*, II, 149–156; W. G. Rumbold, *Bauxite and Aluminium* (London, 1925), *passim;* Schoenebeck, *op. cit.*, pp. 22–26; FTC Docket 1335; BR, pp. 97 ff., *Haskell* v. *Perkins*, Record; BMTC v. ACOA, records of both cases.

vestigate carefully. Only a very small part of these deposits had been purchased by aluminum companies. With the coming of war it was, of course, necessary for the German alumina producers, who now had to assume the burden of furnishing oxide to the new German aluminum works, to use sources of ore within the Central Powers. The same was true of the AIAG, which had been partly financed by German capital and had marketed most of its output in Germany. After sequestration of its French properties it purchased deposits in the Bihar district of Hungary. Other deposits in the same district were acquired by German alumina producers, who also bought ore beds in Dalmatia and Istria. The Austrian government also carried on ore production in the latter provinces. With the peace settlement Istria went to Italy, Dalmatia to Jugo-Slavia, and the Bihar section of Hungary to Rumania. The new Hungary contained large deposits of bauxite hitherto untouched, while Jugo-Slavia also acquired deposits in the interior back of the Dalmatian coast.[20] Of these various bauxites some of the Dalmatian and Istrian ore was comparable to the French bauxite with respect to quality, transport, and development costs. The Istrian bauxite was, in general, more suitable, all things considered, for the production of aluminum at locations near the European or American markets, than most of the remaining European deposits. When it became apparent during and shortly after the war that aluminum was likely to enjoy a great increase in consumption as an important industrial material and a military necessity, there ensued a growing struggle for the possession of the marginal and better extra-marginal deposits.[21] Efforts of the old aluminum and alumina producers and potential competitors were both hindered and intensified by nationalistic governmental policies which ranged from the imposition of

[20] Estimates of the tonnage of ore in these various countries have been reported as follows: Italy, 13,500,000 tons (MR, 1931, p. 33); Rumania, 20,000,000 tons (*Aluminium*, IX, Heft 12, p. 10, June 30, 1927, and Rumbold, *op. cit.*, p. 83); Jugo-Slavia, 30,000,000–50,000,000 tons (Schoenebeck, *op. cit.*, p. 25, and U. S. Commerce Reports, 1928, III, 490); Hungary, 200,000,000 tons (Fox, *op. cit.*, pp. 257 ff). Of the enormous Hungarian reserves only 15,000,000 or 20,000,000 tons appear to be suitable for aluminum reduction at present. The proportions of the reserves in other countries which are suitable for aluminum do not appear.

[21] Before the war bauxite of 60 per cent alumina content and 2 per cent silica was used in Europe. Since the war it has been necessary to exploit ores containing only 52 per cent alumina and up to 7 per cent silica.

onerous conditions of export to attempts to foster domestic aluminum producers.[22] Hungary, Jugo-Slavia, and Rumania appear to have moderated their policies during the twenties, but in the case of Italy just the opposite is true.

In spite of governmental restrictions upon the free play of economic forces an energetic and determined quest during the first few post-war years would probably have provided a new aluminum enterprise contemplating production for the chief European or American markets with bauxite deposits economically capable of justifying the investment requisite for effective production. During the twenties the established firms and the new German and Italian companies did purchase or lease much of the known bauxite remaining in Europe.[23] The Bauxit Trust, the ore company of the Vereinigte Aluminiumwerke, acquired the largest amount of ore. During the first post-war years the German producer relied upon ore from Istria, where deposits were purchased in 1922, and upon importations from France. When the Italian government restricted the export of bauxite a few years later with the purpose of fostering a domestic aluminum industry, the Germans set out upon a determined campaign resulting in the acquisition in 1925 and 1926 of about twenty million tons of ore in Hungary, and smaller deposits in Jugo-Slavia and Rumania. The desires of the Italian government were met by erection of a reduction plant at Mori, Italy, partly financed by the German concern, where the Germans could use their Istrian bauxite. It was not until 1924 and 1925 that the Aluminum Company of America acquired its Istrian and Jugo-Slavian ore. The holdings of the French and British firms were also extended about the middle of the twenties.

The question of the extent to which the bauxite situation was changed through purchases of ore lands in the middle twenties by the established firms has produced conflicting evidence. The leading bauxite official of the Aluminum Company of America, testifying in the late twenties, stated that at that time there were many

[22] See above, p. 91. Unsettled political conditions and the embryonic stage of law in the new states were also hindering factors.

[23] Established firms were doubtless in a better position than a new company to meet the desires of local governments for assured supplies of metal or coöperation in promotion of domestic enterprise, but it would seem that many if not most of the acquisitions of bauxite in the twenties were made without such conditions.

commercially available bauxite deposits not owned by aluminum companies which could be used as profitably in the manufacture of alumina as the ore then being worked by the company.[24] Testimony of this official in 1935 in the Baush case set forth the known occurrences of large quantities of bauxite in various parts of the world, but did not deal with the question of ownership of economically suitable ore.[25] L. T. Emory, who in the course of investigations for different employers made a considerable study of the bauxite situation, stated in 1927 that he knew of no available bauxite deposits anywhere which would justify a new venture in aluminum production in the United States.[26] An officer of a European aluminum company expressed the opinion to the author that most of the better bauxite deposits then known had been removed from the market by the end of the twenties.

It is unquestionable that there exist at present large tonnages of good bauxite in various parts of the world which are not owned by aluminum companies or producers of alumina. Some of these deposits are accessible to cheap transport. The difficult question which it has been impossible to resolve concerns the amount of ore remaining outside the control of established firms at the end of the twenties which was economically suitable in quality, location, and so on for substantial ventures contemplating immediate production of aluminum for sale in the chief markets of Europe and America. Deposits in India, the Gold Coast of Africa, Australia, and other regions, are known to contain many million tons of high-grade ore, but a large part of them are evidently too inaccessible to use at present for this purpose. Government provision of

[24] FTC Docket 1335, pp. 5305 ff.

[25] BMTC v. ACOA appellant, fols. 5169 ff. The statement of another officer of the company in the same proceeding that "bauxite is no longer a problem," owing to the opening up of extensive foreign deposits since 1912, may perhaps give the implication that there is plenty of economically satisfactory ore which new entrants could acquire today (fol. 6110).

[26] Letter to EMJ, CXXIII, 771 (May 1927). See also his testimony in FTC Docket 1335, Record, p. 2212. Cf. Leon Henderson, director of the research and planning division of National Recovery Administration, Report on Aluminum Industry, p. 5, where it is concluded that the dominant position of the Aluminum Company of America rests partly on control of sources of high-grade bauxite. It is said (p. 18) that one company which is now considering the establishment of alumina and reduction plants in the United States says that it owns adequate supplies of high-grade domestic bauxite. No indication is given of the size of the projected enterprise.

adequate transport facilities in some sections, such as Herzegovina, might, of course, render economical the exploitation of good bauxite which cannot now be cheaply transported. The known facts suggest that ever since the war acquisition of ore reserves suitable for new firms would have required considerable pains, and that by the end of the twenties the remaining known deposits economically suited for competition with the established producers may have been distinctly limited. It is very probable, however, that large quantities of excellent bauxite in fairly accessible locations in the tropics still remain to be discovered. It would not appear that the costs of exploration are very great, since the ore usually occurs on the surface.

In conclusion, it does not seem that the almost complete lack of new firms in this industry in Europe and America since the war can be explained entirely by the acquisition of large ore reserves by the established firms; but it is evident that these acquisitions greatly enhanced the difficulty of the problem facing potential entrants.

It is not clear whether the power problem of an independent venture in America or Europe could have been solved more easily than the problem of acquiring satisfactory ore. As will be explained in a subsequent chapter, the number of power sites which are economically suitable for aluminum production is quite limited.[27] A new aluminum enterprise would require cheap energy which was well located with respect to ore and markets. Unless power could be obtained which would enable a new firm to lay down aluminum in a given market at a cost, all things considered, which was not far above the cost to old producers of reaching that market with metal made by developing a part of their power reserves, the new firm would not be able to enter the industry with a reasonable expectancy of profits.[28]

[27] Below, pp. 185–186.

[28] The important comparison is between the relative costs of meeting additional increments of an expanding demand. For wherever overhead costs are substantial and capital equipment quite durable, the marginal cost for a projected new firm, which includes the capital costs, is not likely to be lower than the marginal cost incurred with the existing investment of old firms, which does not include costs of capital already invested, unless striking improvements in equipment or process can be introduced. Hence when such improvements are not possible a new firm can ordinarily enter only by capturing an increment of an expanding demand. See below, p. 145.

Relevant information upon power costs is meager. The following passage by an electrical engineer represents a generalization applicable to the years just prior to the recent depression.

In the early days of the industry it was possible to develop hydroelectric power at an investment of around $100 per horsepower installed. Since that time not only has the cost of machinery and construction advanced, but the most desirable of the world's water-power sites have been taken up. For this reason at the present day the capital cost is of the order of $250 to $350 per horsepower.[29]

Two or three years earlier an officer of the Aluminum Company of America stated that investment per horsepower for new hydroelectric developments in the United States varied between $150 and $200, averaging nearly $200.[30] A recent paper by a representative of the Aluminum Company discusses the unfavorable location of the large water powers remaining to be developed in the United States.

Practically all of the large water powers remaining to be developed in the United States, except those on the St. Lawrence River, are located in such a rugged and inaccessible terrain as to preclude the establishment of reduction works at or even near the power sites. Thus, long distance transmission becomes a significant and costly factor.[31]

St. Lawrence power has not yet been made available for aluminum production, nor can it be predicted when and on what terms it will become available. It is not clear whether the statement quoted is intended to apply to the undeveloped power in the South owned by the Aluminum Company as well as to other sites. Presumably, however, the company selected those sites which appeared to be the most economical of all which might have been obtained. Hence, if conditions have not been markedly altered since their purchase, it would appear that the Aluminum Company has possessed some advantage over potential entrants on the score of power costs. In 1926 the Department of Justice designated as one of the chief reasons for the absence of serious attempts to enter the industry the apparent difficulty in the acquisition of adequate power in advantageous locations.[32] Whether the situ-

[29] E. V. Pannell, *Metal Industry,* XXVII, 72 (February 1929).
[30] Below, p. 213.
[31] T. J. Bostwick, *Transactions,* World Power Conference (1933), II, 328.
[32] BR, p. 51.

ation will be modified by government development of power in the eastern part of the country cannot be definitely known at present.

The cost of energy per horsepower year at the Saguenay development of the Aluminum Company of America is said to be under $5.[33] This figure may give some indication of the cost at which the reserve power on the Saguenay can be made available. A new firm in Canada which intended to enter the markets served by Aluminium Limited would doubtless need to have power at something under $10 per horsepower year at least. In fact, it might require somewhat cheaper power than that at the disposal of Aluminium Limited in order to offset a disadvantage in the bauxite cost.

Available information upon relative costs is insufficient to determine whether it would be profitable to found a new firm to sell aluminum in the United States which was produced with Canadian water power. Unless it was believed that the average cost of the new firm, including the United States import duty, would not greatly exceed the average cost of an additional increment produced by the Aluminum Company in the United States, there would be an appreciable risk of poor returns for the investment of the new entrant. The import duty of the United States is equal to $88 per metric ton of aluminum. Assuming that four horsepower years are used in producing one ton of metal, the duty is equivalent to $22 per horsepower year. It is doubtful if new energy has cost the Aluminum Company more than $20 per horsepower year.[34] It seems plain that the existence of a substantial duty has meant that a new firm established in Canada to sell in the United States would have required some advantage over the Aluminum Company in other departments as well as in power expense in order to compensate for the duty. For the future the problem is complicated by the uncertainty as to whether the

[33] Testimony of an officer of the Aluminum Company, *Haskell* v. *Perkins et al.*, Record, p. 1036. The investment was given as about $65 per h.p. The price which will be paid by Aluminium Limited, if it uses any of this power, does not appear. It pays $12 per h.p. year for the energy taken from the Saguenay Power Company (BMTC appellant v. ACOA, fol. 612).

[34] This is inferred from evidence that power could be rented in the United States during the early twenties for a minimum of about $20 per horsepower year (*Haskell* v. *Perkins et al.*, Record, pp. 801, 1421, 1575); and testimony upon investment per horsepower.

Canadian dollar will follow the pound or the United States dollar.

Authoritative opinions assert that, with the exception of Norwegian sites, nearly all the European water power which can be considered for aluminum reduction in the near future has already been taken up and developed in part or in whole. Many of the best sites in Norway are controlled by the French, British, and American aluminum firms. The use of brown coal in certain sections of Europe would probably enable effective competition if suitable ore could be secured. It is said that the cost of energy made from brown coal in Germany is now in the neighborhood of $20 per h.p. year.[35] The investment per horsepower required for a new hydroelectric development in Norway is given as $110, equivalent, perhaps, to about $10 per h.p. year.[36] Investment per horsepower for a new hydroelectric development in Germany is estimated in the same source as just double that for Norway. Evidently these estimates refer to present costs for developments which might be considered typical. Their only value is to afford some rough comparison of the costs of new power in various locations. There are no data obtainable which show with any precision the relative costs of new firms and of added increments produced by the established companies.[37]

The known facts about the bauxite and power situation seem to indicate that a new firm established to sell in the principal markets of the existing producers might have somewhat higher costs than the cost to the latter of expanding their output by drawing on ore and power reserves. It cannot be concluded, however, that a new firm which suffered some cost disadvantage could not have an expectancy of good profits. This matter requires brief analysis.

If a new entrant felt that he must be prepared for purely competitive price making he would need, of course, to be sure that

[35] MW, XIII, 380 (May 18, 1934). Bannert (*op. cit.*, p. 51) gives about $16.00 per h.p. year for 1926. The cost figures given here in dollars have been converted at the pars of exchange prevailing prior to devaluation of the United States dollar in order to facilitate comparison with figures for earlier years and the figures for Canada expressed in United States dollars of the old parity.

[36] MW, XIII, 380. It appears that power can now be purchased at something less than $10 per h.p. year in Norway (*Transactions*, World Power Conference, 1933, II, 319).

[37] Investment per horsepower of the first stage of the Lochaber development appears to be about $340. It is estimated that this will be reduced to an average investment of about $185 per horsepower for the whole development (*Engineering*, CXXXV, 636, June 9, 1933).

his unit cost would be as low as the lowest marginal cost at which the old firms could meet an expanding demand.[38] In later chapters it will be shown that competitive rather than oligopolistic price making has in fact occurred from time to time in this industry. It is manifest that the entry of several new firms, if that were possible, might of itself precipitate competitive price making. Were the new firm certain that it could rely upon duopolistic or oligopolistic price making, it might enjoy an expectancy of good profits even though its cost were somewhat higher than the marginal cost to the old producers of meeting additional increments of demand. It would need to insert its capacity ahead of the contemplated expansion of the established firms. Its relative success or failure in catching the existing incumbents off their guard, as it were, by invading the field while they were still *planning* extensions would affect the length of the period in which its profits would be small or nonexistent. The length of time required to secure the properties necessary to enter this industry and the difficulty of maintaining secrecy militate against success in this respect.

If the old firms tended to keep in existence such an investment that marginal revenue in all markets would equal their marginal costs of serving those markets, the new firm would reap normal profits as long as its average cost for some rate of output was no greater than the price (average revenue) at which marginal revenue was equal to the marginal cost of the old companies. In other words, the difference by which the average cost of the new entrant could exceed the marginal cost of the established firms would be determined by the spread between price and marginal revenue. In all markets in which the price was ordinarily set toward the lower end of a quite elastic portion of the demand curve — which was followed by a section of unitary elasticity or inelasticity — the spread between price and marginal revenue

[38] This would be true even if the established firms introduced more effective resources — bauxite deposits, power plants, and so on — as demand expanded, provided they engaged in purely competitive price making. For the result of that would be that the resources of lesser productivity, which had been brought in earlier, would be utilized less intensively, so that marginal cost with the new and the old would be equalized in so far as technological obstacles did not prevent that. It is exceedingly doubtful, however, that the old firms would practice competitive price making when introducing more effective resources, for at such a time they would be made particularly aware of the unprofitability of price cutting.

would not be large, and the average cost for a new firm could not much exceed the marginal cost to the established producers of supplying additional increments.[39] It seems likely that in some important markets for aluminum, such as the markets for cable and airplane materials, the differential between price and marginal revenue is small because demand is quite elastic at prices higher than those charged.

If it were shown that this condition of a small differential between price and marginal revenue existed in most important markets, it could not be concluded, however, that the average cost for a new firm could not much exceed the marginal cost to the old companies of expanding supply, *because we cannot be sure that the established producers are able to equate marginal cost and marginal revenue in most markets.* Under certain circumstances the marginal cost curve for a firm will not intersect the curve of marginal revenue. This is possible when demand, instead of being continuously elastic, exhibits stretches of unitary elasticity or inelasticity following a section of elasticity. When marginal cost is always either below or above the positive portions of the marginal revenue curve corresponding to the elastic parts of the demand curve, it is impossible to equate marginal cost and marginal revenue. A simple situation of this sort is illustrated in the diagram on the next page.

In an industry characterized by perfect ease of entry such a condition could not long endure. An influx of new firms would move the demand and marginal revenue curves of all producers to the left until marginal cost and marginal revenue were equal. In an industry such as aluminum, however, there is no reason to suppose

[39] The case in which the cost disadvantage of the new firm would be just equal to the full spread between price and marginal cost of the old firms is the limiting case. It is interesting to note that new firms with cost disadvantage approaching this amount might be unable to gain more than normal profits with any rate of output. Normal profits would require nearly full utilization of the investment appropriate for the price at which marginal revenue was equal to marginal cost for the old firms. Should the new firm restrict output by an amount X in an attempt to equalize its marginal cost and marginal revenue, an old firm would take over this amount X as soon as demand had expanded enough to permit it to bring in the smallest practical "lump" of added investment, for until it did so marginal revenue would be above the marginal cost at which it could expand. The new firm would then find itself producing for less than normal profits. Hence if its officers used foresight they would from the beginning content themselves with normal profits only.

that this would necessarily occur. Only the aluminum producers themselves can tell us whether marginal cost and marginal revenue have been so related in most markets that they could be made equal. No evidence on this point has appeared, so it is impossible to know whether the possible gap between the average cost of a new firm and the marginal cost of additional supply from old firms is large or small.

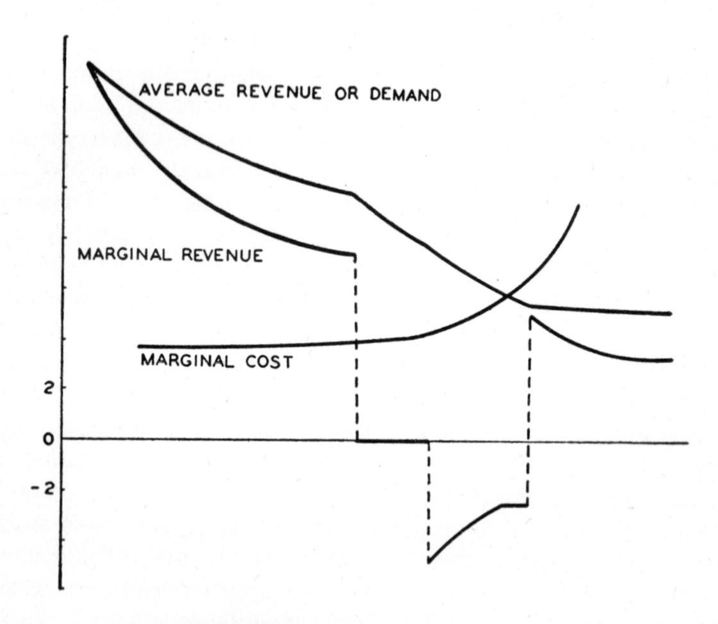

Whatever this spread, a new entrant would have to take account of the possibility that the difference between price and marginal revenue might be diminished by an increasing elasticity of demand at lower prices. Furthermore, the average expectancy for the spread would be narrowed by the probability of occasional periods of competitive price making and the possibility of economic warfare. It goes without saying that a new firm would need to be provided with ample reserves in case its power of resistance were subjected to a searching test.

In view of the considerations surveyed in this section and the strength and organized market control achieved by the principal companies (which will be examined next), it would not be surprising if new aluminum ventures depending on bauxite were limited for some time to come to countries having power and access to ore which are removed from the chief markets of Europe and America, or to countries whose governments foster domestic producers. The recent developments in Russia, Japan, and Hungary seem to illustrate these possibilities.

It is impossible to predict the success of attempts to utilize lower-grade ores commercially. The Blanc process of obtaining potassium chloride and alumina from leucite, a mineral containing 23.5 per cent of alumina, has been operated in Italy for several years. Apparently, however, no large amount of alumina has yet been produced by this process. In the past thirty years many attempts to discover an economic method for recovery of potash and alumina from alunite have failed. In the fall of 1934 it was announced that the Bohn Aluminum and Brass Company of Detroit had nearly perfected a process which they intended to operate commercially, using the alunite which exists in abundance in Utah. According to report the Tennessee Valley Authority has also been experimenting upon the reduction of aluminum from low-grade ores.[40] It is said that one of the European processes for extracting alumina from low-grade ores is quite as economical as the Bayer process, although it would not yield any decisive saving as compared with the latter. The fact that it has not been used is attributable to its possession by one of the established aluminum firms, which found no advantage in employing it. Altogether a very large amount of effort has been devoted to experimentation upon methods of using lower grade ores.[41] It is to be hoped that some process may be discovered which will appreciably lower the cost of production of this useful metal. Should that occur the world might indeed experience something like the Aluminum Age envisioned by the founders of the industry, and the problems incident to market control might become simpler.

[40] HR, p. 18.

[41] See *Aluminum Industry,* I, chaps. V–VII; United States Bureau of Standards, *Acid Processes for the Extraction of Alumina;* Czimatis, *op. cit.*

CHAPTER VII

NATIONAL MONOPOLIES AND INTERNATIONAL RELATIONS

1. Continuing Monopoly in America

Inasmuch as a new aluminum enterprise in the United States could probably have obtained economical bauxite in Europe or the Guianas and power in Canada, if not in the United States, during the first part of the post-war decade, control of ore and power by the old companies cannot completely explain the failure of competition to arise under the stimulus of rapidly expanding markets. As in the pre-war period, the expansion policy of the Aluminum Company of America has had a weightier significance than merely increasing the difficulty of others in securing these essential materials under circumstances enabling effective competition. War prosperity and the return to a higher duty in 1922 contributed to the striking increase in size and financial strength of the Aluminum Company. After a period of hesitation during which rapid growth of consumption here could not be cared for even with the addition of Norwegian properties, expansion was resumed in 1925 when ground was broken for a dam at Santeetlah on the Little Tennessee River. In the same year the vast development on the Saguenay River was launched with the building of a reduction plant at Arvida, which began operation in the middle of 1926 with 100,000 horsepower from the Duke-Price development at Isle Maligne. The year 1926 also witnessed the commencement of construction to enlarge the capacity at Badin, North Carolina, and develop power at Chute-à-Caron on the Saguenay. Facilities in use and under construction were again ahead of demand at the current price, and potential power had been more than doubled. Moreover, during this period an interest in several foreign enterprises had been acquired, more attention had been paid to the cultivation of foreign markets, and at home the company had strengthened its outlets by taking over Aluminum Manufactures, improving its foundries, and waging a determined campaign which soon made it once more a prominent factor in the castings division of the industry. Semifabricating facilities kept step with reduction capacity.

The promoter who launched a new aluminum company in the post-war years would have faced a stronger business organization than the Aluminum Company of the earlier period. Perhaps the cost of ingot production in the United States had been somewhat lowered by the larger power plants in the South; evidently the Saguenay plant became the lowest cost unit in America and possibly in the world. This suggests that some power sites which would have enabled an independent to compete successfully before the war later became extra-marginal. It is to be doubted that the company has gained any considerable economies of combination since pre-war days. A well-managed firm of much smaller size than the Aluminum Company could probably produce and sell quite as efficiently, provided it were well integrated and possessed suitable ore and a well-located site capable of cheap, large-scale power development.[1] Yet the initial investment necessary to create a substantial firm has doubtless been greater since 1920 than it was before the war.[2] Not only has the financial strength of the Aluminum Company become more impressive; the investment in ore reserves and potential power of the two North American companies also increased. The introduction of operating capacity by outsiders would, if it resulted in capturing any substantial part of the market, retard the future operation of this idle investment. Even if that part of the Saguenay investment now belonging to Aluminium Limited is to be used chiefly to serve markets outside the United States, the idle reserves of ore and power of the Aluminum Company of America are a factor of significance. Since it would be impossible to predict exactly what policy the Aluminum Company would follow should a new firm enter the field, the latter would need to be prepared for a searching test of its purse.

Although scientific knowledge of the several stages in the production of aluminum ingot has become more widespread than it was twenty-five years ago, the complexity of the process has probably remained a deterrent element of some force. In 1912 the Aluminum Company was one jump ahead of potential competi-

[1] See below, Chap. IX, for analysis of economies of large scale investment.
[2] Greater after allowance for the higher prices of the factors of production.

tors. The growth in specialized skill of its scientific and operating staff and the resultant improvements in various parts of the complicated process—particularly at the alumina stage — may have more than offset the wider dissemination of knowledge about the process, and placed the company two jumps ahead in recent years. The creation of an operating organization of comparable efficiency might require several rather than a few years. Finally, uncertainty incident to the possibilities of economical use of low-grade ores for aluminum has probably exercised some discouraging influence upon potential competition.

Whatever the views of promoters, there has apparently been little desire among bankers to assist in creating new aluminum enterprises. Neglect of this industry in a world where investment bankers increasingly engage in the promotion and operation of industrial enterprises argues weighty deterrent elements. One wonders whether the mere prestige and strength of the Mellon banking group has not influenced the situation.

The absence of attempts by more than a few of the larger immediate consumers of aluminum to integrate backwards may be explainable on the hypothesis that it would have been difficult to obtain satisfactory ore and power, or upon the supposition that it was felt best not to do anything which might possibly antagonize a powerful competitor.[3] Or perhaps the complexity of the process in itself was enough to deter many efforts in this direction. In any event, successive abandonment by prominent business men of plans to produce aluminum must have had a cumulative effect. Besides the Uihlein venture and the Duke-Haskell episode, two other projects received publicity. In 1917 it was reported that the Anaconda Copper Mining Company was negotiating for bauxite properties in Georgia, and for a few years this company experimented upon aluminum production.[4] About the time that Anaconda dropped its plans Henry Ford apparently set his metallurgists

[3] It is an obvious fact that the existing organization of the industry confers upon the Aluminum Company the power to affect in large degree the fortunes of those competitors in the fabricating branches who are largely dependent on it for supplies of virgin ingot. The nature of testimony before the Trade Commission some years ago and conversations with independent fabricators indicated their awareness of this. This feeling may have been strengthened by what has seemed to some the failure of government agencies to settle satisfactorily the complicated problems incident to the situation. (See Part IV, *passim.*)

[4] EMJ, CIV, 539 (September 1917); MI, XXIX, 14 (1920).

to work upon aluminum reduction. In 1923 he bought bauxite lands in Georgia and it was said that he bid for Muscle Shoals with an eye cocked toward aluminum production.[5] Yet to date he has produced none of the metal except for experimental purposes. It would appear that these plans were abandoned before they had reached a stage requiring the accumulation of ore reserves; yet such reserves could probably have been obtained at the time.

Nearly fifty years have elapsed in the life of the aluminum industry without the production of a single market-minded virgin ingot in the United States outside the plants of the Aluminum Company. The company emerged from the war twice as large and much more formidable. Since 1915 it has played a leading role in the international competition of existing aluminum producers for the better ore and power sites which has substantially diminished the opportunities open to potential competitors. In two instances the company acquired bauxite or power which was apparently being considered as a basis for establishing a rival aluminum producer on this continent. Although capacity was unequal to demand at high prices during the years 1922–1924, resumption of expansion in 1925 introduced additional facilities rapidly thereafter, and added an enormous reserve of potential power. The history of this industry demonstrates that the existence of the antitrust laws has not been sufficient to create conditions under which competitors would arise. A complicated process requiring an investment of several millions, limited natural resources, vigorous expansion by the existing firm, uncertainties of obsolescence, disinclination of bankers to provide financial assistance — some or all of these elements have raised effective barriers to entry of a sort with which the antitrust laws as interpreted and administered were not designed to cope. From the considerations surveyed up to this point it would not seem likely that several new firms will enter this field in America in the near future unless, indeed, some cheap process of using low-grade ores is developed.

[5] *New York Times,* August 3, 1921, p. 7; MI, XXXII, 30 (1923); Anderson, *op. cit.*, p. 61. The Hooker-J. G. White-W. W. Atterbury bid for Muscle Shoals was reported to contemplate production of an alloy of duralumin to be used in railroad-car construction. Whether the reduction of aluminum was intended is not clearly apparent, but it would seem so.

2. National Monopolies in Europe

Since 1914 there have been only seven new aluminum enterprises launched in Europe outside of those established with the participation of the leading producers. Of the seven new firms four, or perhaps five, owe their existence to the activities of governments. The two most important are the state enterprises in Germany and Russia. A small plant was built during the war at Steeg, Austria, by the Stern und Hafferl electric firm at the instigation of the Austrian war department.[6] It has continued as a very small producer. Aluminiumwerke, Bitterfeld, owned jointly by I. G. Farbenindustrie and the Metallgesellschaft, represents the outgrowth of participation by these two corporations in the development of a war-time industry in Germany. It is not clear whether the Aluminiumwerke Manfred Weiss, which has recently established a small plant in Hungary, received government encouragement beyond the imposition of a protective tariff. The remaining two of the seven new enterprises were promoted in Norway by private interests who shortly sold control to established producers. Det Norske Nitridaktieselskap, which began to produce aluminum in 1914, was purchased in the following year by the Compagnie Alais. Norwegian interests formed the Norsk Aluminium Company to take advantage of war profits, suffered financial reverses during the post-war depression, and sold a half interest to the American firm, as we have seen.

Of the new companies which have remained independent of old producers the Vereinigte Aluminiumwerke alone has exercised a significant influence in the European market. Its rapid growth to an established position among the leading producers has been sketched in an earlier chapter. From its inception this state corporation has received a large measure of government protection against domestic and foreign competition in Germany, which has ordinarily been the largest consumer of ingot aluminum in Europe. By a decree of the *Bundesrat* in 1917 the erection of new aluminum plants or expansion of existing ones was made subject to approval of the *Reichskanzler*.[7] At the same time a qualified

[6] Debar, *op. cit.*, p. 35.

[7] Günther, *op. cit.*, p. 22; Kupczyk, *Wirtschaftsdienst*, XVI, 281 (February 13, 1931).

Einfuhrverbot was passed, under which the import of aluminum would be sanctioned by the government only when the German companies could not or did not wish to deliver metal with the same conditions as to quality, price, time, and so on.[8] Since a government permit was required for importation, and the rules referred to were merely an expression of general policy, dumping could easily be controlled. The first of these decrees has continued in force. When Germany lifted the ban on imports in 1930 in accordance with an international agreement obtained by the League of Nations, it was replaced by a protective tariff of 0.25 M. per kilogram (equal to about 15 per cent ad valorem). It is obvious that the monopoly position of the Vereinigte Aluminiumwerke could not be threatened if private enterprise so desired, unless the government were willing to share the business; while the restrictions on importation gave the government an opportunity to exploit consumers if it wished to do so.

The situation in France since the war has been quite similar on a smaller scale to that in the United States. Consolidation of the Alais and Froges concerns left only one outsider, who coöperates in control of the market with the Compagnie AFC through l'Aluminium Français. This effective market control was protected from foreign invasion by a tariff of 2 fr. per kilogram, which was raised in 1926 to 3.40 fr. per kilogram (equivalent to 20–30 per cent ad valorem during the past decade).[9] The Compagnie AFC and its associate produce principally for the home market, although plant expansion in the latter half of the twenties was intended to produce an export surplus. The chief difference between the postwar situation in France and in the United States lies in the greater significance of bauxite control in the former. A new French producer of aluminum would have had to obtain bauxite abroad. Much of the ore available in the first five years after the close of hostilities was definitely inferior to the French bauxite in quality. Obviously, transport costs would be somewhat higher even for bauxite which was equal in quality. Unless a new French venture could have obtained an assured supply of this material on economical terms from some alumina firm which owned large reserves

[8] Günther, *loc. cit.*

[9] This is the minimum tariff which applies to imports from the other European producing countries. A special tariff of 5.10 fr. was laid upon ingot aluminum from the United States some years ago.

in France, it would probably have been at a disadvantage with respect to ore costs. Needless to remark, concentration in the industry and expansion at a rate at least equal to growth in demand have enhanced the formidability of l'Aluminium Français.

The disadvantage of a small home consumption in a world where most of the largest markets are protected has been shown in the description in an earlier chapter of the experience of the Neuhausen concern since the war. It is manifest that there has been little incentive to promote a new aluminum venture in Switzerland. Until 1931 the Swiss imposed only a nominal import duty of 5 fr. per 100 kilograms. In that year the duty was raised to 65 fr. per 100 kilograms, equal to about 25 per cent ad valorem. The disadvantages of more expensive power and labor in England have doubtless discouraged fresh capital and enterprise from entering this industry while the home market remained open to foreign sales. Support for this view exists in the fact that the high power costs at the new Lochaber plant of the British Aluminium Company were reduced somewhat by guarantee of bonds of this company by the British government. The English market was unprotected until 1932, when a 10 per cent ad valorem duty was set up for aluminum as England joined the move toward protection.

Thus in each of the four countries domiciling the leading producers of Europe there existed during the post-war decade factors becoming progressively stronger each year which acted as deterrents to the entrance of independent enterprise in spite of the very large increase in consumption. Acceptable ore may have been obtainable for only a few years after the peace settlement. In France semimonopolistic control of superior bauxite was fortified by the strong organization and impressive size of the two producers. In Germany the government decreed legal protection for its own offspring. In Switzerland a home market much too small to satisfy the ambitious Neuhausen firm would have necessitated precarious reliance on those foreign markets which still remained open. High costs in England might have meant small returns with the ever-present possibility of invasion of the unprotected home market. And it was well known that the producers of aluminum acted in concert for market control and might oppose their combined front to any invasion of the field.

Norway, Italy, Spain, and some of the successor nations of the old Austro-Hungarian empire were the only countries where independent enterprise might have found it feasible to establish aluminum plants. Since power costs are high and free enterprise restricted in Italy, a combination of government interest and aid from established producers was required to obtain a producing industry there. The successor nations offered small home markets, unsettled financial conditions, and barriers to free trade in neighboring lands. The Spanish market was too small to act as an anchor for an ambitious venture. Although the Norwegian law restricting ownership of power to Norwegian citizens doubtless added to the difficulties attending promotion of a new aluminum enterprise in that country by foreign capital, it did not make it impossible. Whether the failure of independent enterprise to take advantage of the possibilities of Norway for aluminum has been due to inability to obtain suitable bauxite, control of all the better power sites by established interests (in other industries as well as aluminum), exercise of power and influence by existing producers, or simply a lack of interest in attempting to promote an enterprise which would require such an exacting combination of elements and involve such large risks, cannot be definitely known.

Before passing to consideration of international agreements among producers it must be emphasized that throughout the twenties more than half of the consumption of aluminum ingot in Europe occurred in the markets of France and Germany. Imports into the protected market of France were negligible. The protective device in Germany was administered so as to permit imports which ranged between a quarter and a third of German consumption. The German and French companies developed capacity for export only in the middle or latter twenties. The world picture presented immediately after the war showed five national monopolies, of which three, producing perhaps three-quarters of the total output, sold almost altogether at home. In the ensuing decade, however, this situation was materially changed. With the rapidly expanding expectancies for aluminum, which were partially verified by the swift increase in demand beginning in 1922, there occurred a growing competition in the form of investments in ore and power reserves, as well as reduction capacity, which began to threaten a heightened competition in price and sales.

3. International Competition and Coöperation

For six or seven years after the end of war there seems to have been little active competition between the leading companies of the world except in sales in the United States, particularly during the depression of 1921–1922. Outside of the Swiss, who were forced to devote more attention to foreign markets, each company was engaged chiefly in satisfying the home demand. During this period there existed informal price understandings between some, if not all, of the European companies.[10] In 1923 a definite price agreement was made between the French, the British, and the Swiss. In their sales in Europe the Americans seem, in the main, to have respected the European price.[11] After the depression it was probably understood among the European producers that home markets would not be invaded in any determined fashion. It would be wrong, however, to conclude that there were no competitive influences at work. Aggressive price competition in the sale of ingot aluminum does not seem to have occurred, except in the development of new markets such as the Orient, unless it was induced by business depression. Until the latter twenties, at least, the same appears to have been true of competition in sales at a given price. Nevertheless, within the limits of the kind of market control which existed at various times, there have always been competitive forces which expressed themselves partly through a race in expansion to benefit from the increasing demand in new uses and countries where aluminum consumption had been low,[12] and to prevent increasing imports or withdrawal of government protection as a result of failure to meet the home demand; partly through exporting more metal to markets formerly served by other countries; and partly through attempts to influence the nature of agreements on price, markets, and, when they existed, quotas. It is difficult to determine whether the European com-

[10] Benni, Lammers, Marlio, and Meyer, *Review of the Economic Aspects of Several International Industrial Agreements* (League of Nations, Economic and Financial Section, Geneva, 1930), p. 26.

[11] Cf. *Wirtschaftsdienst,* XIII, 210 (1928), and XIV, 939 (1929).

[12] Cf. remarks of Dr. Gustav Naville of the board of directors of the AIAG at the general meeting in 1924 (*Aluminium,* VI, Heft 9/10, p. 8, May 25, 1924), who stressed the necessity of expansion by the company in order not to lose its relative share of the world market.

panies have felt these competitive influences more strongly than the Aluminum Company of America, but it seems probable that they did until the latter twenties. The British market stood open to foreigners, while the Neuhausen firm was forced to develop sales abroad. The German import restriction was so drawn, and apparently so administered, that potential imports were a factor for consideration. Until about 1925, when it began to develop a capacity destined to produce an export surplus, the Compagnie AFC probably felt these competitive elements the least. By the middle of the twenties the growth of American investments in Europe had introduced a new element into the European situation.

By 1925 it was apparent that the capacity then under construction and projected was capable of producing more aluminum than could be sold at the current high price. The world prices of copper had been falling since 1923. Formal cartel organization was again set up in the following year, perhaps with the chief purpose of insuring an orderly reduction of price. When oligopolists, believing that a change to a lower price is required to maximize profits, make reductions independently, there is often a danger that the reductions of some will be misinterpreted as acts of price competition. Such a possibility would exist whenever opinions differed as to the amount or time of the best change. Under such circumstances agreement may prevent the outbreak of price competition. The new cartel included the four chief European companies. Three small firms — Aluminiumwerke, Bitterfeld, the Steeg concern in Austria, and the Giulini enterprise which operated a small plant at Martigny — were affiliated with it. The Aluminium Corporation of England remained outside the organization. At the expiration of the two years to which the original agreement was limited it was renewed without material change for a period of three years more.

Neither the Aluminum Company of America nor its Canadian subsidiary joined the cartel, and apparently there were no agreements between the American and European producers until Aluminium Limited made agreements with European firms relating to Asiatic markets.[13] However this may be, it appears that in their

[13] It has been asserted by some that understandings relative to the American market have existed between the Aluminum Company of America and the Europeans at various times since the war. See, for instance, the testimony of Mr.

sales in Europe the cartel price was in the main respected by the Aluminum Company before 1928 and by Aluminium Limited after its formation. The Europeans did not always reciprocate in their sales in the United States, although it appears that since 1922 foreign metal has not as a general rule been sold here at prices substantially below those charged by the Aluminum Company. Many sales have been made at the prices quoted by the domestic firm. Apparently from time to time sales have occurred at prices equivalent to a discount of ½ cent or 1 cent a pound from the list price of the Aluminum Company, and sometimes at a larger concession.[14] There is no indication that the European firms cut prices vigorously in order to build up a substantial export market in the United States. The bulk of the exports to this country, which have ordinarily represented less than 15 per cent of the total sales of virgin aluminum in all forms here, seem to be explained by distress selling during depression, attractive profits owing to temporary shortage here, and disposal of temporary surpluses incident to the abrupt introduction of a large "chunk" of new capacity in Europe.[15]

The failure to adjust investment in Europe to a part of the lucrative American market must be ascribed partially to the tariff in the United States. It is to be doubted that oligopolistic forces alone would have produced this result, for cost in some parts of

---

Haskell that an officer of the Aluminum Company told him of such understandings (FTC Docket 1335, Record, pp. 2546 ff. and 2636 ff., and BMTC v. ACOA appellant, fol. 1754). See also *Kartell Rundschau,* XXV, 588 (1927); *Wirtschaftsdienst,* XIII, 210 (1928), and XIV, 939 (1929). This has been repeatedly denied by representatives of the company. See, for example, BMTC v. ACOA appellant, fols. 5765 ff. My own interviews have produced conflicting reports, but those who might be expected to have better knowledge on this point have stated that the Aluminum Company did not have any understandings with the cartel.

It has been said that Aluminium Limited had understandings relative to the European market with some or all of the cartel members. An officer of Aluminium Limited has testified, however, that prior to organization of the Alliance Aluminium Cie. there were no such understandings (BMTC appellant v. ACOA, fol. 1170.

[14] See open market quotations, p. 242 below, and BMTC v. ACOA appellant, exhibits 193, 270–280, 270A, 324–334, 344, 347–400, 425–427, 432–446, 526–528, 556. Cf. HR, pp. 4, 8. After 1929 foreign metal was sold in indeterminate amounts at larger concessions from the list prices of the American firm. It appears, however, that the Aluminum Company itself was making substantial concessions on some of its sales in these years.

[15] See Chaps. XI and XIII.

Europe has probably been somewhat below cost in the United States, owing to cheaper power and labor and lower transport expenses. In the absence of a protective tariff here it is likely that European capacity would have been adjusted for a substantial export surplus to be sold in this country. Doubtless oligopolistic price making would then have governed the market, but price would have moved at a lower level and European sales would have been larger. Removal of the duty after acquisition of the Norwegian interests, and particularly after development of the low-cost Canadian capacity, whose large potential output could be used to enter markets served by the Europeans, might not have produced the result which could earlier have been expected, if the Europeans believed that reprisals would follow any determined cultivation of the American market.

During the twenties imports from European producers failed to increase in the same proportion as the sum of output in the United States and imports by the Aluminum Company.[16] Sales of the Europeans in the United States declined relative to sales of the Aluminum Company after 1926.[17] Evidently the European producers did not share proportionately with the domestic firm in the expanding consumption here. Existence of the duty, which did not change between 1922 and 1930, may not entirely explain the failure of the European firms to adjust their investment for increasing exports to the United States as demand increased here. In the absence of agreement that result could be explained by any of the following conditions. (1) The cost of producing additional increments of metal in America may have declined relative to the cost of additional amounts in Europe.[18] While information on cost reductions is unsatisfactory, nothing has appeared to suggest that the cost of additional metal declined in the United States relative to Europe during the twenties. It would appear, however, that the

[16] Output figures are given in Table 38, and statistics of imports in Table 15. (Imports of fabricated aluminum were negligible.) Annual percentage ratios of imports from European producers to the sums of output in the United States and imports by the Aluminum Company were approximately as follows:

1923 — 19	1927 — 15
1924 — 11	1928 — 7
1925 — 18	1929 — 7
1926 — 22	

[17] Below, p. 325.

[18] Cf. the reasoning above, pp. 144 ff.

cost of the added increment from the Saguenay in Canada was lower than the cost of additional metal in most places in Europe. (2) The Europeans may not have felt confident that the Aluminum Company would pursue oligopolistic rather than competitive investment, output, and price policies here. In so far as the marginal cost here was below the sum of marginal cost abroad and the duty, Europeans would invest for export to the United States only if they were reasonably sure that price here would tend to exceed marginal cost by some appropriate margin. (3) European producers may have feared that maintenance of a certain proportion of total American sales might lead to more aggressive cultivation of foreign markets by the American companies.

The purposes of the cartel were announced in 1926 as follows: (1) exchange of experience and patents; (2) coöperative cultivation of markets — an international bureau was created with the function of stimulating interest in aluminum; (3) reduction of costs and prices, and stabilization of the latter.[19] The apparatus for market control (which was not made public for several years) appears to have been as follows.[20] There was no attempt to control production directly, but sales quotas were fixed in 1926 and again, perhaps with slight changes of 1 or 2 per cent, in 1928. Of the aggregate sales of ingot aluminum, including alloys in ingot form, by members of the cartel, each company was allotted a certain percentage. This arrangement covered all sales of ingot, in the home market as well as abroad, except sales in the United States. The actual quotas were apparently somewhere in the neighborhood of the following: French, 30 per cent; Germans, 27 per cent; Swiss, 23 per cent; British, 20 per cent. The mechanism of money fines for exceeding the proportionate sales quota was not employed. Instead, it was agreed that a company which had exceeded its allotted proportion of the total sales during the preceding three months should purchase from the companies which had sold less than their quotas sufficient aluminum to bring the sales of the latter up to their quotas. Such purchase was made at the regular cartel price, which was determined quarterly.

The price was fixed for all markets and included transport expenses. Except for the fact that the buyer in a country levying a

[19] *Kartell Rundschau,* XXIV, 615 (December 1926).

[20] Benni, *loc. cit.,* several magazine articles, and personal interviews.

protective tariff on aluminum might be charged the amount of the duty,[21] aluminum sold everywhere in Europe for the same price. Each producer was free, however, at least during depression, to make special prices in his own home market to meet particular conditions of demand. The exclusive right to sell in his own home market was reserved to each producer, but this does not mean that he could supply all the metal sold in his market from his own plants. On the contrary, in the event that demand increased more rapidly in one producing country than in the others, the latter were enabled through the quarterly adjustments to participate in supplying the former. Since the company in country A would, in the course of satisfying the increased demand there, have exceeded its percentage of total cartel sales, it would be forced to buy appropriate amounts of metal from the other producers at the next adjustment date. Thereafter it would, in effect, be acting as sales agent for them in its own country.[22]

Although this cartel probably strengthened monopolistic market control as compared with the agreements which it supplanted, competitive forces were by no means eliminated. In so far as the quota arrangements were effectively enforced, competition in sales of ingot aluminum at home or abroad would not exist. Each company benefited in the proportion fixed by the agreement from every sale anywhere — except in the United States — no more, no less. Its owners would have exactly the same incentive to develop new markets as if they owned a proportionate interest in a single monopolistic company equivalent to their percentage quota in the cartel. To the extent that the quota arrangement could not be rigidly enforced, however, competition could be expressed through evasion. It is said that the situation in Norway, where two plants were owned by the British, one by the Americans, and two jointly by all three producers, always created difficulties in the administration of the scheme. Apparently the arrangement was, in fact, evaded to some extent by some of the companies. Stronger competitive elements existed as well. The cartel made no attempt to control the price or sales of half-products and finished goods; and

[21] Evidently this provision was not applied in those producing countries which had tariffs.

[22] The most striking instance of this happened in Germany during 1927 when the Swiss and French were enabled to send in several thousand tons. See *Berliner Tageblatt,* January 11, 1928.

integrated concerns were free to use as much ingot as they cared in their rolling, fabricating, and casting plants. Members of the cartel preserved independence as to investment policy. It has been said that relative capacity could have no influence in changing quotas. Nevertheless, inasmuch as the door was open to unregulated use of additional capacity when the product was carried beyond the ingot stage, and the possibility existed of a breakup of the cartel, competition in investment was not altogether ruled out. Furthermore, as in every cartel, competition between the members was expressed through attempts to influence price policy and rearrangement of quotas.

Other disturbing elements created difficulties. After the installation of new plant on the Saguenay, American sales in markets formerly served by the European producers grew markedly. In the Orient price competition in the sale of ingot and half-products broke out about 1927,[23] evidently due in large part to an attempt to build up markets for the Canadian metal. It was reported that the European producers of half-products had joined together in order better to meet the competition of the Americans.[24] When the depression became serious and stocks accumulated rapidly, it was realized that a stronger organization was required to prevent a breakdown into outright competition. The first move resulted in subjecting the Oriental markets to control. About the beginning of 1931 Aluminium Limited and the French, German, Swiss, and British companies made an agreement pertaining to the Japanese market.[25] A few months later the British Aluminium Company and Aluminium Limited reached an agreement with respect to sales in India.[26] The European cartel was to expire in the fall of 1931. During the earlier part of that year Aluminium Limited took a prominent part in the negotiations leading to the formation of a new cartel, which was incorporated in October as the Alliance Aluminium Compagnie of Basle.

All the European members of the earlier cartel were included in the Alliance with the exception of the two new Italian companies. Acting apparently under the influence of the Italian gov-

[23] *Kölnische Zeitung,* October 12, 1927, and February 17, 1928; *Kartell Rundschau,* XXVI, 369 (June 1928).

[24] *Verein der deutschen Ingenieure Nachrichten,* March 19, 1930, p. 13.

[25] BMTC appellant v. ACOA, fols. 1160 ff.

[26] *Ibid.,* fols. 1169 ff.

ernment, they refrained from formal attachment in order to develop their business as rapidly as possible both at home and abroad.[27] At that time the tariff on aluminum ingot was raised from 42 pre-war gold lire per 100 kilograms to 260 lire per 100 kilograms (then equivalent to an increase from 3.67 cents per pound to about 6 cents per pound). The Aluminium Corporation, which had never belonged to the cartel, sold out to the Alliance in December 1931. The new Japanese company coöperates with the cartel. Since Aluminium Limited has at last joined the international organization, the only important producers left outside are the Aluminum Company of America, the Russians, and the Italian firms. It will be recalled that one of the latter is partially controlled by the AIAG and that the other was partly owned by the Vereinigte Aluminiumwerke until 1935. The life of the Alliance Aluminium Compagnie is fixed at ninety-nine years. However, any one or several shareholders who control 200 shares, or one-seventh of the total share capital, may demand dissolution at any time upon six months' notice. The fact that 1,400 shares were issued when current production of the member firms was nearly 140,000 tons per annum suggests that participation in stockholdings may have reflected relative outputs.[28] If that is so, it means that any one of the five chief firms can dissolve the organization at any time. However, since the cartel was incorporated in Switzerland, where organization fees are more than nominal, and a total capital of 35,000,000 francs was subscribed by members, it is reasonable to suppose that the associate companies have all made investments sufficiently large to preclude capricious desertion.[29]

Few details of the provisions relating to control of the market have been made known.[30] An arrangement was adopted for financ-

[27] *Metall und Erz,* XXVIII, 536 (November 1931).

[28] Figures of participations of Aluminium Limited, the German state corporation, and the French have been published. Lack of precise output figures for these producers makes it impossible to test the hypothesis suggested in the text. Aluminium Limited took between 20 and 30 per cent of the stock of the Alliance (BMTC appellant v. ACOA, fol. 860). L'Aluminium Français apparently subscribed 25 per cent of the capital (J. L. Costa, *Le Rôle économique des unions internationales de producteurs,* Paris, 1932, p. 121). It is reported that the Vereinigte Aluminiumwerke took about 17 per cent of the shares (*Handbuch der deutschen Aktiengesellschaften,* Ausgabe 1934, I, 293).

[29] The authorized capital is 65,000,000 francs.

[30] The following account has been made up of bits of information drawn from various sources believed to be well informed.

ing the large inventories of aluminum. Owing chiefly to its lack of control of production, the preceding cartel had not exemplified the maximum degree of monopolistic regulation of the market of which a federated instrument is capable. On paper at least this deficiency was evidently remedied by the new agreement, which, according to authoritative information, provided for definite regulation of all production of ingot aluminum by its members, whether for sale or for use at higher stages, through fixation of output quotas.[31] It is said that relative changes in capacity were to have little or no influence upon quotas. Development of a new alloy or product, which opens up a new market, has been mentioned as the chief criterion for alteration of quotas. The possibility of evasion of quota restrictions was reduced by employment of a prominent firm of accountants to make periodic audits of the books of each member and check its inventories. In the absence of a central sales agency each member was to be free to dispose of his contingent where and as he pleased. Sales quotas appear to have been adjusted to production quotas. It is understood that a uniform price of ingot was to be fixed for all markets in which sales were regulated. If the provisions for market control were substantially as sketched here, competition within the cartel would appear to be limited to the development of new products, alloys, and processes, and to influence upon price and quota determination — except in so far as members were led to engage in competition in investment through fear of dissolution.

Evidently the policies of the cartel are, however, subject to important modification at the behest of national governments. One would infer from the markedly disproportionate increase in German production and capacity in 1934 and 1935, and a separate price reduction in 1934 in that country, that the German government simply announced to the other members of the Alliance that it must be free from restrictions in meeting its domestic needs as it saw fit. Just what internal difficulties this may have engendered in the cartel is not known; nor can it be certain that the Germans will restrict their new-found independence, if such it be, entirely to the home market. Similar policies may be adopted by other governments in the future. It is quite possible, of course, for

[31] Output at higher stages is evidently not controlled except indirectly through limitation of ingot production.

governments to cajole or force domestic private enterprises into representing what is conceived to be the national interest, provided that profits do not suffer severely.

Unless the Germans have decided for complete independence, it appears that monopolistic elements have, for the time being, been strengthened in all markets of the world. Unified control does not indeed extend to all markets. However, excluding consideration of the Russians, whose influence in European and Asiatic markets in unpredictable, there seem to be four separate areas — the United States, Germany, Italy, and the rest of the world — in each of which market control is in large degree coextensive with the activities of producers. How long this situation will endure cannot be foretold. The Italian companies seem to have been chiefly interested in the domestic market. The participation of cartel members may earlier have exercised an influence against unregulated exports. One of the Italian companies is now apparently free to do more or less as it wishes. It remains to be seen whether the Italian government will desire maintenance of a large domestic capacity through foreign dumping. It will also be interesting to see what use is made of the enormous addition to German capacity if the armament demand diminishes. National self-sufficiency is rarely accomplished completely, and necessary imports are often purchased by dumping the product of excess domestic capacity. If new aluminum plants should be established through government encouragement in countries of eastern Europe, the total amount of capacity and the number of firms ready to produce for low export prices might grow to such an extent that coöperative market control would become extremely difficult.

In the years 1929–1931 exports of ingot by the Aluminum Company of America were negligible. In the following three years they averaged about 2,500 tons per year, nearly all of which went to Japan. Estimated exports of European producers to the United States have dropped markedly since 1927.[32] The evident reduction in imports from European producers during the depression, instead of an increase such as occurred in 1908–1909 and 1921–1922, could be explained by the existence of one or more of the following circumstances. (1) It may have been more profitable to

[32] See Table 15.

sell for the most part in markets other than the United States. (2) The Europeans may not have cared to utilize excess capacity by exporting large amounts to this country as they have done in former depressions. By facilitating the holding of stocks off the market, the Alliance has, of course, tended to diminish exports to the United States. An indisposition to sell larger tonnages here might be due to the fear of reprisals threatening both the stability of markets outside the United States and the successful functioning of the cartel agreement. (3) An understanding that European producers would not sell any large tonnages here.[33]

It is questionable whether larger European sales in this market would not have brought, apart from indirect consequences, greater revenues above variable expenses. If differences in the list prices were roughly symptomatic of differences in actual sales prices on opposite sides of the Atlantic, the average price charged declined less in the United States than in Europe.[34] It may be, of course, that the European producers considered that lower prices than those which they actually charged here would have reduced the amount by which their revenues exceeded variable expense, because they believed that demand would show little elasticity at lower prices or because they feared that the Aluminum Company would cut below them, thus precipitating a war which would carry price down to marginal cost. The latter hypothesis is weakened by the fact that the foreigners evidently did cut prices here by a few cents without the result of driving them 'way down.[35] The truth of the former supposition cannot be determined. There is some reason to think that demand at lower prices may have been less elastic in the United States in 1932, when the depression was apparently more severe here than in Europe, but one cannot be sure of this. Unless the Europeans felt that, as it shrank, demand

[33] Officers of the Aluminum Company have denied that there has been any such understanding (BMTC v. ACOA, *passim*).

[34] In 1931 and 1932 the list price here exceeded the official cartel price by more than the United States import duty. See below, p. 241. During 1930–1932 the proportion of total sales of virgin ingot in this country enjoyed by the foreigners seems to have increased somewhat as compared with 1928–1929, although it remained below the proportion for 1926–1927. The important question, however, is whether revenues above variable expenses could not have been enlarged by much greater sales here.

[35] BMTC v. ACOA appellant, Exhibits 347–397. Cf. HR, pp. 4, 8, 15. See also below, pp. 326 ff.

became much less elastic on this side of the Atlantic than on the other, their failure to test it out with lower prices is hardly to be explained on the grounds that their revenues could not have been increased in the absence of indirect repercussions.

It is clear that the above reasoning applies if metal sold in the United States is not counted as part of the output quotas. According to testimony given in 1935 by the chairman of the British Aluminium Company, sales in the United States were not so counted.[36] A moment's reflection will show that the reasoning would be equally valid if aluminum exported to the United States were to be counted as part of the output quotas. In the absence of an understanding or a fear of reprisals, the output quotas of all companies would be sold in the most profitable proportions as between Europe and America, and metal sold here would in fact be exempted from output restrictions.[37]

Evidently the Americans have attained a position of greater power than they formerly enjoyed. The aluminum tariff has in the past given to the Aluminum Company a large measure of freedom to plan its own price, investment, and output policy. Unhampered by quota restrictions, it has not ordinarily had to share the American market with the foreigners to any great extent. It would now seem that the domestic firm will be even more free to pursue its own policies without encouraging imports, unless the instrumentalities of market control abroad should be greatly modified. At the same time Aluminium Limited has gained a substantial footing in the markets formerly served chiefly by the European producers and appears to have become an important factor in European aluminum politics.

In summary, entry into the field of virgin aluminum production has never been easy. The number of producers has remained

[36] BMTC v. ACOA appellant, fol. 1172.

[37] If there were substantial sales here, and if the quotas had been set with regard to the other markets only, output quotas would almost certainly be enlarged in order to permit the most profitable rate of output for the other markets — which would be equivalent to exempting metal sold here from the output restrictions. If it were profitable to sell large tonnages in the United States, it would seem that (in the absence of agreement or fear of reprisals) the total output of Europeans would not be restricted to the rate appropriate for sales in markets outside the United States, unless the Americans were in a position to dictate to the other cartel members, or unless Aluminium Limited took a smaller percentage quota in the cartel than it could otherwise obtain.

small. With the exception of a few enterprises promoted or fostered by governments, only two new firms — in Hungary and Japan — have been established since 1910 which have remained independent of the old companies. In addition, one Italian company has recently become independent. In unprotected markets oligopoly with or without agreements has obtained; in markets sheltered by governments the degree of monopolistic control has been high. The growth in government encouragement of national monopolies has recently threatened the international cooperation of private producers. Competitive forces have taken the form of rivalry in the acquisition of ore and power sites, in the expansion of capacity, in the development of new alloys and products, in sales in new markets, and, finally, in attempts to influence the determination of prices and quotas when agreements existed. Producers of virgin aluminum have enjoyed a greater measure of freedom from the compulsion of the "free market" and hence more opportunity to plan the relation of investment and output to demand than has been characteristic of most other industries. Governments which espouse crude theories of *laissez faire* have permitted or encouraged a degree of power over market forces which is hardly consistent in theory with a rational philosophy of economic liberalism or *laissez faire*. The next section will be devoted to an examination of the consequences of the various mixtures of competitive and monopolistic forces which have obtained in this industry at different times and places.

PART III

SOME CONSEQUENCES OF VARIOUS KINDS OF MARKET CONTROL

CHAPTER VIII

STRUCTURAL EFFECTIVENESS — INTEGRATION

1. Issues and Terminology

Although explanation of the existence of monopolistic elements is often difficult, it is usually a less formidable problem than assessment of their consequences. The aluminum industry presents no happy exception to this rule. What we want to know is, of course, what have been the actual consequences of existing mixtures of monopolistic and competitive forces, and whether some alternative mixture could be expected to yield better or worse results. In Part III an attempt is made to ascertain the consequences of the various mixtures of monopolistic and competitive forces which have existed at different times or in different places in this industry, and to decide whether alternative combinations of these elements would have produced more or less desirable results. In this appraisal it will be inquired which of the alternatives would approach most closely to the ideal results.

Considerations of effectiveness in production and marketing limit the desirable alternatives in most industries to a certain portion of the whole range of alternatives between pure competition and the type of market control in which monopoly elements are likely to be strongest, single-firm monopoly. In the case of aluminum it is apparent at a glance that pure competition[1] has not been and is not likely in the near future to be an advantageous type of market control. A comparison between the actual results of the monopolistic forces in this industry and the probable consequences of pure competition would have no significance for public policy. To anticipate the results of the analysis of structural effectiveness, the desirable alternatives for the aluminum industry are limited to the monopoly end of the scale. Comparisons will need to be made only between the results of single-firm

[1] The necessary conditions for this type of competition are (1) a sufficiently large number of buyers and sellers so that no one of either can affect price; and (2) homogeneity of product.

monopoly, simple oligopoly, and oligopoly with different sorts of agreement. Wherever it seems possible and fruitful, direct comparisons will be made between the actual results of the alternating types of control which have existed in Europe, and also between these results and the consequences of single-firm monopoly of production in America. However, since the various types never operate side by side under exactly the same circumstances, it is manifest that the actual results of one alternative must often be compared with the hypothetical results of another.[2] At the end we shall raise the question whether government control of one kind or another offers potentialities of improvement in results.

In their simplest aspects the principal matters at issue between the various alternative types of market control are these. (1) Effectiveness in production with a given technique of capital goods and administration. Unless some reason develops for believing otherwise, it may be assumed that the grade of business ability in the industry would not depend upon the type of market control which existed. (2) Progressiveness, or the discovery and adoption of improvements which lower cost, improve quality, or develop the adaptation of a basic material to new employments for which its inherent qualities make it admirably suited. (3) The relations between investment, output, prices, and earnings. Which alternative will bring the closest approach to ideal investment, output, and price? Divergence from ideal relations may take the direction of underinvestment, or underutilization, with monopoly profits, or overinvestment with profits above or below normal. The first sort of divergence was emphasized by the older theory of monopoly. More recently it has been urged that in many industries monopolistic control is necessary for the prevention or cure of overinvestment, for what I shall call rationalization.[3]

The more one reflects upon these matters, however, the greater the number of interrelationships which intrude themselves upon

[2] Actually this is what must be done in nearly all attempts to judge the relative benefits of alternative economic arrangements. Usually the alternatives are mutually exclusive. Where they happen to coexist at the same time, the controlling conditions are usually sufficiently different to require the use of hypothesis.

[3] This has to do with what is usually called "stabilization" in the United States. The use of the term "stabilization" is unfortunate because, as usually employed, it has implied that it would be desirable to stabilize something or other, usually price. Under some conditions stabilization of any element is bad, and under most actual conditions stabilization of certain elements seems to be bad.

his neat categories. Under the actual conditions of the present dynamic world price and output policy is closely linked to investment policy — these are but two aspects of the same set of considerations. Further, the investment and price policy of a corporation seems to be the kernel or focal point to which all of the various elements referred to in the questions just raised stand in important reciprocal relation. The old issue of exploitation through restricted output and high price is partly a question of restriction of investment. The newer issue of rationalization concerns the least wasteful adjustment of investment to changing demand, the question of overinvestment and its elimination. While the short-run aspects of both of these involve output policy expressed through degrees of utilization of existing investment, short-run policy in itself is always influenced to a greater or lesser extent by the past and prospective investment programs — and vice versa. The degree of progressiveness may be affected by, and certainly has a determining effect upon, the investment policy. One aspect of this appears in the fact that the ideal investment will not be reached without progressive discovery and adoption of new forms, shapes, and variations of the basic product — in the case of metals, new alloys especially. The profitableness of the investment and price policy of a given period influences the investment policy of the succeeding period, not only in the obvious way, but also by making available greater or lesser sums which may be invested without going to the savings market. The nature of the best structural firm at a given time is also influenced somewhat by the past investment policy and its causes and results, while the actual structure in existence helps to shape the investment policy of the future.

These close interrelationships, of which those mentioned constitute some of the more important, make it evident that a neat division of issues into separate categories should not be allowed to simplify the picture of consequences. In treating several issues separately we must take account of these interrelations in order to apprehend as clearly as possible the more important aspects of the adjustment of investment, output, price, and demand.

Clarity in subsequent analysis will be aided by definition of terms which recur frequently. The size of a firm may be defined by the size of its output or by the number of units of any one of the factors of production of which it is composed. For the problems ahead it seems most useful to refer to the scale of investment in capital equipment and natural resources.[4] As a matter of convenience in exposition the term "scale of investment" will be restricted to the size of investment at one stage of an industry; changes in scale will mean horizontal contraction or expansion. The terms "economies" and "diseconomies" will be employed to characterize only such differential cost advantages and disadvantages as are occasioned by a change from one scale to another. "Integration" will have its generally accepted meaning of vertical combination of operations at more than one stage of an industry under a single business management. It will also be helpful to have one word to connote both the size of a vertical combination and the internal adjustment of scales of production between the various stages. For this I have adopted the word "structure." The term "rationalization" is often used to connote, in part, parcel, or whole, what is deemed by European industrialists to be American application of scientific method to business (or should one now say what was once so deemed?). It has also been employed to mean an ordered adjustment of supply to demand for a whole industry. "Rationalization" will be used here in the latter sense, to mean diminishing or preventing waste in the adjustment of supply to changing demand. Optimum rationalization will achieve the smallest degree of overinvestment consistent with difficulties of adjustment incident to technological conditions. In order to avoid misunderstanding, the application of this term will be restricted to the relation of supply and demand at one stage in the industry. Thus rationalization will concern only horizontal control and not integration. Integration may, of course, aid rationalization. "Best utilization" of a given capacity or scale of production will be used to mean that degree of utilization at which average cost (including all elements of cost) per unit is lowest. Maximum effectiveness is obtained by best utilization of the best structural firm which can exist with a given state of the arts.

[4] Differences in efficiency between firms with different scales of investment are obviously not to be explained solely by the latter. In comparing the efficiency of two scales it will be assumed that for each scale the factors are combined in the most efficient proportions for that scale, and that the sorts of equipment and labor appropriate to maximum efficiency at each scale are employed.

2. Theoretical Relations between Integration and Monopoly

The most economical use of a community's resources will exist when production in each industry is carried on by business units of the most effective structure. In this chapter and the following one we endeavor to determine, as nearly as possible, what is and has been the best structure for a firm engaged in manufacturing and selling virgin aluminum ingot, and thus to decide whether national or international single-firm monopoly or cartel organization has been required for maximum effectiveness in any markets. In the absence of adequate quantitative data, which are unobtainable, much reliance must be placed upon inference from facts of which the quantitative importance is known roughly rather than with precision.

The best structural firm is one which, when working at best utilization, has a lower cost per unit (including all elements of cost to society) than would obtain in the production of any quantity of the same good under any other arrangement. Single-firm monopoly can only be advantageous to consumers on this score if one business unit is able to produce more cheaply than two or more units the total amount demanded at a price equal to cost including normal returns to all factors of production. Offhand this seems to be entirely a question of the economies of horizontal extension of control, of the best scale for one stage of an industry. In the absence of material advantage from integration, that is so. However, when some degree of integration improves efficiency, that may indirectly result in justification of monopoly [4a] at a stage where it could not otherwise be justified. The greatest conceivable advantages of integration cannot, of themselves, cause a single monopolistic unit to be more effective than two or more firms. But it is evident that vertical combination necessitates adjusting the scales of investment between stages. If the best scale happened to produce at a rate which just absorbed the output of the best scale at the next lower stage, and supplied just the right output for the best scale at the next higher stage, there would be no problem of

[4a] Throughout the next few chapters it will often be convenient to abbreviate "single-firm monopoly" to "monopoly." It should be clear from the context when the word is used in this narrow sense rather than in its broader sense.

fitting scales to each other. Where the best scale of any step was adapted to produce most efficiently more or less than could be used in the same period by the best scale at stages above, when they were operating most efficiently, or was adapted to absorb more or less than the most efficient output of the best scales below, such a problem would exist.[5]

If there were an open market in the materials for, and the products of, all the stages,[6] the best structure would include an investment of the best scale at each stage, unless purchase of materials from outside markedly diminished any important advantages of integration — e.g., that of insuring suitability of materials. But if no open market existed, or if purchase of materials would prevent substantial gains from integration, then vertical control would involve fitting together scales of investment some of which would have to diverge from the optimum. The most economical adjustment would yield the best balance of advantage from vertical control and diseconomies incident to exceeding or falling short of the best scales at various stages.[7] This might require a monopolistic scale at a stage where, in the absence of advantages from integration, it would have no justification, and vice versa.[8] For instance, the economy of a monopolistic scale in smelting and the advantages of integration might require monopoly of

[5] In its simpler aspects this problem resembles that of combining machines of differing capacities. Underutilization can be eliminated by adjusting the number of each machine appropriately to the lowest common multiple of their capacities.

[6] An open market should exist wherever efficiency of converting any portion of these goods into other commodities would not be increased by integrated control — i.e., if such operations were carried on by a "different industry" — and also whereever efficiency would be improved if other firms in the "same industry" bought from or supplied the integrated concern.

[7] There might be more than one adjustment which would minimize cost.

[8] Mr. John Jewkes's interesting article, "Factors in Industrial Integration" (*Quarterly Journal of Economics,* XLIV, 1930, p. 621), appeared when I was beginning to think about the relation between integration and monopoly. His ideas were partly responsible for crystallizing my belief that the relation was important, and they were helpful in suggesting some lines of departure for my problem. However, his article seems to be an endeavor to explain the growth of integration and its results to the individual vertical firms. Hence he does not need to make the distinction between private and social advantage which is essential here.

In *The Structure of Competitive Industry* (London, 1931), which came to my notice after the main relations discussed in this section had been formulated, Mr. E. A. G. Robinson deals with the relations of size and efficiency in terms of different sorts of optima and their reconciliation. This type of analysis is similar in many respects to that presented here.

mining for maximum efficiency in the whole process; or a monopolistic scale in smelting might be justified by the efficiency of monopoly in mining plus the benefits of integration. Clearly, monopoly at any stage could not be thus justified via integration if any of its product could be more efficiently used outside the integrated structure — in "another industry" — or in the opposite case, if any of its materials could be more efficiently supplied from the outside. Where these conditions did not obtain, there would be two possibilities. A monopolistic scale at stage *X*, justified on grounds of economy at that stage, might be matched with monopoly at other stages, where the best scale was considerably smaller, with the result that one firm monopolized the whole output at all stages. Or smaller scales at the other stages might be complemented by a smaller scale at stage *X*, and production divided between two or more firms. The first result would be the more likely the larger the proportion which the cost at stage *X* made of the total cost of the final product, the greater the diseconomy of a scale at that stage below the most efficient, and the smaller the diseconomies of exceeding the best scales at other steps. The latter result would be the more probable when the cost at stage *X* would be a small part of total cost whatever scale were adopted there, and the diseconomies of exceeding the best scales at other stages were marked. Finally, when integration brings substantial advantages the abilities of management should be distributed between horizontal and vertical control in the most effective way. A monopolistic scale may not be warranted which would be desirable in the absence of gains from the extension of control vertically. Without laboring the point further it is clear that integration may sometimes bear an important, if indirect, relation to monopoly.[9]

3. Integration in the Aluminum Industry

The remainder of this chapter will be devoted to an examination of the advantages of integration and an assessment of their

[9] It need hardly be mentioned that integration often stands in other relationships to monopoly which are quite important for the strategy of business — as was shown in Chaps. VI and VII — and must be understood if we are to have intelligent social control. Purely strategic advantages of integration do not concern the determination of the best structural unit as here defined.

importance in the aluminum industry.[10] In the following chapter horizontal economies and the problem of combining the best scales in an integrated structure will be considered.

As they realistically occur under dynamic conditions the more important possible cost advantages of a change to integrated organization of production and marketing may be classified as follows: (1) a closer and less wasteful coördination of requirements between successive stages; (2) the location of plants at points which are, for the particular industry, more favorable (because of transportation or production advantages) than the locations already used or likely to be used in the given economic conjuncture; (3) the imposition, at one or more stages, of a more effective scale of investment, or more efficient technique; (4) the promotion of rationalization at one or more stages; (5) the initial extension of able business management to the control of more than one stage, or the imposition of more able management than that already in control. The first class of advantages may be elaborated in this way.[11] By integration the suitability or reliability of raw materials, and, indeed, materials at each stage, may be improved for the specific purpose of the next higher stage. Also, where much effort is devoted to marketing between stages when they are independent, integration yields a distinct saving. The traditionally emphasized gain from greater regularity in the flow of materials through the various stages, or to put it more accurately, better quantitative adjustment to the requirements of succeeding steps, needs no more than mention. This possibility applies also, of course, to the provision of capital equipment as well as materials. Savings from elimination of unnecessary operations may also be included here.

It frequently happens that the existing equilibrium or conjuncture has placed independent plants producing at one stage of an industry in locations which are not the most economical from a transportation or production point of view for the industry in

[10] The reasons for extended treatment of a matter which is but indirectly related to monopolistic elements are: (1) its relationship to monopoly has been neglected by most students; (2) integration is of considerable importance in the aluminum industry, and it is not difficult to confuse some of its advantages with those of horizontal expansion.

[11] Some of the following ideas are found in J. M. Clark, *Studies in the Economics of Overhead Costs* (Chicago, 1923), pp. 136–141.

question. Especially is this likely to happen if the plant is producing materials for several other industries also, or if the industry with which we are now concerned has in its development relocated its own plants (i.e., plants at stages other than the one not yet controlled through integration). Futhermore, it may occur that lack of knowledge or indisposition to assume the risk of tying up altogether with one industry, particularly if the latter is in the pioneer stage, will preclude establishment in the best locations unless it is done by integration.

The third class of advantages concerns improvements in the technique of capital equipment or processes and, equally important, improvements in the technique of administration, or scientific management in the broadest sense. Betterments of either type may be so radical as to constitute "improvements in the arts," or may consist in the adoption of more or less well-known applications of general principles or techniques to the particular purpose of the industrial operation concerned. Opportunity for the imposition by integration of "improvements in the arts" of capital or administration probably exists only when a given stage has had conspicuously unenterprising management, unless changes have been or can be developed by persons associated with the company contemplating the acquisition of vertical control. With such a condition it is quite conceivable that the potential improvement would not be translated into actuality except by integration and that it would become the dominant motive for such a move. The opportunity to impose lesser improvements by integration is undoubtedly more widespread and may exist wherever management, without being inept, is none too alert, well-informed, and ambitious. Where the operating firm is typically smaller than the best size, extension of vertical control may close the gap. Further, when the firms at a given stage are in a demoralized condition owing to serious overinvestment, integration may aid in rationalizing the adjustment of supply to demand at that stage.

The last type of advantage which may be gained by integration overlaps all the others. There may, however, be additional savings which are not included in any of the other classes. The initial extension of able management to more than one stage will clearly be advantageous to the point of best utilization of the business ability involved. An important qualification is that the "excess capacity" of the management may be more advantageously used in expansion of the single plant or in horizontal extension. Furthermore, the problems of the various stages may differ so widely in nature that the utilization of "excess capacity" through *any integration* would result in lessening efficiency. Although the efficiency of the business unit might be increased by placing under control of a given management double the amount of factors in the same stage, it might be increased less or even diminished by adding the control of any factors engaged in the very different work of another stage. This qualification also holds when it is a question of the imposition of more able management upon a stage where less able management is already in command. On the other hand, vertical control may present much greater possibilities for specialization and coördination in administration (the true advantages of "full utilization" of business ability) than are promised by the extension of horizontal control. To some extent at least, the types of specialized abilities which are included within a given managerial group will depend upon the relative advantages to be gained by using this or that sort of ability in the kind of work to which it is best adapted.

The advantages of integration in the aluminum industry may not be comparable to those found in the steel industry, but they are undoubtedly of substantial importance. The chief gains seem to be those relating to the suitability of materials, the adjustment of requirements, and the location of plants. Chemical analysis determines the composition of different bauxites and hence the use to which they are best suited. Perhaps the mining and grading of bauxite by an aluminum company results in enhanced suitability of the material used in the alumina plant. Preparation of its own alumina aids greatly in insuring the reduction of metal of high quality, for the purity of the aluminum tapped from the reduction furnace depends almost entirely upon the purity of the aluminum oxide fed into the furnace. Any metal elements contained in the oxide will also appear in the aluminum because they are less difficultly separated from oxygen. Since the purity of the metal is of controlling importance for many uses an aluminum firm can hardly afford to entrust the quality of its alumina to the disinterested efficiency of an independent concern.[12] The same may

[12] All of the important aluminum firms possess their own alumina works.

be said of the advantage in manufacturing its own carbon electrodes and furnace linings. Impurities in these will appear in the aluminum or interfere with the efficient functioning of the electrolyte. Furthermore, the butt ends of used electrodes may be added to the mix from which the new electrodes are produced. The saving in this regard may amount to 10 to 30 per cent of the total electrode consumption.[13]

Closely connected with the foregoing, although probably of less importance, are the savings incident to elimination of marketing between the various stages. The fact that specifications are so important means that in the absence of integration buying would involve constant testing. The amount of effort spent in this activity is undoubtedly somewhat less in the integrated company. Although integration may bring some saving in the elimination of "sales effort," on account of the grading of materials and half-products it is doubtful if this kind of gain is large in the aluminum industry.[14]

Ownership of power by an aluminum producer results in better coördination between this stage and others. With or without integration the power cost of making a pound of aluminum will be large because of the very great energy consumption (something like 10–15 h.p. hours for each pound of metal). One of the most substantial savings from vertical control proceeds from adjusting the operation of the reduction cells to the generation of a base load at the power plant. Two sorts of gain accrue. The cost of equipment for providing a base load is less than that of the facilities of a plant which must handle peaks of various sizes. And best utilization of equipment is not possible for most plants which sell power — that is, for the sort of plant from which the aluminum producer would have to purchase.[15] In order to utilize generating equipment most efficiently it is imperative to assure regularity of the flow of materials for the reduction process. Ownership of bauxite mines, alumina plant, and electrode factory not only reduces the uncertainty of obtaining material when it is required,

[13] Anderson, *The Metallurgy of Aluminium*, p. 122.

[14] Bauxite and alumina are graded according to chemical composition. Ingot aluminum is graded on the percentage of pure aluminum contained; sheet and wire are graded by the B. and S. gauge, and so on.

[15] It may be objected that a power company would be willing to generate and sell its whole output to an aluminum producer. But see below, p. 185.

but obviously enables a less wasteful scheduling of production and flow in adjustment to the base load of power.

Carrying production through the stage of semifabrication, and to that of finished goods in the instance of wire, castings, cooking utensils, and other manufactured articles, may bring some gains of the same sort. Since these stages succeed the reduction process it would seem at first sight that the gains might not be nearly so great. But any activities at the later stages which aided in regularizing demand would tend to enhance the regularity of the reduction process. The more direct knowledge of the consumption markets which is likely to result from integration may help, at least, to bring about a nearer approach to the best schedule for the power plant and reduction works permitted by the state of the market. By enabling better planning of the flow of materials, vertical control of operations at several stages should diminish underutilization, not only of the costly power plant, but also of equipment and labor at the other stages.

In addition to savings from adjusting reduction to a base load of energy, ownership of power sometimes eliminates the necessity of converting from alternating to direct current. Purchased power usually arrives in the form of alternating current which must go through converters before use in the reduction cells. If the hydroelectric plant is owned, and the reduction works situated close by, direct current generators may be employed which feed the energy directly into the cells. Transforming from high voltages to low is also avoided if the two plants are adjacent. Most of the aluminum companies have secured savings of this sort at some locations. However, these gains are not always possible, for many power sites are not close to suitable plant and town sites, and frequently the reduction plant uses only a part of the power generated or needs interconnection with an adjacent alternating current system to assure a steady load. Frary and Edwards say that the more common arrangement at the larger plants is to generate alternating current.[16]

One of the most conspicuous advantages of integration occurs in situating power plants at locations which, considering all

[16] *Aluminum Industry,* I, 319–320. See also T. J. Bostwick, in *Transactions,* World Power Conference, 1933, II, 323 ff. In the United States only the Niagara Falls and Massena works use D. C. generators.

relevant factors, are the most economical for the aluminum industry. We have seen that cheap power is of vital importance for cheap metal and that an aluminum producer will use a large base load of power. The sites capable of developing power at the cheapest cost fall into three classes in relation to aluminum. Some would be uneconomical because large transportation expenses on raw materials and finished or semifinished goods would more than outweigh the cheapness of energy. In America the water-power sites in the state of Washington will probably be in this class for several years. Many other sites, such as Niagara Falls, which develop power at a very cheap cost, can sell it at rates which include a substantial segment of economic rent, owing to the existence of a demand from many industries and municipalities.[17] Lastly, there are others, well located with respect to aluminum ore and markets, for which there is no general public demand sufficient to make their development profitable. It is these sites, which are out of the way of industrial and population centers, that provide the most economical power for aluminum reduction.[18] This seems to present an instance where an industry must integrate in order to have operations carried on at the most economical locations. It is exceedingly doubtful whether the best sites for aluminum would be developed by independent enterprise which would have to assume, in large part if not entirely, risks connected with this one industry.[19] In any event the other savings from joining the power and reduction stages render it more economical for an aluminum producer to develop his own power. The resort to out-of-the-way sites for cheap power is exemplified in

[17] Even after public regulation of electric rates has been imposed, much rent is included in the charges. Some portion of its capitalized value is usually injected into the valuation at the time when regulation is first imposed, and subsequent revaluations usually reflect in some measure the growth of demand by their consideration of "present value."

[18] This is explained as follows by one of the representatives of the Aluminum Company. "Our plants are located primarily or solely from the standpoint of where there is power available, and we must realize that in practically every case they must be in those districts where there are no large public utilities taking off power, or power is too expensive. We must have cheap power, so that except for . . . our plants in Niagara Falls, our plants are in small communities, what might be termed out-of-the-way communities" (FTC Docket 1335, Record, p. 3188).

[19] The Duke development on the Saguenay seems to be the exception which proves the rule. The aluminum industry may develop and use there far more power than would have been utilized (even if developed) for many years by anyone else.

the unsuccessful attempt by the Aluminum Company to develop the Long Sault section of the St. Lawrence. Even at the present time this site seems to fall in our third class — among those admirable for aluminum but for which no general public demand exists.[20] Since the war the Aluminum Company has developed power in Tennessee and Canada for which no other demand was likely to arise for many years. The British Aluminium Company early resorted to Norway for cheap energy. The advantages of developing out-of-the-way power sites have markedly reduced the proportion of total cost ascribable to energy. While the Aluminum Company was using Niagara Falls current alone, power expense accounted for nearly half the cost of production.[21] Since then the part of total cost made by energy expense has been progressively lowered until at present it is probably not more than 20 per cent at most plants.[22]

Much of the most economical bauxite is also situated in out-of-the-way places. As in the case of power, this might not be used at all if it were not mined by the aluminum producers; or it might not be mined as efficiently by outside enterprise, as will be explained in the next paragraph.

Integration in the aluminum industry may also bring advantages from the maintenance of better scale units, more effective technique, and better rationalization at some stages. Both the scale of investment and the type of equipment used in bauxite mining were doubtless improved by integration in several instances.[23] Maximum efficiency in the exploitation of bauxite resources located in the less industrialized, less wealthy areas could be achieved today only through integration. The same may be true of rationalization. Similarly, experience suggests that ownership of alumina and electrode plants is necessary for the greatest

[20] See the survey of Sanderson and Porter, dealing with markets for this energy, contained in *The St. Lawrence Navigation and Power Project* (Washington, 1929), by H. G. Moulton, C. S. Morgan, and A. L. Lee.

[21] See J. W. Richards, in *Aluminum World*, VIII, 257 (October 1902).

[22] The energy required per ton has also been reduced somewhat.

[23] Concentration of deposits in Arkansas invited steam-shovel stripping of the overburden and mechanization of transport between pits, crushing mill, and drying apparatus. The Aluminum Company of America has probably developed the Guiana deposits more efficiently than would have been done by local enterprise. The same may doubtless be said of the operations of the Bauxit Trust in Hungary and Jugo-Slavia.

economy at those stages wherever efficient plants have not yet developed to supply a substantial demand from other industries. Possession of rolling mills and castings plants has resulted in the discovery and application of improvements which might not have been introduced as soon had these units been separately owned.[24] Advantages of the sort mentioned in this paragraph would certainly be of much less importance than those mentioned earlier if it were not for the great possibilities of coördinated research which views the problems of the various stages in their related aspects. Improvements of all sorts are likely to be more significant when the research staff understands the problems of all the stages, and sets its tasks in the light of the possibilities suggested by a broad view of the whole process and the interrelations of the various steps. The history of the introduction of the more notable alloys and final products shows that one-stage fabricating firms or independent inventors make important discoveries and adaptations of this sort. But it is questionable whether the range of variations would be enlarged as much by competition of this sort as by rivalry between several integrated enterprises.[25]

Finally, the very different operations of the several stages in the production of aluminum offer opportunity for specialization of management. An integrated firm can make use of particular ability for coördinating the administration of several phases of a whole productive process. Indeed, the exercise of such specialized ability is a requirement if the other advantages of integration are to be obtained. Vertical control probably enables specialization in the supervision of relations between stages in greater degree than would be advantageous if such relations took the form of market contacts.

In summary, the decisive advantages of vertical control in the

[24] In 1907 the Aluminum Company of America erected a "continuous" sheet rolling mill at New Kensington. At this time continuous rolling was confined to steel and tin plates, so it seems likely that the introduction of this improvement would have been retarded if the rolling of aluminum had been left to firms which rolled brass and copper. (See *Metal Industry*, VI, 4, January 1908.) In 1909 the Aluminum Castings Company built a new foundry in Detroit which was laid out in the "unit system" and was said to be the "largest, best equipped, and most up to date in the world" (*ibid.*, VIII, 162, April 1910). Recently one of the most modern jobbing sand foundries in the world has been installed at Fairfield, Connecticut, and the first blooming and structural mills for aluminum have been erected by the Aluminum Company.

[25] Below, Chap. XV.

aluminum industry have been, and are, provision of the most economical power, neat adjustment of investment and output between the various stages — particularly adjustment to the steady capacity of the power plant — and assurance of satisfactory quality in the materials for the reduction process. These gains are made possible by and require the use of special coördinating ability on the part of some managers. Evidently a high degree of efficiency ordinarily requires as a minimum bringing under one managerial control the preparation of alumina, electrodes and furnace linings, the generation of energy, and the reduction operation. Under some conditions ownership of bauxite enables some reduction in cost. Accumulation of reserves of ore and power sites has, of course, been occasioned in large part by considerations of tactical advantage. Extension of control beyond the reduction process by bringing into the organization various branches of the semifabricating and finishing stages may afford appreciable benefits. The economical limits to extension in this direction will be set by the possibilities of increasing efficiency through extending administrative control horizontally at the alumina, power, and reduction stages. Vertical control of these three stages at least is imperative. The problem of achieving the best balanced structure, which is considered in the following chapter, involves the best utilization of business ability through the extension of control horizontally, vertically, and laterally — i.e., over the making of commodities other than aluminum which are produced with aluminum or electric energy or both — in such a way that its marginal product is the same in each direction.

CHAPTER IX

STRUCTURAL EFFECTIVENESS — HORIZONTAL EXTENSION AND INTEGRATED BALANCE

THE next step is to ascertain whether there is any stage of this industry where the best scale of investment has been so large as to be monopolistic. If this condition has not obtained in the past it cannot do so in the future (barring a fall in demand) without some substantial change in capital or administrative technique. The best scale of horizontal control is determined by the limits of economy in extending the size of the single plant or of the combination of plants. It changes with improvements in capital technique and the technique of management, and varies with different grades of business ability; and it grows with the increasing "capacity" of a management group as the members mature in the problems of their business. While it is impossible to fix definitely the best scale for a firm in this or that industry at any time, or to forecast its future alterations, nevertheless in many instances approximate limits may be set with some assurance. The problem of the present chapter requires the comparison of the best scale with the market. Maximum effectiveness would require monopoly at any stage only if the output of the best scale would meet the whole demand at prices which just covered normal costs. The best scale may be determined by the most efficient size for a single plant or by the economies of combining plants under one managerial control. It will be useful to distinguish between plant economies and economies of combination. The basis for classification is not the type of economy, for many of the economies of combination are of the same nature as those of the other class. The ultimate limits upon the economy of combination are set by the development of techniques of administration which permit management to administer larger aggregates of economic resources efficiently because they enable better utilization of the latent capacities of business men.

The relation of the best scale of investment to the market demand does not, of course, remain fixed in a dynamic society. This aspect of the monopoly question is often realistically presented as a race between the growth of demand and improvements which enhance the economical scale.[1] What is the situation in this respect in the aluminum industry? The best scale for the single plant at several stages will be examined first, after which the question of economies of combination will be raised.

Examination of the best scale for the reduction plant need not detain us long. Electrolytic reduction of aluminum by the Hall process gives no opportunity for the use of large, specialized units of machinery or for extensive specialization of supervision. The process is a simple one operated with simple equipment.

> A striking feature of electrolytic and electric furnace processes is that they are usually demonstrable on a small scale, and that the industrial application consists simply in multiplying indefinitely these small units . . . the production of aluminum is purely an electrolytic operation, one in which the output is proportional to the number of amperes of current used.[2]

There appears to have been no fundamental change in process or equipment since the beginning of the industry.[3] The quantity of current which it is feasible to use in one aluminum cell determines that the best size of the individual cell be relatively small.[4] The output of the largest pots employed is probably not more than 350–450 pounds of metal a day. Hence large production is obtained through the multiplication of cells. A typical reduction plant consists of several pot rooms, each with one or more rows of from thirty to a hundred cells, depending on the voltage which is best suited to power conditions. Units of labor are easily proportioned to units of capital equipment in a reduction plant, so that about the same amount of labor per unit of output must be used in a large

[1] In some of the industries where the trust problem first presented itself the growth of demand in the United States has apparently so far outstripped the best scale of production under very capable management that there no longer seems to be any doubt that single-firm monopoly is not required for the greatest effectiveness, however this may have been in the earlier years of this century — e.g., oil, steel, tobacco.

[2] J. W. Richards, *Aluminum World,* VIII, 131 (April 1902).

[3] Cf. Guillet, *L'Évolution de la métallurgie,* p. 66, and *Haskell* v. *Perkins,* Record, p. 1673.

[4] The upper limit to the amount of current used is given as about 30,000 amperes. This limit is set by the increasing difficulty of changing anodes and breaking in the frozen crust as the size of the cell increases. See *Aluminum Industry,* I, 302.

plant as in a small one. The same is generally true of energy requirements, which are, of course, unaffected by the scale of the reduction works.[5] Often several small pot rooms are employed rather than one large one in order to stimulate rivalry between the workers of each. Whatever economy this occasions would, of course, be available to a fairly large plant as well as a very large one. The reduction plant seems to present a clear instance of broad limits within which unit cost would vary hardly at all due to change in scale.[6] Actually the best size for a reduction works is set by the considerations which determine the practicable amount of power to use at a given location. The same is broadly true of electrode and furnace lining factories.

What then determines the best size for a water-power development? It is certain that the very large powerhouse is more economical than a lesser scale, owing to the greater economy of giant turbines and generators. The matter is not so simple with the dam, waterway, and storage basin. A very large investment can always turn the materials of nature into a more effective steel works than a much smaller one, and the variations in design of two great steel plants are not numerous. But a very large investment in water-power development at some sites would not yield as cheap energy as a smaller investment at other places. This is due to the extremely wide variation in "site characteristics," such as size of head and flow, degree to which discharge can be controlled by storage, nature of the foundations and valley where the dam is built, and the nature of the penstock location.[7] Usually it is these natural conditions which play the dominant role in deciding the cost of hydroelectric energy. A study of seventeen hydraulic developments varying in horsepower from 1,000 to 85,000 and in head from 21 to 1050 feet

indicates, as might be expected, a general tendency toward a lower cost per horsepower as head and size of plant increase. The site characteristics are a

[5] The savings in consumption of power and labor have come from improvements in electrical apparatus and in design and operation of cells rather than from larger reduction plants.

[6] This conclusion has been verified in conversations with several persons connected with the industry.

[7] See H. K. Barrows, *Water Power Engineering* (New York, 1927), p. 252. Cf. also J. D. Justin and W. G. Mervine, *Power Supply Economics* (New York, 1934), chap. VII.

more dominating feature, however, and wherever the cost of dam or waterway, or both, is high, the cost per horsepower of the plant will be high, irrespective of cost of powerhouse and equipment.[8]

It has been explained that the scale of the reduction plant will be determined by the scale of the power plant. Hence monopoly might be justified on the score of effectiveness if development of sufficient power *at one site to* satisfy the whole demand for aluminum would bring a lower cost of producing ingot than when energy was generated at more than one site. But if such a site existed, it would still be necessary to inquire whether the national economy would be best promoted by allocation of it to the aluminum industry. From what has already been said of the development of the several national monopolies, it is clear that (at least since the war) none of them can be justified on the basis of large-scale economies in power; unless, indeed, any of these firms was prevented from obtaining such a superior site. Each of these companies has, in fact, developed power at several different sites.[9] The Aluminum Company of America is the only one of the large producers which has ever attempted to concentrate all its reduction operations at one site. The Long Sault section of the St. Lawrence would have furnished sufficient energy to meet the demand for aluminum for some decades. Nevertheless it has probably been a wise policy to prevent the use of this power by the aluminum industry, for at some time in the future a general public demand will probably arise for it, and it will be worth more to the public in other forms than aluminum, which can be produced with power that would otherwise not be used.[10] It seems most probable that maximum

[8] Barrows, *op. cit.*, p. 626.

[9] Given the actual development, monopoly could not be justified by the existence of a site at which the total output of the existing plants could be more cheaply produced unless full cost per unit at the one site (including the maximum rent offered by alternative uses of that power) would be lower than marginal cost with the existing plants.

[10] The most economical arrangement would probably have been to allow the aluminum industry to use some of this power under a contract permitting government purchase whenever a sufficient general demand for the energy should arise. But the uncertainties of prognostication, and the probability that the initial type of development for aluminum would have to be scrapped, because not in accord with the best plans for developing the section as a whole, would very likely have prevented agreement upon terms satisfactory to both the government and the Aluminum Company.

effectiveness at the power and reduction stage has not for many years required a monopolistic scale except, perhaps, in countries with very small demand, such as Italy and Spain. But it may be concluded that the best structural firm will possess a hydroelectric development which is likely to be fairly large, and which will in any case have a favorable combination of site characteristics and cheap transportation. Both the number of dams and the number of powerhouses will depend upon the site characteristics, which thus determine indirectly the number and size of reduction plants, each designed to utilize the full base load of energy from one or more power plants.[11]

In the conversion of bauxite to alumina there is much opportunity for mechanization.[12] The Bayer process, which is in general use at this stage, is a complicated method involving digestion of the ore under pressure with caustic soda, filtration to remove the "red mud," precipitation, calcining, and cooling. Large tanks keep the materials in solution for about two days, and an elaborate pumping and piping system conveys the solution from one step in the process to another. Very great requirements of steam for digesting, and heat for calcining and evaporation allow the employment of large boilers and rotary kilns. But after a plant has become large enough to utilize fully the best size and type of boilers, kilns, and pumping system, further enlargement results only in duplication. A plant producing but 10,000–15,000 tons a year (which would yield roughly 5,000–7,000 tons of aluminum) will have many rather than several digesters, precipitation tanks, and filter presses, several boilers, and one kiln.[13] Thus a much larger plant would not seem to be required for a correlation of the capacities of mechanical units at successive steps in the process such

[11] If production of other electrochemical products is undertaken, as in Europe, the plant for aluminum reduction will, of course, be somewhat smaller than the scale suitable for the maximum load.

[12] Information upon operation at this stage has been obtained chiefly from *Aluminum Industry*, I, chaps. III and V; a description of the Salindres plant in *Engineering*, CVI, 163 ff. (August 16, 1918); and a personal visit to one plant.

[13] In 1918 the Salindres plant, which produced about 12,000 tons of oxide a year, had 20 autoclaves (digesters), 24 precipitation tanks, 2 batteries of boilers, 1 rotary kiln, and 1 cooling drum. Neither the account of the operation of the Bayer process given in *Aluminum Industry*, nor personal observation at a plant much larger than the Salindres works, suggests that there have since occurred any changes in equipment which would alter markedly the proportions indicated in the text.

that underutilization may be almost eliminated.[14] According to one authoritative opinion enlargement beyond the scale designed for an output of fifty tons of alumina a day (15,000–18,000 tons per year) brings no savings in cost of capital equipment per unit of output, and offers very little opportunity for economizing labor. Labor costs amount only to about 10 per cent of the final cost of alumina, of which the cost of materials — bauxite, soda, coal — makes up a large part. It appears that cost would not vary appreciably on account of differences in size between factories designed for fifty tons a day and those producing several hundred tons a day.

Both the large-scale savings and the limitations upon them are illustrated by the number of alumina plants in operation. The two companies in France operated thirteen reduction plants in 1930, but only three or four works making alumina. In Germany the Vereinigte Aluminiumwerke has but one oxide factory, while the Swiss concern has three alumina works and four reduction plants. The two English companies coöperated for many years in the manufacture of alumina. The Aluminum Company of America has never operated more than one plant; but, in so far as the above conclusions are correct, few if any savings would be foregone if it had several. Since two to two and one-half tons of bauxite must be used to produce one ton of alumina (which in turn yields only one-half ton of metal), transportation expenses absorb an appreciable part of the cost of the oxide. In 1916, after the acquisition of its Guiana ore properties, the company began the erection of a large oxide plant at Sollers Point on Chesapeake Bay. This plant was never completed, since it was found that the pressing war demands could be filled more expeditiously by enlarging the East St. Louis facilities.[15] But an alumina plant at this location is contemplated for the future. In the absence of any decisive change in the capital technique of alumina purification, the past and present plans of the company for a tidewater plant, when considered with the foregoing analysis, may be taken as additional evidence that the limit of productive economy at this stage is

[14] Professor Myron W. Watkins has theoretically explained the better correlation of productive capacities from enlarging the scale of production. See *Industrial Combinations and Public Policy* (New York, 1927), p. 54.

[15] BR, p. 117.

attained long before the present size of the East St. Louis works is reached. Apparently there are no substantial economies or diseconomies with enlargement to a scale producing several hundred tons a day. It follows that considerations of plant effectiveness at this stage have not required monopoly in America since the date of patent expiration; nor in the more important producing countries of Europe at any time since the last few pre-war years. Evidently the size of those plants which produce alumina by a "dry process" in an electric furnace may easily be accommodated to the amount of energy available for this purpose at a given site.

The possibilities of large-scale operations in bauxite mining depend in the first instance upon the degree of geographical concentration of deposits. Where the ore beds are confined in a small area, as in Arkansas and some parts of southern France, operation by a single company would obviously avoid duplication of crushing and drying equipment, rail connections, and so on, but it is manifest that these savings would not bulk large. The drying facilities are simple, and on account of the large amounts of clay in all bauxite deposits, hand methods of mining which combine shoveling and sorting are the most economical.[16] At this point it becomes apparent that from bauxite to ingot there is no stage at which the economies of the best-scale plant have, for some time past, necessitated monopoly in order to exert their full force upon the cost of aluminum.

The relation of plant effectiveness in the semifabricating and finishing stages to monopoly may be quickly disposed of. Plainly monopoly at the ingot stage could not be required indirectly through advantages of integration unless large plant economies extended so far in most of the higher branches using large amounts of ingot that a nearly monopolistic scale would be best in *each* of them. In Chapters XVIII and XIX reasons are shown for belief that relatively small-scale production in castings and utensils is quite economical. In all the important producing countries of Europe and America there are several plants in each of these branches. The most economical scale for a rolling mill is somewhat larger.[17] But the experience of the Aluminum Company of

[16] *Tariff Information Survey, C-16* (1921), p. 14.

[17] A much less wasteful utilization of the capacities of rolls, shearing machines, annealing furnaces, and so forth, can be secured in a large mill. One substantial item

America indicates that the best scale is not so large as to offset the advantage of having sheet mills located at the reduction works or close to markets. This company rolls sheet at Alcoa, New Kensington, Niagara Falls, and Edgewater, New Jersey. In 1925 there were twenty-eight rolling mills in the German Aluminium-Walzwerksverband,[18] while Switzerland boasted eight mills in 1929.[19] Although some mills may be adapted for somewhat different products the evidence indicates that the economical size of plant in this branch has fallen short of a monopolistic scale, except perhaps in the case of the new hard alloys, demand for which is probably still in its infancy. Evidently single-firm monopoly has not been required for maximum efficiency in the manufacture of most products from ingot except, perhaps, when they were quite new.

Elements of effectiveness incident to horizontal combination of plants in one business unit may be conveniently divided into (1) economies in production and transport, (2) economies in marketing, (3) savings in administration. It is doubtful that combination would yield any significant economies in production, as distinguished from administration and marketing, at stages below semifabrication. It might seem that combination would result in less underutilization of resources used in the preparation of bauxite and alumina, since the aggregate power fluctuations at several sites might be of lesser amplitude than the variations at one site. However, in so far as fluctuations at individual sites could be predicted, as they probably could be, within broad limits at least, the high utilization of resources at other stages could be accomplished by individual firms through variation in inventories. Secondly, there is no opportunity for specialization of plants at stages preceding fabrication. Finally, the steady pace of the reduction plants, ordinarily interrupted only during depression, minimizes the possibility of savings from concentrating fluctuations of production in one of several plants.[20]

---

of economy is in cutting down on the expense of changing rolls; the large mill really consists of several "mills" with different-sized rolls.

[18] *Aluminium,* VII, Heft 16, p. 11 (August 31, 1925).

[19] *Schweizerische Betriebszählung,* August 22, 1929.

[20] This economy has been rather glibly bandied about in the literature on combination. There are good reasons for holding it suspect, not the least of which is that the one trite example of the sugar trust is repeatedly cited. If this saving is to be gained through a reduction of capital costs, less costly capital units must be used

Specialization of plants in semifabrication undoubtedly brings some savings. The Aluminum Company of America has one wire and cable factory at Massena, a rolling mill for the strong alloys and "Alclad" sheet at Alcoa, and the new mill for making blooms and structural shapes at the former location. About 1928 a special mill for rolling, drawing, and working the strong alloys was erected by the AIAG at Chippis. The Germans also have a special alloy mill at Bonn. At present additional rolling or blooming mills for the strong alloys might represent wasteful duplication even in the United States, where consumption is much larger than in any country of Europe. But if the great future predicted for aluminum in building should materialize, there would, of course, be room for other mills. Furthermore, it need hardly be said that the most effective structure for an aluminum producer does not need to comprehend all varieties of products.

The hackneyed saving which combination may bring through the elimination of cross freights can be transferred from the text-book to the account book only if the best scale of plant is so large that more than one plant for each market area would involve the sacrifice of efficiency.[21] For many years the demand for sheet in the United States seems to have allowed enough efficient rolling mills so that three or four firms could each have operated one or two mills without much, if any, resulting increase in transport cost. In Europe, where intensity of demand is greater per square mile, savings in cross freights in this industry are probably of negligible importance. It is obvious that horizontal combination in the important finishing branches can bring no marked manufacturing or transport economies.

---

to carry the peak and suffer the various degrees of idleness. In some industries, this would be impossible; in others such units could be attached directly to the other plants instead of being segregated in one plant, and this might lower the managerial costs. Perhaps there might be a significant saving in labor through a lessening of the costly item of turnover; but there is also the possibility that marginal direct costs would be lower with two plants operating than with one.

[21] It might seem that this economy could always be secured because the size of the market area tributary to each plant could be made that which would keep the plant busy without invading a market tributary to another plant. But the extent of the market in any geographical area is not determined by the square miles of the region —whereas freight rates are determined partly by mileage—but rather by the number of purchasers and their intensity of demand. Obviously, if the extent of demand in a fairly small geographical area is great, the savings in cross freights would be negligible.

The foregoing analysis leads to the conclusion that unless horizontal economies in marketing or administration have been substantial, maximum efficiency has not required single-firm monopoly in the leading markets in Europe or America at any stage of the aluminum industry, with the exception of a few branches of semifabrication and finishing. It is evident that conclusions about these two sorts of economies, particularly those of extending administrative control, are difficult. The earlier chapters have portrayed the encroachment of aluminum upon the well-established markets of the older metals. Since this invasion has been strongly resisted by the producers of the older metals and by the forces of ignorance, inertia, and habit, selling expenses have always been high. The development of the strong alloys will have to make its way in many fields (aircraft is a notable exception) where these forces are even stronger than in some of those which aluminum has already captured. Under such circumstances the opportunities for combination to effect a substantial saving in advertising and other sales expenses would not appear to be marked, unless there is reason to think that each of a few firms would spend large sums upon sales campaigns designed to attract business from the others. With regard to standard products whose quality is easily tested, such as ingot, sheet, and cable, each of a few firms might refrain from any such campaign because of a belief that it would provoke similar expenditures by the others with no other result than an increase in expenses all around.[22] In so far as this was so, there would be no sales expenditures of the sort which clearly could be saved by combination. If the few firms did engage in wasteful advertising against each other the appropriate remedy would be an agreement upon a code of economic selling practices. Merger for the sole purpose of eliminating wasteful advertising may be considered in the same category as marriage to avoid expensive hotel dining. Combination might, indeed, reduce the expenditures on advertising of competing differentiated alloys or forms of the basic product, but perhaps at the expense of reducing their variety unduly.[23]

Economy from the extension of managerial control over larger aggregates of labor and capital depends upon the opportunities for

[22] See Chamberlin, *The Theory of Monopolistic Competition,* pp. 150, 170.

[23] Cf. below, p. 347.

better utilization of managerial abilities and for specialization in administration and upon the feasibility of coördination.[24] Both vertical and horizontal extension may afford opportunities of increasing the efficiency of the firm through increasing the efficiency of management, and both require coördination if such savings are to be realized. For the sort of operations which present little opportunity for specialization in management, enlarging the scale of investment results first in utilizing the management better, and thereafter in multiplication of the number of "fully utilized" managers. Sooner or later increasing difficulties of coördination will cause higher unit cost.

What can be said of the question of increasing efficiency of management by horizontal extension of control at the several stages of the aluminum industry and by vertical extension, and of the best balance between them? The nature of the industrial and marketing processes in this industry suggest that opportunities for specialization in administration are quite limited at most of the stages. Mining and preparation of bauxite, manufacture of electrodes, generation of electric energy, and electrolytic reduction are all very simple processes yielding simple products. They cannot be broken up into a great variety of operations, as can the manufacture of automobiles and of many kinds of machinery. No amount of specialization can appreciably alter the energy required to reduce a pound of metal or the amount of transportation from the mines to the smelting plants. Much the same may be said of nearly any one of the fabricating and finishing branches, considered by itself. At each of the stages the limits of specialization seem to be reached with a few managers. Owing to the simplicity of operations, however, these few can evidently administer a fairly large aggregate of resources just as efficiently as a much smaller one. The conclusion must be that beyond the point where further specialization of managers fails to increase efficiency, the scale of investment at each of these stages may be enlarged to a great extent without appreciable effect upon unit cost through increasing or decreasing efficiency of management. While the few managers are becoming better utilized unit cost will diminish but

[24] Specialization may be considered as one way of better utilizing the abilities of business men. Upon economies of large-scale management cf. D. H. Robertson, *The Control of Industry* (New York, 1923), and Chamberlin, *op. cit.*, Appendix B.

slightly because the total cost incurred for management is small; with the addition of more administrators unit cost will remain constant for some time because the work of coördination remains relatively simple. Although the Bayer process of preparing alumina is more complex than the work of other stages, it does not seem sufficiently complex to induce a high degree of managerial specialization.

Wide opportunity for specialization in productive and marketing operations is presented when several branches of fabricating and finishing are brought under one administration. The most efficient degree of specialization in the work of developing new forms, alloys, and products, and in direction of the personal selling organization [25] cannot be used by the very small firm. However, a firm which is much smaller than some of those now operating can specialize upon the manufacture and progressive variation of a few products rather than many. The success which the European concerns have shown in these departments, as exhibited by new products and alloys and by the growth of consumption, seems to compare favorably with the results in America, and it is by no means certain that a firm needs to be as large as the French and German companies to achieve these results. There is no reason to think that maximum economy would require monopolization of all fabricating operations by a single firm.

As far as the preceding considerations go it now appears that at every stage of this industry there is a wide range within which unit cost is not appreciably affected by the scale of investment. Two important corollaries are evident. The scale of investment at the several stages can be easily fitted into an integrated structure without incurring substantial economies or diseconomies unless the firm is very small or very large. With a given state of the arts, effectiveness would probably not differ markedly whether the scale of investment at each stage were monopolistic or were considerably smaller than that. Hence, the relative desirability of oligopoly and single-firm monopoly must be tested largely on other grounds. Since the conclusions reached here with respect to single-firm monopoly apply a fortiori to agreement and cartel organization, the desirability of the latter must also be determined upon other grounds. The conclusions summarized in this paragraph have

[25] Printed advertising is given over partly to outside specialists.

probably been justified by the conditions obtaining in all except the smaller markets of the world for some time past.

Considerable coördination is required by extension of control over several fabricating branches and also by integrated control of several stages below, which, as we saw in the last chapter, is necessary for maximum efficiency. Sooner or later as control is extended horizontally at all stages the difficulties of coördinating the work of direction of operations at all stages will grow. This is so because the number of contacts or relations between members of the administrative forces of the several stages will be increased. After a point they will probably increase in greater proportion than the growth in investment. Since a larger "amount" of coördination cannot be obtained simply by adding more business men, extension of horizontal control at all stages must at some point bring diminishing efficiency. It is impossible to ascertain whether any of the existing aluminum firms has expanded horizontally beyond the point where a marked drop in efficiency begins to express itself through rising unit cost or a poorer quality of product or service. Should demand continue to expand rapidly in the future, however, it is plain that the present firms will not be able to grow indefinitely year after year at a corresponding rate without injuring efficiency, unless improvements in the technique of administration or equipment occur. The powerful position of the established firms not only makes it impossible to infer from the lack of new entrants that they are not yet too large for maximum efficiency; it also suggests that newcomers may not enter until the older concerns have gone some distance along the path of increasing cost.

One of the other issues between the different types of market control is progressiveness. Some of the foregoing analysis has suggested the artificiality of separating the two issues of effectiveness and progressiveness, for the degree of progressiveness may be related to the size of the firm as well as to any differences in incentives, afforded by the several types of market control, to discover and apply improvements which reduce cost, better quality, or widen the range of adaptations of the basic product. Integrated research on the problems of the various stages may, as we have noted, bring advantages. The question here concerns the best size for that part of the research organization devoted to the problems of each stage. At what point do the advantages of greater specialization and increased contacts between minds of different make-up and training cease to be significant? We need the opinions of those who know most about research as to the relation between the horizontal size of the research organization and its productive results.

The analysis of these two chapters will now be summarized briefly. The most economical structure for an aluminum firm has always required vertical control of several stages. It must include the cheapest possible power as determined by the best balance of site characteristics, large-scale development, and transportation expense. It does not seem probable that horizontal expansion beyond a point reached many years ago by the leading firms yields any substantial increase in effectiveness, although it does not soon tend to raise unit cost. This conclusion is substantiated for the past decade by the statements of officials that cost reductions have been due chiefly to technical improvements in cell design and operation, and the like. It appears that there is a wide range of scales of investment at all stages within which unit cost remains approximately constant. Thus there is no difficulty in fitting together an integrated structure composed of efficient scales at all stages. Continued horizontal expansion, which is, of course, necessary to maintain monopoly power as demand increases, encounters diminishing efficiency sooner or later, however, because of the growing difficulties of vertical coördination incident to more numerous contacts between stages.

It seems unlikely that the cost of producing aluminum in the United States would have differed markedly in the last twenty years or so had there been three or four business units instead of one, unless the rate of progressiveness had been much less.[26] In Europe, where there are several small, cheap water powers, the existence of two or three times the actual number of the leading companies might not have increased cost much, if any. But the existence of a large number of firms would have resulted in much lower efficiency. Oligopoly and single-firm monopoly are the desirable alternatives.

This conclusion would have been nearly as evident twenty-five years ago as it is now. Unless there was reason to believe that

[26] This matter is discussed below, p. 348.

single-firm monopoly in this industry would give a higher rate of progressiveness or would bring no net disadvantage, all things considered, there was no economic reason for permitting it to exist even at a time when it was temporarily justified from the standpoint of structural effectiveness. For in the actual world, competitors do not spring up full grown (especially in an industry of this nature) when economists say, "Presto," even where government agencies exist for the purpose of protecting them from trial by purse rather than efficiency.

NOTE:

It was with much reluctance that I finally decided, during my original study of the American monopoly, to abandon several attempts at quantitative measurement of changing effectiveness with expansion which, I had hoped, could be used to supplement the analysis given in this chapter. The difficulties which seemed insuperable were partly the lack of data and partly the presence of too many factors which have exerted a real but imponderable influence.

Figures of average cost of production for ingot in each year from 1920 to 1925 appeared in the Benham Report. As would be expected, they reflected the inflation, depression, and succeeding pickup of business at a lower price level. Cost figures for the years 1925–1931 were computed in the Baush litigation. It is not clear whether they are comparable with the earlier figures, and it is known that they were affected to some undetermined extent by improvements in apparatus. It is impossible to compute a series of cost figures from data on ingot output and prices and earnings because the amounts of earnings from the sale of ingot alone are not distinguished. Even if we had a continuous cost series, conclusions about changing effectiveness relative to expansion would require precise information upon all other influences which had affected cost. A lower apparent rate of earnings upon capital investment in the post-war decade as compared with pre-war years can be explained by several factors other than decreasing effectiveness, as will be shown in a later chapter. Professor Watkins suggests (*Industrial Combinations and Public Policy,* p. 136) that when return to capital makes an increasing part of "value added by manufacture" this indicates an enhanced rate of progressiveness unless the use of monopoly power explains the increase. When monopoly power is appreciable, such increase might also reflect economies of horizontal expansion or of more effective integration. We should like to know the results of this test applied to the Aluminum Company, even though the interpretation might be difficult. But the Census Bureau has never published "value added by manufacture" for aluminum ingot production. A usable substitute for this figure could be computed by multiplying a moving average of ingot production by a similar type of average ingot price, upon the grounds that the amount of ingot carried through later stages was worth the market price to the company. But it is not known what part of the earnings come from sales at the higher stages. The change in the ratio of earnings per dollar of capital to earnings per pound would be meaningless in the absence of precise measurements of changes in proportion of capital invested in the finishing stages and in idle reserves of power and bauxite. Many other difficulties might be mentioned, but enough has been said to show why precise quantitative study of the problem was not feasible; for much the same reasons it has not been practical in the case of the European concerns either. A large amount of statistical data appearing in the Baush litigation seems to be of dubious value to the economist for a variety of reasons.

CHAPTER X

INVESTMENT, PRICE, AND DEMAND — INTRODUCTORY

1. Issues

THE next few chapters present an examination of questions connected with the relations of investment, output, prices, earnings, and demand. In this chapter a discussion of the nature of the issues involved will be followed by a description of certain important conditions of supply and demand in the aluminum industry and consideration of some features of the aluminum price structure.

What we wish to know are the actual relations of investment, output, prices, earnings, and demand under the types of market control which have existed from time to time, and whether alternative types could be expected to approach more or less closely the ideal relations. The ideal relationships would exist when new investment equaled the largest amount which would show an expectancy of normal earnings [1] on the average over a period of years during which such an output was produced at every point of time as could be sold at prices equal to marginal direct cost; and when output was in fact regulated at all times so as to equate price and marginal direct cost. Divergences in either direction from these ideal relations are equally bad. In the one case the excess increment of resources is invested where it has less value to consumers as a whole than in some other employment; in the other case investment is restricted to an amount less than the ideal, with the result that a certain increment of resources must be invested in another industry where its social value is less, if it is invested at all. The terms "overinvestment" and "underinvestment" will be used in this book to mean divergences from ideal investment. Under-

[1] Normal earnings may be conceived as the returns which tend to be earned in those "industries" where entry is not appreciably hampered by barriers of any sort and where the products of the various firms are not effectively differentiated by any device, such as brand, trade-mark, or advertising slogan, the attractive power of which cannot readily be duplicated.

investment enables prices greater than average full cost, including normal earnings, and hence profits above normal. With overinvestment price may be above, below, or just equal to average full cost, depending upon the particular quantitative relationships existing and the degree of underutilization. If price is set equal to marginal direct cost under conditions of overinvestment, it will necessarily fail to cover average full cost.

Under certain conditions the adjustment of investment to changing demand will involve underutilization of capacity [2] even though no mistakes are made in estimating the movement of demand and no unpredictable changes in demand occur. In other words single-firm monopoly or oligopolists would, under certain conditions, plan to keep in existence greater capacity than they expected ordinarily to employ at the rate of best utilization; or would find underutilization more profitable, under other circumstances, than operation at the rate of best utilization, as intended. Any one of the following conditions might occasion underutilization persisting over several years. (1) The introduction of more efficient equipment before the retirement of older facilities. If the entry of new firms were relatively easy, new instruments would be put into operation whenever they promised to produce at an average full cost below the expected price. With effective barriers to entry, oligopoly would produce the same result in so far as it appeared that some oligopolists would be unable to follow the lead of others within a short time. A monopolist similarly protected would have no incentive to introduce new appliances before older instruments had worn out, except when average full cost with the new would be less than marginal direct cost with the old. The ideal relationships are represented by an equilibrium in which price tends to equal average full cost of production with the more efficient equipment best utilized, and in which underutilization of older facilities is at a minimum.[3] Such an equilibrium should not be considered to

[2] Throughout the analysis of the next several chapters estimates of ingot capacity in physical terms will be used as well as monetary figures of investment. Lack of information makes it impossible to discover the rates of output corresponding from time to time to best utilization of the investments of the several companies. It seems probable that the estimates of capacity in terms of rates of production per year bear some approximate relation to best utilization of investment.

[3] In other words, an equilibrium such that the older equipment is used up to the point where marginal cost is equal to the average full cost with the more efficient facilities when they are best utilized.

exhibit any overinvestment; the true social value of the older equipment is equivalent to a capitalization at the normal rate of earnings of what it can return in that equilibrium. In so far as investment is not written down to such a figure there will be an appearance of overinvestment, while as long as older instruments are not retired there will be underutilization of existing capacity, although not of capacity which is truly economic. Manifestly, introduction of more efficient equipment may occur more rapidly than is consistent with the ideal equilibrium, with the result of true overinvestment and underutilization. (2) Strategic advantage from accumulation of large idle reserves of materials. (3) Such pronounced "lumpiness" of capital instruments that for many years the growth of demand is not sufficient to permit best utilization. (4) Conditions of diminishing cost for the individual firm such that the rate of output which best utilizes any given scale of investment A can be produced at lower average full cost with a larger scale than scale A. This situation would obviously exist through a range of output within which the average cost curve of some larger scale lay below the average cost curve of some smaller scale at every point. It would also exist as long as the average cost curve of some smaller scale was intersected to the left of its point of best utilization by the average cost curve of some larger scale. Long-run underutilization would not necessarily occur, however, when any rate of output between zero and the rate x which best utilized a given scale A could be produced more cheaply with scale A than with any larger scale, even though some rate of output greater than x could be made with a larger scale at lower unit cost than the cost of x with scale A.

But if entry into the field is quite unrestricted long-run more or less permanent overinvestment and underutilization may develop with this last type of diminishing cost, or with long-run constant cost combined with considerable lumpiness. Hence another set of circumstances leading to underutilization is found in (5) a combination of unrestricted entry and cost conditions which would not in themselves induce a larger scale of investment than it was ordinarily intended to operate at about best utilization. When the permanent existence in the industry of capital and enterprise invested by newcomers has not been considered by the old firm, or firms, in formulating their investment policies, the most profitable program for all may involve operation at less than capacity. When entry is quite unrestricted the result may be pronounced overinvestment combined with normal competitive profits, high prices, and a large degree of underutilization of equipment.[4] However, if entry is unrestricted it may well happen that the original incumbents, realizing the impossibility of obtaining any profits above normal, will decide to invest and operate the ideal amount of capacity.[5] It is when entry is not impossible but is opposed by considerable barriers that the holders of the field may be inclined to act as if outsiders will not intrude or upon the hope that incursion can be summarily quelled. Under such conditions, in so far as newcomers do become well-established, the result is quite likely to be moderate profits attended by high prices, overinvestment, and underutilization of existing capacity. The condition of underutilization will only endure, of course, if the idle equipment is so specialized that it cannot easily be turned to the production of other commodities which will yield some surplus over direct costs. With highly specialized capital and considerable uncertainty shrouding the prospects of new competitors, the condition of underutilization may exist continuously for many years during which investment may be less than, equal to, or greater than the ideal amount, while profits exceed or remain below normal returns. Furthermore, the course of events may revolve around a vicious circle without ever proceeding along a stable path. If monopoly or oligopolists restrict investment farther than is necessary to prevent or cure overinvestment, high profits may attract newcomers whose entrance creates overinvestment. On the other hand, monopoly may, of course, eschew extra profits and attempt to keep in existence the ideal investment with more success than competitors would have.

Finally (6), a monopolist or an oligopolist may decide to create overinvestment continuously over several years and underutilize it, with the purpose of building up an industry, or some part of it, more rapidly than would otherwise be accomplished. Such a policy may be based upon strict commercial calculus with the

[4] See Chamberlin, *op cit.*, pp. 104 ff.

[5] Such a decision might be influenced by other motives, such as a wish to be known as leaders of a dominant firm, a desire to develop an industry in a particular way, and the like.

expectancy that present sacrifices will be recouped in the long run; or it may proceed from other motives, such as the desire to be considered the leaders of a basic industry, preoccupation with the development of engineering technique, and the like.

The foregoing considerations indicate that ideal investment may be created without being best utilized, that underutilization may or may not represent exploitation, and that the fact of underutilization of existing physical capacity is no proof of true over-investment. Each situation must be analyzed to discover the elements responsible for it.

With effective barriers to entry neither a monopolist nor oligopolists would have any reason connected with rational profit calculus, except those noted under (2), (3), or (6) above, to follow a policy of long-run underutilization of capacity if they operated under long-run constant cost or diminishing cost of the third type explained above — the condition in which some larger scale of investment *B* can produce more cheaply some amount greater than the number of units which best utilizes a smaller scale *A*, *but cannot produce as cheaply as scale A* any amount of output from zero to the number of units which best utilizes scale *A*. Under such circumstances the issue of long-run exploitation involves only the question whether investment has been restricted to less than the optimum amount with the result of prices which yield high profits. We have noted that underutilization may or may not represent exploitation. It is now evident further that best utilization of the capacity in existence constitutes no proof that exploitation is absent, for existing capacity may be far less than the ideal amount.

Furthermore, it is evident that long-run rationalization — the prevention or cure of overinvestment — of which so much has been heard recently, does not operate upon a different plane from monopolistic restriction of investment for purposes of exploitation. These constitute merely two aspects of the relations of investment and demand, and the benefits of rationalization may be mixed in varying degree with the disadvantages of exploitation.

When restriction of output below the rate which best utilizes capacity is undertaken as a long-run policy, it will also characterize the short-run adjustment, of course. In the absence of long-run underutilization, curtailment of output in order to maximize profits may occur when a contemplated increase in demand does not materialize, when new knowledge about the state of demand indicates that a smaller output would be more profitable, or when the degree of monopoly control is enhanced; but it is not likely to continue as a long-run policy (except under the conditions already discussed) if the demand schedule is moving ahead. Short-run rationalization may also involve temporary reduction of output.

Later analysis will be clearer if the nature of ideal output and ideal investment is theoretically examined somewhat further. In general, ideal output is that rate of output for which price is equal to marginal cost. In the short run marginal cost includes only those items of expense which vary with short-period changes in output — items which are commonly referred to as direct or prime costs. With large changes in output over a long period nearly all elements of expense become variable.[6] The marginal cost incident to a substantial change in output which requires the enlargement of capacity will include the cost occasioned by the additional equipment, as well as direct expense for labor, materials, and so on. Under conditions of constant cost long-run marginal cost will be equal to average full cost. With diminishing cost long-run marginal cost will, of course, be less than average full cost of all units produced until the most effective scale of investment is best utilized. Hence, in a profit-economy business men cannot in general be expected to sell goods at a uniform price equal to marginal cost as long as they are in the stage of growing up to the best scale.[7] These principles, familiar to economists, occasion no difficulty. It is otherwise with the nature of the demand to which investment should be properly related.

As usually explained, the essential theorem of economics, that with a given set of demands in a community the particular allocation of resources which gives equality of marginal products in all uses will best meet those demands, says merely that investment in each industry should be ideally adjusted *to the actual effective demand* of the present or near future. If we could take the position

[6] The reader who is unfamiliar with cost theory may consult J. M. Clark, *Studies in the Economics of Overhead Costs.* Professor Clark's term "differential cost" refers to the concept here called marginal cost.

[7] This seems to be another way of expressing part of the reason for Professor Pigou's proposition that the allocation of resources would be improved by granting bounties to industries operating under conditions of increasing return.

that the aim of economic activity [8] should be merely to meet in the most economical fashion demands as they happened to be; if, in other words, demands as they chanced to exist might be regarded as given data, the matter could be left without further discussion. If we are really interested, however, in maximizing the satisfactions obtainable (with a given distribution of income) from scarce resources, we must inquire whether any wants which could be better or more cheaply satisfied with products already developed, or with variations of these, are in fact being more expensively or less satisfactorily supplied because consumers do not have adequate knowledge of the relative qualities of alternative goods or because a basic product is not being adapted to many uses for which it would be better fitted than something else. A concept of ideal demand to which investment should be properly related would be useful.

One would like to say that ideal demand for a commodity would depend upon perfect knowledge of the true cost-utility ratios between it and all other commodities. This would not, however, be a useful concept, because the prices of many commodities do not equal their costs, and because neither the acquisition and spreading of true knowledge about the properties of a good nor the development of variations of a basic product is costless. The ideal demand curve will here be defined to represent the amounts of a commodity which would be demanded at various prices at any particular time, given the prices and forms of other commodities, if resources had been expended in acquiring and imparting true knowledge about the commodity, and in developing all possible variations of it, up to the point where the full costs of those resources were just returned.[9] Ideal demand for a commodity may be said to reflect the choices which consumers would make if, in addition to the possession of the incomes, preferences, and knowledge which they would have apart from the activities of its producers, they were provided with true information up to the point where the expenditures on this account would just be repaid; and if they were confronted with all the possible variations of the prod-

[8] Not the only aim from a broad sociological standpoint, of course.

[9] Not necessarily returned immediately. Resources should be devoted to these uses in so far as their present value in these uses is no less than their present value in any other employment.

uct — in composition, alloy, form, or other property — for which the costs of development would be fully returned. Although the position of the ideal demand curve at any time cannot be precisely defined, because it depends partly upon adaptations of understood principles of composition, design, and so on, which have not yet been made, and discoveries not yet achieved, the concept represents something of significance. If it is important that every commodity be pushed into all possible uses up to the point where, at a demand price equivalent to its average full cost, including expenses of true advertising and of developing variations, it ceases to have any net advantage over substitute or alternative goods, then the social worth of the various alternative types of market control for an industry is to be tested partly by their relative success in pushing effective demand closer to ideal demand, and in bringing into existence and operating an investment which approaches the ideal investment appropriate to ideal demand. Existence of ideal investment with respect to actual effective demand at any time is desirable, but it is also important that effective demand be moved ahead to coincide with ideal demand, or, in other words, that it be pushed as far forward as can be accomplished economically. The same amount of monopoly net revenue[10] might be earned under different types of market control, one of which would maintain a larger investment and output than the others because more aggressive cultivation of markets had moved effective demand closer to ideal demand than the position which it would have assumed under the other alternative types.[11]

[10] From the long-run standpoint monopoly net revenue means the excess of aggregate receipts over aggregate costs of production, including normal earnings as defined above.

[11] Critics have objected to the concept of ideal demand as here defined upon the ground that it is indeterminate, since the position of the curve is made to depend partly on something not yet known. (In so far as the curve is determined by advertising known truth, the criticism does not apply.) With respect to discoveries of principle, the curve does depend upon something not yet known. Resort to the familiar simplified distinction between discovery of new principles and adaptation of known principles in new fields suggests that the position of the ideal-demand curve would be determined to a substantial extent, however, by activities which could be achieved without anything that might be considered in any important sense to constitute discovery of new knowledge. As commonly used, the concept of the most effective scale of production or investment assumes that known general principles of technology and organization will be adapted for particular application. I hesitate to limit ideal demand to determination by advertisement of existing

2. Some Characteristics of Supply and Demand Conditions

The relations of investment and demand in the aluminum industry have not been complicated by numerous striking innovations in process and equipment. The Hall-Héroult electrolytic process has not been replaced, nor does it appear to have undergone substantial variation.[12] Adoption of "dry processes" of preparing alumina, the efficiency of which is still in question, has been limited to a few new plants. Alterations in equipment at the stages from bauxite through ingot seem to have taken the form of minor improvements, with the exception, perhaps, of some improvements in apparatus at the alumina stage.

Evidently the individual aluminum firm has operated, since the earlier years of the industry at least, under conditions of long-run constant cost or diminishing cost of the type with which no rate of output up to and including that which best utilizes any given scale of investment is produced more expensively with that scale than with a larger one. Hence it appears that maximization of profits has not required continuous maintenance of idle excess capacity. The history of this industry indicates that entry has not been easy. Growing overinvestment and underutilization as a result of continuous entry of newcomers has not been a feature of significance.

During the post-war years the strategic investment of the leading firms in reserves of ore and power has grown appreciably. A study of the relations of investment and demand requires consideration of the question whether these idle holdings, which do not represent current capacity to produce aluminum, should be regarded as a part of current investment.

The adjustment of capacity to changing demand must often be somewhat jerky in the aluminum industry, because of the length of time required to create new capacity, some "lumpiness" of capital instruments, and the perplexing uncertainties of demand.

---

knowledge and adaptation of known principles, because business men appear to act upon the belief that expenditure of resources will yield discoveries of new utility-giving properties and new principles for product variation, and it is highly important that they should do so.

[12] Above, p. 190.

A description of the first two elements is contained in the following paragraph.

> The first period of time which is usually required in building an hydroelectric plant is the time required to get together the various necessary properties. For instance, in this group of properties to which I referred yesterday on the Little Tennessee River, it was necessary for us to acquire something over 5,000 separate titles. When you consider — like buying a right-of-way — it is necessary to acquire continuous strips on both sides, one is always confronted with the fact that certain properties take a long time to acquire, properties that are in the hands of minors, etc. It took us, I suppose, five years to acquire properties on the Little Tennessee River. After the property is acquired, four or five years are required for the building of the ordinary hydroelectric development. The expense varies in this country now from $150 to more nearly on the average $200 per horsepower, so that a project with an installed capacity of 50,000 horsepower costs about $10,000,000. Our Cheoah property cost us over $10,000,000. It is the very large amount of money involved and the great length of time entailed that makes it impossible in the aluminum business to have a productive capacity follow along after the demand, rising and falling as the demand rises and falls.[13]

It should be noted that the passage quoted refers to an initial acquisition of power sites. Once the riparian rights necessary for an extensive development, such as that of the Little Tennessee region, have been acquired, this potential capacity may in many instances be brought into existence by several stages. The time factor is cut to the number of years required to complete a dam and power plant. Recently this seems to have averaged about a year less than the period indicated in the statement. Since the war the larger holdings of potential energy in the form of power sites have tended to reduce the time needed to bring fresh capacity into operation. When a power plant is underutilized by existing reduction facilities, capacity can be enlarged within about a year by expanding electrolytic plants and equipment at lower stages. Many of the European power developments used for aluminum have been so small as to occasion very little unevenness in the adjustment of investment to demand. In some instances the significance of the time element or of "lumpiness" may be lessened by purchase or sale of power. It is impossible to generalize with respect to the extent and significance of these two supply characteristics, beyond saying that they have often exercised some ap-

[13] Testimony of an officer of the Aluminum Company, FTC Docket 1335, Record, p. 714. Cf. annual reports of the Compagnie AFC in the last ten years.

preciable influence upon the relations of investment and demand. Everything depends upon the characteristics of particular power developments, the opportunities for purchase or sale of power, and the rapidity with which demand moves forward.

At all times during the life of the aluminum industry it has probably seemed nearly certain that effective demand was bound to move ahead in the future, chiefly because a large gap separated effective demand from ideal demand. Great uncertainty must have obtained, however, with regard to the rate at which effective demand would move or could be made to move ahead and the degree in which its slope at different points would change. Uncertainties are due largely to the great variety of potential uses for the new metal and to the fact that in nearly every employment it must make its way against the competition of substitute materials, such as other metals, wood, and bakelite and other synthetic products.[14]

The shape and position of the curve of effective demand from nearly every use is determined by a series of price-utility ratios between aluminum and its substitutes. It has already been remarked that there is little market for aluminum transmission cable when the contained aluminum sells for more than double the price of copper. Many of the ratios are determined in more complicated fashion. For instance, aluminum alloys and various steels compete for use in bus and truck bodies. The price ratio which leaves the builder upon the margin of doubt as between the substitutes is a resultant of relative mechanical qualities, machinability, amount of paying load per unit of vehicle weight, and so on. The price-utility ratios between aluminum and enamel cooking utensils depend upon relative heat conductivity, corrosion resistance, appearance, and prejudice. Ratios for pistons are influenced by relative heat conductivity, expansion, weight, and strength, among other factors. As the curve of effective demand moves ahead, its forward progress must exhibit a wriggling or undulating motion as a result of changes in price-utility ratios between substitutes in the several markets. Alterations in these relationships come from better knowledge, variations in composition, form, or other attri-

[14] The term "substitute" carries no implication of necessary inferiority; inferiority or superiority of the substitutes varies with a number of circumstances, as will be explained.

butes, disproportionate price changes, and altered conditions of supply and demand in the industries which use the competing substitutes. The intensity of demand for hard alloys by the railroads will be affected by the growth of understanding of their properties by railway men, the degree to which suitable alloys have been or can be developed, the success with which producers of other metals can improve their products, and changing conditions in the railroad industry itself. Elements making for economical use of aluminum vary between railways with different grades, different climatic conditions, different kinds of freight or passenger traffic, different wage conditions, and so forth. The changing demand for railway transport may favor those roads which can use much aluminum economically, or otherwise. Although the demand from the aviation industry may be fairly inelastic for a considerable stretch above a price which would throw aluminum out of several other uses, the development of tubular steel and alloys of magnesium has tended to flatten this curve.

The following example illustrates the adverse influence of improvements in a substitute and changed conditions in a consuming industry. In 1922–1923 nearly 20,000 tons of aluminum were used in crankcases for passenger automobiles in this country. By 1930 only a third of that amount was employed for this purpose.[15] During the intervening years two factors combined to move demand to a lower position. More intensive cultivation of the lower reaches of the demand for automobiles was accomplished by reduction of costs in every item possible. Cast-iron crankcases were at first substituted for aluminum in many cars. Then improvements in electric welding enabled the economical and satisfactory use of pressed steel for this motor part.

Other influences upon demand are capable of more accurate measurement. Since the war the output of secondary aluminum recovered from scrap has increased markedly relative to primary or virgin metal. Import duties and the character of international relations between producers affect the position and shape of the demand curve for the individual producer.

Evidently the aluminum firms have enjoyed a considerable measure of freedom to plan the relations of investment and demand. It appears that they have been under no necessity of creat-

[15] *Aluminum Industry,* II, 633.

ing long-run overinvestment or of steadily underutilizing capacity, although the time required to provide additional capital equipment, "lumpiness," and uncertainties of demand tend to hinder neat adjustment of investment to changing demand in this industry. The next few chapters will examine the results, beginning with the case of single-firm monopoly in the United States. The available data upon investment and earnings of the integrated firms represent in each case the total result of all their activities. For this reason it is advisable to survey briefly some leading features of the aluminum price structure.

3. Some Features of the Aluminum Price Structure

It does not appear that the production of different grades of aluminum ingot, or the production of aluminum and other electrochemical products, with the fixed plant described above constitutes a case of joint supply that justifies differences in price between the several grades or products which are not based upon cost differences.[16] Since the proportions in which the different grades or products can be produced are apparently not unalterably fixed by technical conditions, they may be and will be varied in response to changes in the relative intensities of demand for the several grades and products. In general it appears that the quality of ingot aluminum can be controlled by care in the selection of materials and the operation of reduction cells. It is said that the immediate product of each cell differs to some extent in composition from that of every other cell.[17] For this reason and because the aluminum as it comes from the cells generally includes some impurities from the electrolytic bath, it is standard practice to remelt the original pigs in order to refine them and combine them in such a way as to produce ingots of grades adapted for different demands. It would appear that the various grades can be produced in any desired proportions. Evidently the adjustment of their output to their respective demands so that the price of each equals its

[16] The conditions of joint supply are well explained by Professors Taussig and Pigou, in *Quarterly Journal of Economics,* XXVII, 693–694, 691 (August 1913). I have elaborated the theoretical point presented in my discussion here in an article entitled "Joint and Overhead Cost and Railway Rate Policy," *Quarterly Journal of Economics,* XLVIII, 583 (August 1934).

[17] FTC Docket 1335, Record, pp. 5621 ff.

cost, including normal returns to capital and enterprise, is not prevented by fixity of the proportions in which they can be produced. Continued difference in the relations of cost and price between several grades or products does not seem to be explained in this industry by conditions of joint supply.

The extent to which the cost of producing different grades of aluminum ingot varies is not clearly shown by the available information. Aluminum of exceptional purity, which can be produced only with the aid of a special refining process, evidently costs more than ordinary 99 per cent ingot. The cost of production of 99 per cent and of 98–99 per cent ingot was said to be substantially the same.[18] Whether this statement applied to production in any proportions does not appear.

The power to discriminate between different consumers by charging prices which yield different returns per unit of productive factors — in other words, prices which include different margins above or below total cost per unit — is conferred by the existence of monopolistic elements or by imperfect knowledge on the part of consumers. In so far as it is not a result of monopoly the latter condition accounts for sporadic discrimination in most industries and has no bearing upon the problems examined here. The possible extent of monopolistic discrimination depends upon the degree of transferability of units of a commodity from one market, one set of consumers, to another, and of units of demand from one market to another.[19] Discrimination is profitable when elasticities of demand in separable markets are different.[20] Conditions enabling the exercise of monopolistic power to discriminate evidently exist in the aluminum industry, and the data upon prices, costs, and earnings indicate that discrimination is practiced in this industry, although they do not afford a basis for precise estimate of its range and degree.

The term "price discrimination" is ordinarily employed to signify that different prices are charged for identical units of the same good, or that commodities which, although differing in some

[18] Testimony of an official of the Aluminum Company, FTC Docket 1335, Record, p. 576. Cost was defined to include materials and labor of plant administration and superintendence, taxes, and interest on borrowed money.

[19] See A. C. Pigou, *The Economics of Welfare* (3rd ed., London, 1929), pp. 275 ff.

[20] See Joan Robinson, *The Economics of Imperfect Competition* (London, 1933), p. 181.

respects, possess a common element, are sold at prices equivalent to different amounts per unit of the common component. The influence of monopolistic forces upon the price structure of aluminum and its products can be most easily understood if we think of a certain grade or quality of aluminum ingot, the least pure grade that is made, which is a basic component of all other grades of ingot and all aluminum products. In so far as the prices of other grades of ingot and of semifinished and final products diverge unequally from the sums of the price of this basic component and the respective conversion costs, the result is that consumers of the final products are paying different prices per unit for the basic component. Such differences in prices may exist when the basic component is made into various final products — by converting it into different alloys[21] and then into various shapes and forms adapted for different uses — provided there are differences in the elasticities of demand for the aluminum contained in the several finished articles.[22] Such differences in prices paid for the basic aluminum component are, then, likely to be expressed in the prices of all those aluminum products in the sale of which monopolistic forces are dominant and for which the possibilities of resale or transference of demand to other aluminum products are slight.

Resale of the basic ingot by those who could purchase it more cheaply to those from whom higher prices were asked would tend to prevent discrimination in its price above the margins allowed by costs of resale. The integrated firms which do a part of the work of converting the basic component into all final products, and carry out all the conversion operations for some products, are in a position to practice a greater measure of discrimination. Those final or semifinished articles which they monopolize can be sold directly at prices which, after subtracting conversion costs, yield prices for the basic component which differ from those that it brings in other markets. It is said that, in order to compete successfully with copper, the Aluminum Company of America, which had no competitors in the fabrication of aluminum cable,

[21] Since no aluminum of 100 per cent purity is made, all grades of ingot used constitute different alloys.

[22] The demand for a final product is a joint demand for the basic component and the factors of production required for converting it into the final product. The demand for the basic component is derived by subtracting the costs of conversion from the demand for the final product.

often sold this commodity at prices yielding lower revenues per pound of aluminum than were gained in other markets. According to testimony the price of aluminum cable varies with the price of copper cable.[23] Recent testimony suggests that from time to time in the last fifteen years the price of pure aluminum cable may have been below the price of 99 per cent ingot from which it was fabricated.[24] If a policy of this sort were to be effective, investigation of inquiries for cable would be necessary in order to obviate resale to demanders of ingot in general and to prevent them from purchasing cable directly. Variations in the differentials between the price of ingot and the price of various rolled products have occurred in several countries since the war. If, as seems likely, these variations are not to be fully explained by changes in cost, they evidence discrimination in the price of the basic component.[25] It appears that the sums received by the Aluminum Company of America for sheet during part of the period 1925–1930 were equivalent to less than the market price of the contained ingot plus the full conversion costs, including normal earnings.[26] Duralumin products seem to have been sold in this country at prices below the sums of the market price of metal and full conversion costs during much of the time between the war and the depression beginning in 1929.[27]

In general, it is probably true that discrimination in the price of the basic component cannot be carried into effect to any large degree through control of the prices of semifinished or final prod-

[23] BMTC appellant v. ACOA, fol. 1656; BMTC v. ACOA appellant, fols. 5713-5714.

[24] BMTC appellant v. ACOA, fols. 3139 ff., 6691 ff. Cf. HR, p. 14.

[25] See Chaps. XVI and XVIII for discussion of changing differentials. An officer of Aluminium Limited explained its price policies as follows: "If we are speaking, as I assume we are, about aluminum in all its forms, raw aluminum, sheet and cables, there is in the instance of some of these products another factor which comes in. I hardly know how to describe it except to say that we sometimes desire to take on a marginal load, as for example in the form of electrical cables, which we would not otherwise get, and sometimes make a concession in the price for that particular commodity. This marginal load is on the fabricated articles principally. We were speaking, as I understood it, about the tendency which Aluminium Limited desires to follow with respect to the price of its articles. We desire to get a fair price and a uniform price. But I have been explaining the obstacles which sometimes arise toward getting a uniform price, and among them I may include the fact that some traders are closer buyers than others" (BMTC appellant v. ACOA, fols. 1177–1179).

[26] Below, pp. 387–389.

[27] *Ibid.*

ucts unless it takes the form of setting some prices lower than the sum of the market price of ingot and the costs of conversion. Prices appreciably above such an amount would be likely to attract newcomers in most branches of the industry above the ingot stage. For instance, should an integrated firm achieve a monopoly of the sale of cooking utensils, and raise the prices of the latter, consumers would be enabled, through the action of independent enterprise, to change their demand for the integrated firm's ware into a joint demand for its sheet and independent capital and enterprise to convert this material into utensils. This sort of transference of demand would be possible in most of the finishing branches and could be carried back as far as ingot. For the most part, final products cannot be sold for long at prices which are considerably in excess of the price of ingot plus the outside conversion costs.

Nevertheless, for those semifinished and final products in the sale of which competitive elements are controlling, the same sort of thing may be accomplished in some degree through discriminatory prices of ingot or half-products. Monopolistic forces are strong at the ingot stage. Hence the chief obstacles to discrimination are resale and transference of demand. The effectiveness of these can be considerably diminished by grading ingot according to qualities broadly adapted for different uses. Aluminum of 99 per cent purity can be used interchangeably with aluminum of 98–99 per cent purity in most of the uses to which the latter is put; but the less pure grade is not as satisfactory as the 99 per cent metal in some employments. If a higher price is charged for the purer metal, all demanders of this grade will not be able to substitute 98–99 per cent metal. For many years the Aluminum Company of America charged one cent more for 99 per cent ingot than for its 98–99 grade. This margin was reduced to two-tenths of a cent in 1926 and advanced to four-tenths of a cent the following year. "Metallurgical 94–99," which used to sell at the same price as 98–99 per cent, was divided into two grades, 94 plus and 98–99. The former was quoted for several years at about a cent and a half below the price of the latter. Alloy ingots have been priced at varying differentials from 98–99 per cent. Price differentials between different grades of ingot have existed also in Europe. It is questionable that variations in grading and relative prices have simply measured changes in the relative costs of producing different grades.[28] Further, it appears that during the latter twenties the prices of rolled products tended to bring the Aluminum Company of America less revenue per unit of the basic component than was received for the metal which entered the other broad outlet for aluminum, the castings market.[29]

Direct price discrimination in the sale of the same grade of ingot or other product has also occurred to some extent in this industry. Prices in certain foreign markets, particularly in the Orient, have been from time to time much lower than those charged in protected home markets.[30] The European cartel inaugurated a uniform delivered price for all points within producing countries.[31] The Aluminum Company of America appears to have followed a policy of freight absorption.[32] Evidently consumers close to mills have paid more for a unit of aluminum plus a unit of transport than those located farther away. Finally, the same aluminum product may be sold at varying prices to different sets of consumers (in the same regional market) whose elasticities of demand differ because of a difference in available substitutes. Instances of this sort of discrimination seem to occur especially during depression.[33]

Although it is impossible to determine accurately the extent of discrimination in the aluminum industry, the following conclusions seem to be valid. The sale of some of the semifinished and final products can be controlled to an extent sufficient to permit prices corresponding to discriminatory prices for the basic component. Many, perhaps most, of the semifinished and final products are sold at prices which do not diverge markedly from the sums of prices of materials and outside conversion costs. This does not mean, however, that consumers of most final products pay exactly

[28] There is some indication that an increasing amount of 99 per cent ingot has been sold as 98–99 per cent since the war. If this has been so, it may have represented an attempt to discriminate between different consumers of the purer grade, to sell it at a lower price in new markets while maintaining a higher price in older markets. This may have exerted some influence toward a reduction in the differential between the two grades.

[29] This inference is suggested by the analysis of Chaps. XVI–XIX.

[30] For instance, before Aluminium Limited and the foreign producers made an agreement in 1931 relative to the Japanese market the price of 99 per cent ingot in Japan was apparently about 14 cents per pound, while it was 3 or 4 cents higher in Europe (BMTC appellant v. ACOA, fols. 1162–1163).

[31] Benni, *op. cit.*, p. 26.

[32] BR, p. 309.

[33] Cf. above, n. 25, p. 219, and below, pp. 327, 477.

the same prices for the basic component of aluminum. Some discrimination seems to exist in the sale of ingot of different grades and of the same grade; and half-products may be sold at prices which contain different amounts for the basic component. The multiplication of patented special alloys adapted to particular uses widens the limits of discrimination, because increased monopoly power is conferred, while special suitability and the costs of turning one alloy into another hinder resale and transfer of demand. Finally, although changes in the pattern of discrimination occur from time to time it seems clear that any major changes in the relations of investment, prices, earnings, and demand will involve changes in the prices of ingot. In subsequent analysis changes in the price of 98–99 per cent ingot, sales of which seem to have exceeded those of other grades, will be used as a rough index of changes in the level of prices. While the prices of all grades of ingot and all other products do not always vary by the same amount, a marked change in the basic ingot quotation is usually accompanied by shifts in the same direction in most other aluminum prices. Broad changes in the fundamental relationships to be examined in the next few chapters are likely to be attended by movements in the general level of aluminum prices, although they may also involve alterations in the discriminatory pattern.

We have seen that technological conditions do not necessitate long-run overinvestment in the aluminum industry — circumstances are not similar to those exhibited by some public-service industries in which "lumpiness" is so great that normal costs cannot be returned without discrimination. The general economic objections to discrimination apply in the case of this industry. A discriminatory price structure not only forces some consumers to pay higher prices than others; it apportions the consumption of different sets of consumers in amounts which represent unequal percentages of the quantities which they would take at prices equal in every case to full cost. Furthermore, provided no consumers are served at prices below full cost, discrimination will ordinarily result in a smaller output and consumption than the quantity appropriate to prices that equal full cost.[34] This is so because every

[34] The statements in the text refer to divergence from an ideal long-run equilibrium. The same points are relevant to the short-run problem, for which ideal output is that amount which equates price and marginal direct cost.

price above full cost will, under typical long-run conditions of demand in the real world, restrict consumption to some amount less than the quantity which would be taken at prices equal to full cost.[35] There is one qualification to the general objection to discrimination which should be noted. When the output of the best-scale firm is large relative to consumption in a sizable geographical area, both efficiency and desirable competitive activity may be facilitated by some degree of freight absorption.

In closing, another point concerning the aluminum price structure should be mentioned. Since the degree of monopoly power at the stages above ingot is very much less than that exercised in the production and sale of ingot, it is to be expected that whatever monopoly profits may be secured will be chiefly contained in the price of ingot, and that profits at most of the stages above will not greatly exceed normal earnings.[36] Control over the adjust-

[35] It is sometimes maintained that discrimination will result in the same output as uniformity, provided every point on the demand curve can be tapped. Professor Pigou has pointed out several reasons why markets cannot usually be separated in such fashion that this can be accomplished. A more fundamental obstacle inheres in the fact that the sort of demand curve which is appropriate for problems of price determination in a market where price is uniform is not adapted for problems of price discrimination, if the same consumers are represented at different points on the curve, as is ordinarily true. Every point on the conventional demand curve used to deal with problems of markets in which price is uniform should be understood to represent the total amount which will be demanded at a given price if each consumer pays just that price for all that he buys, no more or no less for any part of it. (Cf. A. A. Young in R. T. Ely, *Outlines of Economics,* 5th revised ed., New York, 1930, p. 180, and E. H. Chamberlin, *The Theory of Monopolistic Competition,* p. 27.) The potential demands of a given consumer at different prices must be regarded as alternative choices which are mutually exclusive. The demand curve of an individual consumer signifies that he will purchase 100 units at 10 cents apiece, or 120 units at 9 cents, or 150 units at 8 cents; it does *not* mean that he will buy 100 units at 10 cents, plus an additional 20 units at 9 cents, and a further increment of 30 units at 8 cents. It follows that when he is sold 100 units at 10 cents apiece his potential demands for larger amounts at uniform lower prices are automatically canceled, with the result that the total market demand at points below 10 cents is smaller than the amounts shown in the conventional curve. (He would, of course, be willing to purchase simultaneously 100 units at 10 cents and 20 additional units at 4 cents apiece, which would be equivalent to 120 units at 9 cents. This seems to be what happens in the case of step rates for electric energy.) Without laboring the point further it should be clear that it is impossible to tap every point on the conventional demand curve. It is evident further that the conventional curve is of no service in dealing with problems of discrimination. Several different curves must be constructed which represent different possibilities in division of consumers into separate markets. One of these will be appropriate to the best discriminatory price structure from the standpoint of maximizing profits.

[36] Some evidence tending to confirm this will be found below, p. 543.

ment of investment and output to demand is exercised by the integrated firms directly at the ingot and lower stages, and hence indirectly at higher stages. One would expect that they would set the price of ingot with reference to the more profitable of the markets for final products in which they cannot or do not care to achieve considerable monopoly power. In so far as ingot used for the final products whose manufacture is not monopolized by the integrated firms cannot be sold at different prices, some compromise will be necessary. The maximum profits implicit in demand conditions will not be obtained from any of these markets.

CHAPTER XI

INVESTMENT, PRICE, AND DEMAND IN THE UNITED STATES

1. Earnings

THE next four chapters are devoted to an examination of the relations between investment, output, demand, and price which have existed in the chief aluminum markets of the world since the expiration of the basic patents. Long-run adjustment of investment to demand is emphasized in the first three chapters. The following chapter presents an analysis of short-period relations, with particular reference to depression. In Chapter XV an attempt is made to compare the probable results of different types of market control operating under the same market conditions and to set forth the best alternatives for public policy.

We begin with the American company. The discussion of the preceding chapter suggests that there has been no need for long-continued underutilization of power plant and reduction capacity in the United States in the period here surveyed (1909–1935), unless it has been occasioned by great "lumpiness" of equipment. No new firms have ventured into the field. Cost conditions have evidently not been such as to require creation of more capacity than it was intended ordinarily to utilize. "Lumpiness" has not, as a matter of fact, occasioned the existence of idle plant, for the Aluminum Company has operated its power and reduction facilities at the capacity permitted by stream-flow conditions in all years except those of intense business depression. During the years 1909–1929 its plants were markedly underutilized in only three years.

Annual financial reports were not regularly published by the Aluminum Company of America until 1927. Estimates of investment and earnings in the years 1909–1926 have been made on the basis of fragmentary data supplied to government agencies or appearing in financial advertisements. Table 12 exhibits the results of this study.[1]

[1] The nature of the available data, the methods used in making estimates, and the factors which suggest understatement of the rates of return are described in some detail in Appendix C.

TABLE 12

ESTIMATED RATIOS OF EARNINGS TO INVESTMENT OF THE ALUMINUM COMPANY OF AMERICA, 1909–1935 *

Year	Investment† (*Thousands*)	Net Earnings (*Thousands*)	Rate of Return (*Per Cent*)
1909	$ 25,500	$ 3,600	14.1
1910	28,500	4,590	16.1
1911	28,150	5,100	18.1
1912	28,150	4,463	15.9
1913	33,380	7,500	22.5
1914	39,990	7,500	18.8
1915	47,015	9,000	19.1
1916	65,450	20,000	30.6
1917	85,880	14,000	16.3
1918	101,200	11,230	11.1
1919	114,830	10,500	9.1
1920	127,500	12,500	9.8
1921	150,220	def. 10,000	−6.7
1922	139,760	3,000	2.1
1923	139,600	14,000	10.0
1924	150,510	13,425	8.9
1925	170,000	22,892	13.5
1926	203,260	19,747	9.7
1927	248,100	18,160	7.3
1928	235,000	23,390	10.0
1929	223,220	27,330	12.2
1930	235,940	13,630	5.8
1931	242,240	6,495	2.7
1932	240,500	def. 510	−0.2
1933	234,660	3,400	1.4
1934	229,610	8,100	3.5
1935	220,500	10,820	4.9

AVERAGE RATES OF RETURN IN VARIOUS PERIODS

Pre-war	(1909–1914)	— 17.6
War	(1915–1918)	— 19.3
Post-war	(1919–1920)	— 9.4
	(1921–1922)	— −2.3
	(1923–1929)	— 10.2
	(1923–1925)	— 10.8
	(1926–1929)	— 9.8
	(1930–1934)	— 2.6

* Sources of data are given in Table 37, Appendix C.
† Average total assets during year.

Investment represents the average of total assets of the Aluminum Company and its consolidated subsidiaries, less depreciation and depletion during each year. Annual earnings are equivalent to net earnings after operating expenses, including depreciation, depletion, taxes, and interest on current debt. All figures given by the company have been used. There is some reason to think that the investment figures of the table understate true total investment, in the latter part of the whole period at least, for it is questionable whether the sums accumulated in the depreciation and depletion account are not much larger than amounts which would represent true accrued depreciation and depletion from the social standpoint. Several considerations suggest that earnings are appreciably understated in the table.[2] The earnings figures there given, taken in conjunction with records of additions to assets through security sales, do not account for the full growth of investment during the period 1909–1929. It does not appear that any substantial revaluations of assets upward occurred in this period.[3] Furthermore, the earnings figures presented in Table 12 do not include interest on current debt, or the full earnings of those subsidiaries whose dividends, rather than total earnings, appear in the consolidated income statement. The annual charges to depreciation and depletion may have been too great in many years.[4] On the whole, it does not seem at all probable that the average ratios of earnings to invest-

[2] See Appendix C for details.

[3] Below, pp. 540–541.

[4] It may be urged that even though this reserve account exceeded true depreciation and depletion, the excess would not be greater than reasonable provision for obsolescence. From the social standpoint the question of who should bear the costs of obsolescence presents a knotty problem. Furthermore, the case of aluminum appears to be somewhat unusual. Obsolescence has evidently been very small in the past forty years; but nearly the whole of the investment from mines up through the reduction stage may become obsolete for the production of aluminum all at once. I am not yet prepared to express myself with finality upon this problem. It may be said, however, that savers may not consider obsolescence except in those instances where the rate can be predicted with some assurance. Furthermore, with respect to the aluminum industry, it is obvious that in so far as obsolescence reserves are reinvested in plant designed for the electrolytic process, they afford no provision (from the private or the social point of view) against obsolescence of facilities adapted to that process. This is subject, of course, to the qualification that power plants at some locations, alumina works, and electrolytic plants may be useful to some extent for other purposes than aluminum reduction. In so far as they would not be employed in making other commodities, it would appear that sums invested in such equipment are not to be considered as reserves for the sort of obsolescence which would render that kind of equipment useless.

ment were less than the figures shown in the table; and it is likely that they were somewhat higher, perhaps much higher.[5]

The figures of the table fall short in other respects, also, of providing exactly the sort of information which is desirable as a basis for decisions on public policy. Nothing is known about payments other than dividends to executive officers or directors. Most of the stock has been closely held by directors, officers, and their families, and estates of deceased officials. Payments of this sort may have been so large as to absorb some part of the true earnings of capital or much smaller than the usual rewards. Even if the figures of the table could be regarded as a correct statement of earnings and investment from the economic standpoint they would not in all probability show the real earnings and earnings ratios for the investment used for aluminum. A not inconsiderable part of the company's investment represents holdings in subsidiaries which sell power and transport services to the general public. The rate of earnings on this part of the investment may have been greater or less than the rate of earnings on the part used for aluminum.

Further, we have seen that there is reason to think that the investment from ore up through reduction plants which is devoted to the production of ingot receives higher rates of return than the investment at some later stages, and that returns vary somewhat between these later stages.[6] To the extent that this is so, correct rates of return upon total investment used for aluminum and its products would understate the rates of return upon that part of investment (from ore up) used to produce some articles, and so tend to indicate that it had been carried closer to ideal investment than was in fact true, and vice versa.

Again, the rates of return in the table do not, of course, express the rates of earnings upon operating investment. At all times since 1909 the company has possessed some idle reserves of ore and power. Between the last pre-war years, when the first acquisitions

[5] Understatement of investment would not result in overstatement of the rate of return if earnings were understated by about the same amount at about the same time. Under such circumstances understatment of investment is consistent with understatement of the rate of return.

[6] Normal returns have evidently been earned in several branches much of the time. But several million dollars' worth of equipment devoted to hard alloys and their products apparently received no net earnings in the period 1916–1929. (See below, p. 389.) It is questionable whether sheet of all classes has always returned sums equivalent to normal returns upon the equipment at this stage (*ibid.*).

of ore were made in South America and Europe, and the latter twenties the proportion of investment represented by idle reserves may have increased. If so this would account for some fall in the rate of earnings upon total assets. In so far as a reduced rate of earnings is due to growth of idle investment at a more rapid rate than operating investment it is not to be regarded as a symptom of a better adjustment of output and price. In order to ascertain whether the relations of investment, output, and price have improved over a period of years we need to know the change in the ratio of earnings to operating investment. No figures of operating investment of the Aluminum Company are available.

Moreover, it is clear that a normal return on total assets at any given time cannot be considered symptomatic of ideal relations between investment, output, price, and demand if any part of the assets are idle. The view is sometimes advanced, however, that in instances where materials are sufficiently scarce to make it imperative for firms to acquire large reserves in order to ensure continuous operation, steady maintenance and inflow of capital cannot be obtained unless normal returns on total investment — and so larger returns on operating assets — are gained. This is not true. The ordinary expectations of capital require only normal returns on original cost of operating investment.[7] Presumably reserves of materials are purchased at prices no greater than the best possible estimate of the present value of the annual net additions to revenues which will begin when they go into operation at some point in the future. No return upon this idle part of investment is expected during the intervening years, and output and price policy would, in general, be the same in the intermediate period whether or not such investments were made. It follows that during a time in which large reserves were being accumulated, the normal expectancies of savers would be fulfilled even though the rate of return on total investment were below normal. After reserves purchased in the past went into operation they would earn more than normal returns on their purchase price unless the expected additions to revenue failed to materialize.[8] In a period

[7] Unless at the time of acquisition it was not intended to use it until a later time. See the following discussion in the text. The qualifications concerned with changing price levels and improved equipment are beside the point here at issue.

[8] The analysis of this paragraph assumes that depletion is charged at a fixed amount

during which no new acquisitions of reserves occurred, and during which the materials secured in the past were being rapidly exhausted, it might happen that a rate of return greater than normal on original cost of total investment would not be inconsistent with the ideal relations of output, price, and demand.[9] Inasmuch as the Aluminum Company possessed substantial reserves of ore and power, which were apparently increased from time to time, during most of the period 1909–1935, it would appear that the ideal relations between investment, output, price, and demand would have been indicated by rates of return upon total assets which were less than normal.

Finally, the information presented in corporate reports does not, of course, make it possible to distinguish monopoly gains and economic rent.[10] If the book valuation of natural resources differs from the capitalization at a normal rate of the true economic rent, the rate of return upon total investment will not reveal the relations between rent and monopoly profit. Assume that natural resources are carried on the books at their purchase price. Let us first suppose the purchase for immediate use of an additional power site or increment of ore lands that was expected to yield, with the exercise of monopoly power, an increase in revenue which, after returning all other normal costs, exceeded the economic rent of this unit of natural resources. The price paid for this property might fall anywhere between the capitalization at the going normal rate of the full addition to revenue (rent plus monopoly profit) and the maximum value of the property in another use.[11] If the price paid was near the capitalization of the full addition to revenue and if the estimate of added revenue proved to be ap-

---

per ton mined. Depletion charges can, of course, be made in such a way that the annual rate of return on total assets tends to be equal throughout a period of years.

[9] The following scheme might be useful as a device of social accounting if it could be made practicable. No asset would be included in investment until it was actually being used. When included it would not be reckoned at the actual original cost, but at current capitalization of the annual addition to net revenue expected from it at the time it was purchased — i.e., the additions upon which its purchase price was calculated. The actual additions to revenue should not be capitalized, for such a process would always show a normal rate of return.

[10] Economic rent is the additional revenue above all other costs which would accrue to owners or users of natural resources when price was equal to long-run marginal cost exclusive of rent.

[11] Provided, of course, that the latter was lower than the former.

proximately correct, the rate of return shown upon original cost would be little above normal; but a substantial amount of monopoly profit would have been concealed at the outset by capitalization in the purchase price. Indeed, if the addition to revenue were less than had been estimated but greater than true rent, a return below normal might be shown, although monopoly profit was actually being paid by consumers. Monopoly gain might also be consistent with a showing of a normal rate of return if true rent diminished, on account of changing conditions, while book value of natural resources remained the same. Thus absence of monopoly profit is not to be inferred from a normal rate of return upon book valuation of natural resources unless it is known that the book value is equal to capitalization of true economic rent.

On the other hand, a greater than normal rate of return upon book value of total assets might be consistent with normal earnings including rent if the purchase price of natural resources was less than the capitalization of true rent; or if, over a period of years, rent increased considerably and the book valuation of natural resources, whatever its original basis, remained unchanged. Finally, the same would be true when use was made of reserves of materials acquired earlier and now carried on the books at a figure equal to their present value at the time of purchase, if such value was lower than the current capitalization of true rent.

Since available information does not disclose which of these various possibilities has characterized the history of the accounts of the Aluminum Company, it is impossible to assess the significance of this aspect of the matter.[12]

Even when data are fairly satisfactory, comparison of rates of return in different industries must be interpreted with caution. In the present instance comparisons will be used simply as one of several methods of testing the conclusion suggested by the analysis above. According to Professor Crum's figures the rate of earnings in metal manufactures during the three years 1924–1926 exhibited a stability which was characteristic of all branches of manufacture. In 1926 the percentage ratio of net earnings before interest (and after taxes in so far as they could be eliminated) to total assets of all corporations in the metals group which reported any net

[12] This is also true in the case of the European companies whose earnings ratios are studied in the next two chapters.

income was about 8.8.[13] Apparently the rate of return was about the same in the two preceding years.[14] Corporations included under metal manufacture comprise integrated smelting companies and firms which make highly specialized metal products, such as automobiles and electrical equipment. Returns of the latter group would in general tend to be higher than those of companies whose assets contained substantial reserves of materials or other idle investment, if depletion were charged according to a fixed rate per ton used. Furthermore, profits of firms producing specialized manufactured articles may have been higher on account of more effective differentiation of product.[15] One would infer that the rate of return upon operating investment used for aluminum of the Aluminum Company appreciably exceeded the comparable average rate earned by many firms engaged in similar processes in other metal industries, unless idle reserves of the latter made a negligible proportion of their total assets and they

[13] W. L. Crum, *Corporate Earning Power* (Stanford University, 1929), p. 186. This study is based on analysis of income-tax statistics. The figure 8.8 does not appear in Crum's tables. He gives 8.03 as the percentage ratio of net income after interest to total assets. On p. 253 the percentage ratio of interest paid by all metal corporations to total assets is given as 0.756. There is no reason to suppose that the percentage ratio of interest to total assets for those companies reporting net income would differ sufficiently from 0.756 to permit any substantial error in the rate of 8.8.

[14] Strictly comparable figures are not obtainable, since total assets were not reported for 1924 and 1925. Crum estimates total assets from data reported on capitalization, surplus, and other items, and computes percentage earnings ratios of net income after interest to the estimated total assets. After adding percentage ratios of interest paid to total assets we get the figures of 8.25 and 8.66 as the percentage ratios for 1924 and 1925 in the metals group as a whole. These figures overstate the rate of return since the estimate of total assets was perhaps 30 per cent too low (Crum, *op. cit.*, pp. 188–189). But these figures refer to the group as a whole, including corporations reporting no net income, whereas the figure in the text represented the rate of return of companies reporting net income. Since these two factors would, in part at least, offset each other, there is reason to think that the rate of return for those firms reporting net income did not differ greatly in 1924 and 1925 from the rate in 1926.

[15] Data given in L. H. Sloan, *Corporation Profits* (New York, London, 1929), at pp. 140 ff., tend to support this hypothesis. Average percentage ratios of net earnings before interest to total assets, less current liabilities, were as follows in specified branches of metal manufacture in 1926 and 1927:

	1926	*1927*
17 copper firms	7.93	7.47
22 steel firms	7.26	5.63
6 electrical equipment firms	12.05	11.73
7 agricultural implement firms	10.36	10.35
14 miscellaneous mining and smelting firms	9.83	9.56

earned in fact considerably higher rates of return than the average received by the whole group of enterprises engaged in metal manufacture.[16]

In summary, there are several reasons for belief that the rates of earnings upon total assets of the Aluminum Company used in the aluminum business were higher than those shown in the table during much of the period under survey, and that the returns upon operating investment used for ingot were considerably higher than the rates of return on total assets indicated by the table. Inasmuch as the rates exhibited by the table are hardly to be considered less than normal, it would appear that investment and output have been appreciably below the ideal amounts and price correspondingly above the ideal price. Apparently the average rate of earnings of similar enterprises in other metal industries was much lower in the normal years 1924–1926 than the rate earned by the Aluminum Company. Comparison with the average rate of earnings upon a part of total investment received by a sample of large firms in the nonferrous metal industries in 1924, 1926, and 1927 is inconclusive.

It will be noticed that the average rates of return shown by the table have been much less in post-war periods than before the war.[17] Furthermore, according to the figures of the table the rate of return declined slightly after 1925. The lower rates in 1926

[16] Professor R. C. Epstein gives percentage ratios for the years 1924–1928 of net earnings before interest to total capital at the beginning of the year of forty-eight large nonferrous metal corporations (*Industrial Profits in the United States,* New York, 1934, p. 292). Total capital is defined to include only assets represented by funded debt, capital stock and surplus. Similar ratios may be computed for the Aluminum Company for the years 1924, 1926, 1927. (Large changes in capital in the middle of each of the years 1925 and 1928 render corresponding figures for those years useless.) The average rate received in these three years by the forty-eight companies included in Epstein's sample was 10.7 per cent, while the similar average rate for the Aluminum Company was 9.9 per cent. Since the proportion of idle fixed capital may have been much greater for the Aluminum Company than the average proportion for the group of forty-eight firms, and the rates of return gained by integrated smelting companies similar to the Aluminum Company may have been lower than those received by manufacturers of more highly differentiated articles, this comparison does not demonstrate that the operating investment of the Aluminum Company earned lower rates of return in these years than the operating investment of similar enterprises in the nonferrous metals group. That may have been so, or just the opposite may have been true.

[17] Perhaps rates of return on cost of reproduction have been even less since the war. Inability to estimate cost of reproduction prevents computation.

and 1927 are to be explained partially by the abrupt increase in idle investment incident to the Duke merger. In the next two years the proportion of operating to total investment of the Aluminum Company may have grown with the transfer of much idle investment to Aluminium Limited and the completion of new dams and power plants in the South. It will be recalled that the water rights and partially constructed dam of the Alcoa Power Company on the Saguenay were retained by the Aluminum Company, however. The average rates of return shown by the table are about the same in 1924–1925 and 1928–1929. It is not clear whether the proportion of operating to total assets was larger in the latter of these two periods, as a result of the transfer to Aluminium Limited of foreign properties, including a substantial part of the idle ore and power reserves. If that was so, the rate of return on operating investment was a little lower at the end of the twenties than it had been five years earlier.

It is interesting to estimate the changes during the period 1923–1929 in monopoly net revenue as computed from the data of Table 12. For this purpose 8 per cent has been used as a normal return. According to this calculation monopoly net revenue increased to about $9,350,000 in 1925, declined to a minus quantity in 1927, and rose again until it stood in 1929 at almost the identical figure of 1925.[18] The average indicated monopoly revenue is somewhat greater in 1928–1929 than in 1924–1925, however. If total assets exceeded operating investment by the same proportion in both periods, monopoly net revenue with respect to investment in use must have been increased from the earlier period to the later by a greater amount than these figures indicate. We have seen, however, that the proportion of total investment in use may have been greater in 1928–1929 than in the middle twenties.[19] One might

[18] The figures are as follows (in thousands of dollars):

1923 —	$2,790	1927 —	–$1,740
1924 —	1,355	1928 —	4,700
1925 —	9,350	1929 —	9,375
1926 —	3,455		

The decline in 1926 and 1927 is to be attributed partly to the increase in idle investment. When 6 per cent is used as a normal rate of return monopoly net revenue increases slightly between 1925 and 1929; when 10 per cent is employed, it diminishes slightly.

[19] During the twenties there evidently existed some unused fabricating capacity. Changes in the proportion of idle to total fabricating capacity are not ascertainable.

infer that monopoly net revenue was somewhat larger, perhaps not a great deal larger, in the later period. Evidently aggregate monopoly revenue did not, however, grow in the same proportion as operating investment in the years 1923–1929. Unless this conclusion is vitiated by inaccuracies in the figures or assumptions, it is to be regarded as a symptom of a somewhat larger investment relative to effective demand at the end of the decade.

The period 1923–1929 was undoubtedly characterized by a forward-moving effective demand. Under such conditions a failure of monopoly revenue to grow as rapidly as operating investment might be explained by an increase in the elasticity of demand at horizontal points, by a policy of sacrificing some immediate profits in the expectancy of building up demand for the future, by an overestimate of the rapidity of movement of demand, or by policies appropriate to objectives other than maximization of profit. Alteration in the elasticity of demand might be due to changes in the competitive influence of substitutes or foreign aluminum or to growth in the strength of potential competition. In the examination of prices and the influences bearing upon price policy, to which we turn in a moment, an attempt will be made to discover the explanation of the apparent improvement in the relation of investment to demand.

One more question should be introduced at this point. The significance attaching to the fact that the capacity of power plants and reduction works of the Aluminum Company was utilized to the full permitted by operating conditions during the period 1923–1929 must not be overlooked. Upon certain assumptions this would necessarily mean that investment had not tended to exceed the amount appropriate to maximum monopoly net revenue, except temporarily, at any time during the period. These assumptions are that the company endeavored rationally to maximize profits and that European sellers did not "compete" in this country in the sense of acting with neglect of the ultimate consequences of their policies. If the investment of the Aluminum Company had exceeded that proper to maximum profit, and if the domestic firm and the Europeans had acted as rational oligopolists (without any agreement), the full capacity of the Aluminum Company would

---

Capacity of power plants and reduction facilities was well utilized during the period 1923–1929.

not have been utilized. This may be explained as follows. A certain amount of capacity x is appropriate to maximum profits with particular demand and cost conditions. If x capacity is exceeded, monopoly revenue must be absolutely less than it would have been with just x capacity. The problem is to determine what rate of output will maximize the excess of aggregate revenue over aggregate variable expense, or in other words equate marginal revenue and marginal cost, with the capacity $x + y$. Reflection shows that marginal cost and marginal revenue will not be equated either at the rate of output which best utilizes $x + y$ capacity or the rate which would leave idle all the "excess" capacity y — i.e., the rate which best utilizes x capacity. At the rate which best utilizes x capacity marginal revenue will exceed marginal cost for some rates of output of y facilities. If it be assumed that marginal cost at the rate of output which best utilizes $x + y$ capacity will be about the same as marginal cost with x capacity alone, operated at best utilization, then marginal revenue will of course be less than this. Hence some rate of output between that which best utilizes x alone and that which best utilizes $x + y$ will be appropriate to maximum profits under the circumstances, and the full capacity will not be utilized.[20]

There are other possible explanations of the full utilization of the Aluminum Company's capacity. Price competition between the Europeans and the domestic company, with neglect by each of the effect of his actions upon the policies of the other firms, would keep price low enough to utilize any size of capacity up to the ideal amount. It does not appear, however, that this sort of competition prevailed generally throughout the period 1923–1929. The assumption of oligopolistic rationality seems closer to actual conditions. A strong threat of competitive price making by new domestic firms might have led to full utilization of capacity larger than that proper to maximum profit. Or again, "excess" capacity, created purposely or by mistake, might have been fully utilized

[20] If marginal cost with the new equipment (after the investment is made) is much lower than marginal cost with the old equipment, the most profitable adjustment will result in underutilization of the older facilities alone. An amount of investment $x + y$ of the domestic firm could, of course, be best utilized profitably if the foreign producers withdrew from this market an increment of product equivalent to the output of y facilities when best utilized. That would simply mean that the domestic producer was operating at best utilization the amount of capacity appropriate to maximum profit with the larger demand now available to it.

in order to test out the elasticity of demand at lower prices and familiarize a greater number of people with aluminum, with the hope that the immediate profits sacrificed would be more than offset as a result of more rapid transformation of latent into effective demand. Finally, full utilization of any amount of capacity greater than that consistent with maximum profit might be explained by motives other than strict pursuit of profit.

The chief questions to be examined in the rest of this chapter are, then, the explanation of the apparent improvement in the relation of investment and demand between 1923 and 1929, and explanation of the full utilization of domestic ingot capacity. The simultaneous existence of these two conditions is consistent with any of the following hypotheses. (1) Demand became more elastic as it moved ahead, and investment tended to equal an amount appropriate to maximum profit. (2) The threat of entrance by newcomers likely to pursue competitive investment and price policies was sufficient to induce a more liberal adjustment of investment to demand and full utilization of existing capacity. (3) A larger capacity than that appropriate to maximum monopoly revenue was created by mistake or as part of a program to convert latent demand into effective demand by lowering price and advertising; [21] and was then fully utilized at lower prices in order to build up demand. (4) Investment and price policy were controlled by objectives other than maximum profit. It should be recognized that the "development" policy noted under (3) would probably be undertaken only if it were believed that as the demand curve moved ahead it would become more elastic. Unless lower prices were to be ultimately advantageous it is questionable whether they would be put into effect.

The above discussion has proceeded upon the tacit assumption that cost conditions remained unchanged during the period 1923-1929. Provided demand did not decrease through the relevant range, cost reductions should occasion a different adjustment resulting in somewhat larger investment and output and greater monopoly revenue. But changes in cost would not invalidate the propositions advanced above.

[21] In so far as consumers with a latent demand at prices below those currently charged will not pay much attention to advertising until "the price is right," price must be lowered in order to command a hearing.

We now proceed with an examination of the course of changes in prices of aluminum and its substitutes, in imports, capacity, and costs.

2. Capacity and Price

After recovery from the depression of 1908–1909 the Aluminum Company's list price of ingot approximated 20 cents a pound. With the reconstitution of the European cartel and better business abroad in 1912 it was raised to 22 cents.[22] A lowering of the import duty by 5 cents in the fall of 1913 occasioned a reduction of 3 cents in price. In the next year and a half price fluctuated between 18 and 20 cents. Earnings were evidently high in the short period between 1910 and the outbreak of abnormal war demand in 1915. Expansion of plant was not begun until demand had revived from depression and moved ahead markedly. Evidently investment lagged behind the increasing demand from the growing automobile business. The additional facilities of 1914–1916 might have resulted in capacity for an output large enough to bring returns down to or below normal for some years if the war demand had not appeared to tax it as soon as completed. Or they might have yielded large profits in capacity production for a bustling peace-time automobile industry.

The post-war years present the first normal period of some duration in the mature life of the Aluminum Company. The course of prices here and abroad during the years 1920–1932 is shown in Table 13 and Chart II.[23] During 1920–1921 the world aluminum price structure crumbled under the impact of dumping of government inventories and general business depression. In the United States a sudden pickup in demand for aluminum in 1922 became so intense that by the end of the year there was a general shortage of the metal. Until the tariff was settled in September the Aluminum Company of America made no change in the price of ingot, with the result that it remained below the foreign price plus the 2-cent duty. A few days after the 3-cent tariff increase went into effect the company raised its price 2 cents; and within two months had added 3 cents more in order to take the full advantage

[22] Prices for 1912–1915 appear in BMTC v. ACOA appellant, Exhibit 344.

[23] Price data for the years 1933–1936 are given below, pp. 321 ff.

allowed by the new duty.[24] By early 1923 price at home and abroad had advanced another cent. During the three years 1923–1925 the Aluminum Company's price followed the foreign price

CHART II

PRICE DATA FOR 98–99 PER CENT PRIMARY ALUMINUM INGOT IN THE UNITED STATES AND EUROPE, 1920–1932

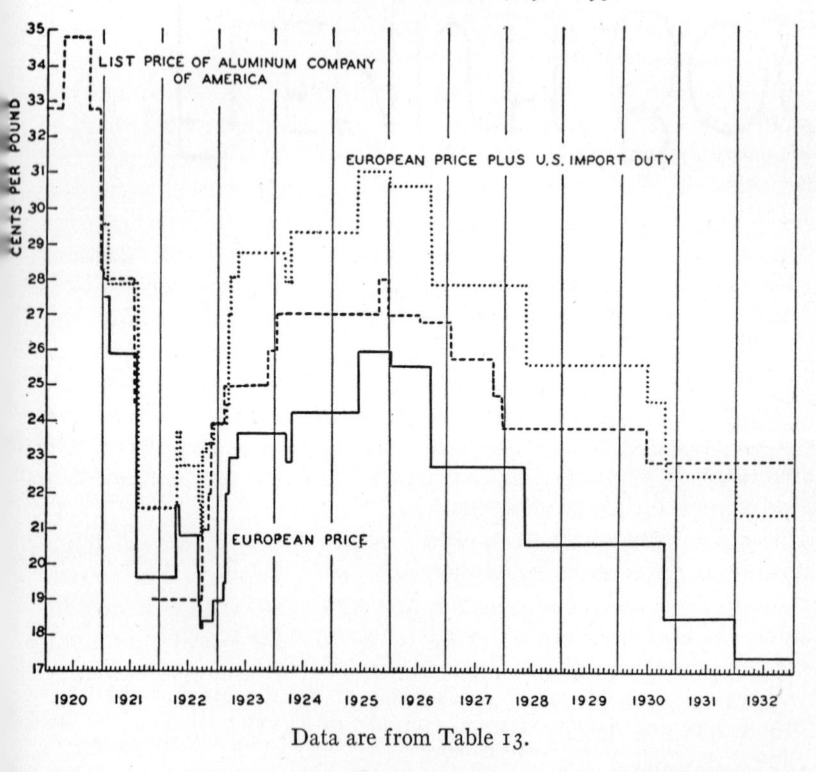

Data are from Table 13.

upward but remained somewhat under the margin allowed by the duty. The New York open-market quotations (Table 14), which afford some indication of the price of imported metal, indicate that

[24] It is said that when the new duty went into effect the company announced that price lists would no longer be issued and quotations would be made privately. It was several months after each change before the trade journals were even able to say definitely what the price had been for some time past. This policy was evidently continued for more than a year. See *Metal Industry* and *Automobile* for these years.

TABLE 13

PRICE DATA FOR 98–99 PER CENT VIRGIN ALUMINUM INGOT, 1920–1932 *

(Cents per Pound)

Date of change	List Price of Aluminum Company of America	European Price	U. S. Import Duty	European Price plus U. S. Import Duty
1920				
January 1	32.8			
April 19	34.8			
October 1	32.8			
December 20	28.3			
1921				
January 10	28.0	27.5	2.0	29.5
February 5		25.9		27.9
July 15	24.5 †			
August 3		19.6		21.6
November 15	19.0			
1922				
April 22		21.7		23.7
May 6		20.8		22.8
September 2		19.8		21.8
September 16		18.8		20.8
September 23		18.2	5.0	23.2
September 26	21.0			
September 30		18.4		23.4
November 1	22.0			
November 22	24.0			
December 2		19.0		24.0

* List prices of the Aluminum Company of America are from BMTC v. ACOA appellant, Exhibits 344 and 556. List prices evidently mean the price quotations given by the central office to salesmen of the company, for the Aluminum Company has not ordinarily circulated published price lists among customers. European prices are the London quotations (from *Mining Journal*, London) until September 1926, converted at average monthly exchange rates. Thereafter they are the official cartel prices in gold pounds converted at par. Before 1926 slight differences in prices prevailed from time to time in the chief European markets. Between 1926 and 1934 the officially announced price was the same in all producing countries of Europe.

† Between August 15 and November 15, 1921, schedule prices of ingot were suspended, owing to unsettled market conditions.

TABLE 13 — *Continued*

Date of change	List Price of Aluminum Company of America	European Price	U. S. Import Duty	European Price plus U. S. Import Duty
1923				
February 7	25.0			
February 24		19.5		24.5
March 3		22.0		27.0
March 17		23.0		28.0
May 12		23.7		28.7
November 13	26.0			
1924				
January 10	27.0			
March 8		22.9		27.9
April 12		24.3		29.3
1925				
June 6		26.0		31.0
October 22	28.0			
December 17	27.0			
1926				
January 1		25.6		30.6
July 1	26.8			
September 17		22.8		27.8
1927				
January 15	25.8			
October 20	24.8			
December 21	23.9			
1928				
May 18		20.6		25.6
1930 ‡				
June 26	22.9		4.0	24.6
October 18		18.5		22.5
1932				
January 1		17.4		21.4

‡ There were no changes in prices in 1929 or in 1931.

TABLE 14

Monthly Average Prices of 98–99 Per Cent Virgin Aluminum Ingot in the New York Open Market *

(Cents per Pound)

Month	1920	1921	1922	1923	1924	1925	1926	1927	1928	1929	1930	1931	1932	1933	1934†	1935†
January	32.00	22.86	17.74	22.75	27.61	27.00	27.00	26.37	23.90	23.90	23.90	22.90	22.90	22.90	23.30	20.50
February	31.83	24.50	17.33	23.25	27.71	27.00	27.00	25.83	23.90	23.90	23.90	22.90	22.90	22.90	21.65	20.50
March	31.50	23.44	17.52	24.95	27.57	27.00	27.00	25.55	23.90	23.90	23.90	22.90	22.90	22.90	21.65	20.50
April	31.61	23.25	18.07	26.00	27.46	27.00	27.00	25.55	23.90	23.90	23.90	22.90	22.90	22.90	21.65	20.50
May	31.95	23.06	17.92	26.24	26.43	27.00	27.00	25.55	23.90	23.90	23.90	22.90	22.90	22.90	21.65	20.50
June	32.00	22.75	17.87	26.25	26.27	27.00	27.00	25.55	23.90	23.90	23.76	22.90	22.90	22.90	21.65	20.50
July	32.00	22.62	17.87	26.25	26.37	27.00	27.00	25.55	23.90	23.90	22.90	22.90	22.90	22.90	21.65	20.50
August	32.21	20.22	17.87	26.07	26.52	27.00	27.00	25.55	23.90	23.90	22.90	22.90	22.90	22.90	21.65	20.50
September	31.44	19.02	18.26	25.50	27.24	27.00	27.00	25.55	23.90	23.90	22.90	22.90	22.90	22.90	21.65	20.50
October	29.12	17.85	20.32	25.50	27.16	27.24	27.00	25.29	23.90	23.90	22.90	22.90	22.90	22.90	21.49	20.50
November	27.80	17.50	20.87	25.80	27.00	28.00	27.00	24.30	23.90	23.90	22.90	22.90	22.90	22.90	20.50	20.50
December	23.83	17.50	22.52	26.31	27.00	28.00	26.85	24.26	23.90	23.90	22.90	22.90	22.90	22.90	20.50	20.50
Year	30.61	21.11	18.68	25.41	27.03	27.19	26.99	25.40	23.90	23.90	23.39	22.90	22.90	22.90	21.58	20.50

* Prices are reported in the *American Metal Market*, and reprinted in the *Yearbook of the American Bureau of Metal Statistics*, by whose permission they are reproduced here.

† 99 per cent ingot. Prices of 98–99 per cent ingot in 1934 and 1935 not reported.

TABLE 15

IMPORTS OF ALUMINUM INGOT INTO THE UNITED STATES, 1919–1933
(*Metric Tons*)

Year	Total Imports	Estimated Tonnage from European Producers
1919	8,003	2,360
1920	18,178	12,600
1921	13,870	12,390
1922	18,122	14,700
1923	19,534	12,500
1924	13,333	8,100
1925	19,690	12,520
1926	33,965	18,400
1927	32,744	13,830
1928	17,189	7,020
1929	21,961	8,020
1930	11,112	6,050
1931	6,261	5,090
1932	3,631	2,720
1933	7,580	6,080

Import figures represent general imports of aluminum and alloys in crude form, including scrap. Figures of imports from European producers have been taken from the following sources. For the years 1919–1922 they represent total imports minus the tonnage received from Canada. The figures for the years 1923–1924 are based on statistics of imports by the Aluminum Company given by an officer of the company in the case of *Haskell* v. *Perkins* (Record, p. 1020). The figures for the years 1925–1931 appear in Exhibit 126, BMTC v. ACOA appellant. (Whether these figures represent "imports for consumption" or tonnage entering United States ports is not explained.) Six thousand seven hundred tons purchased by the Aluminum Company from the Vereinigte Aluminiumwerke in 1925 and 1926 have been added to the figures of imports from foreign producers given in Exhibit 126. Since most of this amount was imported in 1926, 6,000 tons have been added to the figure for that year and 700 tons to the figure for 1925. Figures for 1932 and 1933 appear in Table 5 of the Henderson report of NRA. Since scrap and virgin are not separated in import statistics, the figures of imports from European producers probably represent some slight overstatement of their imports of virgin aluminum into this country.

European producers were making some sales here at quotations somewhat above those of the domestic company during 1923 and 1924.[25] In the next two years the New York price was the same as the Aluminum Company's list price much of the time. Additional evidence shows, however, that European companies made some sales at prices below the list price of the domestic firm from time to time throughout the twenties.[26] Estimates of imports from European producers exhibit an increase from 1919 to 1922 and a decline in the ensuing two years (Table 15). The fact that imports did not drop farther in 1923 is probably to be ascribed to the procedure of contracting for sales a year ahead, while the failure of imports from Europeans to disappear almost entirely in 1924 under conditions of world shortage is doubtless to be explained by premium prices here.

Inspection of the ratios of Table 16 shows that the price of aluminum rose proportionately higher than the prices of the other nonferrous metals in the years 1922–1924. In 1925 the price of aluminum increased relatively less than the price of copper and an index of nonferrous metal prices. The price ratio of aluminum to copper reached a high in 1924 greatly exceeding the ratios for 1919 and 1920, which in turn surpassed the ratio for 1913. In the next three years the prices of aluminum and copper declined at about the same rate. A marked reduction in the aluminum quotation at the end of 1927 combined with an increasing copper price in the ensuing two years lowered the ratio to a point substantially below the average for 1923–1924. The ratio of the price of aluminum to a nonferrous-metals price index also rose to a high in 1924, which, however, remained below the ratios of 1919 and 1920, although much above the 1913 ratio. This ratio declined abruptly in 1925 owing to a marked rise in the nonferrous index, and remained about constant (near the 1913 ratio) until 1930, except for an increase in 1927, when the index declined relatively more than the price of the light metal. In the latter twenties aluminum was offered at prices equivalent to lower ratios to the prices of the other nonferrous metals than had existed in 1923–1924 or in the short post-war boom. The ratio of the aluminum price to the

[25] The New York open-market quotations on virgin aluminum seem to apply mainly to spot sales of foreign metal and of small amounts of domestic metal offered for resale.

[26] Above, p. 159.

TABLE 16

RATIOS OF YEARLY AVERAGE LIST PRICES OF ALUMINUM INGOT TO YEARLY AVERAGE PRICES OF COPPER AND A WEIGHTED NONFERROUS METALS INDEX IN THE UNITED STATES, 1919–1934 *

	AVERAGE PRICES (*Cents per Pound*)			RATIOS	
Year	Aluminum	Copper	Nonferrous Metals Index	Aluminum to Copper	Aluminum to Nonferrous Index
1913	21.18	15.27	9.98	1.39	2.12
1919	32.83	18.69	12.93	1.76	2.54
1920	33.58	17.46	12.88	1.92	2.61
1921	— †	12.50	8.22	—	—
1922	19.90	13.38	9.34	1.49	2.13
1923	25.04	14.42	11.13	1.74	2.25
1924	26.98	13.02	11.29	2.07	2.39
1925	27.15	14.04	12.63	1.93	2.15
1926	26.90	13.80	12.59	1.95	2.14
1927	25.63	12.92	11.31	1.98	2.27
1928	23.90	14.57	10.89	1.64	2.19
1929	23.90	18.11	11.11	1.32	2.15
1930	23.38	12.98	8.81	1.80	2.65
1931	22.90	8.12	6.23	2.82	3.68
1932	22.90	5.56	4.78	4.10	4.80
1933	22.90	7.03	6.61	3.26	3.46
1934	20.50 ‡	8.43	7.70	2.43	2.66

* Average list prices of 98–99 per cent virgin ingot of Aluminum Company of America are computed from the data of Table 13. Averages prices of electrolytic copper at New York are taken from the *Yearbook of the American Bureau of Metal Statistics.* The weighted nonferrous metals index comes from *Metal Statistics.*

† Owing to suspension of list prices of aluminum during part of 1921 it is impossible to compute average price for that year.

‡ Approximate estimate of average price quoted made from figures of *American Metal Market.*

nonferrous index was approximately equal to the 1913 figure in three of the five years 1925–1929. The aluminum-copper price ratio, however, did not sink to the 1913 level until 1929.

The price reductions of the Aluminum Company occurred within a space of two years. At the end of 1925 the quotation was lowered by one cent. After a reduction in the European price equivalent to 3 cents in September 1926 when the cartel was formed, the domestic firm cut its figure by about 3 cents in successive stages during 1927. The timing of these price reductions may have been influenced by temporary increased foreign offerings at a little less than the domestic price. (See Tables 14 and 15.) Study of the relations of supply and demand in European markets seems to indicate that the increased exports from European companies to the United States in the years 1925–1927 in large part reflected a temporary surplus with respect to current prices at home.[27] The temporary European surplus was incident to a trend toward a new and better adjustment of investment and demand. Fundamental forces operating in this direction were at work on both sides of the Atlantic in the latter half of the twenties, as will be explained. By 1928 when the cartel lowered its price about 2.2 cents a pound exports to the United States by European producers had shrunk greatly. The Aluminum Company did not follow this price reduction.

It seems evident that the American company would have reduced its price considerably some time between 1926 and 1930 even in the absence of lower quotations for imported metal. Lower price ratios between aluminum and other metals than those obtaining in the middle twenties were better suited for rapid development of the new markets. To meet the new demands great increases in operating capacity occurred in the years 1927–1929. Imports by the Aluminum Company diminished after 1926.[28] Cost of production was being appreciably cut at the same time. After 1927 no further change occurred in the list price of the Aluminum Company until the middle of 1930, when a reduction of one cent in the import duty was accompanied by an equal drop in price.

[27] Below, Chap. XIII. In these three years nearly 13,000 tons, or about 30 per cent of the metal imported from European firms came from Germany, where a large power plant had just gone into operation.

[28] BMTC v. ACOA appellant, Exhibit 126.

A depressing influence upon the price of primary aluminum was probably exerted by the growing production of secondary aluminum recovered from scrap. Secondary aluminum, which competes with the virgin metal in many uses, constitutes an economic substitute which makes the demand for virgin more elastic than it would otherwise be. The exercise of monopoly power by one or a few producers of primary aluminum would be rendered impossible if (1) secondary metal were a perfect substitute for virgin in all uses; (2) any quantity of virgin aluminum which would bring a price just equal to its marginal cost if there were no secondary aluminum in the market could be entirely replaced by secondary metal at the same marginal cost; and if (3) the number of firms selling secondary aluminum were large enough to result in purely competitive market relations. Although the third condition may be met in practice it would appear that the other two are not completely fulfilled. An increase in the supply of secondary which could be produced at a marginal cost equal to current price would ordinarily encourage some reduction in the price of virgin, however. During 1923 and 1924 the estimated output of secondary metal was equivalent to about a third of the tonnage of primary aluminum produced. In 1925 the production of secondary aluminum jumped to 40,000 tons or about 60 per cent of the year's total for virgin metal.[29] The tonnage of secondary recovered apparently increased but little thereafter, with the result that the proportions increasingly favored virgin after 1927.

In 1920 the Aluminum Company evidently possessed a capacity of a little over 300,000 h.p. for production in the United States, which was equivalent, perhaps, to a maximum output under ordinarily favorable conditions of 70,000 metric tons of metal per year. During the years 1922–1925 the demand for ingot moved forward substantially. In 1922 the company sold about 21,000 tons in the form of ingot at an average price of about 18 cents per pound.[30] During the following year nearly 32,000 tons were sold for about 21.5 cents per pound. About the same amount brought 25 cents per pound in 1924, and in the first six months of 1925, 25,000 tons were sold at an average price of 25.5 cents. An officer of the company has testified that capacity was insufficient to meet

[29] Estimates of production of secondary aluminum are given in Table 39, p. 572.
[30] Sales data for these years are given in BR, pp. 117–118.

demand in the latter part of 1922 and in 1923.[31] Yet new domestic construction was not begun until 1925.[32] In the meantime the domestic shortage was alleviated somewhat by importation from Norwegian firms in which the American company took an interest. Ten years earlier the company had waited until sales exceeded its output to begin extensions which would take a year or two to complete. Now it waited three years after revival had started — during which the purchase of foreign capacity apparently increased supply by less than 15 per cent — before beginning construction of dams in the South requiring two or three years to finish. Even with the addition of foreign capacity the company's sales exceeded its output until 1927, the difference being made up by purchases of primary aluminum from foreign companies and of scrap in this country.[33] Substantial delivery delays occurred during the upswings in 1912 and in 1922–1924, as well as in the peak years 1906–1907 and 1919–1920.[34] These delays probably tended to retard the growth of consumption in some uses in which aluminum was economically superior to other materials. The absence of serious delays in the late twenties is another symptom of a better relation between investment and demand.

Between 1925 and 1930 domestic capacity was nearly doubled by the building of three new dams in the South and the purchase of additional energy for Massena.[35] The great Saguenay development was also initiated in this period. Although some metal from the Saguenay was imported by the Aluminum Company for a few years, the Canadian properties now administered by Aluminium Limited have produced chiefly for markets other than the United States since 1928. As long as this continues, the relations of investment and demand in the domestic market are likely to be influenced principally by changes in the capacity within this country.

No thoroughly satisfactory cost data have been published. Two sets of figures of the plant cost of producing ingot in three of the four domestic plants appear in the records of the Baush litiga-

[31] FTC Docket 1335, Record, p. 5257. Data submitted in this case show marked declines in inventories and increases in unfilled orders during these two years.

[32] Between 1920 and 1925 an increase of 18,000 h.p. from additional dynamos and improved technique was offset by the relinquishment of 10,000 h.p. of purchased energy and the sale of over 10,000 h.p. generated at Badin (*ibid.*, pp. 697 ff.). Canadian capacity was not extended until 1926.

[33] Annual report, 1927.

[34] The delays of the twenties and extenuating factors involved are discussed below, pp. 421 ff.

[35] Above, p. 77.

tion, both prepared from answers of the Aluminum Company to detailed interrogatories submitted by the Baush attorneys. The figures of plant cost given in the exhibit introduced by attorneys of the Aluminum Company, and data on other expenses and on sales appearing in exhibits introduced by the other side, are used here to compare relative changes in the price and cost of ingot. The figures of plant cost of ingot seem to include wages, alumina, power and other materials, repairs, depreciation, taxes, and plant overhead, interest on working capital, and perhaps some interest on fixed capital. The average yearly list price of 98–99 per cent ingot fell by about 3.25 cents between 1925 and 1929, and that of 99 per cent ingot by about 3.9 cents. The average price received for ingot of all grades (sold by the company in the form of ingot) appears to have declined from 26.20 cents per pound in 1925 to 22.73 cents per pound in 1929, a drop of 3.47 cents.[36] The cost figures indicate that in the same period plant cost was reduced by 7.6 cents, 5.8 cents, and 4.9 cents respectively at three plants.[37] Since these figures are presented in order of relative outputs it would appear that cost for the three plants as a whole diminished by an average of about 6 cents. Total general expense of the Aluminum Company was slightly less in 1929 than in 1925.[38] Aggregate sales expense, including advertising, doubled between 1925 and 1929.[39] Allocation of about one-half of selling expense to ingot — probably a generous proportion — in each year would give selling expense per pound of about 1 cent in 1925 and 2 cents in 1929. It would appear that the cost of producing and marketing ingot was reduced by at least 5 cents per pound between 1925 and 1929.[40] The figures indicate that price fell by at least a cent less than cost. This would suggest a higher rate of return on investment in 1929 than in 1925, but just the opposite seems to have been true. Unless the explanation is to be found in a larger proportion of idle assets in 1929, of which there is no clear evidence, the prices of some commodities other than

[36] Average price received is derived by dividing total receipts from ingot sales by number of pounds sold as given in BMTC v. ACOA appellant, Exhibits 289 and 294.

[37] *Ibid.*, Exhibit 117. [38] *Ibid.*, Exhibit 58. [39] *Ibid.*, Exhibit 258.

[40] There is no reason to think that the change in cost at the fourth plant, for which no figures were given, was so strikingly different as to cause a variation in the average of more than ½ cent. The fourth plant produced about one-seventh of the total output in 1925 and about one-fifth in 1929.

ingot must have fallen somewhat more than their costs,[41] or revenues from investments declined. About the latter we have no information.

We have seen that the record of changes in investment, operating capacity, and earnings evidence a tendency toward a better relation between investment and demand in the latter twenties. It now appears that this was not true of the investment used to produce that part of the total output sold in the form of ingot. In so far as the price of ingot was lowered by less than the diminution in cost a more restrictive relation is indicated. Unless the lower rate of return at the end of the twenties is illusory, or is wholly explained by a drop in the rates of return or dividends of firms in which the Aluminum Company held investments, that part of the company's investment used to make commodities sold in forms other than ingot must have shown the less profitable adjustment to demand. The information available does not demonstrate clearly where this relation occurred. It may have obtained in the markets for bauxite, or alumina,[42] or in the markets for any products made from ingot. If it occurred in the markets for products made from ingot there was in all probability a change in the discriminatory price structure. This may be explained as follows. Evidently the relation between demand and the whole of the investment, from ore through finishing plants, used to produce articles made partly or solely by independent fabricators who purchase ingot from the Aluminum Company was less rather than more liberal. Ingot was apparently sold at prices higher relative to cost. There is no reason to suppose that the investment of the finishing firms who fabricate ingot purchased from the Aluminum Company was much more or much less liberally adjusted to demand. In many of these fields the economical scale of investment is evidently small relative to total demand, and entry is not opposed by substantial barriers.[43] With expanding demand, at least, it is reasonable to suppose that there is a strong tendency in each of these branches of the industry for investment *in fabricating plant* to be kept close to that amount which tends to yield normal returns. Changes in the investment in fabricating plant in these

[41] Cf. below, pp. 379 ff. It does not seem likely that the relatively small quantities of ingot sold at lower prices in the Orient influenced the total result much.

[42] The Aluminum Company sells bauxite and alumina to firms making chemicals, abrasives, and other commodities.

[43] See Chap. XVIII.

branches would probably keep step roughly with changes in the amount of capacity used by the Aluminum Company in producing ingot to be used in these branches. Prices of the final products made in these fields would tend to equal the price of ingot plus normal cost of conversion. If the adjustment of investment to demand were less restrictive in other markets, the prices charged for final products in those fields would be less than the sum of the company's market price of ingot plus normal expenses of conversion. In other words, the rate of return upon the total investment from ore through finishing plant used to serve these latter markets would be less than that obtained in the former fields. The lower rates of return would be likely to occur upon those parts of investment devoted to products sold chiefly by the Aluminum Company itself. Table 17 shows the distribution of sales of the Aluminum Company between ingot, sheet, and all other products made from aluminum.

TABLE 17

SALES OF VARIOUS PRODUCTS BY THE ALUMINUM COMPANY OF AMERICA, 1923–1929

(*Thousands of Pounds*)

	1923	1924	1925	1926	1927	1928	1929
Ingot	73,276	71,739	89,592	71,857	65,167	89,689	81,030
Sheet	60,551	44,409	58,177	61,123	63,905	71,305	64,147
All other products from aluminum	51,504	44,073	48,649	53,235	57,268	93,188	97,142
Total	185,331	160,221	196,418	186,216	186,340	254,181	242,319

Figures for the years 1923 and 1924 are taken from BR, p. 118. Data for the other years are found in BMTC v. ACOA appellant, Exhibits 290–294.

The absolute totals and the proportions of the three classes of products shown in the table were nearly the same in 1926 as in 1923. In the interim ingot sales had increased temporarily, while sales of sheet and other products were decreasing proportionately. After 1926 metal sold in forms other than ingot or sheet accounted for an increasingly larger proportion of total sales, while sales of ingot and sheet both decreased proportionately. In both 1923 and 1926 the company sold nearly 40 per cent of its metal in the form of ingot. By 1929 this had dropped to less than 34 per cent. In 1926 only about 29 per cent of total sales was made up of aluminum in forms other than ingot and sheet — scarcely more than the 1923 percentage. This proportion grew to 37 per cent in 1928 and about 40 per cent in 1929. The fact that the Aluminum Company was carrying a larger part of its total output through semi-fabricating or finishing stages would afford opportunity for more discrimination, particularly if the additional product took the form of articles made to a small extent, if at all, by other firms.

With an expanding demand for aluminum during the years 1922–1925 the ingot capacity of the Aluminum Company was enlarged but little. Rising prices of substitutes permitted increases in the price of aluminum which at first took full advantage of the higher tariff imposed in 1922. In 1925 the company embarked upon an extensive program of expansion which almost doubled domestic ingot capacity in the ensuing five years. The ratios between the price of aluminum and an index of nonferrous metal prices were lower in the period 1925–1929 than in the post-war boom or the years 1923–1924. The rate of return on all operating investment of the company appears to have declined between 1925 and 1929. Evidently the relation between total operating investment and the demands for its products was more liberal toward the end of the twenties. Comparison of data on reduction in the price and cost of ingot between 1925 and 1929 seems to indicate, however, that investment used for ingot sold directly did not earn a lower rate of return in the latter year. If price had been reduced as much as the apparent saving in expense, the ratios of aluminum prices to the prices of other nonferrous metals would have been much lower at the end of the decade. Evidently certain parts of the investment only were more liberally adjusted to demand. If the less restrictive adjustment occurred in the markets for aluminum products, rather than in the markets for bauxite, alumina, power, or transportation, the pattern of discrimination in prices received for the basic component aluminum was evidently altered.

3. Changes in Demand

Assuming the same degree of utilization of operating investment, a change to a lower rate of return on operating investment might result from the movement of prices closer to costs. This might be explained by changes in the character of demands. In the next few pages "demand for aluminum" should be understood to mean the derived demand prices for the basic component, except where it clearly refers only to the demand for ingot sold to outside fabricators. In the first place, availability of the same or greater quantities of a substitute at lower prices than formerly would diminish the demand prices for aluminum throughout the portion of the demand curve affected by this substitute. Lower prices would be necessary in order to keep the same volume of sales. Evidently such a change in demand for aluminum occurred after 1925 on account of growing quantities of secondary ingot offered at lower prices. With rising prices the output of secondary aluminum increased rapidly up to 1926. In spite of falling prices thereafter, it grew slightly in the next three years. The same sort of change must have been occasioned by declines in the prices of other metals after 1925. Now, if at the same time that demand prices for some quantities of aluminum tended to be depressed by these influences, the total composite demand for aluminum and its substitutes was moving forward, the demand for aluminum might sooner or later assume a new position which showed (as compared with its position prior to any of these changes) sections of greater elasticity at horizontal points ahead, representing lower prices than those which had been charged before these changes began.[44] It seems probable that the demand for aluminum underwent changes of the sort just described during the years 1926–1929.

[44] The outcome might be something like that portrayed in the following diagram, in which D^1 represents the demand curve before any of the changes, and D^2 the new position after the various tendencies have worked themselves out.

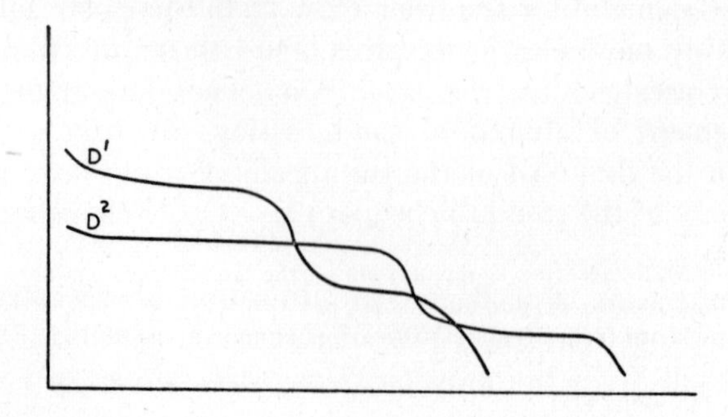

In the second place, the demand for aluminum would be depressed within certain ranges by unfavorable changes in price-utility ratios occasioned by improvements in the quality of substitutes. After 1923 the demands from the automobile industry for sheet and sand castings sank markedly. About 1920 large tonnages of aluminum sheet had been used in the manufacture of automobile bodies, although its price was between seven and eight times the price of steel body stock.[45] When relative area per pound is considered this means that the price of aluminum was 2.3–2.7 times the price of steel. By 1924 very little aluminum sheet appeared in the form of automobile bodies, although the price ratio was then slightly more favorable to the light metal. Two factors seem to have played the chief part in reshaping the demand from this market. Exploration of the lower reaches of demand for automobiles encouraged elimination of the more expensive items. It appears that the steel companies were more successful in the progressive development of their product.[46] In the early twenties large-scale production of aluminum bodies was not entirely satisfactory, owing to the softness of the metal and the difficulty in developing suitable equipment. Nor was steel sheet very satisfactory at that time. Then the steel companies devised a type of thin sheet suitable for bodies and capable of taking paint well. They also coöperated in research with the automobile firms to develop satisfactory machinery and technique of manufacture, with the result that steel became superior to aluminum in every respect save weight. A few years later the Aluminum Company began to make strenuous efforts to regain this market with the strong alloys, which are in many respects admirably suited for automobile bodies. With the much larger output of automobiles the derived demand for the basic ingot from this potential market was probably more elastic at prices below those prevailing in the middle twenties.

Replacement of aluminum sand castings by iron automotive castings in the first half of the twenties appears to have been the result chiefly of the trend to cheaper cars. Other elements tending

[45] The Budd Manufacturing Company, one of the largest body builders, consumed 10 to 12 million pounds of aluminum sheet per year (BR, p. 145). Price data for steel body stock are from *Metal Statistics;* for aluminum sheet from FTC Docket 1335.

[46] BR, pp. 145 and 148.

to diminish demand prices for aluminum in some markets were the development of stainless steel and thin-walled tubular steel and improvement in the welding of steel sheet.

It would appear that the derived demand for ingot contained in some of the products fabricated by one-stage firms from ingot *sold* by the company became more elastic at lower prices during the latter half of the twenties. This condition and the necessity of lowering prices to hold the same volume of business in some markets may help to explain the reduction in the price of ingot. It must be recalled, however, that the price of ingot was not apparently lowered as much as the cost of ingot. As already pointed out, the lower rate of return on operating investment toward the end of the twenties would be explainable by greater reductions in the derived price of the ingot contained in other products, chiefly those made largely by the Aluminum Company itself.[47] To some extent changes in demand of the sort described up to this point may have occurred in the case of such products. But the most decisive influence contributing to greater elasticity of demand in the latter twenties seems to have been the conversion of latent into effective demand through the development of new alloys and products. When the curve of effective demand was pushed forward it undoubtedly became more elastic at lower prices as aluminum became a capable substitute for various alloys of iron, copper, and zinc in heavy-duty employments. That is, the demand for the basic component in these new uses was more elastic at prices below those charged in the years 1924–1925 than its demand in older applications. It is to be doubted that these new influences affected the actual demand or the price policy to any great extent in the first half of the twenties. But during the latter half of that decade the aluminum industry was undergoing a process of transition from a condition of limited markets to one of diversified markets, in many of which the new alloys had to make their way in competition with the established metals, in particular the basic product of the "iron age," to a much greater extent than ever before. This may be briefly explained as follows.

A large part of the rapid growth in investment during the latter half of the twenties both here and abroad was induced by conversion of latent into effective demand through the development of

[47] Cf. the analysis in Chap. XVI.

new alloys and products, and determination of rules, standards, and formulae to assist utilization.[48] The immense potentialities of industrial application of strong aluminum alloys had always appealed to the imagination. The practical possibilities of alloy development must have been vaguely perceived before the war as a result of the work of Wilm and Rosenhain. Both the potentialities and the feasible achievements were made somewhat more concrete by experience in the war and the studies of heat treatment in government laboratories. The campaign to exploit these possibilities was begun during or shortly after the war by the leading aluminum firms. In 1919 the research staff and equipment of the Aluminum Company of America were greatly enlarged. During the next few years several promising alloys were developed by the company's scientific men. The same thing was happening in Europe. Commercial application of the pioneer alloys of the duralumin class and of the Pacz silicon-aluminum alloys was already yielding fruit. By 1925 some of the possibilities of 1920 must have appeared as probabilities for 1928 or 1930. It was reasonable to suppose that as demand from these new markets increased in the next few years it would exhibit greater elasticity at lower prices. It appears that prices of products destined for the new markets were lowered markedly and energetic attempts were made to reduce costs.[49] At the same time an intensive advertising campaign was initiated to aid in developing the new demands. Many informative pamphlets, beautifully designed, have been distributed. The number of articles about aluminum products and their uses appearing in the trade and scientific press multiplied several times in the latter half of the decade. The two-volume work, *The Aluminum Industry,* written by members of the research staff and other divisions of the Aluminum Company falls partly into the category of educative advertising. In addition to technical analysis of various processes of producing aluminum and its products, the book contains an extended exposition of the properties of aluminum, its alloys, and the products made from them, and a detailed and commendably informative explanation of the countless

[48] A description of these developments has been given in Chap. III.

[49] See below, pp. 381 ff. It is probable that prices of new alloys in various forms and of old alloys in new forms contained a lower price for the basic component than it had returned in several older markets.

uses to which they are adapted.[50] Before the depression broke, it was clear that the campaign of research, demonstration, lower prices, and advertising had moved the curve of effective demand ahead appreciably.

There seems to be an interesting similarity in spirit between the program of the last dozen years or so and the campaign in which the Aluminum Company cultivated new markets during the first fifteen years or so of its life. Inferences from the number and nature of journal articles and the meager historical evidence contained in the handful of books on aluminum suggest a marked slowing down in this kind of aggressivenes during the first decade of the century. In the earlier campaign outstanding discoveries of principle were few, doubtless because the adaptation of existing principles borrowed from the fund of experience with other metals provided a difficult task and impressive returns. Only within the last fifteen years or so does it appear that the majority of significant developments in new aluminum alloys and products have come from the producers of ingot here and abroad.[51] Thermit, aluminum bronze powder, and duralumin were discovered in Germany by fabricators or government scientists. The Y-alloy came from the National Physical Laboratory in England. In this country the Doehler die-casting process and the Pacz silicon alloys were the result of independent invention.[52] Prior to the twenties the major discoveries of representatives of the Aluminum Company seem to have been the steel-cored transmission cable, an aluminum-manganese alloy, and the perfection of a continuous rolling mill.

[50] It is unfortunate that the book seems to fall below this high standard in certain other respects. One has an uneasy feeling that the impression is conveyed that nearly all of the wisdom on things aluminum in this country resides with the Aluminum Company and a few noncommercial research organizations. Little space is devoted to the alloys and products of other domestic fabricators. Apparently the names of an engineer who designed one of the most successful types of piston and of the person who developed the die-casting process are not mentioned. The historical description of patent litigation seems to present a somewhat one-sided view. Treatment of economic issues concerned with competition and monopoly and public relations is superficial and displays an inadequate grasp of the nature of the issues.

[51] Cf. Chaps. I and III, *passim*.

[52] An article published in 1923 entitled "Twenty Years in the Metal Industries" listed four significant achievements in progressiveness in the aluminum industry. These were the die-casting process, the Pacz alloys, duralumin, and the introduction of continuous rolling of aluminum sheet (*Metal Industry,* XXI, 1, January 1923).

After the eye of science descried the latent demand for aluminum far ahead, it must have been apparent that wider opportunities were extended to competitive influences. Knowledge indicating the possibilities of greatly enlarged consumption of aluminum grew during and after the war. It would have been strange if the ranks of potential competitors had not been augmented. Threat of the entrance of newcomers might have made the expected future demand for the products of the Aluminum Company potentially more elastic because of the possibility that new firms would engage in price competition. Although the likelihood of entry in the United States may have been less in the latter half of the twenties,[53] it is possible that a somewhat larger operating investment for the later twenties was planned about 1925 than would have been undertaken in the absence of this contingency. Perhaps it is more likely that the influence of potential competition was exhausted in stimulating acquisition of additional idle reserves of ore and power and in inducing more strenuous efforts to develop latent demand before others did.

After 1925 effective demand was pushed closer to ideal demand, and operating investment for some uses was apparently adjusted to immediate effective demand in such a way that its rate of return was less than that received on many parts of investment earlier. Operating equipment for ingot was used to normal capacity until the advent of the depression. Some of the company's fabricating plant, however, was probably not operated continuously at full utilization.[54] Evidently the changed relation between some parts of investment and the demands for their products is explained partially, at least, by increased elasticity of demand. Potential competition from new integrated firms here and existing enterprises abroad in the new alloy and product development probably exerted some impetus towards a greater investment here by presenting an incentive to acquire a larger amount of the limited ore lands and power sites and by making it desirable to capture the markets for new alloys before others did. It may be that the possibilities of converting latent into effective demand would not have been explored so energetically if it had not seemed probable that others would do it in the event that the Aluminum Company did not. This is not to say, however, that the lower rates

[53] See Chaps. VI and VII.

[54] See Chap. XVI.

of return on some parts of operating investment were caused by threat of price competition. It seems more probable that operating investment was affected by potential competition only in so far as the latter force helped to induce activities resulting in pushing effective demand ahead more rapidly and enlarging the amount of investment appropriate to maximum monopoly revenue. Further, it is likely that actual competition from foreign producers was similarly limited in its effect. Again, it should be recognized that the increase in competition in the development of alloys and their products by one-stage firms could influence investment used for ingot only in so far as it converted latent into effective demand. For such competition would not tend to bring more investment in the stages succeeding ingot than would be appropriate to the circumstances of demand and the prices charged by the Aluminum Company for ingot or half-products. It other words, the Aluminum Company roughly controls the total amount of investment in the competitive finishing stages (wherever virgin ingot is a significant material) appropriate with any given demand to normal returns at the finishing stages.

The widened opportunities for competition by one-stage firms in the alloy development, and perhaps the influence of potential competition from integrated firms, are attested by the increased prominence devoted to the trade name "Alcoa" in advertising of the Aluminum Company. The range of variation in the characteristics of pure aluminum ingot, sheet, wire, and the few standard alloys made before the twenties was doubtless not great. But specific alloys are by nature differentiated products. An invitation is extended to attempts to convince consumers that those produced by a particular firm are generally superior. That trade names and advertising will be able to make the individual demand curve of a given firm very much less elastic is, however, to be doubted in the case of aluminum alloys because of the ability to compare by scientific testing the degree to which the desired qualities are present in the products of different firms.

If it may be concluded that the relation of operating investment to demand was influenced by potential competition and by actual competition from European ingot producers and domestic one-stage fabricating firms only in so far as these forces induced activities which moved effective demand farther ahead, it follows that for those parts of the investment which earned lower returns toward the end of the period but continued to be fully utilized (if there were any such) investment had not been carried beyond the amount appropriate, when fully utilized, to maximum profit — unless motives other than maximization of profit were dominant or mistakes had been made. In the absence of the latter possibilities the improved relation between these segments of investment and their demands must have resulted wholly from increased elasticity of demand. Unfortunately no one but the executives of the company can tell us whether considerations of maximum profit were controlling, or whether mistakes were made.

The case is somewhat more complicated with those parts of investment which earned lower returns in the latter twenties but were not used to normal capacity; for here two other possibilities enter. Investment may have been pushed beyond the point proper to maximum immediate profit, or, indeed, beyond the ideal amount, and price set rather low with the purpose both of preparing for future increase in demand and helping to develop that growth. Or underutilization might be explained simply by cost conditions such that the scale of fabricating equipment which would produce most cheaply the output appropriate to maximum profit was capable of a much larger output. It is impossible on the basis of the available evidence to be certain that either one of these elements was at work, but considerations presented in Chapter XVI suggest that both exercised some influence upon the adjustment. Unfortunately, the depression intervened before there had been enough experience with the newer demands to assess with assurance their elasticity relative to older demands. However, it does not seem improbable that in general the demands for the strong alloys and their products will turn out to be somewhat more elastic within the decisive price ranges than was true of many older demands for the metal, with the result of a lesser degree of restriction of investment devoted to serving these markets and lower rates of return upon that part of total operating investment. In proportion as this is so, one would expect that much of the fabricating plant underutilized in 1929 would come to be well utilized, with lower returns to total operating investment used for these markets than the returns received in other markets characterized by less elasticity of demand.

Finally, in the case of those parts of the investment which earned the same or higher returns in the late twenties as they were receiving four or five years before, there are three possible explanations. With much the same elasticity of demand at both times, or less elasticity in the later years, investment might have been roughly equal to the amount proper to maximum monopoly revenue at both times. With an increase in elasticity of demand, investment might have approached more closely, towards the end of the decade, the amount appropriate to maximum profits than it had earlier done. Or investment might have been less or more at each time than the respective amounts proper to greatest profit at each time. There seems little reason to think that investment for these markets was much less during the twenties than the amount appropriate to maximum profit with current effective demand, in so far as this could be apprehended.

One other aspect of the matter should be mentioned. Up to 1919 nearly all of the increase in the Aluminum Company's investment represented reinvestment of earnings. There is no reason to suppose that the amount of surplus earnings bore any close relation, during any part of the period, to the amount of additional investment required for maximum profit. The equivalence of these two quantities at any time in any industry would be sheer coincidence. However, one cannot infer that prior to 1919 the investment of the Aluminum Company exceeded or fell short of the amount suited for maximizing profit from time to time. Earnings might have exceeded the reinvestment necessary to maintain the best size of investment from the standpoint of the company, and the excess might have been distributed in dividends, invested in other enterprises, or used to acquire idle reserves of materials. Dividends were small; and the available material shows no indication of large investments in other enterprises except subsidiaries engaged in the aluminum business or companies holding or operating sources of power. But substantial acquisitions of idle reserves of ore and power were made. Since the quantitative relations involved in this problem cannot be determined from the available evidence, it is impossible to decide from these considerations whether investment was kept about equal to the amount appropriate to maximum profit or not. The fact that the company could doubtless have obtained substantial additions to capital through sale of securities without voting power which carried maximum income limitations would suggest that operating investment was not kept much below the amount that the executives thought necessary for maximum profit with current effective demand. However, with great uncertainties about the consequences of attempts to develop latent demand, single-firm monopoly might pursue a cautious, unaggressive policy which resulted in a smaller investment and smaller profits than could have actually been secured by more rapid and energetic development of variations of the basic product. On the other hand, for one reason or another the executives of the company may have maintained an investment larger than that proper to maximum profit. We cannot tell.

In the period 1919–1929 a total of $111,000,000 of new securities appears to have been sold by the Aluminum Company. When account is taken of retirement of securities out of earnings, by refunding, or through operations incident to the transfer of some properties to Aluminium Limited, the indicated net addition to assets of the Aluminum Company through sale of securities in this period becomes about $40,000,000. Investment used or intended to be used in serving markets within the United States appears to have grown by more than $100,000,000 in the same period.[55] Unless the executives were moved chiefly by motives other than maximization of profit, one would infer that earnings were not sufficient during the twenties to provide for the growth in investment necessary for maximum profits.

It would appear that during the twenties investment for most markets did not fall much short of the amounts proper to maxi-

[55] Since the estimates of investment at the end of 1918 and 1919 (Table 37) probably represent understatement as compared with figures for later years, the true increase in total assets in the period 1919–1929 cannot be determined. (See Appendix C.) The figure in the text is reached by the following process. During the years 1921–1929 total assets of the Aluminum Company increased from about $158,000,000 to $233,000,000, a growth of $75,000,000. In the two years 1919–1920 $24,000,000 of notes were sold, and some earnings were reinvested. Furthermore, the growth in the period 1919–1929 of that part of the assets to be used for serving markets in the United States must have been greater than the indicated growth in total assets, because a substantial portion of the company's assets during the years 1919–1927 represented foreign properties turned over to Aluminium Limited in 1928 in exchange for stock of the latter, which was immediately distributed to stockholders of the Aluminum Company. The Aluminum Company's surplus was reduced about $23,000,000 by this transaction, some part of which represented assets acquired before 1919.

mum profit with current effective demand; and that investment for some markets tended in the latter twenties to exceed the amounts that would have been immediately most profitable. Evidently the Aluminum Company has successfully rationalized its long-run adjustment of investment to demand in most, perhaps all, markets. Except during depression underutilization has been limited to some fabricating capacity. It seems likely, however, that overinvestment has been avoided through a policy of underinvestment [56] for some markets at least.

[56] With respect to ideal investment.

CHAPTER XII

INVESTMENT, PRICE, AND DEMAND — THE FIRST EUROPEAN CARTEL

1. Issues and Facts

We now turn attention to the results of the first European cartel and the growth of outsiders. It will be recalled that this agreement reserved the home markets to the domestic firms and divided the competitive market (chiefly Germany) among the members. A minimum price was agreed upon from time to time. There was no central sales agency, and apparently no strong centrol control.[1] It is plain that this cartel did not entirely eliminate competition between its members. The possibility of evasion of quota restrictions in the competitive market may not have gone unappreciated. As in most cartels, competition existed in the form of attempts by individual members to exert a controlling influence upon the determination of price and quotas. To say the least, there was no single-minded coöperation — the existence of which is necessary if a cartel is to function like single-firm monopoly — between Neuhausen and the French companies. Moreover, each member was free to do as he would with his own investment policy. During the life of the cartel the French companies increased their capacity relative to that of the AIAG.

The significant questions are the following. (1) Did the large increase in investment between 1905 and 1914 represent a better or worse approximation to ideal investment? (2) Were the increase in capacity and attendant fall in price due chiefly to the policy of the cartel — i.e., what it would have done anyway in the absence of outsiders — or were they occasioned chiefly by the influx of new firms? (3) Did the entrance of new competitors result in a large degree of underutilization without much reduction in price? (4) Was curtailment of output involving underutilization practiced by the cartel to increase profits? (The price and

[1] Above, p. 36.

production policies of competitors in Europe during depression will be compared with those of the Aluminum Company of America in a later chapter.)

The facts seem to have been as follows. Before the formation of the cartel, price in Germany had been 2 M. per kilogram.[2] According to Kossmann the companies were receiving good profits with this price.[3] In 1902 the cartel raised price to about 2.50 M. and curtailed production somewhat for two years.[4] The resulting surplus energy was used to some extent for the production of other commodities.[5] During the boom which started in 1904 the price of aluminum was raised until it touched 4 M. in 1907. In 1906, if not earlier, existing capacity was hardly able to fill the demand. After the boom was well under way in 1905 the old companies embarked upon programs of great expansion, and new firms entered the industry. By 1908 the installed power of three years earlier was almost tripled, while 1914 saw in existence nearly five times the capacity of ten years earlier.[6] After the crisis of 1907 the cartel reduced price several times before its dissolution at the end of September 1908. In the next three years reported prices fluctuated between 1.05 M. and 1.60 M. With reconstitution of market control and enhanced general prosperity in 1912, quotations ranged between 1.50 M. and 1.80 M. until the outbreak of war. These prices were much below the price charged during the first years of the cartel and were equivalent to about half of the amount secured during the boom.

Production expanded in about the same proportion as capacity between 1905 and 1913. Using the capacity figures of Table 11, it appears that for every ton of aluminum produced in Europe in 1905 there existed 9.3–10.6 h.p. This ratio fell to 7.8 the following year — obviously because more power was devoted to aluminum and less to other commodities — rose to about 15 in 1908

[2] Prices in Germany are given by the Metallgesellschaft (reported by Günther, *op. cit.*, p. 14). The trend of price movements in other European countries was undoubtedly the same, although the prices were apparently not always exactly equivalent in all countries. Figures of average yearly prices in France somewhat lower than the German prices are reported in AG, 1913, Partie sup., 192. Escard (*op. cit.*, p. 55) gives French prices roughly equivalent to the figures of the Metallgesellschaft for Germany.

[3] *Op. cit.*, p. 114. This was not true of the British Aluminium Company, but see below, pp. 269.

[4] Bannert, *op. cit.*, p. 33. [5] Gautschi, *op. cit.*, p. 47. [6] See Table 11, p. 123.

on account of the conjuncture of large new capacity and depression, and fell to 10–10.5 in 1912 and to 9.5–10 in 1913.[7] In so far as the estimates of installed capacity and of production are reliable, it may be inferred that about the same proportion of available energy was devoted to aluminum production in 1913 as in 1905. Since there was probably little unused capacity in 1905 it seems reasonable to presume that plant was well utilized in 1913. Hence, unless the lower level to which price had fallen yielded less than normal profits in 1912–1913 there is no reason to believe that the tremendous expansion carried investment beyond the ideal amount with reference to the state of the world market as a whole. European investment may have been somewhat greater than the ideal amount for European demand alone, since large exports were being sent to America. Their ready reception here supplements other indications that the Aluminum Company's capacity was too small.[8]

Although the financial data are very unsatisfactory for our purposes, they afford some indication of the course of profits of the European companies. By 1906 the AIAG had accumulated an amortization account, through annual debits to profit and loss,[9] which represented just about 40 per cent of the 26,000,000 francs standing on the books as original cost of fixed capital. In 1910 the amortization account was equal to about 29 per cent of fixed capital, which had increased to a little more than 53,000,000 francs. Three years later amortization had grown to about 38 per cent of fixed capital, which had remained approximately the same. Evidently there were lumped together in this account true depreciation, reserves for obsolescence or anything which might impair the value of the old capital, and reinvestment of earnings.

[7] The higher figures for 1912 and 1913 result from inclusion of 20,000 h.p. ascribed to the two plants at Beyrède and Venthon, which were not used for aluminum after their purchase by l'Aluminium Français.

[8] Above, p. 238.

[9] Annual reports of the AIAG for the years 1901–1910 are reprinted in Dux, *op. cit.*, pp. 49 ff. Data for the years 1911–1913 have been taken from the *Schweizerisches Finanz-Jahrbuch*. In 1906 and 1907 a premium of 16,480,531 francs resulting from sale above par of 10,000,000 francs par value of new shares (50 per cent paid up) was charged to a special "Bauabschreibungskonto" which was later merged into the general amortization account. Since it is desired to show only the amounts accumulated out of gross income this premium has not been included in the amortization account in computing the percentages given in the text.

(No surplus was accumulated; earnings not distributed in the form of dividends or *tantièmes*, or charged to a very small reserve account, went into the amortization account. The annual charges to the latter reflect the ups and downs of gross income.) Furthermore, it was admitted on several occasions by officers of the company that substantial secret reserves existed.[10] Hence it is quite impossible to discover, even approximately, the actual earnings record of this firm. That it was unusually profitable can scarcely be doubted, however. In addition to accumulating the large amortization account and indeterminate secret reserves, the company distributed dividends averaging 10 per cent of paid-up share capital in the nineties and 17 per cent in the decade 1901–1910.[11] During the depression and recovery, dividends equaled the following percentages of paid-up share capital, which represented perhaps one quarter of total assets less real depreciation: 1908—18, 1909—12, 1910—14, 1911—14, 1912—20, 1913—20. This company suffered no losses in 1908 and 1909. By reducing markedly the annual charges to amortization it was able to pay out of gross income nearly the whole of the interest charges, dividends, *tantièmes*, and similar disbursements in these two years.[12] During 1910 and 1911 earnings remained small. The record of 1912 showed appreciable improvement, which was followed by a striking increase in the next year.[13] One receives the impression that the AIAG gained very large profits during the years 1901–1907, enjoyed some earnings in the depression, and made good returns in the few years before the war.

Earnings data for the two large French companies indicate that their records were similar to that of Neuhausen, although somewhat less profitable.[14] It appears that during the years 1902–1907 the Froges concern received earnings — after charges to a renewal fund which evidently represented a reserve for depreciation and obsolescence — which were equivalent to an average return of

[10] See Dux, *op. cit.*, p. 40, and Gautschi, *op. cit.*, p. 59.

[11] Annual dividend percentages are given by Dux, *op. cit.*, p. 34.

[12] See data in Dux, pp. 38–39, 84–89. The division of earnings between aluminum and calcium carbide cannot be ascertained.

[13] Accepting the actual amortization charges, which may have been inadequate, the rates of return on undepreciated book value of total assets were about as follows: 1910 and 1911 — 4 per cent; 1912 — 5 per cent; 1913 — 8 per cent.

[14] Financial reports for 1901–1910 are reproduced in Kossmann, *op cit.*

about 10 per cent on undepreciated book value of total assets.[15] By 1909 reserves equal to about 30 per cent of fixed investment had been accumulated. Since the method of computing rates of earnings probably overstates investment (no depreciation is deducted) and understates earnings, it would seem that returns could not have been less than normal and may have been much greater than normal. If secret reserves were accumulated, the latter is more likely. In the two depression years the company reported no losses, although no charges were made to the renewal fund. It paid an average dividend in these two years equal to about half the average annual amount distributed in the years 1905–1907. In 1909 the dividend was paid from revenue. The indicated earnings of the Alais firm were slightly less than those of the Froges enterprise during prosperity and slightly more in depression.[16]

Available data on earnings of the French companies during the years 1910–1913 are fragmentary. Although prices in France seem to have been lower in 1910 and 1911 than during 1909, it appears that the companies fared at least as well during these two years as in 1908 and 1909, owing to a larger volume of sales. Annual reports indicate that in both 1910 and 1911 the Alais firm earned approximately 5 per cent on book value of assets, or about the return received in 1909.[17] It was said that the French companies came through the bad years 1909–1911 without serious damage.[18] The earnings of five firms, including the AIAG, the Compagnie Alais, and three of the French independents, were reported to be appreciably greater in 1912 than in the preceding year.[19] The higher prices following upon reconstitution of the in-

[15] Presumably book value of capital assets represented original cost. The figures of book value used as a base for computing rates of return were gross values without deduction of the amounts in the renewal account. The rates of return for individual years, expressed in per cents, were as follows:

1902 — 6.2	1905 — 12.4	1908 — 1.8
1903 — 6.5	1906 — 22.1	1909 — 4.4
1904 — 10.4	1907 — 6.0	1910 — 4.9

[16] No amortization account appears on the balance sheets of this enterprise, although it is shown in the profit and loss account. Unspecified reserves on the balance sheets remained about the same during the years 1904–1909. They equaled 20 per cent of book value of capital in 1909.

[17] AG, 1911, Partie sup., p. 240; *ibid.*, 1913, pp. 2367–2368.

[18] AG, 1912, Partie sup., p. 371.

[19] AG, 1913, Partie sup., p. 191. In 1912 the Compagnie Alais earned more than 7 per cent on book value of assets.

ternational cartel could have exercised little influence upon the revenues of 1912, inasmuch as the majority of sales in that year had been contracted for at lower prices. Hence in the light of general business conditions we may presume that earnings of all companies improved markedly in 1913. It might be inferred that the old firms were earning at least normal returns.

The British Aluminium Company presents a different record. It was forced to reorganize about 1901 and again during the next depression. Its reports show earnings during the four boom years 1904–1907 equivalent to about 6 per cent on book value of total assets.[20] No reserves or depreciation account appear on the balance sheet. Apparently this company was not obtaining normal profits. This result should not be regarded as a symptom of general overinvestment in the industry, however, for it seems very questionable whether aluminum reduction in Great Britain was economically justifiable. Kossmann reports that the Kinlochleven power development, where a very large dam was required for collection of water in the lake, was too costly for economical production.[21] Investment per horsepower at Kinlochleven was about $300, while the original cost of power developments at Neuhausen, Rheinfelden, and Lend-Rauris seems to have been something less than $200, $235, and $100 per horsepower, respectively.[22] It is probable, too, that British wage costs were above those on the Continent. The fact that this company turned to Norway for expansion before any of the others seems to confirm the view that production in Great Britain was not economical.

It has been impossible to secure much information about the financial condition of the new entrants. It was said that the Italian firm, the Société des Pyrénées, and the Société du Sud-Est operated with losses during the depression, while the Société du Sud-Ouest and the Aluminium Corporation failed.[23] Even if their entrance had not been greeted by depression, it is probable that some of the newcomers would have suffered losses for a few years. Doubtless the capacity of 1908 could not have been fully utilized

[20] The reports are reprinted by Kossmann.

[21] *Op. cit.*, p. 49.

[22] Calculated by dividing horsepower of each into the original cost (as given in 1910 report of the company) of *all* plant at each location. Cf. Kossmann, *op cit.*, p. 80.

[23] Kossmann, pp. 111, 115.

immediately had prosperity continued. More significant than their failure to earn during the first few years of their existence is the fact that none abandoned the field except to sell out to members of the cartel, and that the latter operated all the plants so obtained, with the exception of the two at Beyrède and Venthon, whose capacity was certainly not more than 20,000 h.p. and probably much less.

Consideration of the changes in price and the probable amount and change in cost of production support these impressions of the profits of aluminum producers in Europe. Price in 1912 and 1913 was scarcely half of the average for the boom years. Undoubtedly cost per unit had been decreased somewhat by the larger power developments at Chippis, l'Argentière, and St. Jean, and it does not seem improbable that some reduction in general expenses per unit occurred with the transition from a small to a rather large annual output. Perhaps some economies were obtained from larger alumina works, and from integration. It is not likely, however, that unit cost was cut in half.[24] Much lower profits than those of 1904–1907 seem more probable. While cost estimates by persons outside the aluminum companies may not be considered altogether reliable, it is interesting to note that all the estimates for pre-war cost which I have discovered tend to support the interpretation given above. In 1909 Lodin estimated that the cost per kilogram in France under the most favorable circumstances was about 1.20 fr. (0.97 M. per kg. or 10.5 cents per lb.), while the average cost for all French companies was 1.305 fr. (1.05 M. per kg. or 11.4 cents per lb.).[25] These figures apparently included interest and depreciation on fixed investment, but no general expense or profit. Kossmann, who included general expense as well as interest and depreciation, arrives at an estimate of 1.36–1.50 fr. per kilogram (1.10–1.20 M. or 12–13 cents per lb.) for the older, well-established firms. The *Frankfurter Zeitung* hazarded the opinion that with a price of 1.50 fr. per kg. (1.20 M. per kg. or 13 cents per lb.) the aluminum companies were not incurring losses, and the *Mineral Industry* concluded that cost in Europe (without profit) ranged from 1.20 fr. or 0.97 M. to 1.50 fr. or 1.20 M.[26] In

[24] See Chaps. VIII and IX.
[25] Reported in MI, XVIII, 20 (1909).
[26] *Ibid.*

1910 Professor M. G. Flusin estimated that average cost per kilogram was about 1.50 fr.[27] Two estimates published in 1917 and 1919 were 1.30 fr. and 1.25 fr.[28] During the years 1908–1914 the prices reported for Germany were below 1.20 M. only for a part of 1911. In the following two years they averaged about 1.60 M. The degree of accuracy of the figures published by the Metallgesellschaft cannot, of course, be known with certainty. Actual prices received may have been lower. The lower of the two series of prices in France, which may perhaps register actual prices received more nearly than the higher figures of the other series, shows an average price of 1.50 fr. for 1909, 1.30 fr. for both 1910 and 1911, and 1.40 fr. for 1912.

2. Conclusions

The foregoing analysis suggests these conclusions. Until the end of the boom, capacity was much below an amount which under existing demand conditions would have yielded only normal returns when best utilized. By 1912, although a depression had intervened, a greatly enlarged investment was being well utilized at much reduced prices which enabled almost all of the investment in the industry to earn moderate returns. While the amount of investment which would have yielded normal returns when best utilized may have been somewhat exceeded in the years 1912–1914, the excess was probably not very large.

Two additional considerations support the conclusion that the actual investment of 1912–1914 represented a much better approximation to the ideal investment for that time than was true of the relation between existing and ideal investment of 1904–1906. The trends of production, prices, and consumption through the years 1904–1909 reinforce belief that investment was much too small and price much too high during the boom. Although production of aluminum in Europe was a little greater in 1908 than in any preceding year, and increased much more in 1909, the

[27] AG, 1912, Partie sup., p. 371.

[28] By Guillet and Goldschmidt. See Gautschi, *op. cit.*, p. 58. The rates of output used as a basis for computing the various cost estimates here presented were in no case explicitly stated. Inasmuch as some of the estimates were compared with prices it seems reasonable to infer that those cost estimates related, roughly at least, to existing rates of output.

stocks which accumulated in 1907 and 1908 had nearly disappeared by the end of 1909.[29] Now it is unlikely that the demand schedule in 1909, before recovery had gone far, was located to the right of that for 1907, so it seems most probable that much of the increase in consumption in 1909 was due to the fall in price. It may be inferred that prices considerably lower than those actually prevailing during the preceding boom would have carried off the output of a much larger investment without bringing less than normal returns to capital and enterprise. Secondly, according to Kossmann, the four old companies had not troubled themselves much to develop new applications for aluminum.[30] It was not until 1908 that they entered upon an aggressive marketing campaign to push the demand schedule to the right. This explains part of the increase in consumption which enabled the investment of 1912–1914 to return good earnings when operating at somewhere near best utilization. Thus it appears that the investment and price policy of 1904–1907 was still farther removed from the ideal than has already been indicated. It follows also that, although the capacity of 1908 might have represented temporary overinvestment even had prosperity continued, it was not too large to yield profits when markets were energetically cultivated. Of course, it was too large an investment for the actual demands of the depression years, but it is obvious that the ideal long-run adjustment of investment to demand inevitably results, in most industries, in some overcapacity during depression. When attention is turned away from this to the question of adjustment of investment to increasing demand it becomes apparent that the trouble was not too much capacity in 1908–1909 but too little in 1905–1907.

The objection may be raised that, since facilities for aluminum production cannot be expanded quickly, it is impossible for aluminum producers to keep up with increasing demand during a boom.[31] It is manifest that, on account of the inelasticity of facilities which has been described above,[32] aluminum capacity must

[29] *Statistische Zusammenstellungen,* XVI, 16 (1915). Very little of the increase in tonnage sold in 1909 is to be accounted for by exports to America. The total of imports into the United States in 1908 and 1909, of which some part probably came from the Canadian subsidiary of the Aluminum Company, was only about one-twelfth of the European production in these two years.

[30] *Op. cit.,* pp. 67, 73–74.

[31] Cf. Gautschi, *op. cit.,* pp. 45 ff.

[32] Pages 212–213.

often either anticipate or lag behind rapid bursts in growth of demand. Plainly, the best balance of interests of producers and consumers requires more expedition in adjustment to increasing demand than occurred in the instance at hand, where expansion lagged far behind demand before and during the boom. But in 1905 and 1906 plans were launched whose fruition in 1908 enabled investment to anticipate somewhat the expected demand of the next year or so — i.e., the probable demand without consideration of depression. It would appear that depression and dissolution of the cartel resulted in a greater fall in price than would have occurred otherwise; and that these circumstances spurred the producers to aggressive cultivation of new markets, some part of which would seemingly have been required anyway to utilize the additional capacity created by the next wave of expansion, which may have resulted in some temporary overinvestment in the last two years before the war.

From the standpoint of rationalization the results of cartel control were not salutary. It is hardly likely that substantial overinvestment would have developed had the old companies remained independent. Evidently cartel control was not really necessary for rationalized coördination of supply and demand. If it be contended that the rise of outsiders, which certainly helped to bring about desirable results with respect to investment and consumption, spoiled the proper adjustment of investment by the cartel members to the expected demand of 1908–1910, it should be recognized that the cartel directly invited the invasion by its investment and price policy. Without cartel control, investment and consumption might have expanded earlier. If it be assumed that that would not have occurred, it follows that market relationships were not appreciably influenced by the existence of the cartel, and that more advantageous relations between actual and ideal investment and price were, in any event, to be obtained only by the influx of new firms attracted by large profits.

Until about 1908 investment and price policy in Europe and America were quite similar. The price increase in America was more restrained, and perhaps investment was expanded about a year sooner here, but capacity and price were evidently far from the optimum on both sides of the Atlantic. After 1908, however, the activities of producers in Europe and America seem to have exhibited a decided contrast. It seems clear that there was a much closer approximation to ideal investment in Europe than in America, where the Aluminum Company waited until 1912 to begin new construction. The evidence suggests that it was not until 1915 that completed capacity in America would have been sufficient (perhaps somewhat more than sufficient) to care for actual demand at prices yielding only normal profits. The better investment and price situation in Europe must have had an influence, expressed through imports into the United States,[33] upon the situation here.

It cannot be concluded with assurance that, simply because the better approximation to ideal investment and price came after the rise of outsiders, it was chiefly due to replacement of cartel control by competition between several concerns; or that in the absence of cartel control, investment, production, and price policy in the years 1901–1907 would have been better. It is quite clear that after the formation of the cartel the four old European companies "sat back" and enjoyed the generous profits yielded by price control.[34] Apparently little effort was made to push their relatively unknown product into new uses. In fact, the price increases during the life of the cartel were greater than the average increase in prices of other metals.[35] Would the situation have been any better had the four not joined in the loose cartel? Certainly it must have been difficult to assess the probabilities of the growth of the infant automobile industry in 1902–1904, when, as hindsight shows, extensive enlargement of plant should have been started. Nevertheless, it is not unlikely that if the four had remained altogether independent, some would have been more venturesome and more aggressive in marketing than they were as members of the cartel. The investment and price policy of the cartel constituents before 1905–1906 was appropriate for large profits with a static or slowly increasing demand and exhibited at the same time a method of "playing safe" with respect to the possibility of a large and rapid growth in demand. Although the

[33] See below, p. 317.

[34] Cf. Kossmann, Bannert, Gautschi, Günther.

[35] See Bannert, *op. cit.*, pp. 33–35. Price increases during a boom may be desirable rather than otherwise from the social point of view. But both the level from which they start to rise and the degree of the increase may be, as seems to be true in this case, undesirable.

cartel was not capable of a definite common or central policy, and although it did not, of course, control the investment programs of its members, it would not be surprising if the fact of concert in market control and the reservation of home markets [36] tended to lessen venturesomeness.

Certainly the growth of outsiders was not required to show the older companies that demand was in fact increasing with great rapidity during 1904 and 1905. In any case the latter would have expanded their plant considerably, as the single firm did in America. In the absence of the twin plagues, outsiders and depression, the cartel members might have enlarged total capacity in the industry to the point which it actually reached in 1914 even if the cartel had hung together. If this had happened it would probably have been the result of energetic rivalry for the largest share of the growing demand. But if market control, even of the weak sort exhibited by the cartel, had been preserved throughout, it is extremely doubtful that price would have been allowed to fall so far. Although, in an attempt to deal with the outsiders, the cartel reduced price several times before its demise late in 1908, it tried to keep price much above the depth to which it fell in 1909, which was, if anything, a better business year. The outsiders, the depression, the large increase in investment, and competition between the old firms themselves all contributed to the great fall in price. Given the low price of 1909–1911, it was imperative to cut costs and widen markets. The active competition of these few years was both a cause and a result of the low price. Furthermore, after three years of lower prices and expanding markets it might have been impossible, without seriously reducing demand, to raise price upon reconstitution of market control to the level at which the cartel would very likely have held it in the absence of internal or external difficulties. The demand curve in 1912 was certainly shaped somewhat by the experience with the low prices of the preceding years.[37] Hence it seems unlikely that the cartel, if left untouched by internal or external forces of disruption, would have brought into existence the actual capacity of 1913–1914. And it appears still more unlikely that this capacity, if introduced, would

[36] At this time only the French market was protected by a high tariff.

[37] Both the president of the AIAG and the board of directors of the Alais firm explicitly recognized that the great increase in consumption had been stimulated by the low prices (AG, 1913, p. 2365; *ibid.*, 1912, Partie sup., p. 180).

have been operated at fairly high utilization with such low prices. Uninterrupted price control would doubtless have resulted in higher price, less expansion of demand, in both the market and schedule senses, and either smaller capacity or a lower degree of utilization, or both.

The third and fourth of the questions set at the opening of this section may be answered from what has already been said. The entry of new firms did not result in substantial underutilization with little reduction in price. They engaged in price competition with the older companies, and after the dissolution of the cartel all of the producers acted, much of the time at least, like competitors rather than as a monopolist, with the result that price fell to a point which permitted good utilization. This may have resulted as much from a belief that substantial gains would attend a lessening of the broad span between actual and latent demand as from the impact of depression forces, under which monopolistic agreements, to say nothing of monopolistic tactics pursued independently by a few firms, often crumble.

With a slowly increasing demand in 1902–1903 the cartel members devoted less of their horsepower to aluminum and more to other products in order to raise the price of the former and gain greater profits. The same thing may have been done in 1912–1913. Whether underutilization of reduction works occurred at either period cannot be known, because there are no reliable figures of capacity of reduction works.

CHAPTER XIII

INVESTMENT, PRICE, AND DEMAND IN EUROPE DURING THE POST-WAR DECADE

1. Earnings

Monopolistic influences in the post-war European aluminum industry have existed in the form of national monopolies with varying degrees of government protection, subsidy, and operation, and instruments of international market control which, until very recently at least, became increasingly stronger every few years. Nevertheless, elements of competition were probably somewhat stronger in Europe than in the United States throughout most of the post-war period.[1] In this chapter some of the results of market control in post-war Europe will be examined. Trends in earnings, prices of aluminum and competing metals, capacity, and proportionate utilization of equipment will be studied for evidence bearing upon changes in the relations of actual to ideal investment, output, and price.

Table 18 exhibits the results of study of the earnings of the three leading European aluminum companies.[2] These will be considered in the order in which they appear in the table.

The Compagnie AFC evidently shows the original cost of its operating equipment on the balance sheet. Reserves for depreciation, depletion, and obsolescence are apparently accumulated in an account labeled *Amortissement Général*. Annual charges to this account are deducted from gross earnings. Other reserves, unspecified as to purpose, are taken out of net earnings.[3] The amounts credited to *amortissement général* probably represent a substantial amount of reinvestment as well as true capital charges,[4]

[1] See above, Chap. VII, *passim*.

[2] The data have been taken from the published annual reports of the companies.

[3] No surplus account is used. What would be called surplus in American procedure is distributed through the accounts mentioned.

[4] In the report of this company for 1931, it is said (p. 5) that the reserves include reinvestment.

TABLE 18

RATIOS OF EARNINGS TO INVESTMENT OF THREE EUROPEAN ALUMINUM COMPANIES, 1922–1935

COMPAGNIE AFC

Year	Undepreciated Investment * (*Fr. 1,000*)	Net Earnings † (*Fr. 1,000*)	Rate of Return (*Per Cent*)
1922	347,910	13,904	4.0
1923	363,830	17,040	4.7
1924	421,750	21,123	5.0
1925	504,270	30,867	6.1
1926	622,300	48,303	7.8
1927	718,693	53,912	7.5
1928	782,643	63,434	8.1
1929	925,000	67,823	7.3
1930	1,093,557	61,792	5.7
1931	1,241,832	37,286	3.0
1932	1,305,301	12,867	1.0
1933	1,325,357	29,668	2.2
1934	1,349,845	27,543	2.0

AIAG — NEUHAUSEN

Year	Undepreciated Investment * (*Fr. 1,000*)	Net Earnings † (*Fr. 1,000*)	Rate of Return (*Per Cent*)
1922	137,375	4,150	3.0
1923	139,130	8,020	5.8
1924	144,835	11,635	8.0
1925	155,750	12,280	7.9
1926	159,990	10,905	6.8
1927	161,730	11,755	7.3
1928	168,515	10,935	6.5
1929	183,665	11,575	6.3
1930	206,980	7,970	3.9
1931	216,255	6,555	3.0
1932	215,000	5,130	2.4
1933	213,500	4,175	2.0
1934	215,000	4,755	2.2
1935	214,000	5,115	2.4

TABLE 18 — *Continued*

VEREINIGTE ALUMINIUMWERKE A. G.

Year	Depreciated Investment * (*RM 1,000*)	Net Earnings ‡ (*RM 1,000*)	Rate of Return (*Per Cent*)
1924	34,900	3,570	10.2
1925	42,580	3,870	9.1
1926	48,660	4,170	8.6
1927	51,240	4,335	8.5
1928	55,750	4,355	7.8
1929	55,870	3,460	6.2
1930	55,400	2,295	4.1
1931	59,530	1,530	2.6
1932	66,825	1,495	2.2
1933	65,670	618	0.9
1934	58,040	1,650	2.8
1935	62,215	2,998	4.8

* Average of total assets (undepreciated or depreciated as indicated) at beginning and end of year adjusted for sales of new securities and retirement of funded debt when information was at hand. Funds held for current dividend payments have been eliminated.

† Net earnings before interest.

‡ 1924–1930, net earnings *after* interest. Funded debt of this company has been negligible, but current obligations have always represented a substantial part of total liabilities. It was impossible to discover annual interest payments before 1931. 1931–1935, net earnings *before* interest.

for the total of this account was equivalent to about 23 per cent of the gross value of fixed assets plus investments in 1923, 33 per cent three years later, 37 per cent in 1929, and 36 per cent in 1933.[5] For this reason computation of investment by subtracting from total assets the balance sheet total for *amortissement général* each year would understate investment appreciably, and so tend to overstate the rate of return. Hence the ratio of net earnings to

[5] It is not clear from the reports that this account has been debited at the time of retirements. It is increased each year by exactly the amount charged to it in the profit and loss account. Since it would be absurd to suppose that it has not been debited for retirements (the reports mention credits to plant account for retirements) it is reasonable to assume that the retirement debits to *amortissement* have been made before closing the account to profit and loss.

undepreciated investment (i.e., the total gross value of assets as given each year) has been taken. The resulting percentage ratios are, of course, much less than the true ratios.

The figures for net earnings of the Compagnie AFC given in the table represent reported gross earnings less the annual charges to *amortissement*. Thus they include interest charges. They could be regarded as true net earnings of capital if the annual charge to *amortissement* were not so large after 1923. In each of the following two years, more than 5 per cent of original cost of fixed investment was charged against gross earnings to this account; in each of the four years 1926–1929 more than 10 per cent of fixed investment was deducted. The percentage fell to about 8 per cent in 1930 and 3–4 per cent in the next three years. It has already been remarked that the life of a large part of the fixed investment is very long. The rates of return shown in the table are much smaller than the true ratios of net earnings to investment throughout, because the undepreciated investment base is somewhat larger than true investment. For the years 1926–1929 at least the degree of understatement is probably magnified by the large charges to *amortissement*.[6]

On the surface there seems to be a contrast between the trend in rates of return for this company and the trend for the two other European companies and the Aluminum Company of America. Studies of the earnings records of the latter all suggest falling ratios after 1925, while the French company appears to have earned larger returns in the latter half of the twenties. This is to be ascribed chiefly to improved utilization of capacity and higher prices incident to currency depreciation. In the period 1922–1924 the company was evidently unable to utilize its equipment to the full. In order to meet the rapidly expanding demand it was

[6] If 5 per cent of fixed investment plus investments, which is, perhaps, quite liberal, had been charged to depreciation in each year 1925–1929 the indicated net earnings and the ratios would have been as follows:

	Earnings (*Fr. 1,000*)	Ratio (*Per Cent*)
1925	33,000	6.5
1926	61,100	9.8
1927	68,300	9.5
1928	79,500	10.2
1929	81,500	8.8

forced to purchase metal abroad, doubtless at high prices.[7] It was not until 1926 that inventories began to approximate a normal requirement. French prices of ingot rose rapidly between 1922 and 1926. The subsequent decline left the ingot price 25 per cent higher in 1929 than it had been in 1924.

In 1926 price (in paper francs) stood 60 per cent above the average for 1924.[8] Production had grown by about a quarter, and gross earnings (gross sales minus operating expenses) by about 120 per cent, or 20 per cent more than the increase in gross sales indicated by the changes in output and price. If gross sales of fabricated aluminum and of other products of the company kept step roughly with the changes in gross sales of aluminum ingot, operating expenses must have failed to increase quite as much. The higher rate of return indicated for 1926 is explained by the fact that net earnings were about 128 per cent larger than two years earlier, while investment was only 48 per cent greater.

After the formation of the cartel in 1926 the price of aluminum was reduced. In 1929 it was 25 per cent below the average for 1926 in France. Output had been enlarged only about 15 per cent by 1929. Evidently gross revenue from aluminum ingot declined somewhat. Gross earnings from all operations were reported as 107.8 million francs in 1929 as compared with 77.3 million francs in 1926 — an increase of 40 per cent. Net earnings showed about the same percentage increase. Investment was expanded by nearly 50 per cent, with the result that the rate of return declined slightly. The larger gross earnings might be explained by different factors. They may have been due in varying degree to increasing profitableness of commodities other than aluminum, to a policy of carrying more ingot through later stages, to decreases in the prices of fabricated goods which were less than the drop in the ingot price, or to reductions in the expenses of producing aluminum or its products. It is known that the cost of ingot declined somewhat in these years. Since aluminum is the company's principal product, it is scarcely probable that the whole amount of the increase in gross earnings was occasioned by increasing profits on other commodities. It would not appear that the rate of return upon investment devoted to aluminum was markedly higher or lower

[7] See annual reports.

[8] Prices are given in the *Revue de l'aluminium*.

at the end of the twenties than three years earlier. The information available is too meager for even a sophisticated guess as to whether operating investment was a smaller or larger proportion of total investment in 1929 than in 1926. In both years appreciable amounts of investment must have been locked up in power developments in process of construction and in idle bauxite reserves.

The AIAG of Neuhausen also states gross value, presumably original cost, of its equipment on the balance sheet. No surplus is accumulated, and reserve accounts remain small. Depreciation, depletion, obsolescence, other reserves, and reinvestment are evidently lumped together in *Amortisation*. Inasmuch as *Amortisation* equaled 80 per cent of fixed investment at the end of 1921, and 75 per cent of the gross value of participations in subsidiaries was also amortized,[9] the same procedure has been followed in calculating rates of return as in the case of the French company. The gross stated value of all assets is taken as the base. To the reported figure of net earnings (after general costs, taxes, and *Amortisation* charges) have been added the annual amounts of interest paid, calculated approximately from data on obligations outstanding and their rates of interest. In proportion as the undepreciated investment is larger than true investment the ratios are smaller than the true rate of earnings. Although plant cannot have depreciated to the extent that it has been written down, actual accrued depreciation may be greater than in the case of the Compagnie AFC because some of the plant is older. It is not true in the case of the AIAG, however, that the indicated rate of earnings is still further understated by large annual charges to depreciation. Deductions from gross earnings for *Amortisation* averaged only 2 per cent of gross value of fixed capital in each year during the period 1925–1929.[10] The balance sheet and income account of this company do not consolidate the operating properties and returns of subsidiaries. Hence the ratios shown in the table

[9] The *Amortisation* account is split into two parts, one of which applies to fixed capital, the other to participations.

[10] The fixed plant investment was maintained at a net depreciated value of around 15 million francs, while gross value grew to more than 80 million francs. Large depreciation charges in the twenties would have entirely wiped out the net value of fixed capital. In the past five years gross value of plant has been about 90 millions and net value around 20 millions.

represent understatement according as the earnings of subsidiaries were larger than their dividends to the AIAG, or overstatement according as the securities held by the latter were carried at a gross value much below the true value of the property they represented.

The lower rate of earnings of the AIAG after 1925 seems to be unquestionable. Evidently it was due chiefly to the lower prices of aluminum. It may also reflect increased difficulty in finding markets abroad except at still lower prices than those charged in the leading European markets. Between 1925 and 1929 investment increased by about 18 per cent. Production estimates show a growth of 13 per cent. It is not clear whether more ingot was being carried through later stages in the latter year. Gross earnings fell in 1929 to a figure equivalent to about 7 per cent less than four years earlier, while net earnings were reduced by about 6 per cent. As in the case of the French company it is quite possible that the return to investment devoted to aluminum diminished somewhat more or somewhat less than the figures of return on the whole investment, because this enterprise also makes other products. It does not appear that the proportion of operating to total investment changed much between 1925 and 1929 in the case of this company.

The books of the Vereinigte Aluminiumwerke are kept in somewhat different fashion. The original cost of capital goods in use does not appear; this is written down progressively, and the net figure for depreciated investment is given each year. Hence a percentage ratio of net earnings to book value of total assets has been computed. There is no way of determining whether the book value of assets, which was set up after the depreciation of the mark, was high or low. For two reasons the figures of net earnings cannot be considered accurate. Since the profit and loss account of this corporation for many years did not consolidate the results of its operating capital and the operating capital of its subsidiaries (chief of which is the Erftwerk A. G.), the figures of net earnings include only the dividends received from subsidiaries, rather than the full earnings of the latter.[11] Secondly, until 1931 the interest charges were included with various expenses in an item

[11] Reports of the Vereinigte Aluminiumwerke consolidate the investment and income of the Erftwerk beginning in 1932.

called *Handlungsunkosten und Zinsen*. It has been impossible to discover the rate at which interest was paid on the current debt of this corporation, which was always relatively large. The figures of net earnings in the table represent net earnings *after* interest for the years 1924–1930; thereafter, net earnings *before* interest. Interest charges were probably a million marks or more during some of the years in the earlier period. Whether the earnings ratio has been understated in the table on account of large depreciation charges is not certain. These were about 10–12 per cent of the depreciated book value of fixed assets in each year between 1923 and 1929. Since this corporation owns no electric plants itself,[12] this rate of depreciation may not be too high.

There is no altogether clear explanation of the apparent falling rate of earnings. Between 1924 and 1928 output grew by nearly 60 per cent, and gross earnings increased by about 50 per cent.[13] Although the average price fell from 2.24 M. in 1924 to 1.98 M. per kilogram in 1928, gross earnings per kilogram remained just about the same.[14] The fall in price per kilogram was manifestly offset by an equivalent reduction in operating expense per kilogram. The lower rate of net earnings must then be due either to a larger investment per unit of output or greater deductions of one sort or another from gross earnings. Evidently it was chiefly the result of the latter influences, because total assets, which increased from 34.9 million marks in 1924 to 55.8 million marks in 1928, showed about the same proportionate growth — 60 per cent — as output. All of this increase occurred in inventories, accounts receivable, and securities of subsidiaries (which reflected increased participations).[15] Aggregate net earnings were only 22 per cent larger in

[12] It will be recalled that they are owned by the VIAG.

[13] Gross earnings were as follows (in thousands of marks):

1924	6,413
1925	7,510
1926	8,412
1927	9,604
1928	9,614
1929	8,722

[14] This is of more significance in the case of the German firm, which has not entered the fabricating branches or the production of *Nebenprodukte* to the extent that the other two leading continental firms have done. The year 1928 is taken for comparison instead of 1929 because the beginnings of depression were marked in Germany in 1929.

[15] The totals of these accounts for 1928 were not extraordinary compared to other

1928 than four years earlier, although gross earnings increased by 50 per cent. The explanation is not found in proportionately larger depreciation charges. This item absorbed about 2 million marks in 1924, 2.5 millions in each of the next three years, and 2 million again in 1928 and in 1929. The increasing proportion of gross earnings was taken by *Handlungsunkosten und Zinsen* and by taxes. The larger share of this increase must be ascribed to unspecified costs and taxes, for the credit obligations did not grow substantially during the years in question.

It may be concluded that the rate of earnings was appreciably higher than the figures indicate. The lower price of 1928 would not have resulted in a lessened rate of earnings in the absence of an increase in working capital and participations [16] coupled with higher general expenses and taxes. The diminished rate of return in 1929 is to be explained by lower gross earnings reflecting the beginnings of depression, combined with increased inventories and participations. The true rate of earnings may not have fallen at all. A tax item of about 1 million marks appears in the profit and loss account for the first time in 1927. It increases to about 1.5 million marks in the following years. No explanation of the sudden appearance of this item is offered. When taxes are included in net earnings the rate of return becomes 10.9 for 1927, 10.8 for 1928, 9.2 for 1929, and 6.7 for 1930. It appears that the proportionate utilization of total investment of the Vereinigte Aluminiumwerke increased appreciably between 1924–1925 and 1928–1929, and hence that the correct rates of return on total investment would represent less understatement of the rates of return on operating investment in the later years than in the earlier years.

The inadequacy of the data severely limits conclusions upon comparative trends in the rates of earnings of the three companies. The Swiss enterprise evidently suffered a substantial decline in the rate of return after 1925, which reflected lower prices and the increasing difficulty of finding markets abroad. Internal factors

years. Accounts receivable were somewhat larger at the end of 1928 than in the year after or the year preceding, but not enough so to account for the fall in the ratio as compared to 1926 and 1927.

[16] Upon which dividends were received at a rate lower than the rates of earnings of subsidiaries.

peculiar to France, particularly currency inflation, apparently accounted for a higher rate of return for the Compagnie AFC after 1925. No definite trend in the next four years seems established.[17] In the case of the German corporation unsatisfactory data indicate that the true rate of earnings did not change much between 1924 and 1928, but one cannot place great confidence in this.

An examination of the earnings records of these three companies suggests that they were not receiving earnings less than normal during much if any of the period 1923–1929 and that they may have been gaining rather large returns. It should be borne in mind that another element, in addition to those mentioned above, causing understatement of rates of return on operating investment is the inclusion of idle resources in investment. An inspection of the figures of investment and net profit of the British Aluminium Company in these years indicates that this corporation received moderate returns.[18] Investment increased through the twenties, and net profits were smaller per year in the period 1926–1929 than in the two years 1924–1925. Hence the rate of profit declined in the latter part of the decade. It is probable that rates of return varied between different parts of the investment of each company.

Although the financial reports of the three continental companies are in some respects quite detailed, they tell but a part of what we really wish to know. A successful study of the sort here attempted would require as a minimum the precise distinction between idle and operating investment, clear separation of true depreciation, depletion, and the like from reinvestment, consolidation of the results of operations of all controlled subsidiaries, separation between the results of the aluminum business and those of other departments, and separation of the returns earned at different stages or branches of the production of aluminum and its products.

2. Capacity and Price

The trends of aluminum prices were the same in Europe as in America during the twenties. Table 19 gives approximate European prices in cents per pound. As in America the lower prices

[17] The slightly lower rate earned in 1929 cannot be regarded as significant in the face of the various inadequacies in the data mentioned in the text and the fact of beginnings of depression. [18] Brief figures are reported annually in the *Economist*.

after 1925 represented a proportionately greater reduction than occurred in the prices of competing metals such as copper and tin until the depression. Price trends of the nonferrous metals were substantially the same in the London market as in New York. Table 20 shows the changing relation between prices of aluminum and copper in London and Berlin.

TABLE 19

APPROXIMATE AVERAGE YEARLY PRICES OF 98–99 PER CENT ALUMINUM INGOT IN EUROPE, 1922–1930

(Cents per Pound)

Year	Price	Year	Price
1922	19.0	1927	22.7
1923	22.5	1928	21.4
1924	24.0	1929	20.7
1925	25.5	1930	20.1
1926	25.2		

These averages are derived from the price data of Table 13.

Informal agreements between some of the European companies existed from time to time between 1918 and 1926, when a formal cartel including the four chief producers was set up. In the years 1922–1925 prices rose to a high level. The cartel started its life in the fall of 1926 by lowering price, which had been about £118 per long ton, to £105.[19] Successive reductions were made in the next six years, giving prices as follows: May 1928 — £95, October 1930 — £85, January 1932 — £80.[20] To some the fact that the cartel has several times lowered price without once raising it has seemed paradoxical; others regard the price reductions as proof that the cartel has served the best interests of consumers. It would appear, however, that the cartel had no choice but to lower price in 1926 if the extensive additions to capacity just completed or under construction were to be well utilized and the markets of its members substantially protected from the American company, whose foreign capacity was increasing. It does not need to be repeated that growing competitive influences were beginning to

[19] It is said that for a time preceding the formation of the cartel concessions had been given from the price of £118 established by a loose price agreement.

[20] European prices in cents per pound and dates of change are given in Table 13 on p. 240.

TABLE 20

RATIOS OF AVERAGE YEARLY PRICES OF ALUMINUM INGOT TO AVERAGE YEARLY PRICES OF COPPER IN EUROPE IN CERTAIN YEARS

Year	98–99% Aluminum	Electrolytic Copper	Ratio Aluminum to Copper
		London (*£ per Long Ton*)	
1923	111	72 11*s*	1.53
1924	123	68 6*s*	1.80
1925	123	67	1.84
1926	114	65 14*s*	1.74
1927	105	62 6*s*	1.69
1928	98	69 9*s*	1.41
1929	95	85 8*s*	1.11
1930	92 10*s*	62 3*s*	1.50
1931	93 3*s*	42 13*s*	2.19
1932	96 5*s*	35 19*s*	2.68
1933	100	36 7*s*	2.75
1934	100	33 6*s*	3.00
1935	100	35 9*s*	2.82
		Berlin (*R M per Kilogram*)	
1924	2.24	1.28	1.75
1925	2.37	1.356	1.75
1926	2.288	1.335	1.71
1927	2.10	1.266	1.66
1928	1.98	1.406	1.41
1929	1.90	1.739	1.15
1930	1.86	1.274	1.46
1931	1.70	0.825	2.06
1932	1.60	0.548	2.92
1933	1.60	0.521	3.07
1934	1.57		
1935	1.44		

Sources of average price statistics:

Aluminum

London — 1923–1930, derived from quotations in *Mining Journal;* 1931–1935, from *Metallwirtschaft.*

Berlin — 1924–1931, from *Statistische Zusammenstellungen;* 1932–1935, from *Metallwirtschaft.*

Electrolytic copper

London — 1923–1931, from *Statistische Zusammenstellungen;* 1932–1935, average of bid prices as given by *Engineering and Mining Journal.*

Berlin — 1924–1931, from *Statistische Zusammenstellungen;* 1932–1933, from *Metallwirtschaft.*

make themselves felt by 1925 and 1926. We pass immediately to a consideration of the relation between capacity and demand in Europe during the post-war decade.

As far as cost conditions and circumstances governing entry to the field are concerned, there would be no reason to expect the development during the post-war decade in Europe of a situation in the aluminum industry characterized by continued high price and increasing underutilization of an expanding capacity.[21] Evidently expansion to a larger scale was not spurred by knowledge that marked reduction in cost per unit would result. With the exception of the two Italian companies no outside enterprise entered the industry, although potential competition existed in the form of expressed interest by governments, and doubtless some private groups, in launching new ventures. Since newcomers were very few, unless other elements tended to cause a growing underutilization, one would expect to find that investment and price were so adjusted to demand that little productive capacity was idle. Then the only question of importance, concerned with long-run policy, would be whether the relation between actual and ideal investment was becoming better or worse.

However, the possibility of increasing underutilization of expanding capacity must not be ruled out, because two other important factors existed, either one of which may lead to such a situation. Although a larger scale of plant and organization of the same general kinds would not have brought marked savings, the development of new power sites may have. On the continent of Europe and in Great Britain the sites developed since the war have probably not lowered cost. But with expansion of demand it has become profitable to produce more and more metal in Norway and Canada, where power costs are substantially cheaper. It is not certain whether the full costs of producing in these two countries and selling in markets served by producers with plants in Europe have been considerably less than the marginal costs of the latter. The existence of such cost conditions might have resulted in a growth of idle plant on the Continent and in Great Britain as new equipment was introduced by European and American producers, unless protection to home markets sufficed to keep the older facilities in operation.

[21] Cf. what is said above, pp. 205 ff.

In the second place it is possible that the existence of cartel control may result — even when diminishing costs do not act strongly — in enlarging capacity, through reinvestment of monopoly profits, to a point such that the most profitable price and output policy involves underutilization.[22] Various reasons may account for such action. Members of the industry may prefer to enlarge their own companies even though investment in other industries would be more profitable. Ignorance or mistakes in judging demand may contribute. Competition in investment between cartel members may express differences in judgment and optimism or attempts to redistribute quotas. For these reasons, examination of the relations between investment and demand must consider the possibility of idle capacity.

At the outbreak of the war capacity for aluminum reduction in Europe amounted to something between 40,000 and 50,000 tons per annum. (Table 21 shows changes in estimated capacity in Europe and America between 1914 and 1932.) No substantial additions were made in France, England, and Switzerland during the period of hostilities, but new plants in Norway and Germany added 45,000 tons or more, raising total European capacity to 90,000 tons at least. Det Norske Nitridaktieselskap, which belonged to the French, had facilities for an annual production of about 10,000 tons. Shortly after 1920 (if not before) the Norsk Aluminium Company reached a capacity of about 6,000 tons. After the abandonment of two of the temporary war plants, the Reichswerke in Germany were still equipped to produce at least 25,000 tons a year. European capacity as a whole had roughly doubled during the war, while in America more rapid expansion had carried capacity from about 35,000 tons to something like 85,000 tons. During the first few years following the war the European companies were prevented by chaotic political and economic conditions from employing their full productive power.[23]

With recovery in 1922 and 1923 from the business depression, the demand for aluminum shot upward to a position where most of the productive capacity was fully taxed. The annual reports of

[22] Cf. E. Lederer, "Monopole und Konjunktur," *Vierteljahrshefte zur Konjunkturforschung,* 2. Jahrgang (1927), *Ergänzungsheft* 2.

[23] Railroad embargoes and abnormal conditions in the dismembered Austro-Hungarian empire hindered the flow of bauxite to the Swiss and German companies. See Czimatis, *op. cit.*

the Compagnie AFC announced that in order to meet the demand, at rather high prices, the company was forced to buy metal abroad, in addition to importing the few thousand tons it produced in Norway. Indeed, there was a shortage of aluminum for two or three years in all countries of the world. The available supply of metal could scarcely satisfy demand even at the higher prices charged, although plants everywhere were worked to the full capacity permitted by existing conditions.[24] In its report for 1923 the Neuhausen company announced that it had found new markets, to replace its sales in Germany, which took all the metal it could produce under political conditions that still hampered its raw material supply. In Germany, the invasion of the Ruhr and difficulties incident to the final stages of currency depreciation and the *Goldumstellung* kept utilization down to about two-thirds of capacity.[25] It was apparent that demand was likely to move much farther forward, particularly under the influence of new alloys and products which were being developed every year. The margin of European capacity unutilized on account of abnormal conditions would not augment supply greatly. Under these circumstances all the companies of Europe enlarged their capacities.

TABLE 21

ESTIMATED CAPACITIES OF ALUMINUM PLANTS OF THE WORLD IN CERTAIN YEARS *

(*Thousands of Metric Tons*)

	1914	1919	1926	1931	1935
Europe †	40–50	90–95	125	182	222
America	35	85	105	165	165
European companies †	40–50	90–95	114	164	202
American companies	35	85	116	183	185
Total world	75–85	175–180	230	351‡	422§

* Capacity at end of year. Estimates have been made from study of published estimates of capacity and output.

† Excluding Russia.

‡ Includes 4,000 tons capacity in Russia.

§ Includes 30,000 tons capacity in Russia and 5,000 tons capacity in Japan.

[24] During 1924 the companies were sold out for several months ahead (*Aluminium*, VI, Hefte 15/16, p. 2, Aug. 28, 1924).

[25] The increased German demand was met partly by importation.

In Germany the hydroelectric development on the Inn River, which had been started toward the end of the war, began to deliver power to a new reduction plant of about 10,000 tons capacity in 1925. In 1922 or 1923 the Compagnie AFC embarked upon an extensive program of expansion which for some time resulted in an increase in capacity in nearly every year.[26] Between 1923 and 1929 four new aluminum plants added about 9,000 tons to the capacity of this company,[27] while enlargements of older plants increased productive power by a few thousand tons more. In 1925 additional plant was completed by Det Norske Nitridaktieselskap in Norway, raising its capacity from 10,000 to 15,000 tons. The AIAG finished new power developments in Switzerland in 1925 and 1926 which resulted in an enlargement of the Chippis reduction works equivalent to about 5,000 tons a year. By the end of 1925 the capacity of European aluminum works had evidently increased to 115,000 tons, and perhaps 10,000 tons more went into operation in France and Switzerland during 1926. In 1925 small stocks accumulated for the first time since 1922, but in 1926 a total European output estimated at 112,000 tons must have been very close to operating capacity. Plans for further large additions to investment were already under way. Concessions for two more power developments had been obtained by the Compagnie AFC in 1924 and 1925, and construction was begun in 1926. Plans for the new Italian plants were, at least in the case of one, beyond the project stage. The Swiss producer was expanding its plant in Austria, and building with the French a small plant in Spain, while the British Aluminium Company was engaged upon development of another Scottish water power which would add 12,000 tons or so to its capacity. Altogether the developments under construction or in contemplation in Europe about 1926 involved an accession of 25,000–30,000 tons within three years or so. While these plans were being carried out, futher expansion was undertaken, with the result that by 1930 or 1931 the total capacity of 1926 had been increased by nearly 60,000 tons to a little over 180,000 tons per year.

The capacity actually in existence by 1926 was probably not capable of producing more aluminum than would be consumed at

[26] See above, p. 88.
[27] The other French producer added one small plant to its capacity in 1928.

the lower price set by the cartel when it was formed in September (£105 per long ton). The stocks which accumulated during the business recession in the latter half of 1925 were not large.[28] With enhanced business activity in 1926 and 1927 they were evidently reduced to normal.[29] It seems clear that no further price reductions would have been necessary in 1927–1929 to keep the capacity of Europe well utilized if it had not been increased. Between 1926 and 1929 European output grew by 20,000 tons. It has been impossible to discover how much of the new plant was ready for operation in 1929, but one gathers the impression that during 1928 and 1929 (until the start of depression) new plant was operated at capacity as soon as it was available. Press reports indicate that the companies were disposing of all their product with little difficulty in 1928 and 1929.

Comparison of many of the published estimates of capacity and output suggests that there was considerable idle capacity in Europe nearly every year. I believe this inference is wrong, except for Germany, as will be explained in a moment. Familiarity with capacity estimates for this industry teaches that they usually represent capacity under ideal climatic and operating conditions rather than under typical conditions. The figures presented in this book probably err somewhat in the same direction, although an attempt has been made to bring them closer to the probable results under typical conditions. Furthermore, since most of the producers have not given out information on capacity, the estimates are to be regarded as approximations only. It does not appear, as a result of careful examination of estimates of capacity and output and study of market conditions, that there was any appreciable voluntary underutilization of facilities in France, Switzerland, England, and Norway between 1923 and the beginnings of serious business recession in 1929 and 1930. A drop in output in England in 1926 and 1927, when production averaged about 2,000 tons per year less than the total for 1925, may have represented voluntary curtailment incident to the cartel quota

[28] Information upon market conditions in Europe has been secured from comments in *Aluminium, La Revue de l'aluminium,* and annual reports of the companies, and from personal interviews.

[29] A part of the increased output of Germany incident to the operation of the new plant in Bavaria was disposed of in America during the business recession. Exports to America from the European companies declined sharply after 1926.

arrangement, or it may simply have reflected poor climatic conditions.[30]

It appears that the plants built in Germany during the war were not operated at full capacity at any time during the post-war decade. During 1919–1920 the output of the Reichswerke represented less than half the capacity, and in 1922–1923 it amounted to but two-thirds of capacity. Although the percentage improved thereafter, production seems to have remained at least 15 per cent below capacity on the average in the latter half of the twenties. Doubtless the low utilization of the first few post-war years is explained by abnormal political and economic conditions. Failure to reach capacity output in the second half of the post-war decade has been attributed to high cost — in particular, to increases in the cost of electric energy to the Erftwerk — and to lack of correspondence between capacity and the sales quota given the Vereinigte Aluminiumwerke by the cartel.[31] Inasmuch as the amount of German capacity during the post-war decade reflected the influence of war needs rather than commercial principles, the fact of underutilization in Germany does not represent a significant qualification to the general conclusion that European ingot capacity was operated at or very close to full capacity (permitted by climatic and political conditions) during the period 1923–1929. Fabricating equipment, particularly that installed to make products from the new alloys, was probably not fully utilized all the time.

Continuous operation of ingot facilities at a rate close to capacity cannot be regarded as a sure symptom that the total investment of the European firms did not exceed an amount appropriate to maximum monopoly revenue after 1924.[32] Substantial sales in the United States during 1925–1927 at net prices lower than those received at home[33] probably indicate that some companies — particularly the Vereinigte Aluminiumwerke — regarded their existing capacity as greater than that which would maximize monopoly revenue if its full product were sold in Europe (and any

[30] Analysis of changes in imports from cartel countries and from British plants in Norway does not resolve the question.

[31] The German quota appears to have been lower in relation to capacity existing when the cartel was formed than the quotas of the other members.

[32] Cf. what is said above, pp. 235 ff.

[33] Cf. Alfred Marcus, *Grundlagen der modernen Metallwirtschaft* (Berlin, 1928), p. 213.

other customary markets).[34] After 1926 the total sales of ingot outside of cartel countries by European producers diminished but remained about one-fifth of total output in each year 1927–1929. Some part of these exports brought lower net prices than those received in Europe. This in itself does not demonstrate that European investment as a whole was larger than an amount designed to yield maximum profit through geographical discrimination. Reports of price wars in the Orient at this time suggest, however, that the part of the investment used for products sold in the East might not have been fully utilized with oligopolistic rather than competitive price making.[35]

Up to this point we have found no indication that the European companies may have devoted to European consumers an amount of investment greater than that which, when finally utilized, would return maximum profits. The apparent absence of price competition on ingot in Europe and the evident scarcity of potential entrants does not permit the inference that capacity used was greater than the amount appropriate for maximum profit because of the operation of such forces. European press reports indicate, however, that the substantial price reductions of 1926 and 1928 may have been motivated partly by a wish to forestall possible price competition from the Aluminum Company of America or Aluminium Limited.[36] Further, in the latter half of the twenties price competition occurred in the sale of rolled products, and perhaps other semifabricated or finished articles, where monopolistic elements were less strong than at the ingot stage.[37] Rolling capacity was greatly enlarged during the post-war decade, particularly in Germany, under the influence first of the reconstruction demand, and later of the development of foil and the new hard alloys. It

[34] It is possible, indeed, that European firms, after creating just about the right investment for maximum profit when fully utilized by sales in Europe, decided that they could make larger profits temporarily through sale of some metal in the United States, although such a policy would not permanently be more profitable.

[35] See above, p. 163. That price competition in eastern markets took the form partly of price cutting on half-products or finished goods is indicated by the giving of export rebates to manufacturers of these goods in Germany, where the ingot producer had not integrated forward to the same extent as the aluminum firms in other countries.

[36] See for example JFE, XXXVII, 173 (June 1928), and *Wirtschaftsdienst*, XIII, 1059 (1928).

[37] Personal communication.

appears that from time to time the integrated firms cut prices of rolled products below the sum of the market price of ingot and the full conversion cost.[38] Finally, the cartel price reduction of May 1928 seems to have been part of a program to develop new applications and new markets.[39] It is said that some of the companies experienced a fall in gross revenue from sales of aluminum as a result of the lower price. Evidently European ingot investment in 1928–1929 was somewhat larger than the amount which would have currently yielded maximum profit, but was fully utilized as part of a policy of building up demand.[40] In conclusion, it would seem that during most of the period 1925–1929 total European aluminum investment was somewhat greater than the quantity appropriate to maximum monopoly revenue.

That the relations betwen the investment of European firms and the demand for their products improved during the twenties is evident. In the years 1923–1925 aluminum was scarce, price high, and profits evidently quite good. We cannot be sure whether, in the absence of abnormal difficulties, the capacity existing in 1923–1924 would have constituted great underinvestment or not. Expansion was begun before, or as soon as, shortage of stocks developed. As new capacity came into operation in 1925 and 1926, increased exports temporarily removed an added strain which might have broken the high price. The relations between investment and demand were much better in the late twenties because effective demand was being moved closer to ideal demand. For the rest, it is not clear whether the adjustment of investment to *current effective demand* improved during the twenties. Unsatisfactory earnings data suggest that European investment as a whole experienced a slight drop in the rate of return and that monopoly revenue did not grow as fast as investment. However, it is impossible to determine definitely whether or not the rate of return on operating investment devoted to aluminum fell. It appears that reductions in the cost of producing ingot between 1925 and 1929 were just about equal to reductions in price in the case of two of the

[38] In Germany, where rolling mills were in large part not integrated with ingot producers, a rolled products cartel found it impossible at times to control price.

[39] Edwin Kupczyk, in *Wirtschaftsdienst,* XVI, 283 (February 1931).

[40] It is said that some members of the cartel were quite willing to reduce price farther in 1929 (*Magazin der Wirtschaft,* V, 1210 ff., August 1, 1929).

companies.[41] We have seen that price reductions on some fabricated products apparently resulted in net prices for ingot below its market price and that low prices existed for both ingot and fabricated goods from time to time in eastern markets.

We may conclude with some assurance that the movement of effective demand much closer to ideal demand conferred a net benefit on consumers in general. While the adjustment of investment to current effective demand became more liberal in the markets for some articles and in some geographical areas, it is questionable whether this relation was improved in all markets. It would appear that the benefits of a larger investment and output relative to existing effective demand were restricted to consumers of certain products — particularly those involving new adaptations, such as the strong alloys, where the endeavor to cultivate demand with great rapidity necessarily involved larger capacity relative to existing demand — and consumers located outside Europe. Insufficient information about changes in the discriminatory price structure precludes judgment as to whether the relation of composite investment to total effective demand became better or worse. It should be reiterated that throughout the period 1923–1929 returns on operating investment and monopoly revenue to the European companies as a whole may have been quite large indeed and underinvestment quite pronounced. Or total investment may have been not far from the ideal amount proper to current effective demand, with a discriminatory price structure which brought large losses in some markets. That large returns accrued from some particular markets or sections of demand seems scarcely to be doubted.

The better relation between effective demand and ideal demand, and the more liberal adjustment of investment to some parts of effective demand, resulted from the influence of two elements continuously making prospective demand for the products of each individual firm more elastic at lower prices. In Europe as in America exploration of the possibilities of new alloys, new products, and new uses for old products was producing a transition to larger and more diversified markets, in several of which demand was

[41] No cost data have been published by the companies. The statement is based upon information about proportionate reductions in cost secured from informed sources.

more elastic at lower prices. Development of the new markets depended upon the movement of effective demand, while the greater elasticity at lower prices inevitably meant a lower rate of return per unit of investment devoted to these markets, even apart from a development price policy. The growth in output associated with the transition also tended indirectly towards lower prices by increasing the supply of scrap aluminum. In the second place, a tendency to purely competitive rivalry seems to have characterized the acquisition of bauxite and power, the expansion of equipment, and the development of new alloys and products. In the matter of short-term price and output policy one's total effect upon the market may be easy to predict; hence rational, considered oligopolistic price policies should easily prevail in the absence of very considerable uncertainties, shortsighted behavior, or nonprofit motives. When it comes to introduction of new capacity, acquisition of reserves of materials, and development of new adaptations of a basic product, however, uncertainties surrounding the effects of one's activities must certainly be much greater.[42] Policies proper to maximum joint monopoly revenue for all are less likely to obtain. Evidently a tendency toward competitive rather than strictly oligopolistic adjustments with respect to these long-run elements helped to push effective demand farther ahead and also kept total investment used for European and Asiatic markets somewhat beyond that appropriate to maximum profit. Rivalry of this sort did not only exist between the Europeans as a body and the American companies. There are also evidences of competition in investment and new adaptations within the fold of cartel members.[43] And, finally, it may have been felt that with the transition to larger markets potential competition from new processes using lower grade ores would become a significant threat if price were not substantially lowered.

We should like to know whether the new facilities introduced between 1929 and 1932 could have been operated at capacity without lowering the price of £95 established in 1928 had depression not intervened while it was being built. Whatever the answer to this question, the available data on capacity, price, markets, and earnings do not suggest that the European companies

[42] See below, pp. 337 ff. for analysis.

[43] Above, pp. 47 ff. and 157; below, p. 304.

were adding an amount of operating investment which could not have been fully utilized, in the absence of a marked decline in the rate of increase of demand, without reducing earnings below normal. This is subject to the possible qualification that true over-investment might have existed for a time if the energy from the 260,000 h.p. plant of the Alcoa Power Company on the Saguenay River, which has not been used for aluminum since its completion in 1931, had been employed to make aluminum for the markets of European producers.

In closing this section some discussion of the significance of developments in the past five years for the general long-run relations of investment, output, and demand in Europe would be appropriate. Economic self-sufficiency, tariffs, quotas, and maladjusted currencies have, however, created a situation which has much reduced the significance of consideration of such general relations for Europe as a whole. Only a few observations seem pertinent. Price is everywhere much lower and capacity much larger than at the beginning of 1930. One price reduction has occurred in Germany and one in France since 1932. It is doubtful, however, if the average price received for aluminum, which has probably been well under official prices, is relatively as low as the prices of many other commodities, including competing substitutes. Programs of self-sufficiency and armament building have created large demands in Germany and Italy, part of which may prove to be temporary. On the whole, it would not be surprising if in the next few years geographical discrimination, and perhaps discrimination between different commodity markets, assumed more significance, and the average degree of utilization of capacity were at the same time lower, than was true in the twenties.

3. Oligopoly and Cartels

We now turn our attention to the connection between market relations and the sorts of market control which prevailed during the twenties.[44] Evidently the loose price agreement existing between some of the European producers in the years 1923–1926 exercised no appreciable influence on the adjustment of investment to demand. Investment policies in these years reflected ex-

[44] Oligopolistic relations, agreements, and cartels are described in Chap. VII.

pansion to meet prospective increases in demand in home markets, the beginnings of rivalry in sharing old markets and new foreign markets, and the wishes of governments. While European capacity was rapidly extended it apparently failed to reach the point to which purely competitive tactics would have carried it. This seems to show how a few producers, untroubled by newcomers, can, even in the face of large uncertainties of demand and supply conditions, restrict investment on a rising market below the ideal amount without any coöperation in the adjustment of capacity to demand.

With regard to short-term relations, it is obvious that the agreement may have facilitated rising prices, but it is quite as plain that price would have risen substantially with the capacity and demand conditions obtaining in 1923 and 1924 even in the absence of any agreement. The maintenance of the high price during 1925 and the first half of 1926, when stocks were larger, and when existing or prospective conditions must have indicated the advisability of a price reduction in the near future, may have been due in larger degree to the existence of agreement. The uncertainties of prospective demand at this time were considerable, and opinions may well have differed sufficiently so that, in the absence of agreement, independent reductions made only with the purpose of moving to a price more profitable for all would have provoked a wave of price cutting merely because they were misinterpreted.

Under the cartel agreement of 1926–1931 price, which had hitherto shown minor variations between some countries, was the same in all producing countries.[45] No evidence appears of violations of the price provisions, although it is said that all members were not always in accord as to the advisability of the three reductions in price made by this cartel. Since we do not know precisely how price and cost reductions compared for the individual companies, it is impossible to tell whether consumers in some countries benefited more or less under this arrangement than they otherwise might have.

It will be recalled that the cartel agreement provided for sales quotas which were to be maintained by an equalization device of purchases and sales between members. The estimates in Table 22 indicate that after 1926 the proportions in which output was

[45] Special prices were permitted to meet local peculiarities.

divided between the four chief members were less variable. One would infer that when the covenant was renewed in 1928 the British sales quota was increased somewhat and the French quota diminished. According to the estimates of trade between cartel countries presented in Table 23, Great Britain and Germany were net importers from other cartel countries and Switzerland a net exporter to other cartel countries in the years prior to 1926. This relation was continued under the cartel arrangement, and the amounts grew larger in each case. Except for exports of 5,000 tons in 1927, most of which went to Germany, the trade of France with other cartel countries was negligible. The figures show, however, that the French production in Norway — 4,000–5,000 tons

TABLE 22

DIVISION OF CARTEL OUTPUT OF INGOT ALUMINUM BETWEEN LEADING COMPANIES, 1922–1930

Year	Vereinigte Aluminiumwerke		L'Aluminium Français		Aluminium-industrie A.G.		British Aluminium Co.	
	1,000 Metric Tons	Per Cent	1,000 Metric Tons	Per Cent	1,000 Metric Tons	Per Cent	1,000 Metric Tons	Per Cent
1922	14.8	*30.1*	11.8	*24.0*	16.2	*32.9*	6.4	*13.0*
1923	15.8	*24.8*	17.2	*27.0*	17.7	*27.7*	13.1	*20.5*
1924	18.8	*23.7*	23.9	*30.1*	22.4	*28.2*	14.3	*18.0*
1925	26.0	*27.7*	25.5	*27.1*	25.2	*26.8*	17.3	*18.4*
1926	29.4	*29.2*	29.9	*29.7*	25.2	*25.0*	16.2	*16.1*
1927	27.2	*27.6*	30.8	*31.2*	25.2	*25.5*	15.5	*15.7*
1928	29.7	*27.3*	32.6	*30.0*	27.4	*25.2*	19.0	*17.5*
1929	32.0	*27.2*	34.2	*29.1*	28.6	*24.3*	22.8	*19.4*
1930	29.9	*27.0*	30.8	*27.8*	28.2	*25.4*	21.9	*19.8*

These figures have been computed from the basic data of Table 38. Figures for the Vereinigte Aluminiumwerke include the production of the Bitterfeld plant, which cannot be distinguished. Figures for the AIAG include output of the small plants at Steeg, Austria, and Martigny, Switzerland, which cannot be distinguished. It has been possible to separate the output of the Rheinfelden, Germany, plant of the AIAG and include it in the figures for the latter. Figures for the British Aluminium Company include output of the Aluminium Corporation. The production in Norway, Italy, and Spain has been allocated to the various companies on the basis of proportionate ownership of capacity. Since this method can be expected to yield approximately correct results only when all plants are operating near capacity, the figures have not been carried beyond the year 1930.

per year — was not imported for home consumption. Whether the metal was sent to cartel countries or others cannot be determined. According to consumption estimates of the Metallgesellschaft (based on production estimates and statistics of foreign trade), consumption grew much more rapidly in Germany and Great Britain in the latter twenties than in France and Switzer-

CHART III

ESTIMATED EUROPEAN PRODUCTION OF PRIMARY ALUMINUM BY LEADING COMPANIES, 1910–1930

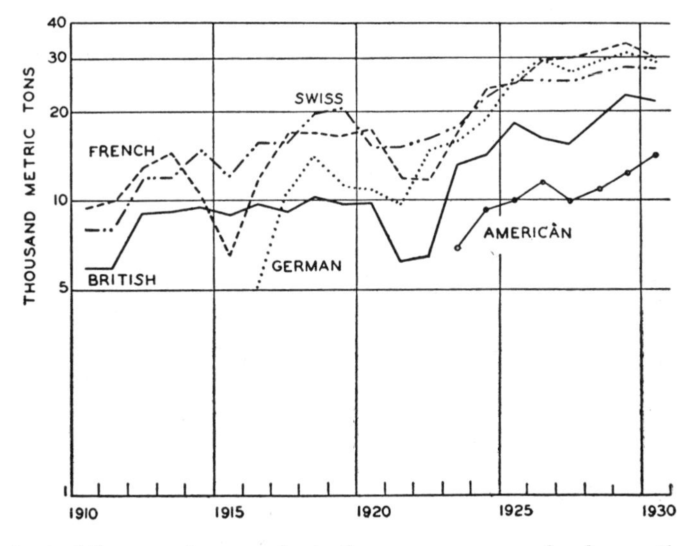

Output of the several companies in the years 1910–1921 has been estimated from the basic data of Table 38 according to the method explained in the note to Table 22. Data for the years 1922–1930 are given in Tables 22 and 25.

land. It would appear that the British were given a small increase in their quota in 1928 because of this fact and the imminent completion of the Lochaber plant; while the French suffered a reduction in their quota on account of the lag in French consumption. Under the cartel arrangement the Swiss and perhaps the French, through shipments from Norway, shared in the rapidly increasing consumption in England and Germany. Without any such device of market control the shares might have been quite different. It

TABLE 23

Statistics of Foreign Trade in Aluminum Ingot of Cartel Countries, 1923–1929

(Thousands of Metric Tons)

	Great Britain							Germany						
	1923	1924	1925	1926	1927	1928	1929	1923	1924	1925	1926	1927	1928	1929
Imports														
From foreign plants of domestic company	1.2	4.2	3.9	6.7	7.5	8.3	8.9	—	—	—	—	—	—	1.3
From others in cartel countries	0.9	2.9	3.2	4.8	6.3	2.7	8.2	4.4	4.8	8.0	5.0	11.0	12.2	10.8
From all others	0.7	2.9	4.1	0.7	2.1	2.7	5.7	1.1	0.7	2.7	0.5	1.8	2.6	2.2
Total	2.8	10.0	11.2	12.2	15.9	13.7	22.8	5.5	5.5	10.7	5.5	12.8	14.8	14.3
Exports														
To proprietor companies in cartel countries	—	—	—	—	—	—	—	—	—	—	—	—	—	—
To others in cartel countries	1.2	0.4	1.3	0.4	1.1	1.6	1.6	0.1	0.5	0.9	2.3	1.2	1.2	1.4
To all others	4.0	2.8	3.4	3.8	4.4	4.5	4.3	0.1	0.6	3.4	11.6	3.9	2.4	2.7
Total	5.2	3.2	4.7	4.2	5.5	6.1	5.9	0.2	1.1	4.3	13.9	5.1	3.6	4.1

	Switzerland							France					Norway						
	1923	1924	1925	1926	1927	1928	1929	1925	1926	1927	1928	1929	1923	1924	1925	1926	1927	1928	1929
Imports																			
From foreign plants of domestic company								—	—	—	—	—							
From others in cartel countries								—	1.1	0.3	—	0.1							
From all others								0.1	0.2	0.1	0.1	0.1							
Total			0.4	0.2	0.5	0.7	0.6	0.1	1.3	0.4	0.1	0.2							
Exports																			
To proprietor companies in cartel countries	—	—	—	—	—	—	—	—	—	—	—	—	1.2	4.4	3.9	9.2	7.7	8.3	8.9
To others in cartel countries	2.7	4.8	4.2	5.9	7.0	6.5	6.3	1.7	0.5	5.0	1.3	0.2	0.2	1.7	2.3	0.3	3.8	1.9	7.9
To all others	6.7	7.8	9.7	9.5	6.4	8.5	6.9	0.4	0.4	1.1	1.8	2.0	11.4	13.1	14.3	12.6	10.6	6.6	12.8
Total	9.4	12.6	13.9	15.4	13.4	15.0	13.2	2.1	0.9	6.1	3.1	2.2	12.8	19.2	20.5	22.1	22.1	16.8	29.6

With the exception of the estimates of imports by domestic firms from their foreign plants and of exports to proprietor companies in cartel countries, the figures of this table have been taken from official statistics of foreign trade. For the most part they represent net imports and net exports after elimination of reimports and reëxports. Imports by domestic companies from their foreign plants have been estimated as follows. When the total imports from a country in which there was a foreign plant of a domestic firm were less than or equal to the output of that plant, estimated according to proportionate ownership of capacity in that country, the full amount of the imports from that country was regarded as imports from the foreign plant of the domestic company. When the total imports were greater than the estimated output of the foreign plant the latter was considered as the imports from the foreign plant. The same general method was followed in estimating exports from Norwegian plants to proprietor companies in cartel countries. Cartel countries include Great Britain, Germany, France, Switzerland, Norway, and Austria. Although the figures comprehend scrap as well as virgin, it is believed that the general relations which they exhibit would not be substantially modified by the elimination of scrap. The word "others" signifies other persons. A dash means none, or a negligible quantity.

would depend upon whether the sales quotas were such as to equalize marginal costs of all producers.[46]

It seems likely that the cartel of 1926–1931 exercised some influence upon short-run market relations. The fact that capacity was too large in 1925–1926 for the current price, the price wars in eastern markets, and the tendency to compete in pricing fabricated goods in Europe are symptoms of a situation in which the ingot price might have gone lower in the absence of cartel control. Some redistribution of sales volumes between companies might also have occurred, although the fact that existing ingot capacity was well utilized indicates narrow limits for this. A marked change in the pattern of geographical discrimination might have resulted, however, accompanied by larger consumption in Europe.

That this cartel substantially affected the adjustment of investment to demand is to be doubted. Between 1926 and 1931 changes in the relative capacities of the four chief members seem to have been more disproportionate than between 1919 and 1926. (See Table 24.) When the relative division of capacity in 1931 is compared with the proportions in which the sum of existing and projected capacity was divided at the time of formation of the cartel, the changes seem to be greater than the variations in percentages of existing capacity between 1919 and 1926. It is not denied, of course, that the cartel may have exerted some restraint upon competitive building; but a consideration of the nature of the agreement reinforces the belief that such influence was not great. The aggregate volume of sales was regulated only through price. The equalization measure provided no compensation — as a system of cash fines and bonuses may — for the member whose capacity did not increase as rapidly as total sales. Hence there was some incentive to revise upwards one's estimate of the future increase in demand, if the expansion of others indicated that they held opinions which were pleasantly more optimistic. Neither the quantities of ingot used in the fabricating plants of members nor the prices and sales volumes of fabricated goods were controlled. The facts of disproportionate expansion suggest that the firms

[46] At least in the same market. Discontinuities in the respective curves of marginal revenue in separate markets might have rendered it impossible to equalize marginal costs of serving all markets.

TABLE 24

Division of Aggregate Ingot Capacity of Leading European Aluminum Companies in Certain Years *

Company	Capacity End 1919		Capacity End 1926		Existing Plus Projected Capacity End 1926		Capacity End 1931		Capacity End 1935	
	1,000 Metric Tons	Per Cent	1,000 Metric Tons	Per Cent	1,000 Metric Tons	Per Cent	1,000 Metric Tons	Per Cent	1,000 Metric Tons	Per Cent
L'Aluminium Français	30	*34.5*	33	*31.1*	39.5	*29.8*	39.5	*26.6*	39.5	*23.3*
Vereinigte Aluminiumwerke	25	*28.7*	31	*29.2*	31	*23.4*	38	*25.6*	55	*32.4*
Aluminium Industrie A.G.	20	*23.0*	25	*23.6*	33	*24.9*	42	*28.3*	46	*27.1*
British Aluminium Co.	12	*13.8*	17	*16.0*	29	*21.9*	29	*19.5*	29	*17.1*
Total	87	*100.0*	106	*99.9*	132.5	*100.0*	148.5 †	*100.0*	169.5 †	*99.9*

* Estimates have been made from study of many published estimates of capacity and output.

† Does not include 5,500 tons owned jointly through Alliance Aluminium Compagnie.

acted upon somewhat different estimates of future demand, or hoped that relative changes in capacity or the individual development of new products would bring about alterations in quotas, or anticipated dissolution of the cartel, or intended to fabricate more ingot. Oligopolistic calculations of market relations in the sale of articles beyond the ingot stage were complicated by the different degrees of integration in the organization of the several companies. Also, it must not be forgotten that during the life of this cartel Aluminium Limited[47] was endeavoring to build up its foreign markets. Clearly, the provisions and administration of this cartel agreement were not adequate to discourage rivalry in expansion. In this industry, as in many others, that would require rigorous enforcement of *output* quotas and agreement that quotas would not be altered as a result of relative changes in capacity; or conditions such that all oligopolists believed nothing was to be gained from rivalry in expansion. Complete elimination of all competition in long-run adjustments would also necessitate an arrangement for sharing in unchanging proportions the results of all variations or adaptations of the basic product. In the earlier years of the cartel of 1901–1908 the simpler markets then existing did not provoke the imagination of the members sufficiently to induce energetic rivalry to gain the larger share of an expanding market. In the twenties the great possibilities and the considerable uncertainties of future demand created a situation in which rivalry in long-run adjustments could have been curtailed only by the devices just mentioned.

It is believed that the agreement reached in 1931, when the Alliance Aluminium Compagnie was formed, included provisions for rigid control of output and an agreement that quotas would not be changed upon the basis of disproportionate growth in capacity. Unfortunately, paucity of information and the chaotic conditions prevailing during the last five years make it impossible at present to assess the consequences of this strong organization for coöperative market control. The more striking developments since the creation of the Alliance cartel and a few of the possibilities have been described in Chapters IV and VII.

Under the influence of semicompetitive rivalry in expansion in Europe, ingot capacity was approximately doubled in the twelve

[47] Before 1928 the Aluminum Company of America.

years 1920–1931. (See Table 21, p. 291.) In the United States single-firm monopoly brought nearly the same result, with an increase from about 70,000 tons in 1919 to perhaps 125,000 tons in 1931, the more rapid expansion after 1925 making up the ground lost in earlier hesitation. When Canadian capacity is added to that of the United States it appears that total American capacity was just about doubled in these years. Until 1929 more than half of the Canadian exports of ingot ordinarily came to the United States.[48] As capacity increased in both countries more of the Canadian metal was sent to other markets and less to the United States. Apparently it was the intention that Aluminium Limited, in the beginning at least, would sell chiefly in foreign markets. A comparison of the analyses presented here and in Chapter XI suggests that the relation of investment to demand for ingot tended to be less restrictive in Europe than in America during the later twenties. However, lack of adequate data and the absence of an effective apparatus for measuring changes in demand prohibit definite conclusion. Competitive influences may have yielded a better adjustment in Europe; or the Aluminum Company may have been more successful than the European firms in moving effective demand ahead.

The relative growth in estimated output by European and American producers is shown in Table 25. The war proportions were renewed in 1923–1924. Expansion of investment in these years enabled the Europeans to jump ahead in 1925–1926, after which American output assumed a commanding lead. The proportions in 1928–1929 were not markedly different from those of 1918 and 1923–1924. It will be noticed that the share of total European output produced by the American companies remained roughly stable. The situation had changed appreciably, however, between the earlier and the later twenties. In 1928–1929 the Europeans were selling a much smaller proportion of their output

[48] Canadian exports of ingot (in thousands of metric tons) were divided as follows in the years 1927–1931:

	To the United States	To Other Markets
1927	18.9	4.6
1928	10.5	7.9
1929	12.9	20.2
1930	5.4	14.2
1931	1.1	8.7

in the United States,[49] while a larger part of the production of the American companies was being disposed of in markets served by European firms. In 1923–1924 the Aluminum Company was importing a large part of its Norwegian output, and exports from Canada to countries other than the United States averaged less than 5,000 tons. By the late twenties Norwegian output had grown, imports from Norway into the United States had declined, and exports from Canada to countries other than the United States had increased.

TABLE 25

ESTIMATED PRODUCTION OF INGOT ALUMINUM BY AMERICAN AND EUROPEAN COMPANIES, 1918–1930 *

(Thousands of Metric Tons)

Year	European Companies	American Companies † Total	American Companies † United States	American Companies † Canada	American Companies † Europe	Percentage of European Output Produced by American Companies
1918	62.9	71.6	56.6	15.0	—	—
1919	59.4	73.3	58.3	15.0	—	—
1920	53.1	74.9	62.9	12.0	—	—
1921	43.1	32.7	24.7	8.0	—	—
1922	49.2	43.4	33.4	10.0	—	—
1923	63.8	75.2	58.4	10.0	6.8	9.6
1924	79.5	90.1	68.3	12.5	9.3	10.5
1925	94.1	88.5	63.5	15.0	10.0	9.6
1926	100.7	96.4	66.9	18.0	11.5	10.2
1927	98.7	120.1	74.2	36.0	9.9	9.1
1928	109.8	142.4	95.5	36.0	10.9	9.0
1929	120.4	146.7	103.4	31.0	12.3	9.3
1930	114.1	152.0	103.9	34.0	14.1	11.0

* European output of American companies is estimated as a proportion of Norwegian output equivalent to proportionate ownership of facilities in that country plus a tonnage equal to the capacity of the small Italian plant owned by the Americans. The series cannot be continued after 1930 owing to lack of information concerning utilization of plants during depression.

† Aluminum Company of America alone until 1928, when its Canadian and European reduction works were transferred to Aluminium Limited.

[49] Imports from European producers are given in Table 15.

TABLE 26

RELATIVE POSITIONS OF THE AMERICAN AND EUROPEAN COMPANIES IN FOREIGN TRADE IN ALUMINUM INGOT, 1922–1930

(Thousands of Metric Tons)

	1922	1923	1924	1925	1926	1927	1928	1929	1930
Net imports from Europeans to America	13.5	4.7			11.2				
Net exports from American companies to all markets outside North America			5.3	6.1		1.7	12.6	23.3	21.7
Net exports from American companies to markets other than Japan and North America ...			2.7	3.9			8.6	15.3	12.7
Exports from American companies to Japan ...	1.8	1.5	2.6	2.2	2.2	2.4	4.0	8.0	9.0
Exports from European companies to Japan ...	2.2	2.2	1.6	2.4	4.8	4.7	4.8	5.1	3.9

The figures of this table have been computed from official statistics of foreign trade. Net imports (or exports) equal imports minus exports (or exports minus imports). Output of American-owned European plants, estimated as explained in the note to Table 25, is treated as an export from the American companies in so far as it is not imported into North America. Because virgin and scrap aluminum are not separated in most foreign-trade statistics the true relations with respect to virgin metal may be somewhat different from those shown in the table. It is believed that elimination of scrap would not greatly alter the trends. It might, indeed, result in larger figures of net exports from the American companies, because it is very probable that cartel countries were larger exporters of scrap than Canada and Norway, in which nearly all of the exports from American companies originated. Neither of these two countries has been a large consumer of aluminum.

Some indication of the results appears in the estimates of Table 26. Beginning in 1928, the year in which Aluminium Limited was created, the excess of American sales in markets served by European companies over European sales in America evidently increased greatly. (American sales in foreign markets were, of course, greater than the figures of the table show by the amount of European exports to America.) Nearly all of the American metal sold in foreign markets came from Canada and Norway. In 1928–1930 about 44,000 metric tons of ingot were exported from America to markets outside North America, nearly all of which went from Canada. Of this total about 21,000 tons were sent to Japan, where the sales of European companies were diminishing.[50] Several thousand tons were exported from Canada to England. Because of the impossibility of dividing Norwegian exports according to plants of origin, we cannot determine the total sales of Aluminium Limited in Europe. Reference to Table 23 indicates that the net exports of cartel countries to all other countries were largest in the middle twenties and declined thereafter to a position roughly similar to that prevailing in the earlier twenties. In 1928 and 1929 Great Britain, Germany, and France had very small net exports to countries other than cartel countries. Swiss exports to countries outside the cartel group were but little larger than five years earlier, while Norwegian sales outside cartel countries had diminished. Clearly, Aluminium Limited was obtaining an increasing proportion of the business in markets outside North America, and the sales of the European producers outside cartel countries were not increasing rapidly. We should like to know the extent to which the foreign sales of Aluminium Limited were enlarged without price competition, through additional sales expenditure or quality competition. Apparently price competition took place in Japan and not in Europe. Unfortunately, for reasons explained, it cannot be determined exactly how much metal Aluminium Limited sold in Europe proper and how much in other countries (outside Japan and North America). From 1923 until 1927 the United States and Germany exported nearly equal amounts of half-products to India. Beginning in the latter year exports from the United States grew markedly, while those of Germany fell off.[51] The exports from the United States may have been chiefly metal belonging to Aluminium Limited fabricated in

[50] Perhaps they were declining more than the figures of the table show. It is impossible to discover the origin of exports from Norway to Japan, which amounted to about 1,000 tons per year, 1928–1930. In the table they are credited entirely to European firms.

[51] United States exports to India ranged between 1,200 and 1,600 tons in the years 1924–1926, while those of Germany were from 1,000 to 1,300 tons. In the four years 1927–1930 half-products sent from the United States amounted to between 2,000 and 3,000 tons per year and those from Germany diminished to less than 1,000 tons.

this country. In the past few years the Ottawa agreements have aided England and Canada, which have been the chief exporters of half-products to India.[52] In this instance of oligopoly it appears that Aluminium Limited enlarged its share of the market, in some places by price competition, in others by such methods as increased selling expenditures or quality competition.

It is impossible to ascertain exactly how the Alliance cartel has affected the shares in the world market outside the United States. Aside from the great increase in German output a few tendencies of recent years are evident. In 1933 and 1934 Canadian exports to markets outside the United States were larger than in any previous year with the exception of 1929. Since 1930 Canadian exports to Japan have maintained their lead. Swiss exports in 1933–1935 had not yet reached the level of 1925–1929. English exports in 1933–1934 remained well below the figures of 1925–1931, while German exports have been less than 1,000 tons since 1933. Of the cartel countries Germany and Great Britain have continued to be the chief importers. The relative increase in consumption in England in 1935 may reflect chiefly rearmament demands. Evidently the tremendous increase in demand in Germany in 1934 and 1935 was due to rearmament and the policy of self-sufficiency. Prediction of future relations would be well-nigh worthless.

Detailed analysis of the relations of investment and demand in the several countries of Europe is without the compass of this book. In closing this chapter a few of the characteristics of each national market may be summarized. France has shown the nearest approach to the simplest case of private monopoly producing only for a national market. Here it would appear that in the post-war period effective demand was not pushed forward as swiftly as in some other countries, capacity was expanded less rapidly, and profits were greater. In England high costs and free trade helped to keep aluminum capacity relatively small until slowly maturing plans involving government encouragement brought a marked increase in the late twenties. Burdened with some uneconomic capacity resulting partly from the war, the German government corporation seems to have pursued policies quite similar to those of private business enterprises. Profits were good, perhaps quite high in the years following establishment of

[52] MW, xv, 419 (May 8, 1936).

the new currency. Rapid expansion of consumption in Germany was shared by other cartel members owing to the lack of protection and the relatively low sales quota of the German producers. Until the creation of Aluminium Limited the Swiss company was the most international aluminum enterprise in location of plants and markets. Hence it suffered most from the growth of nationalistic tendencies. Investment devoted to the home market was apparently restricted enough to yield very good profits. Economic nationalism in Italy brought forth a substantial quantity of high-cost production.

CHAPTER XIV

RATIONALIZATION IN THE SHORT RUN

It is manifest that the economic welfare of a community would be improved by reduction of the amplitude of periodic fluctuations in total output, real income, and employment, and by a lessening of any tendency to greater inequality between income groups which may occur with such fluctuations. The sort of adjustments to fluctuating demand by an individual firm or a group of firms which tend to minimize the wastes of underutilization of economic resources may be called policies of short-run rationalization. Rationalization policies will often, although perhaps not always, operate against tendencies toward less equal sharing of income.

Fluctuations in estimated output of aluminum ingot in Europe and America are shown in Tables 25 (p. 308) and 38 (p. 569). Price data appear in Tables 13 and 14 (pp. 240–243), Table 19 (p. 287), and in pages 265 ff. Few figures for changes in employment in the aluminum industry are available.[1] Except during periods of marked business depression the steady growth in output has been interrupted only by slight declines, due probably to variations in stream flow, abnormal conditions, minor business recession, or changes in the circumstances of market control. Except in Germany reduction works seem to have been operated close to practicable capacity at all times other than depression. During prosperity changes in the prices of aluminum ingot seem to have been less frequent and of smaller amplitude than price changes in many other industries. The even growth in production and the relative stability of price are to be explained partly by the steady growth in demand, influenced in increasing degree by the activities of producers, and partly by variations in inventories. It is to be doubted, however, that these results could have been accomplished if the number of new entrants had been much larger, or if the few producers had adjusted investment to demand with less forethought and prudence. From the standpoint

[1] Requests for employment figures were addressed to some of the leading companies, who replied that they were unable to furnish this information.

of general economic welfare the benefits of short-run stability during prosperity are tempered by the disadvantages of restriction of investment. Given appreciable fluctuations in demand[2] some underutilization of equipment from time to time is unavoidable if ideal investment is to exist, and considerable price flexibility is probably necessary to obtain maximum desirable utilization at all times. The rest of the chapter will be devoted to the important questions concerning the adjustment of output and price to reduced demand in depression.

Theoretical analysis of the total consequences of different sorts of price and output policies is complicated by the fact that the policies of any one industry exert an influence of greater or less magnitude upon conditions in other industries. Mr. D. H. Robertson has shown that the course of the business cycle may be appreciably affected by the price and output policies of producers of basic capital goods.[3] As a result of the growing adoption of aluminum for equipment and structures in the last ten or fifteen years its demand has become more susceptible to the fluctuations of the cycle than was true in the earlier life of the metal, and the policies in this industry now have greater significance for the general welfare.

With a given distribution of money income the real income of the community will in general tend to be greater the closer the correspondence between prices and marginal costs for all commodities. Prices which exceed marginal cost in greater degree during depression tend to reduce real income more then than at other times. Prices above marginal cost in depression may also tend to lessen aggregate output and employment in the community, or to prolong depression by inducing more hoarding than would occur with lower prices.[4] Given quite inelastic demand through a range of prices above marginal cost, a price equal to marginal cost will enable slightly larger consumption of the article and a little more employment in its production than any higher price on the

[2] Although producers can often influence the long-run growth of demand it is to be doubted that they can ordinarily affect greatly the degree of fluctuation from the trend.

[3] See especially *Banking Policy and the Price Level* (London, 1926).

[4] In an essay "Monopoly Prices and Depression," in a recent volume entitled *Explorations in Economics* (New York, 1936), I develop briefly the theoretical principles here summarized in part.

inelastic stretch. More important, since consumers as a whole will spend a much smaller sum on this commodity than they would spend if its price were higher, they will be able to spend more on other things, with the probable result of larger output and employment in these industries than would obtain otherwise. Furthermore, the higher the price on an inelastic range the greater the revenue of the producing firm. Some part of this revenue may be retained in liquid reserves which would not be hoarded if it went instead to labor in other industries.

In the case of elastic demand at prices above marginal cost the probabilities are not so clear. In such a market a price well above marginal cost will be attended by much less consumption and employment than a price equal to marginal cost. But if consumers as a whole spend in other markets the full amount of the difference between what they now spend at the higher price and what they would spend at a price equal to marginal cost, total money value of consumption will be unaffected and aggregate employment in the community may not be appreciably different. High prices in such markets may, however, result in more net hoarding in the community. Where substitutes are not sufficiently satisfactory in quality and price, firms may accumulate liquid reserves for replacement and extension, or individuals may postpone replacement of durable consumers' goods and set aside funds for their future purchase. Finally, a disproportionate decline in the prices of raw materials and finished goods in depression, which increases the number of bankruptcies and receiverships among finishing firms or enhances the fear of failure, may lead to larger cash hoards and smaller payments to laborers and investors. In the absence of ability to determine the total consequences of any policy by a given firm or group of companies it would seem that reduction of price to marginal cost would be more likely to benefit the whole economy than any other policy.[5] In the case of industries whose basic product is sold in markets with differing elasticity of demand, there seems to be no more justification for discrimination during depression than at any other time. Although a discriminatory price schedule may be accompanied by larger

[5] In many instances, of course, lowering of marginal costs or government subsidy would be required to prevent any reduction in output and employment below their pre-depression levels.

output and employment than a uniform price yielding the same revenue, it will probably not give as much output and employment as a uniform price equal to marginal cost. Reduction of price early in depression to a figure equivalent to the marginal cost of the estimated amount which could be sold at that price might, if it were accompanied by a definite announcement that no further price reductions could be expected for some time, have the further advantage of materially diminishing the tendency of buyers to hold off in anticipation of still lower prices in the near future. Absence of variation in price over several months may be desirable if the price is at the right level.

We cannot discover the full consequences of the price and output policies of aluminum firms during depressions. Owing to inadequate information and inefficient analytical tools, the results within the aluminum industry itself are not clear, to say nothing of the difficulties of tracing repercussions elsewhere. In the following pages the available information will be summarized. It should be borne in mind that the average prices actually received for aluminum in depression were probably less than the list quotations of most companies. It has been impossible to ascertain the extent of divergence.

The Aluminum Company of America had sold No. 1 ingot at 33 cents during the first five years of the century. Beginning in 1905 price was raised gradually to about 40 cents in 1907. In 1906 and 1907 aluminum was very scarce and the open-market price went above the Aluminum Company's quotations. After the crisis it appears that the company lowered its price to 33 cents and shut down two-thirds of its greatly enlarged reduction capacity.[6] Substantial reductions in the European price in 1908, which the cartel was unable to prevent, were followed by greater cuts after dissolution of the agreement.[7] In the next two or three years price in Europe seems to have fluctuated above and below the average costs (without profit) of most of the companies. We have seen that the low prices in Europe brought a great extension of the market and spurred producers to efforts which further enlarged it. Output and, it may be inferred, employment actually increased in Europe during depression. A comparison of production esti-

[6] MI, XVII, 15, 16 (1908).

[7] Above, pp. 38 and 265 ff.

mates with the capacity estimates of Tables 3 and 11 suggests that utilization of a greatly expanded capacity was around 50 per cent in 1908, about 60 per cent in 1909, 75–80 per cent in 1910, 70 per cent in 1911 after the completion of the large Rhône plant, and perhaps 90 per cent in 1912.

Annual American output in 1908 and 1909 seems to have reached only about half the total for 1907, although it slightly exceeded the average production in 1905–1906. Since American capacity was tripled between 1905 and 1908, it appears that little more than one-third of capacity was used here in 1908–1909. With small offerings of European metal in this country before dissolution of the cartel the Aluminum Company lowered its price to 28 cents. At the end of 1908 it apparently quoted 24 cents, and during the following year, when imports of more than 2,200 metric tons were sold here for about 21 cents, the 7 cent duty included, its quotation went to 22 cents.[8] Imports increased to nearly 5,500 tons in 1910 and continued to grow until the war.[9] At the end of 1911 the price of foreign metal in New York sank to a new low of 18.5 cents, duty included. The Aluminum Company was meeting the prices of foreign metal to some extent, although it appears that it did not cut its formal quotations to such a point.[10] Profits of the leading firms were evidently fairly good during the years 1908–1912.[11] Much lower prices than had existed before 1908 brought large increases in consumption on both sides of the Atlantic. While stocks accumulated somewhat in the United States, they were swept off by a spurt in demand in 1912. Under the influence of bustling demand and reconstitution of cartel control in Europe the price of foreign aluminum in the American open market rose to about 25 cents in the last quarter of 1912.[12] The Aluminum

[8] MI, XVIII, 17 (1909). Prior to 1909 imports had never reached 500 tons per year and had averaged about 270 tons annually. Most of this probably came from Canada.

[9] It is impossible to discover what portion of the imports came from the Canadian plant of the Aluminum Company or what proportion of total sales in the United States went to European firms. Countries of origin are not distinguished in United States import statistics until 1912 and production data afford no reliable indication of sales. Canada exported about 2,700 tons annually, 1909–1911, and three times as much in 1912 (United States Tariff Information Survey, 1921, C-16, p. 45). It does not appear how much of this came to the United States.

[10] EMJ, XCIII, 970 (May 1912); XCIV, 529 (September 1912).

[11] See above, pp. 226 and 266 ff.

[12] Tariff Hearings, 1913, House Doc. no. 1447, pp. 1485–1486.

Company kept its price at 21 cents. It is doubtful that price would have fallen as far or consumption have been as large on either side of the ocean if effective cartel control had existed throughout the depression.

The experience of 1920–1922 was somewhat similar to that of the earlier depression. When demand fell off in the autumn of 1920, salesmen of the Aluminum Company were instructed to revamp contracts for 98–99 per cent ingot for 1921 at the price of 32.8 cents which had prevailed since early 1919, except for a few months in 1920 when it had been slightly higher.[13] In Europe, where effective agreement had not yet been reached, price was equivalent to about 28 cents a pound in 1920 and fell to about 19 cents in 1921–1922.[14] Once again the American market, now accessible over a tariff of 2 cents only, became the recipient of large quantities of European metal. Imports of over 12,000 tons from foreign producers in 1921 amounted to half the estimated output of primary aluminum in the United States. In the following year shipments from foreigners were apparently 2,000 tons larger than in 1921. Imports of ingot from European producers were apparently equivalent to about 35 per cent of total sales of primary aluminum in all forms here in 1921 and about 20 per cent in 1922.[15] The New York open-market price, which gives some indication of the quotations at which foreign metal was sold here, sank from 32 cents in the middle of 1920 to 24 cents at the end of the year and continued to decline steadily to a low of 17–17.5 cents early in 1922. The Aluminum Company's list price followed the open-market quotation downward with a substantial lag. Between August and November 1921 its schedule price of ingot was suspended. List quotation was resumed at 19 cents, where it remained until the duty was raised in September 1922. The open-market price continued to be 1–1.5 cents below the Aluminum Company's list price until 1923.

Total European output declined from a yearly average of about 56,000 tons in 1919–1920 to an average of 46,000 tons in 1921–1922, a reduction of less than 20 per cent. In the United States it is estimated that the yearly average fell from about 60,000 tons

[13] BR, p. 225; BMTC v. ACOA appellant, Exhibit 344.
[14] Prices in London and Switzerland have been taken as typical.
[15] See n. 42, p. 81 and note to Table 27.

in 1919–1920 to about 30,000 tons in the next two years.[16] Estimates of the totals of output in the United States plus imports from Europe show a larger decline in this depression than estimates of European output minus exports to the United States. The payroll of the Aluminum Company of America was reported to have dropped from a total of 21,000 persons on January 1, 1921, to 8,000 within the next eleven months.[17]

The AIAG and the British Aluminium Company both returned earnings in each of the years 1921 and 1922 equivalent to about half the annual earnings in 1919 and 1920. The Compagnie AFC earned a small return in 1921 and a moderate one in 1922. The Aluminum Company of America suffered a loss in 1921 and earned a small return the following year.

The situation in the first few post-war years was so chaotic that it is difficult to determine the extent to which aluminum producers pursued competitive or oligopolistic policies. The price decline was influenced by government sales of stocks acquired before the armistice. Recovery from war materials enlarged the supply of secondary metal. Currency depreciation aided some European firms in keeping output at a high level.[18] It is possible that the smallness of the drop in output in Europe was due chiefly to factors other than price competition, but the history of cartels in this industry affords little basis for the supposition that price would have been reduced as far if effective market organization had existed.

It appears that the European firms engaged in active price competition with the Aluminum Company of America, which did not retaliate by undercutting but attempted to brake the fall of price

[16] According to the estimates the proportionate drop in Canadian production was not so great. Figures for total American production have not been used at this point because there is reason to doubt the Canadian estimates for these years.

[17] United States Tariff Commission, *Digest of Tariff Hearings before the Committee on Finance, United States Senate* (1922), p. 230. In France the average number of employees in aluminum reduction works was given as 1,115 in 1919, 1,100 in 1920, 633 in 1921, and 852 in 1922 (Ministère des Travaux Publics, *Statistique de l'industrie minérale*). These figures are not comparable with those of the Aluminum Company, which include employees at all stages. Unemployment is probably greater in the stages beyond ingot, in many of which operations take the form of production to specification.

[18] This was particularly true in Germany. Although little ingot aluminum was exported from Germany in these years large amounts of German aluminum ware were sold in England and America.

by acting rather more oligopolistically. The lag in the downward movement of its list quotation also suggests that price would not have gone as low here in the absence of price competition. The Aluminum Company could probably have diminished the fluctuation in its own output and employment if it had stabilized price temporarily at a figure around 20 cents late in 1920. Although some part of the imports was doubtless prompted by a desire to accumulate stocks in this country in anticipation of a rise in price following an increase in the duty, it is probable that the low price here in 1921 and 1922 resulted in much larger consumption than would otherwise have existed. The demand from the automobile industry, which then used large quantities of aluminum, was probably quite elastic, owing to the existence of substitutes. In this depression price competition in this country evidently tended to lessen the fluctuation in consumption of aluminum here and the fluctuation in output and employment in the European aluminum industry. If the Aluminum Company had taken the initiative in drastic price reduction the latter result might have obtained here. The benefits of larger consumption of aluminum and more employment in its production, which ensue from price reduction in markets with elastic demand, might have been nullified from the standpoint of the whole economy by corresponding losses in the industries producing substitutes. It should be recognized, however, that when the same commodity is sold also in markets with inelastic demand, a uniform price reduction will tend to confer net benefits upon the whole economy because larger sums may be spent on other goods and hoarding may be less. In the absence of price competition reductions may be confined to some extent to markets with elastic demand, with the possible result of no decided gain for the economy as a whole.

If price had emerged from the depression at a much higher level, it might have moved on a somewhat higher plane for several years. The growth in consumption beginning at the end of 1922 and the consequent scarcity in 1923–1924 might have been much less striking, and hence the marked elasticity of demand at lower prices much more tardily appreciated.

During the recent depression market control by the international cartel was not only maintained, as it had not been in previous depressions, but strengthened. At the beginning of the depression the cartel price was £95 per long ton, equivalent to 20.6 cents per pound. In October 1930 the quotation was lowered to £85; and coincident with the effectiveness of the Alliance agreement at the beginning of 1932 it was reduced to £80 gold per long ton, or about 17.4 cents per pound. According to report, world stocks had increased by this time to about 150,000 tons, approximately two-thirds of which were in the United States.[19] The Aluminum Company's list price had remained unchanged since the end of 1927 except for a reduction of 1 cent when the tariff was lowered from 5 cents to 4 cents per pound in the middle of 1930. Regulation of output by the Alliance cartel and its provisions for financing of stocks which kept them in the hands of producers seem to have enabled fairly effective price control.[20] During 1932 and 1933 the total estimated output of Europe plus Canada was only a little over 100,000 tons per year, or about 50 per cent of an estimated capacity of 205,000 tons. The increase in stocks of Alliance members was halted in 1932, and thereafter they were diminished each year by sales in excess of output.[21] In the spring of 1936 stocks of the Alliance were said to be no more than enough to meet a half-year's demand.[22] Stocks in the United States evidently failed to increase after 1932 and diminished in 1934 and 1935,[23] with the result that world stocks were lowered to 100,000 tons by the end of 1935.[24]

The official cartel quotation of £80 gold per long ton, or 2 Swiss francs per kilogram, remained unchanged, except in England, until 1934. After England left the gold standard in the fall of 1931 the London price was quoted at £95 paper. In September 1932 it was raised to £100 paper,[25] where it remained in the fall of 1936. Stability of the price in paper pounds was equivalent, of course, to a separate fluctuating price in gold as the pound changed its

[19] *Metall und Erz*, XXVIII, 517 (November 1931), and information from the League of Nations.

[20] Cf. a statement in the annual report of Aluminium Limited for 1932: "World stocks of aluminium are not excessively large. They are in firm hands and do not weigh unduly upon the market."

[21] MW, XII, 233 (April 21, 1933), XIII, 268 (April 13, 1934), XIV, 334 (April 26, 1935); AG, XXX, 693 (June 23, 1934).

[22] MW, XV, 423 (May 8, 1936).

[23] This is inferred from comparison of the estimates of output and sales given in Table 27, p. 325.

[24] MW, XV, 312 (April 3, 1936).

[25] MW, XV, 1032 (May 29, 1936).

gold value. Other separate domestic prices emerged later. In November 1934 price in Germany was lowered from 1.60 RM to 1.44 RM per kilogram, where it remained in October 1936.[26] In October 1934 the French price was reduced by 1 franc to 9.50 francs per kilogram.[27] It appears that these quotations were still given in the middle of 1936, when the international price applicable to countries outside the group with special domestic prices was still quoted at the figure of 2 Swiss francs established in 1932. To some extent, however, the international price represented only a nominal quotation throughout the depression. Sales were made in Russia at figures well below the official price. Annual reports of the AIAG indicate that most sales abroad had to be made at much lower prices than sales at home.

The official cartel price of aluminum in 1932 had been lowered only about 20 per cent below the average for the years 1927–1929. This was about the same proportionate reduction as that exhibited by a group of commodities in Germany subject to cartel or trust control, while prices of a group of goods sold in Germany without organized market control had fallen to 50 per cent of their average for the last three years of prosperity.[28] Average receipts per kilogram were less than the official cartel price because of a discriminatory price structure in home markets, as well as foreign sales at low prices. Reference to the ratios of Table 20 discloses that little aluminum could have been sold at the published price in competition with copper during the depression. With a nominal price of £80 gold per long ton in 1932, consumers who could substitute copper were paying only about £55 for the lighter metal in one country. £10 to £15 more was received from the automobile industry and still higher prices from markets in which substitution was less satisfactory, with the result that the average price from all markets was £70–£75. Evidently price was reduced by different amounts in segments of the market having different elasticities of demand. This could not, of course, have occurred with energetic price competition.

Total European output of primary aluminum declined very

[26] *Ibid.*

[27] AG, XXXI, 782 (June 22, 1935). Price in Italy also moved separately.

[28] The index numbers for *Freie* and *Gebundene Preise* given by the Institut für Konjunkturforschung have been reduced to a 1927–1929 base for this comparison.

little in 1930. In the following year it amounted to a little over 80 per cent of the 1929 figure. Curtailment in 1932 and 1933 carried output in each of these years down to about 65 per cent of the 1929 high. Before the great increase in production in Germany beginning in 1934 the proportions in which output was shared between the European members of the cartel had changed markedly. Estimated output in the leading countries in 1931 and 1932 was equivalent to the following percentages of their yearly average production in 1929–1930:

	1931	*1932*
France	65	55
Germany	85	60
Switzerland	55	41
England	101	74
Norway	87	72
Italy	135	180

The dependence of the Swiss on a contracting foreign market, the comparative mildness of depression in England, and the determination of Italy to expand output are reflected in the figures. One might infer that the British received a larger quota when the Alliance was formed and the Swiss a smaller one.

The leading European aluminum companies have shown earnings on capital in every year throughout the depression. It will be recalled that the rates of return appearing in Table 18 tend to understate the return on depreciated investment in the case of the Compagnie AFC and the AIAG. In 1931 the Compagnie AFC reduced its depreciation charge to an amount equivalent to about 4 per cent of original cost of fixed investment. In the following years it averaged a little less. If these charges fell short of true depreciation the rates of return in the table may represent no understatement. The amortization charge of the AIAG, which had been averaging about 2 per cent of original cost of fixed investment in the latter twenties, was reduced to about 1 per cent during the years 1931–1933. If it had been maintained at 2 per cent the rates of return upon investment would have been reduced by only about a half of 1 per cent. In the case of the Vereinigte Aluminiumwerke the ratio of depreciation charges to fixed assets was not much reduced in depression. Earnings reported for 1933 and 1934 would have been much larger if it had not been for special write-offs in those years.[29] In 1931 and 1932 profits reported by the British Aluminium Company averaged nearly 80 per cent of the average profits of the years 1926–1929. This percentage fell to 40 in 1934 but rose to nearly 60 in the following year. In view of the severity of the depression, the profit record of the European aluminum companies is striking.

In its first full year of operation Aluminium Limited seems to have earned almost 6 per cent on total net depreciated assets of about $65,000,000.[30] Average annual investment grew slowly to about $70,000,000 in 1932 and thereafter declined to about $68,500,000 in 1935. In 1930 a tonnage exceeding current output, estimated at 34,000 metric tons or slightly larger than in 1929, was sold at lower prices yielding earnings of a little over 3 per cent of investment. In each of the next two years sales were approximately 27,000 tons. Reduced consumption and lower prices resulted in earnings of only about one-half of one per cent of investment in each of these years. In the four years 1932–1935 output averaged scarcely 50 per cent of capacity. Higher prices incident to appreciation of foreign currencies brought a rate of return of about 1.5 per cent in 1933, although sales were less by 3,000 tons than in the preceding year. With reductions in cost and continued improvement in demand the company showed earnings of over 3 per cent in 1935. Depreciation charges were about 4 per cent of gross value of land, plants, and facilities in 1929 and 1930. In the next three years they were reduced to a little over 3 per cent. In each of the years 1931 and 1932 the difference between the amounts charged to depreciation and sums equivalent to 4 per cent were just about equivalent to the small earnings shown. In 1933 a 4 per cent depreciation rate would have reduced the rate of return on investment to about 1 per cent.

During the recent depression, imports into the United States from European companies diminished instead of increasing as in earlier depressions. Shipments to this country from European producers of virgin may have declined a little more or less than is indicated in the figures of Table 27, which include imports of

[29] In 1933, 3.3 million marks were deducted for equipment partially or wholly abandoned, and in 1934 inventories or short-lived equipment were written down 1.5 million marks.

[30] Information presented in this paragraph has been taken from annual reports of the company and the estimates of output in Table 38.

TABLE 27

STATISTICS OF OUTPUT AND SALES OF ALUMINUM IN THE UNITED STATES, 1926–1935

(*Metric Tons*)

(1) Year	(2) Estimated Output of Primary Aluminum Ingot in the United States *	(3) Estimated Sales of Aluminum in All Forms in the United States by Aluminum Company of America †	(4) Imports from European Producers ‡	(5) Estimated Total Sales of Primary Aluminum in All Forms in the United States § (3) + (4)	(6) Imports from European Producers as Proportion of Total Sales in the United States (4) ÷ (5) (*per cent*)
1926	66,850	77,550	18,400	95,950	19.18
1927	74,200	72,630	13,830	86,460	16.00
1928	95,500	103,760	7,020	110,780	6.34
1929	103,400	101,070	8,020	109,090	7.35
1930	103,900	64,250	6,050	70,300	8.60
1931	80,530	47,490	5,090	52,580	9.68
1932	47,600	31,110	2,720	33,830	8.04
1933	38,600	39,105	6,080	45,185	13.46
1934	33,646	53,370	—	—	—
1935	54,113	72,000	—	—	—

* From Table 39.

† Total sales of Aluminum Company minus (1) total exports from the United States of aluminum ingot, plates, sheets, tubes, castings, etc., and (2) exports of ingot, blocks, bars, etc., from Canada to countries other than the United States in the years 1926–1928. (Since the Canadian plants were turned over to Aluminium Limited in the middle of 1928 only one-half of the Canadian exports of that year were subtracted. Exports of aluminum utensils and some other manufactured articles from the United States cannot be deducted because government figures report only value in these years. For the same reason exports of fabricated aluminum from Canada cannot be subtracted. In neither case would it appear that the tonnage exported was large in the years covered by the table.) Total sales of the Aluminum Company, 1926–1931, are from BMTC v. ACOA appellant, Exhibits 288–293; 1932–1934, estimated from statements in each annual report on the proportionate increase or decrease in sales compared to those of the preceding year.

‡ From Table 15.

§ Includes imported scrap and secondary aluminum and any scrap entering into articles sold by the Aluminum Company. Excludes a small quantity of imported fabricated aluminum.

scrap and secondary aluminum.[31] According to the estimates in the table, imports from European producers amounted to a somewhat larger proportion of total sales of virgin aluminum in all forms in the United States during the years 1930–1932 than in 1928 and 1929, but a much smaller proportion than in 1926 and 1927. A conspicuous increase is indicated for 1933, perhaps explainable as a wager on inflation in this country. No figures for imports from European producers in 1934 and 1935 are available. Inasmuch as allocation of the total amount of imports in 1934 to the European firms would give them a percentage of total sales almost identical with that of 1933, it is probable that their proportion dropped somewhat. Apparently the percentage fell farther in 1935, because total imports increased in less proportion than estimated sales.

The New York open-market price, which in the main seems to be a "spot" quotation, remained equal to the list price of the Aluminum Company throughout the years 1930–1933. (See Table 14.) In 1934 it seems to have fallen by about 3 cents. Although it does not appear that the list price of the Aluminum Company was lowered in 1934, the *American Metal Market*, which had formerly reported the company's list price, began early in 1934 to report lower figures for the company's price. If the average price paid by consumers of ingot was in fact lower in 1934 and 1935 this may have been due to price competition. Evidently the list price of the Aluminum Company was unchanged between the end of 1927 and November 1934 (the last date for which we have definite information), except for a reduction of 1 cent in 1930 corresponding to the reduction in the tariff.[32]

To some extent the stable list price of the Aluminum Company has been merely a nominal quotation during depression.[33] According to testimony the leeway given salesmen to sell at prices below the list to meet particular circumstances has been much

[31] Official statistics do not separate imports of the primary metal. An official of the Aluminum Company testified in 1933 that imports had been largely virgin aluminum (BMTC v. ACOA appellant, fol. 1476).

[32] The list prices of 22.9 cents for 98–99 per cent ingot and 23.3 cents for 99 per cent ingot established in June 1930 were maintained through November 1934, according to data in Exhibit 556 in BMTC v. ACOA appellant.

[33] *Ibid.*, fols. 5725–5730.

greater during depression than at other times.[34] Data are not available to show the extent of actual price reductions during the most severe part of the depression. It appears that the company's average sale price per pound of ingot of all grades fell from about 22.8 cents in 1928 and 1929 to 22.3 cents in 1930 and 21.2 cents in 1931, an indicated decline which exceeded the reduction in list prices by less than a cent.[35] Evidently cuts from the list became larger as conditions grew worse. An examination by representatives of the NRA of the record of ingot sales of the Aluminum Company during a part of 1934 indicated that different prices were charged to different groups of purchasers in accordance with the degree of price competition from substitutes.[36] Price competition from producers of secondary aluminum probably had the same sort of effect as that of other substitutes. The differential between published prices of primary and secondary ingot fell from 1–2 cents in the years 1922–1929 to 5–6 cents in 1931–1932.[37] There was some price competition from European producers.[38] The decline in imports from European firms and the continuous correspondence between the New York open-market price and the list price of the Aluminum Company throughout the years 1930–1933 indicate, however, that it was less severe than in earlier depressions. Existence of different prices to different groups in 1934 shows that neither competition from European producers nor from makers of secondary ingot was strong enough to produce a uniform reduction.[39]

In 1930 and 1931 the output of the Aluminum Company was maintained close to the 1928–1929 level, and stocks accumulated rapidly. (See Table 27.) The company borrowed $27,500,000 to keep a high rate of operations and sustain employment.[40] As sales continued to drop, output was apparently curtailed in 1932 to less than 50 per cent of the amount produced in each of the years 1929 and 1930, and was further restricted in 1933 and 1934. In 1933 about one-third of capacity was being utilized and

[34] *Ibid.*, fols. 1366–1370.
[35] *Ibid.*, Exhibit 449.
[36] HR, p. 14.
[37] Prices from *American Metal Market* as reported in HR, p. 7.
[38] *Ibid.*, pp. 4, 8, and 15; BMTC v. ACOA appellant, Exhibits 347–397 and 556.
[39] Cf. HR, p. 14.
[40] BMTC v. ACOA appellant, fol. 5728.

stocks had grown to 135,000 metric tons.[41] In the two succeeding years an improvement in sales which was not matched by the increase in production carried inventories well under 100,000 tons.

According to estimates based upon annual reports, the Aluminum Company received an average return of 2.6 per cent on total assets in the five years 1930–1934. (See Table 12.) With the exception of a small deficit in 1932, some earnings were shown in every year of the depression. Reductions in the ratio of annual depreciation charges to the gross value of fixed plant explain only a small part of the earnings during depression, if any.[42]

That part of investment actually used during depression may be estimated roughly by applying to the investment figures of Table 12 the ratio of ingot output to ingot capacity or of tonnage sold to capacity in each year. If the ratio of output to capacity is employed, the indicated rates of return upon investment used in the five years 1930–1934 are 5.8, 4.2, –0.6, 4.7, 13.1, making an average of 5.4. Use of the ratio of tonnage sold to capacity gives the rates of return as 8.2, 6.7, –0.8, 4.3, 7.7, making an average of 5.2. Since the figure for "investment used" in each year still contains its proportionate part of the idle ore and power reserves included in total investment, these figures tend to understate the returns upon the real amount of investment used. The rates of return upon that part of investment used in producing aluminum which was *sold* during these five years must have been much larger still, for expenses amounting probably to several million dollars were incurred to produce nearly 50,000 tons of metal which was not sold. The reported earnings represent, of course, revenues

[41] *Ibid.*, fol. 1282.

[42] In 1930 and 1931 the annual charge to the lump reserve labeled "depreciation, depletion, and workmen's compensation insurance" bore approximately the same ratio to gross value of land and plant at the end of the year as in 1927–1929. In each of the following three years charges to this reserve, although larger, bore a somewhat smaller ratio to plant account, which was increased in 1932 by the transfer of assets formerly represented by securities of non-consolidated subsidiaries. If the 1927–1929 ratio had been maintained throughout 1932–1934, the average rate of earnings for 1930–1934 would still have remained above 2 per cent. Beginning in 1932 the account was labeled simply "depreciation and depletion." If no reserve for workmen's insurance was included in it, the ratio of depreciation to fixed investment may not have declined at all during depression.

after subtraction of costs incurred in producing metal which went into inventory as well as metal sold. These considerations suggest that an income sheet which isolated expenses and revenues for the metal sold would show for 1930, 1931, and 1933 earnings at least equivalent to a normal return upon investment used to produce this part of the product, and might show an average rate of return for the five years 1930–1934 which was not below normal. Given less than best utilization of capacity, a normal rate of return upon that part of investment whose product is sold indicates that the average price received is above marginal cost of the output sold.[43] It is impossible to discover whether the company produced in the years 1930–1932 more metal than could have been sold at prices equal to its marginal cost.

Whether price went down to marginal cost in Europe or the United States in 1908–1910 and in 1921–1922 we do not know. One might infer that it approached marginal cost more closely in the earlier depressions than in the years 1930–1934, and that price cuts were more nearly uniform in sections of the market with differing elasticities of demand in the earlier depressions. In the recent depression price reductions in markets with the more elastic demands enabled larger consumption and probably more employment in the aluminum industry than would have occurred with higher prices. Whether prices in these markets fell to marginal costs cannot be determined. It appears that prices in some other markets were not lowered as much. The earnings records of nearly all the leading companies suggest that prices in some markets were above marginal cost unless those in others went below it, which is to be doubted. If prices in markets with quite inelastic demand remained high, there was a tendency to intensify depression because part of the sums spent upon aluminum might have maintained larger output and employment in other industries, without

[43] This can be shown as follows. In addition to the familiar curves of total average unit cost, average variable cost, and marginal cost, let there be drawn a second curve of total average unit cost upon the assumption that aggregate capital costs are reduced in the same proportion as output, or, in other words, that capital cost per unit remains constant at the amount given by best utilization. This curve, which parallels the average variable cost curve at a distance above it equal to the constant capital cost per unit, will lie above the marginal cost curve until it meets the latter at the point of best utilization. Hence any price upon this curve, which would, of course, yield normal returns on the investment used, will be above marginal cost.

diminution in consumption and employment in the aluminum industry, had the prices of aluminum been lower; unless, indeed, these sums were used to produce for inventory in the aluminum industry. In the first part of depression production of most firms exceeded sales appreciably, but this does not seem to have occurred in Europe after 1931 or in the United States after 1932. The financial reports of the aluminum companies give very little indication of hoarding. No further information upon the question of hoarding is available. On the basis of the evidence here surveyed it would appear that lower average prices and greater uniformity as between markets with differing elasticities of demand would have given better rationalization than the policies followed by the aluminum companies in this depression.

CHAPTER XV

CONCLUSIONS AND POSSIBILITIES

1. Theoretical Comparison of Different Types of Market Control

Some of the analysis of the last six chapters may be summarized in the following conclusions, which must be regarded as tentative, owing to limitations of data and method. Since the earlier years of the aluminum industry single-firm monopoly has not been required for a high degree of efficiency in the production and marketing of ingot except in the smaller markets of some European countries. In the United States it appears probable that the investment maintained by the Aluminum Company has not ordinarily fallen far short of an amount proper to maximum profit with current effective demand. It may have been larger at times, particularly in the late twenties. The rate of earnings upon total assets seems to have been much lower in good years during the post-war period than before. This result may be partially explained by the growth of idle investment in reserves of ore and power. For the rest, it was apparently due to the influences which account for a tendency during the twenties towards a progressively lower rate of return upon the investment devoted to some particular markets. Greater elasticity of demand for aluminum in several new uses and in some older employments was a significant factor. Overestimates of effective demand, a policy of selling at low prices to encourage rapid development of demand, unavoidable provision of excess facilities in some fabricating branches, or expansion prompted by nonprofit motives may have contributed in varying degree to these results. The estimates of earnings given in Chapter XI do not suggest, however, that operating investment was as large as ideal investment with respect to current effective demand in all markets. Effective demand was pushed forward markedly.

In the first seven or eight years of the century the expansion of capacity of the few European producers did not keep step with a rapid growth of demand, which they had done little to bring about, with the result that their investment may even have been too small for maximum profits. High prices and large profits may have been facilitated by a rather weak cartel. After the cartel had disintegrated in 1908 as a result of internal friction, influx of new firms, and depressed markets, a much better relation between demand and a greatly expanded investment was brought about by low prices and aggressive cultivation of new markets. The post-war history of this industry in Europe has been characterized by a tendency towards competitive adjustments in expansion of capacity, in accumulation of reserves of bauxite and power, and in development of new adaptations of the basic product, which have interacted to push effective demand ahead rapidly and to keep total European operating investment somewhat larger than amounts appropriate to maximum profit. The relation between investment and current effective demand seems to have become progressively less restrictive during the twenties in those markets where elasticity of demand was increasing and in foreign markets where price competition prevailed part of the time. Whether the same tendency existed in other markets is not clear. It appears that the rates of earnings of the leading European companies were lower in post-war years than earlier, but it is probable that they remained somewhat above normal earnings and possible that they were much above this standard. Until 1932, cartel agreements were not strong enough to exercise a dominant influence upon the long-run relations of investment and demand. Further, neither the nature of technological conditions and industrial structure in the aluminum industry nor the circumstances governing entry to the field are such as to lead to the belief that serious overinvestment would occur in the absence of agreements between producers.

During prosperity nearly all reduction works have always been operated at practicable capacity. Prices have in general exhibited a relatively high degree of stability, both with respect to frequency and amplitude of change. In the depressions of 1908–1909 and 1921–1922 price competition, originating among European companies, resulted in great reductions in prices on both sides of the Atlantic, which probably enabled larger consumption and employment in this industry than would have existed with stronger control of price. The Alliance cartel was evidently able to preserve effective price control during the recent depression. Although some price competition occurred in the United States, it was apparently less severe than in earlier depressions. Both in Europe and in America price reductions seem to have taken the form of varying concessions from list prices in markets with differing elasticities of demand. If prices had been lower in markets with inelastic demand, consumers would have spent less upon aluminum, without discouraging consumption and employment in this industry, and output and employment in other industries might have been somewhat greater as a result of the expenditure of larger sums upon their products.

Analysis of the facts does not in itself enable us to decide which type of market control is most likely to yield the closest approach to ideal market relations. Comparison of the consequences of different types of market organization in Europe is not entirely conclusive because they have operated under somewhat different conditions. Comparison of the results of single-firm monopoly in the United States with those of different mixtures of oligopoly and agreement in Europe is in the main even less satisfactory. But the analysis of the preceding chapters furnishes a background of knowledge for a hypothetical comparison of the consequences of different types of market control functioning under the same conditions. In this operation, to which we now turn, only the more general tendencies will be developed. Different sorts of government participation in market control will be discussed in the following section of this chapter.

Let us make the following assumptions, which seem to accord with the more general technological and business characteristics of the aluminum industry. (1) A homogeneous basic commodity, ingot, is manufactured in various alloys, forms, and shapes for many uses by a firm or firms integrated vertically from ore to markets. In general it is assumed that, owing to scientific testing, differentiation of the products of several firms will consist principally of differences in quality or suitability or in new variations. (The existence of one-stage fabricating enterprises will be ignored until late in the analysis.) Capital equipment must be expanded in substantial chunks, and is highly specialized and quite durable.[1] (2) Total demand is large enough so that several effi-

[1] The influence of improvements in administration and equipment for producing

cient firms may exist and make normal returns when operating at capacity. Average full cost at best utilization will be about the same whether one or several firms exist,[2] although it would be higher if there were many firms, each much smaller. Differences in the size of companies will depend upon differences in the efficiency of management, site characteristics of power developments, locational advantages, and the like. It is further assumed that every regional market can be reached by more than one firm. If enterprises are to have efficient structure and location this may require some freight absorption and hence prices somewhat above cost in territory adjacent to some plants. The problem of the best geographical price structure is not examined in this book. (3) The firms whose activities we shall discuss have already gained most of the economies of large-scale investment and meet constant cost or very gently diminishing cost [3] as they expand further.[4] Hence business men will have no reason to plan the creation of more investment than they intend ordinarily to operate at capacity, except for temporary underutilization due to lumpiness, mistakes in estimating demand, or business depression.[5] The investment appropriate for maximum monopoly revenue (for one or more firms) will be smaller than ideal investment. (4) Conditions governing entry are such that neither underutilization with normal profits nor resort to competitive tactics is caused by the influx of new firms. The policies of monopoly or of oligopolists will not be disturbed by newcomers. (5) Effective demand increases continuously, so that repeated expansion of investment is required to maximize monopoly revenue.

(6) The business executives in the industry differ somewhat in

---

ingot may be disregarded, since they would not require significant modification of the principles developed here.

[2] The possible qualification on account of advantages of a very large research organization, which was noted on p. 202, will be taken into consideration below, p. 348.

[3] Under such cost conditions existence of more than one firm in a given market is possible if newcomers introduce additional investment ahead of the original firm during an upward sweep of demand, or if restriction of investment by the first enterprise provides an opportunity for a new firm to make normal earnings.

[4] Resort to inferior ore or power with expansion may bring increasing cost to all. This would not affect the principles discussed here.

[5] The conditions under which this is possible have been explained above, pp. 205–208.

individual qualities such as imagination, shrewdness, and courage. Each knows that regard for the indirect consequences of his actions, through their effect upon the policies of others, is to his advantage. In order that comparison between the results of the different types of market control may be clear-cut, let us first assume that, whichever type exists, the industry will be in charge of a set of men who possess the same general level of intelligence and who display the same range of differences in personal qualities. In other words, the same differences would characterize the individual executives of a single monopolistic enterprise as would distinguish the managements of the several firms were a condition of oligopoly to exist. It seems likely that in actual practice with either monopoly or oligopoly the executives would be drawn from the same general grade of business ability, for there is little reason to think that differences between oligopoly and single-firm monopoly from the standpoint of entrepreneurial activity are sufficiently marked to attract respective groups of executives which differ substantially in general ability. Our question is whether, given the conditions here assumed, the same sort of men would feel incentives to act in different ways, depending upon the type of market control.[6]

Before laying down the last assumption, which concerns uncertainties of demand, it should be emphasized that there is very little in the above assumptions which would lead to different results as between oligopoly and single-firm monopoly, given the same degree of regard for profit maximization. If demand were known perfectly, its rate of change as well as its position at any given time, and policies proper to maximum profit were pursued in each case, the fundamental market relations would be approximately the same whether oligopoly or single-firm monopoly existed.[7] The only differences implicit in our assumptions seem to be those occasioned by the possibility of slight differences in cost as between the several types of market control, and those incident

[6] To the student of oligopoly it will later be apparent that this question is, under the assumptions made here, nearly identical with the question whether inducements to which oligopolists are subject with a moving demand curve and great uncertainties are different from those with a static and perfectly known demand curve. For, as is explained in the next paragraph, with the latter conditions the results of oligopoly and monopoly would be very similar.

[7] See Chamberlin, *op. cit.*, chap. III.

to possible uncertainties on the part of individual oligopolists about each other's behavior, due to imperfect knowledge of each other's cost conditions and personal characteristics. Oligopoly with uncertainties of this sort might yield investment and output exceeding those of single-firm monopoly.

The business men are well aware of our fifth assumption that effective demand will move ahead continuously as a result of the spreading of knowledge and the introduction of new variations or adaptations of the basic product. They know that a wide gap separates ideal and effective demand. They are, nevertheless, very uncertain as to how rapidly effective demand will move without their efforts, how fast and how far it can be pushed by different amounts of expenditure and different variations of product, how its slope will change, how great is the distance between effective and ideal demand, and how ideal demand itself will move as a result of changes in other parts of the economy.[8]

Whatever the type of market control, there are different possibilities with respect to the way business men may act. Motives outside the range of strict profit calculus may be influential and occasionally dominant. Lack of information on this aspect of the matter, in which personal temperaments are so important, makes it impossible to set assumptions for analysis. Our discussion will be confined to activities designed to yield profit. In the mists of great uncertainty it will be difficult in any case to adjust investment with perfect accuracy to achieve maximum profit. Under which type of market control will investment be most likely to diverge farthest in the direction of ideal investment? The several men in control of a single monopolistic enterprise will place different evaluations upon the uncertainties, but they must reach a common decision. Unless they differ strikingly in intellectual qualities or personal force the policy adopted will probably involve a rate of expansion closer to the arithmetic average of the various estimates of the best rate of expansion than to the estimates of the more optimistic and less cautious or to those of the more pessimistic and less bold. Secondly, without incurring *permanent* loss of profits which it might have had, single-firm monopoly can wait to expand until it is quite evident that effective demand is moving or has already moved in such a way that an increase in

[8] Some demand characteristics have been described above, pp. 214 ff.

investment will enlarge profits. The annual increment of potential profit lost for a short time by such a policy can be permanently added to revenue as soon as the situation is clear. The temporary loss may be deemed worth while in order to avoid the unwelcome condition in which monopoly revenue is less than it might have been because additional investment was added too soon,[9] and investment may ordinarily be kept a little smaller than that which would actually maximize profits.

Would the incentives of oligopolists differ sufficiently from those of single-firm monopoly to induce more or less investment? Differing in imagination, shrewdness, optimism, and courage, oligopolists would, in the shadows of uncertainty, also arrive at different evaluations of the uncertainties and different estimates of the rate of expansion for the industry as a whole which would maximize their joint monopoly revenue. Under such circumstances they might, without agreement, decide that the safest way to achieve the closest practicable approach to maximum profit for each was to follow a "live and let live" policy involving maintenance of a fairly stable proportionate division of total investment, sales, and profits. None would try to capture more than his share of the expanding common market. One who was temporarily ahead, because his more rapid rate of expansion accorded more closely with the growth of demand, would wait for the others to catch up instead of keeping his advantage. Unflinching pursuit of such policies would enable the most pessimistic to control the rate of expansion for the whole industry. As a practical matter, however, it is doubtful that stable shares in the market could be maintained without agreement unless the rate of expansion of each was in the neighborhood of the arithmetic average of their several estimates of the best rate for the whole industry. Given the decision of each oligopolist to forego any possible differential advantage over others, it would not be strange if total investment ordinarily fell somewhat short of the amount which would actually maximize their immediate joint profits for the same reasons which might actuate single-firm monopoly to expand cautiously.

In a mature industry where gradual changes in demand for an essentially homogeneous product may be predicted with considerable accuracy and where there seems to be little opportunity for

[9] Less by part of the capital charges on the "excess" plant.

development of new variations of the basic product, each oligopolist knows that there is little possibility of gaining any greater share in joint monopoly revenue.[10] Under such circumstances any attempt to get ahead of others in expansion would represent disregard of the immediate direct consequences of one's actions and of the later indirect consequences in the form of outright competition which would be likely to ensue; and a policy of sharing the market in approximately stable proportions might prevail. In a market characterized by great uncertainties surrounding a rapidly moving demand curve, where there is a large field for profitable development of new variations of the basic product, it seems unlikely that oligopolists would follow policies appropriate to more or less permanent division of the market in fixed proportions. *For without transgressing the limits of strictly rational pursuit of maximum profit it may still be possible to gain appreciable advantage at the expense of one's rivals.* Let us suppose that each oligopolist, taking careful regard for the ultimate consequences of his actions, endeavors to adjust investment and price in the long run, and output and price in the short period, so as to maximize his own profits. We shall neglect the influence of new adaptations of the product for a time, reserving that for later treatment.

Each oligopolist is well aware that as demand moves ahead he can increase his relative share of the market and his profits by expanding at a rate above that at which capacity as a whole should be extended to maximize the joint profits of all, provided any other lags behind the best rate for the whole industry. And, conversely, each knows that if he fails to keep up with the best rate, others, by exceeding it, can gain what he might have had. Given differences between oligopolists in their estimates of the best rate of expansion for all — that is, the best rate for each if everyone did the same thing — the actual rate at which total investment was enlarged would represent a simple average of these several estimates only if each acted upon his own estimate without regard for the probable policies of others. It is strongly to be doubted, however, that each oligopolist would set his jaw rigidly and follow his own idea of what everyone should do. Indeed, unless he were certain that the others all held the same estimate as he, such a policy would be incompat-

[10] Aside from cost reductions or selling tactics which result in the appearance of differentiation of product, factors which are not under consideration here.

ible with rational pursuit of self-interest involving full regard for the probable policies of others. In so far as each one knows that the several estimates of the best rate for all vary, and has some rough notion, gained from past experience, direct contact, or trade gossip, of the estimates of others and their probable reactions to alternative policies on his part, he will choose for his own policy the maximum rate of expansion which he thinks the policies of others will permit. None will increase capacity at a rate which he thinks will tend directly, or indirectly through inducing others to expand faster, to keep in existence a total investment greater than that which he believes appropriate to maximize their joint monopoly revenue, for that would reduce his own profit below its potential maximum. But each will expand at the maximum possible rate within what he conceives to be this limit, for that is the only way in which he can endeavor to maximize his own profits. The result of these involved calculations by the oligopolists seems very likely to be a larger total investment on the average than that equivalent to a simple average of the several amounts which they individually believe to be best. The reasons for this conclusion are as follows.

Any producer who expands more slowly than the rate which would *in fact* maximize the joint monopoly revenue of all will provide an opportunity for others to exceed that rate to their advantage. In so far as others close the gap left open to them, the more pessimistic or more cautious will have lost a certain annual increment of profit which they might have had. This loss is permanent, not temporary as in the case of the cautious monopolist, because others have installed the additional equipment and the laggards have no better opportunity to gain the advantage in the future than they would have if they had introduced this increment of investment. Knowledge that others have evaluated the probabilities more pleasantly or fear of being left behind may override some part of their caution, with the result that they build at a higher rate than they would if all shared their view of the future.

Consider now the case of the more optimistic oligopolist. Knowing that others have placed a lower estimate upon future probabilities, it is conceivable that he may decide to expand at a rate below that which he originally considered best for the whole industry. But it seems much more likely that he will not. The fact that others are going ahead more slowly than he thought justifiable for all provides an allowance for optimistic error on his part. In so far as he places more confidence in his own judgment than in theirs he will build at a higher rate than he considers best for the whole industry, just because others proceed at a lower rate. The optimist can make the rate of expansion of total investment in the industry equal his estimate of what it should be by exceeding that rate in his own program in proportion as others remain below it, and he has strong incentives to do so. If total investment is not carried beyond the amount proper to maximum joint profit by his activities he is a permanent gainer, as we have seen. Should he go too far, with the result that joint monopoly revenue falls below the possible maximum, he can force his fellows to bear a part of the loss if demand is elastic and marginal revenue positive for prices below that proper to maximum total profit with the appropriate total investment; and he may be able to escape part of the loss if demand is inelastic. With elastic demand he will not need to lose the full amount of the annual capital costs of the "excess" plant, because some additional output can be produced with this equipment at a marginal cost not exceeding marginal revenue.[11] Reduction of price to dispose of this additional volume will enable its producer to gain a larger amount of profit than he would get if the new facilities were not operated at all, even when others follow the cut in price, as they will, of course, do. Sale of this increment of product will make marginal revenue for all less than the amount which would equal the marginal costs of the others if they operated their now enlarged facilities at best utilization, as they had intended, and hence lead them to adopt a lower rate of output in order to reduce marginal cost or raise marginal revenue. If we suppose that they had already begun to operate at best utilization, readjustment in which our annoying optimist expands production while others contract theirs will proceed until all are producing at rates for which marginal costs are identical and equal to marginal revenue — the basic condition of stable equilibrium.

The result will be characterized as follows. Total investment exceeds the amount appropriate with best utilization to maximum

[11] This depends, of course, upon the assumption that marginal cost for some rates of output with the new facilities is appreciably less than marginal cost with best utilization.

monopoly revenue, and total profit is less than this sum. Total output, although larger than the amount which would maximize total profit if investment were smaller, is the right amount for maximum joint profit with existing investment. The investment of each oligopolist is somewhat underutilized. The proportion of output going to the firm which made the mistake is larger than the share it would otherwise have had. The loss of potential profit is shared by all, yet no producer can better his fortunes, as far as the immediate adjustment is concerned, by expanding or contracting output. Once the excessive investment has appeared, this is the best adjustment for each firm which oligopolists acting independently can achieve.[12] Although the fact that a part of the losses can be shifted to others provides no incentive to make mistakes, it may make men more willing to take chances in backing their own judgments than they would be if they had to shoulder the full penalty of mistakes, as is the case with single-firm monopoly.[13]

With inelastic demand at prices below the price yielding maximum total profit with the appropriate investment, output will be no greater than it would have been with the right amount of total investment. The producer with the largest proportionate increase in capacity will not bear the full amount of the loss of potential profit, unless his equipment is completed last. If all introduce new capacity at the same time, no one will operate it at best utilization because in order to sell such a volume he would have to lower price to a point on the inelastic section of the demand curve where marginal revenue is negative. Owing to discontinuity in the marginal revenue curve it will be impossible for all producers to equate marginal revenue and marginal cost, and if they introduce their new equipment one after another there seems to be no sure mechanism by which the marginal costs of all will be equalized and underutilization shared. The plant completed earlier will probably be operated at capacity and the rest at a lower rate.[14] In

[12] It would also be the best adjustment for a single monopolistic enterprise, operating under the same conditions, which had created the same amount of capacity. Power politics within a cartel might produce an arrangement in which the output of some companies was smaller and their profits larger, while others fared relatively worse.

[13] Single-firm monopoly may, of course, be able to shift some of the loss to producers of substitutes.

[14] Those who finish building later may, however, be able to coerce the others into

this case the optimist knows that he will not need to suffer the full loss of potential profit in the event that his estimate proves too sanguine, provided that he can get his new facilities into operation at least as soon as most of the others.

Furthermore, even if the oligopolists in the van of expansion incur substantial losses of potential profits through over-optimistic errors, they have already gained an advantage in the next wave of expansion, for their plant is in existence, and its full operation as demand grows must be considered as data for the estimates of all concerning further enlargement of capacity.[15]

On the grounds considered up to this point those with the more pessimistic view of the rate at which effective demand will grow have reason to expand at higher rates than their estimates of the best rate for the whole industry; while the optimists have little inducement to build at lower rates than *their* estimates of the best rate for all, and a strong incentive to exceed these rates in proportion as others stay below them. Capacity as a whole is likely to grow at a rate considerably above the simple average of the estimates of the several producers. Unless the average of their estimates is substantially below the actual optimum rate, they will collectively tend to exceed the latter.

There is an additional reason which makes this even more likely. Up to this point we have assumed that policies which carried total investment beyond the amount appropriate to maximum total profit represented bona fide mistakes. All are, however, aware not only that it pays to be an optimist provided your fellows are pessimists, but also that mistakes creating excess capacity with respect to maximum profit establish the basis of a permanent future advantage, unless they are made by all, which may more than repay the temporary loss. With the great uncertainties of demand and some uncertainty about the probable course of action of the individual firms, the line between bona fide mistakes within the realm of strictly oligopolistic endeavor and competitive advances which are not too obvious will be very difficult indeed to dis-

---

curtailment of output through temporary price cuts, but there is a danger that this procedure will provoke outright price competition by all.

[15] Up to the point of best utilization, marginal cost with plant in existence, which does not include capital cost, will be lower than marginal cost for additional output in projected plant, which does include capital charges.

tinguish. What more likely than that some will attempt to steal a march by making "mistakes" of a moderate sort in the hope that these will not be interpreted as warlike advances and followed by retaliatory measures? Success may encourage boldness, and repetition is likely to engender suspicion, leading sooner or later to an outburst of true competitive building which carries investment much closer to ideal investment with respect to effective demand, perhaps even beyond it. Even in the absence of any attempts to steal ahead, competitive warfare may be precipitated merely by misinterpretation of expansion programs regarded by their executors as strictly oligopolistic activities.

Upon the assumption that the industry is in charge of the same sort of persons whether oligopoly or single-firm monopoly exists, it is quite probable that a larger investment and output would be maintained on the average with oligopoly even in the absence of outright competitive tactics. Only if the bolder and more optimistic dictated the policy of the single monopolistic enterprise is it presumable that monopoly would achieve the same result. If the executives were not the same sort of persons with one type of market control as with the other, a variety of results would be possible, of course. However, it is reasonable to think that monopoly would not continually overshoot the investment proper to maximum profit, whoever directed its destinies; and it is to be expected that oligopolists collectively, whoever they were, would not ordinarily fall much short of this mark. Hence it hardly seems probable that monopoly would, in any case, do much better than oligopoly, while the chances are good that oligopoly might improve on the results of monopoly. The likelihood of occasional, perhaps frequent, breakdown into competitive building creates a probability that oligopoly would under most circumstances ordinarily give a larger operating investment than single-firm monopoly. In general, oligopolists have stronger incentives to take chances than a single monopolistic enterprise, and they are always tempted to engage in moderately competitive tactics which may from time to time precipitate strenuous competition.[16]

[16] It is pertinent to recognize at this point that in the formation of public policy it is impossible to take into consideration any elements of the problem for which probabilities cannot be estimated at all. Although personalities are of considerable importance in all cases involving small numbers, it is often impossible to form any

At first sight it might appear to some that the waste of underutilization incident to the larger investment likely to be maintained under oligopoly would nullify the advantage. In any industry where underinvestment exists, however, every added increment of capital which is used even partially will, up to the point where true overinvestment begins, be worth more to consumers than that increment is worth when invested where it earns merely normal returns. The specific product of an added increment of investment may exceed, fall short of, or just equal a normal return, but up to the point of overinvestment the average rate of return on the total investment will remain above normal.[17] Although the added increment may earn by itself less than a normal return, this is more than offset by the saving to consumers from a lower price on all other units of the commodity incident to the larger volume sold. Hence, as long as underinvestment exists, consumers will gain more from added investment, even though it is not best utilized, than if it were devoted to uses where it was worth only normal returns.[18] Further, if the circumstances of market control are such that some restriction of output involving underutilization will occur whenever investment exceeds the quantity appropriate to maximum profit, then an amount of overinvestment which results in normal returns on the whole investment represents a better allocation of capital than would obtain if less were invested in this industry and more elsewhere. Moreover, when oligopolists have collectively overshot the mark of investment proper to maximum monopoly revenue, they may redouble their efforts to reduce cost and to make effective demand move closer to ideal demand. Price may be lowered cautiously to test out elasticity of demand.

We must now introduce the probability that effective demand can be moved ahead by adaptations of the basic product for new uses and contrast the incentives to make new developments which

---

idea of the differences in types of persons who would be in charge of an industry with the one or the other type of market control. When this is so, if it is also considered probable that the element of personalities would in any case exercise the dominant influence, public policy would, of course, have no reason to favor one or the other type of market control.

[17] Assuming that investment is not best utilized.

[18] Their gain would, of course, be still greater if it were best utilized.

are felt by single-firm monopoly and oligopolists. Monopoly will, of course, be able to increase profits by discovery of new adaptations. Would oligopolists be likely to produce better variations or a larger number of successful ones? Let us disregard for a moment all questions concerning the most productive scale for a research organization or the supply of funds for research. If it were known that a particular new development could be made by every firm and if the results of making it were perfectly predictable, there would be no reason for any competitive activity, and the consequences would be the same as with monopoly, providing the costs of this adaptation were the same in the two cases. If everyone had complete knowledge of the nature [19] and results of all possible developments, the total results would be identical under monopoly and oligopoly. If it were certain that at every point of time everyone had the same knowledge of all potentially profitable variations and could introduce them with the same facility as any other, but uncertainties concerning the most profitable rate of production of each at every point of time were sufficiently great to permit different estimates of the best rates from the standpoint of the industry as a whole; then the situation would constitute simply a more complicated case of expansion of investment used to make well-known products for which the future movement of demand is very uncertain. The principles developed above would require elaboration rather than addition. Although hampered from time to time by uncertainty concerning the probable policies of others and inability to interpret correctly what they really intended, each oligopolist would nevertheless be able to take account of the indirect consequences of his actions to a substantial extent.

To add one more unreal case in order to throw a clear light upon actual conditions and their consequences, let it be assumed for a moment that each oligopolist was sure that every successful development of a new adaptation would inevitably be followed shortly by introduction through another firm of a product more satisfactory in the same use. With the certainty of a never-ending series of losses as one company after another found its new discoveries almost immediately superseded, none would ever make any attempt at new developments. Over a period of years single-

[19] Including knowledge of how to make the adaptation.

firm monopoly would, however, produce a few of the products making up such a series, spacing their introduction so as to maximize profit with regard for the costs of obsolescence.

With respect to many possible developments in the aluminum industry it would appear that uncertainties are so great that it is nearly impossible to take account of the indirect effects of introducing new variations. The ultimate consequences of a price cut may often be quite evident; but in the case of a fresh adaptation of a basic product it may be impossible to foretell with any assurance whether it can be duplicated or bettered in a month or in ten years. Each oligopolist would have to act upon consideration of direct effects alone. Under such circumstances more improvements in adaptations are likely to be brought in by oligopolists than by single-firm monopoly. Monopoly must bear the cost of obsolescence whenever it replaces one product with another before equipment is worn out. The individual oligopolist has a strong incentive to attempt to better the product of his rivals rather than pay them royalties, for if he is successful the burden of obsolescence falls on them.[20] In the case of those variations for which hazy estimates of indirect results can be made, the amount of investment would probably be less than if it were impossible to estimate ultimate consequences at all, but greater than that which monopoly would maintain. To some extent the reasoning of this paragraph may be applicable also to improvements which lower cost.

There is reason to think that oligopolists would collectively develop a wider range of adaptations, not only because they would continuously make more improvements of existing products, but also because they would discover more products that were new in the sense that they were suited to new uses. Suppose that in a given period oligopolists developed exactly the same set of new alloys and products as monopoly would have introduced. It is hardly to be expected that the gains would have been divided in unvarying proportion throughout the period. The relative lack of

[20] A firm may, of course, accumulate obsolescence reserves out of earnings in anticipation that some of its products will be bettered by others. But that is a different thing from restricting the number of new developments and the amount of investment devoted to producing them as a result of calculation of indirect consequences. This is possible only if it can be predicted which products will be superseded at given times.

success of some, coupled with the knowledge that others had shown greater prizes to be possible, would tend to incite the former to redouble their efforts. Since there is no reason to suppose that the latter would relax their endeavors in greater degree than a monopolist, it appears that oligopolists would tend to widen the range of new products more than monopoly would do.

Oligopolists are subject to incentives to maintain collectively an investment in research which would tend to produce very nearly the best rate of progressiveness attainable in the existing state of science, while single-firm monopoly has no profit incentive to make and use all socially worth-while discoveries at the best attainable rate. In so far as the extent of duplication in research activities and the distribution of the resulting losses are unpredictable, oligopolists have inducements to maintain a total investment in research which would involve some waste. If the individual firm could not tell when it was spending several times an amount which would pay for itself, the competition of several enterprises might occasion considerable waste. It is likely, however, that the line between profitable and unprofitable expenditure can be distinguished with sufficient approximation to avoid the danger of great waste.[21] Furthermore, in research the expenditure of additional effort upon a given problem may often bring better results rather than duplication. It seems more probable than not that whatever wastes might occur with oligopoly would be more than offset by the greater amount of worth-while innovations and the other advantages of oligopoly.

So far we have considered only the difference in incentives to progressiveness between oligopoly and single-firm monopoly. The actual rates of progressiveness yielded by the two alternatives might also be affected by a difference in command of funds for research or a difference in the productivity of research organizations of different sizes. If well-established integrated enterprises could obtain readily and cheaply in the capital market all the funds they desired to invest in research, or if the profits of oligopolists, whether or not as large as would accrue to single-firm monopoly, were sufficient (after payments to investors) for this, the results with monopoly, which has no profit incentive to invest as much in

[21] We exclude from consideration here any "pure" research of which the ultimate results cannot be estimated at all.

research, would tend to be inferior to those with oligopoly. Only if oligopolists were unable to maintain collectively as much investment in research as monopoly could and would support, would the latter tend to bring a higher rate of progressiveness because of better command of funds.[22] It is possible that greater restriction and larger profits for monopoly in one period might later result in the movement of effective demand to positions farther ahead than it would have occupied had oligopoly prevailed. Or, on the other hand, if oligopolists once obtained adequate research funds, the flow of new developments might maintain or increase their aggregate monopoly revenue to amounts which were always sufficient for all the research they desired.[23] The attitude of investment bankers in the matter of supplying funds for research could be discovered, but it is doubtful that the probable relations between profits and the sums desired for research in the two cases could be predicted.

Finally, the rate of progressiveness may be affected by the size of the individual research organization. If research productivity varies markedly with size of the research organization in the range of sizes between those which would be maintained respectively by single-firm monopoly and individual oligopolists, it is possible, but not necessary, that monopoly would give a higher rate of progressiveness. Each oligopolist would not have as large a research organization as monopoly, because of the losses due to waste in duplication of results. On the other hand oligopolists may be less often inclined to reckon that duplication of effort on a given problem brings duplication of results, and they have in any case incentives to make and use more discoveries than monopoly. Hence they might maintain a larger aggregate investment in research which, although organized in less efficient units, would produce more worth-while innovation. Moreover, as a monopolistic corporation reached the stage where the internal difficulties opposing continued economical expansion became more formidable, its rate of introduction of new variations would necessarily slow down.

[22] Perhaps the investment trust device, properly used, would constitute an excellent means for providing funds for research cheaply to many firms in a number of industries.

[23] In so far as monopoly revenue is increased as a result of new variations this is a symptom of better satisfaction of consumers' demands rather than greater restriction.

Any advantage connected with the larger size of its research unit would tend to be diminished because oligopolists could expand their activities more rapidly. We should like to know whether the questions implied in this paragraph can be answered by those who are most familiar with these matters.

The foregoing analysis suggests that under the assumed conditions investment and output would tend to be larger relative to current effective demand with oligopoly than with single-firm monopoly; and that the rate of progressiveness would approach the optimum rate more closely with oligopoly, unless oligopolists were unable to obtain more funds for research than monopoly would use or their research activities were much less productive because their research units were not as large. If it were shown that monopoly had no superiority or only small advantage in the latter two respects, there would be a definite probability of better results with oligopoly, which in addition to its other substantial benefits might be expected to result in greater price reductions during depression.[24]

Up to this point we have spoken as if the results with an integrated single-firm monopoly would not be influenced by competition, actual or potential, of one-stage rivals in the fabricating branches.[25] It is possible that these forces might accomplish the same results as would occur with several integrated firms. But they could not do so if single-stage firms encountered any difficulty in ingot supply, or if the integrated company failed to maintain adequate price spreads between the basic ingot and all of its products,[26] or if there was any substantial advantage in integrating research on new adaptations with the production of ingot. Furthermore, it is probable that the one-stage fabricating firms would not accomplish as much in the way of new variations as large integrated companies. With a few exceptions they remain small, owing to cost conditions and ease of entry, and tend to

[24] If the questions concerning relative supplies of funds for research or the difference in productivity between research organizations of different sizes could not be resolved into probabilities, there would be no basis for giving weight to these aspects of the matter in the formulation of public policy.

[25] Since entry to the ingot stage is not easy, potential competition from inventors of new variations would probably become active through the establishment of new fabricating firms or the sale of processes to existing fabricators.

[26] These matters are discussed in Chaps. XVI–XX.

occupy themselves more with production according to known techniques than with innovation.

It remains to consider the results of oligopoly with agreements (not subject to government control), which really constitutes a whole series of alternatives. For the sake of simplicity let us divide agreements into two classes, weak and strong, according to the criterion of rigorous control of output. It is clear that the simple price agreement, with or without provision for temporary stabilization of shares in total sales of ingot, is incapable of a dominant influence upon the adjustment of investment to demand. Members retain complete independence of action with respect to investment and output. A terminable agreement of this sort, limited in duration to a few years, cannot prevent purely competitive expansion if some members desire to follow policies which will induce that; nor, if they all act rationally, will it deter anyone from trying to gain the utmost advantage for himself when estimates of the best rate of expansion for all differ.[27] Frequent revision of price and shares in the market will probably be necessary to reflect changes in the adjustment of investment to demand. However, association may lead to better understanding of each other's estimates of the future, and more opportunity for some to persuade others. The result might be divergence from the equilibria of simple oligopoly in the one or the other direction. Since there might be much less misunderstanding of each other's real intentions and so less frequent resort to competitive warfare, it is, perhaps, more likely that total investment would tend ordinarily to be somewhat less than with simple oligopoly.

An agreement regulating output of ingot exerts a more deterrent influence upon competitive building because it removes the opportunity for unregulated sales of metal fabricated in one's own plants. It has already been pointed out, however, that control of output cannot completely prevent rivalry in expansion unless it is definitely agreed that output quotas will not be altered because of disproportionate growth in investment.[28] Members must be convinced that this provision will be rigorously enforced and that evasion would cost more than it was worth. When opinions about future demand differ appreciably it is unlikely that such an agreement

[27] For examples see pp. 157 and 300 ff.

[28] Above, p. 306.

will actually be made and adhered to for a considerable number of years. An effective covenant of this sort would scarcely be possible in any event unless the agreed rate of expansion were at least equal to the average of the several estimates of the best rate for the industry. The fact that agreements are often limited to a few years or are terminable upon notice by one or a few members would provide a loophole of which the more optimistic and more courageous might take advantage. Moreover, when there exist great possibilities and great uncertainties with respect to development of new variations of the basic product, it is hardly conceivable that oligopolists would subject themselves to control of competition in introducing new developments. Evidently the strongest practicable cartel agreement might result in keeping total investment substantially below the amount which would exist with oligopoly, but the chances are good that it would be greater than the amount maintained by single-firm monopoly. Since agreements do not seem necessary to prevent undesirable overinvestment in an industry with the characteristics here considered, it may be concluded that the long-run results of oligopoly with agreements are not likely to be superior to those of simple oligopoly, and will probably be inferior.

The fact that oligopolists make agreements to govern short-run market relations must indicate that they think the results would be different and less satisfactory to them in the absence of agreement. Price control usually means prices above marginal cost. No firm would ordinarily sell below marginal cost, except during economic warfare caused by personal animosity or intent to eliminate or discipline others. It is possible, however, that in some circumstances price might be lower and output larger with agreement than with simple oligopoly, even though price would not be reduced to marginal cost in either case, except as a result of the exercise of government authority. When oligopolists have different ideas about the advisability of lowering price at a given juncture or about the extent of profitable reduction, there is danger that individual price cuts not intended as competitive moves may be misinterpreted as such by some, with the result of a general breakdown into price competition. This is the more likely the greater the uncertainties about each other's ideas. If agreement is prohibited or impracticable, all may deem it more advantageous to make no reduction, at least until it is clear whether or not the new conditions are likely to be more or less permanent. With the aid of agreement the lower price might be obtained much more quickly.

It may be concluded that unless it is shown that progressiveness would be much greater with single-firm monopoly because of striking superiority in command of funds for research or in productivity of research, the best results from the social standpoint would probably exist with several producers of ingot integrated back at least through the stage of alumina production and forward into the manufacture of some fabricated products. Existence of one-stage fabricators would not necessarily involve social loss in all branches, however. Wherever there is no great advantage from extending vertical control to the *productive* operations in making alloys or in fabrication, as distinct from extending it to research upon development of new alloys and products, the best arrangement would include single-stage fabricators as well as fabricating departments of integrated companies. Leaving out of account considerations of government control, perhaps the most desirable sort of market control in this industry, from the standpoint of total consequences, would consist of several integrated producers of ingot[29] making also various alloys and fabricated articles, and a number of single-stage fabricating firms in some branches. Such an arrangement might afford the best combination of effectiveness in production, marketing, and research on the one hand, and of competitive forces tending to move effective demand ahead, to enlarge investment and output, and to lower price on the other hand.

2. Possible Alternatives for Government Policy

The results of the analysis in this book do not provide an adequate basis for determination of the best government policy toward the aluminum industry. Although the tentative conclusion is that the fundamental market relations — investment, output, price, and demand — have often diverged appreciably from ideal relations, it has been impossible to determine with precision either the relations between industrial structure and efficiency and progressiveness or the extent of restriction of investment and output which

[29] Except in those small geographical markets where efficiency would be impaired thereby.

has occurred in various geographical markets and in markets for different aluminum products. And we have been unable to decide whether the possibilities of profitable restriction are changing substantially with the newer demands which are developing. Any government which wished to be sure of the best policy toward this industry would need more adequate and accurate information than has been available for this book.

This section will be concluded with a brief consideration of various alternatives which might be employed for the purpose of bettering market relations in the ingot market if it were believed that they could be improved by the action of consumers or government. The analysis here would also apply to any fabricating branches in which monopoly or oligopoly may yield undesirable consequences. In other fabricating branches the preponderance of competitive forces could, under certain circumstances, be relied upon to produce satisfactory results.[30] The discussion will be limited to the economic problems involved. Neglect of governmental aspects of the general problem must not be taken to imply a presumption that they are easy of solution. The success of government activity obviously depends quite as much upon the honesty and ability of public authorities, their efficiency in the organization and conduct of government activities, and the qualities of economic statesmanship displayed by political leaders, as upon the solution of the economic problems involved.

Among the most important possibilities are consumers' coöperation, creation of simple oligopoly in markets where it does not now exist, regulation of investment or price, government competition with private enterprise, and a public monopoly of ingot production. Some of the schemes discussed, particularly under the topic of regulation, will doubtless seem fanciful. They are analyzed because of, rather than in spite of, this, for it is important to recognize the lengths to which certain sorts of control would have to be successfully extended if the most desirable results were to be approximated.

Consumers' coöperation may take the form of a purchasing association or a producing organization. Since a large part of the aluminum output reaches ultimate consumers in products, such as automobiles, airplanes, and washing machines, in which aluminum

[30] Cf. Chap. XX.

is but one of several components, the field for buying coöperatives seems to be limited mainly to manufacturers. Group purchasing by automobile manufacturers, airplane makers, or power companies, or by associations whose membership consisted of firms in a number of different industries, might result in lower prices for aluminum and some of its products. There can be no guarantee, however, that the "negotiated" prices resulting from such a commercial tug of war would contain no element of monopoly profit. Since many purchasers use relatively small amounts of aluminum, it is unlikely that organized buying would be extended to cover most of the aluminum market.

Coöperatives for the production of aluminum would also be limited in the main to manufacturers rather than ultimate consumers. If such organizations were to be instrumental in improving market relations they would have to resemble ordinary commercial firms in many respects. Capital might be raised by sale of securities to members and perhaps to nonmembers, or by loans from government. The not inconsiderable risk of obsolescence, combined with the fact that many users of aluminum shift back and forth from time to time between aluminum and its various substitutes, might make many consumers hesitant to assume a position not unlike that of ordinary investors in a commercial aluminum enterprise. Hence government might have to furnish a considerable part of the capital for such organizations. Government might also need to make available power and bauxite. The complex process would, of course, necessitate very capable management. Furthermore, unless competent research organizations were maintained by the coöperatives, improvements and new variations were introduced boldly, and the difficult problem of adjusting a somewhat inelastic capacity to probable future demand was approached without too much caution, the existence of producing coöperatives would fall short of promoting the most desirable relations between investment and demand. It is questionable whether most consumers would care to assume the risks associated with dynamic progress. Finally, until there existed a probability that nearly all consumers would fill their demands through coöperatives if they could not obtain metal as cheaply from regular commercial producers, the policies of the latter might not be greatly affected. At the present time it does not seem likely that coöperatives for aluminum production would be established without considerable government aid. If government capital is to be used at all its effectiveness in promoting the most desirable market relations would probably be greater when it was employed directly in the establishment of a government corporation which would compete with private producers.

We next consider the establishment of conditions of simple oligopoly of private firms. Unless additional information and study require modification of the tentative conclusions reached in the earlier part of this chapter, there would seem to be a probability that simple oligopoly — i.e., oligopoly without agreement — would bring better results than single-firm monopoly in this industry. In the chief markets of Europe removal or appropriate reduction of tariffs and removal of other restrictions on imports would help to create more effective oligopoly only if there were a strong presumption that governments would not soon revert to protection. Aluminum producers would obviously be hesitant to create investment to serve markets which might shortly be closed again. With the swiftly changing policies of democratic governments in these disturbed times the probability of long-continued access to foreign markets is not great. In any event the military importance of aluminum makes protectionist policies seem imperative to democratic as well as Fascist governments. Moreover, in order to obtain simple oligopoly, admission of foreign producers to domestic markets would need to be complemented by prevention of international agreements. This might perhaps be accomplished if each government prohibited sale within its own country by any party to an agreement affecting the market in that country.[31] Such a provision might, however, have the effect of convincing the oligopolists that all would be better off if each stayed out of the other countries and cultivated his home garden alone and unhampered. As far as Europe is concerned, under present conditions there is little chance that measures of this sort would be deemed practicable. In most countries creation of true oligopoly would doubtless require dissolution of the existing national monopoly or encouragement of new enterprise. In view of the infrequency of periods of low duties

[31] It would obviously be very difficult for a government to prove that a given firm belonged to such an international agreement if its terms were secret and its records were kept in another country.

in the tariff history of the United States and the possibility that European producers might in any case consider it inadvisable to build up an export surplus for sale in the home market of the powerful domestic enterprise, a similar conclusion may apply to the United States.

Successful encouragement of new producers might require that governments take the responsibility of providing them with economically suitable bauxite either by discovery of new deposits, creation of adequate transport where that is now lacking, or acquisition of some part of the reserves now held by established producers. Deposits could be sold to new firms, or governments could undertake to supply their annual requirements on long-time contracts. In the case of power something of the same sort might be necessary. It is also possible that government would need to give assurance that financial assistance in the form of loans or subsidy would be forthcoming in the early years of new enterprises if losses were incurred owing to overcapacity or economic warfare. Since the capacity of an efficient aluminum firm is fairly large relative to the size of the present market in each country, waste of great excess capacity could be avoided only if new companies were added gradually with the growth of demand. Creation of conditions of oligopoly with equally-matched firms might take many years. Prevention of overcapacity during this process might require prohibition of expansion on the part of the existing producer, and of each new firm after it had reached a certain size, until each of the desired number of firms had become well established and had attained an efficient scale of investment. Creation of oligopoly in any of the important national markets in the near future could probably be accomplished only by dissolution of the existing firm into smaller units. The difficulties which may impede effective dissolution are considerable, as the history of trust dissolution in this country shows, but there is no reason to think that they are insuperable. Most of the leading aluminum companies have several power plants and reduction works which could be divided among successor corporations. In order to fashion well-integrated units some new plant would be required at various other stages, depending on the particular circumstances in a given country. Whatever temporary overcapacity was thus created would probably be less than the amount that would be occasioned by the entry of new firms, unless they came in very gradually. It seems quite probable that there are many banking and industrial interests who would readily purchase a set of mines and plants already in operation, even though they would not undertake promotion of a new enterprise from the very beginning.

It is not, of course, to be expected that oligopoly would always maintain the ideal investment or the ideal price structure. With oligopoly or monopoly uneconomic price discrimination might be diminished somewhat by publicity of all sale prices. It could be lessened still more by exclusion of ingot producers from fabrication of all sorts, but this might diminish progressiveness. Prohibition of all uneconomic discrimination would be tantamount to regulation of all the relations in the structure of prices of aluminum in all forms, although it would not require the setting of a base price. It could not be accomplished without formulation of standards for computing costs of all aluminum products.

With oligopoly or single-firm monopoly it is possible that government could obtain a closer approach to ideal market relations through regulation of investment or price. A government board might be empowered to issue orders requiring expansion of capacity of private firms whenever it judged that additional capacity was necessary to approximate the desirable total investment. The results of such a scheme would probably be much less satisfactory than the results with either unregulated oligopoly or public monopoly of ingot production, for it would separate initiative and responsibility with respect to decisions which, owing to the lumpiness, durability, and specialized character of equipment in this industry, would have long-enduring consequences. With price regulation mistakes would be somewhat less harmful because they could be rectified as soon as they were perceived, and executives could follow their own ideas of sound investment policy in the meantime.

Maintenance of somewhere near the ideal investment on the average might be secured by fixing the prices of all products sold by the integrated ingot producers. As long as ingot producers engage in fabrication regulation of the price of ingot sold directly might not suffice. Semifabricated and finished articles might be sold at prices representing a higher derived price for the contained ingot when that was possible, or at prices returning a lower derived price when nonprofit motives ruled. Formidable difficulties would be encountered whether price regulation took the form of setting prices to yield a normal return upon investment or of setting them equal to marginal costs. The latter method seems to be superior. Formulation of standards for computing marginal cost would not appear to present more difficult problems than the evaluation of standards for measuring investment and allocating overhead. The marginal-cost method is applicable to short-run as well as long-run problems, while the traditional return on investment device involves long-run concepts only. Firms would produce the ideal output at any given time if price were set at a figure to which the marginal costs of all companies were equal and at which the aggregate output could be sold.

To induce the companies to expand investment up to an approximation of the ideal quantity, government commissioners would need to forecast the particular prices of ingot and other products which would, under probable demand conditions a year or two hence, be equal to marginal costs with the best possible utilization of ideal investment; and put the companies on notice that these prices would be set at that time if the expected demand conditions materialized. This would be equivalent to telling producers that after such time as additional equipment could be best utilized, because of growth in demand, they would not be permitted to charge a price higher than the estimated marginal cost with such facilities best utilized, even though they had not in fact added such new equipment or did not yet have it ready to operate. In so far as the government authorities achieved a reputation for being correct most of the time, such measures should achieve their purpose, for the companies would lose potential profits or incur losses if they did not expand at the appropriate rate. If the authorities continually made bad mistakes, the companies would, of course, follow their own notions of investment policy. To the extent that the government board perceived its mistakes and refrained from carrying out its prospective orders, the companies would actually determine the rate of expansion. If the mistakes were not recognized, or if price was in any case consistently set too low, desirable expansion would be discouraged, output would tend to remain below ideal output with existing capacity, or the companies would respond to the pleas of frantic customers to bootleg metal at higher prices.

What might at first thought seem a tempting alternative to direct price regulation is a tax per unit on any positive difference between price and marginal cost. With proper standards for computing marginal cost which were efficacious and practicable in application, such a measure would tend to induce the desirable output with existing investment at any given time. At any other rate of output and sales than the one which equated price and marginal cost, a firm would have to pay to the government all of its revenue in excess of aggregate variable expense. At the ideal rate of output it would pay no tax.[32] Needless to say, such a device would have disastrous consequences unless marginal cost were properly computed. For the rest, the most perplexing problem with this scheme, as with direct price regulation, concerns the provision of an effective stimulus to expansion at a rate appropriate for maintenance of the desirable investment. With a tax of this sort companies would probably wait to expand until their facilities were already being strained and price had moved up with increasing marginal cost to a point where it was above average cost. To some extent this sort of price movement is desirable to compensate for low earnings immediately after a new chunk of equipment is introduced. To prevent it from going too far, however, it would be necessary after a point to tax the difference between price and average cost. This would require determination of the amount of earnings to be considered a normal return upon investment and of standards for allocating overhead between the various products.

As compared with public control of investment, direct price regulation or indirect control through the tax device just described would be attended by less separation of private initiative and responsibility and would leave more scope for private initiative. They might, however, have the unwelcome results of diminishing progressiveness, and they would probably increase the pressure to reduce wages whenever that was possible. Some scheme of bounties for progressiveness would probably be needed, and wage regu-

[32] In so far as changes in demand from week to week or month to month were difficult to predict it would be advisable to permit firms to sell stocks at times for prices below the marginal cost at which they were produced and at times of shortage to charge more than marginal cost. This could, perhaps, be arranged satisfactorily by rebates on taxes equal in amount to any negative difference between price and the marginal cost at which the units sold were actually produced.

lation might prove to be indispensable. It is plain that anything like satisfactory results from price regulation, direct or indirect, could be obtained only by a sizable commission, the members and staff of which came to know as much about the aluminum market and the nature of costs in the industry as the executives and staffs of the companies themselves, if not more. Government regulation of aluminum prices might bring better relations between investment, output, and demand than would exist without regulation. However, recognition of the intricacies of computation of costs by outsiders for a complex industrial process, of the difficulties of inducing the desirable rate of expansion, and of the perplexing problems to be met in devising a scheme to minimize any deadening effect of regulation upon progressiveness, does not engender optimism about the results of regulation. The problems of price regulation could probably be rendered somewhat less difficult by the exclusion of ingot producers from all fabricating branches in which competitive forces could be maintained strong enough to keep prices close to marginal costs. But this might slow up the rate of progressiveness, since the competition of a number of one-stage firms with small resources may serve merely to keep prices down to costs and not to produce a high rate of progressiveness.

Government competition with private enterprise possesses significant potential advantages over regulation by commission. It should be more effective than commission regulation because it could change the data of the market in such a way that from the standpoint of their own interests private firms had no choice but to follow policies that would tend toward ideal market relations. With regulation evasion of orders is sometimes possible, and protracted litigation is inevitable. Less litigation might accompany government competition even though the public corporation was, as it should be, subject to legal restrictions. Adverse effects upon efficiency and progressiveness of private enterprise might be less with government competition because business men would feel freer than when they were constantly receiving direct orders from government authorities; and public competition might contribute to the stimulus to improvements. Again, it is to be expected that the officers in charge of a government corporation would acquire a much broader and more penetrating knowledge of conditions in the industry than members of a regulatory board. Moreover, with government competition the work of government officers would result in the production of aluminum as well as control of market relations in the industry. Another possible advantage of government competition is that it might enable somewhat better control of any undesirable results of policies motivated by nonprofit considerations. If private firms were expanding faster than was justified by the rate of growth in demand the public corporation could expand more slowly. Undesirable economic warfare due to personal animosities might be mitigated through agreements promoted by the government corporation. Probably, however, bad consequences of nonprofit motives could not be entirely eliminated by any method short of a change in the philosophy of business men.

It also appears that a government corporation could exercise much more effective control over market relations than government directors on the boards of private firms. Even if the latter method actually brought considerable compromise it might still fall far short of achieving the most desirable results. It would seem much better to have separate public and private corporations, with initiative, control, and responsibility all concentrated in the same hands in each case. Furthermore, public directors on the boards of private companies would have very much less power to alter total investment and output and prices than the management of a government corporation.

Since the aim of government competition should be to achieve the most desirable market relations consistent with the existence of private companies in the industry, the external conditions of production and sale for the public corporation should accord as closely as possible with the conditions under which private companies operate.[33] The government enterprise should have no advantage or disadvantage *because it is a public corporation.* In other words, except for variations incident to differences in efficiency of personnel, the costs of public and private companies should be about the same. Specifically this means several things. The government corporation would need bauxite that was quite as economical in quality and location as the best reserves not yet used. If ore of this description could not be secured except from aluminum producers a government would either have to take by

[33] Cf. E. S. Mason, "Power Aspects of the Tennessee Valley Authority's Program," *Quarterly Journal of Economics,* L, 409 ff. (May 1936).

condemnation some of any such reserves as were under its jurisdiction or acquire inferior ore. In the latter event if costs were not equalized through a low purchase price formal equalization should be made in the accounting.[34] In the case of power, fluorspar, and any other owned materials the same is true. Power or bauxite which would have greater value in any other use should not, of course, be used in aluminum production.

In the second place, in estimating its long-run marginal cost for purposes of planning the initial investment and all subsequent expansion, the government corporation should include an amount equal to the ordinary tax burden of private firms. The public corporation should also plan to earn the normal cost of capital to private aluminum companies even though funds were obtained more cheaply by the government. Again, in planning investment policy the government should include in long-run marginal cost true economic rent arising from the alternative worth of its natural resources in other uses or from any locational advantages. Finally, wages should approximate those paid by private companies.[35]

The government corporation should maintain the amounts of investment necessary from time to time to give the closest approach to ideal investment in the whole industry permitted by imperfect knowledge and human fallibility; and should offer to sell in all markets at prices equal to its marginal costs.[36] Unless ingot producers were excluded from all fabricating branches in which their presence would lead to oligopolistic price making, the government corporation would need to enter these branches also. Since large integrated companies can probably be expected to contribute more variations and new adaptations than small single-stage firms, the latter policy would seem best. With regard to price changes the

[34] The technical adjustment in computing the marginal cost of producing aluminum would be to enter bauxite at the worth which the best ore not yet used would have when used in the ideal equilibrium.

[35] The government might, of course, regulate wages and hours of labor for the whole industry.

[36] Transport expenses should be included in marginal cost except for price making in areas where marginal cost including transport would be lower for another firm. In such instances the government corporation should offer to sell at a price equal to the marginal cost, including transport of the other firm. The freight absorption involved would not be undesirable discrimination because prices would be determined by the lowest marginal costs of serving consumers in such area. The competitive offer to sell would tend to keep prices down to this level.

government corporation might exercise the function of "price leadership." In their endeavor to prevent any large gap between actual investment and the most desirable investment with respect to current effective demand, the officers of the public company should have regard for the balanced expansion of all firms to produce conditions such that no part of the output could be replaced more cheaply at another site. In so far as private firms were able to differentiate their products effectively from those of the government producer they might, of course, be able to obtain higher prices. Since most aluminum products can be tested with some precision for quality in use, it is doubtful that private producers would ordinarily be able to charge higher prices unless differentiation took the form of better quality or service.

Maximum effectiveness in securing the best possible market relations would require that the knowledge of demand and supply conditions obtainable by all producers be as comprehensive and as accurate as possible. Coöperative study of demand would be desirable. The kinds of statistics of capacity, output, stocks, prices, costs, investment, earnings, wages, and employment which would help to produce better market relations should be filed at appropriate intervals by all companies with some government agency or with a trade association composed of all ingot producers. Such statistics should be open to all producers and buyers and probably to the general public.[37] Even with the benefit of comprehensive information, the efforts of the public corporation to assure the maintenance of ideal investment in the industry as a whole might often result in some overinvestment or some underinvestment. Such divergences could, perhaps, be minimized by voluntary agreements between all producers, including the government company, made after thorough discussion and exchange of views as to future probabilities. Expansion policies based on false assumptions as to the actions of others could thereby be avoided. These agreements would specify the quantities of additional capacity to be introduced by each firm within a two- or three-year

[37] Publicity of this basic economic information might, of course, facilitate unfair and unintelligent attacks upon the industry, but it would also facilitate the important work of promoters, investment bankers, financial counsel, and financial institutions. Perhaps compilation of the statistics open to the public in a more general form than that used in presenting the data to members of the industry would diminish the opportunities for abuse in their employment.

period at the end of which a new agreement, perhaps embodying a different proportionate participation in the next wave of expansion, would be made. To be thoroughly effective such agreements would need to be legally enforceable. Provided that the public corporation always endeavored to maintain an approximation to ideal investment in the industry, either through insistence upon appropriate aggregate expansion under an agreement or, in the absence of agreement, by its own investment policy, the private firms would have little reason to attempt restriction, and might welcome agreement as protection against overinvestment. Whenever it was clear that the government corporation was responsible, either without agreement or in spite of agreement, for the creation of overinvestment, it should set its price equal not to marginal cost with existing investment but to marginal cost with proper investment; or, in other words, equal to the marginal costs of other firms when their capacity was well utilized. For as long as considerable reliance is placed in private enterprise for efficiency and progressiveness the financial health of private firms and the initiative of their managements must not be impaired by prices that are too low because of mistakes of government authorities either in estimating demand or in computing costs.

With regard to short-run market relations, particularly the very important matter of prices and output during depression, a government corporation in the market could probably induce better results than would occur with simple oligopoly or with oligopoly plus agreement.[38] The public company could bring about uniform price reductions instead of discriminatory reductions, and it could keep prices in the neighborhood of marginal cost. If this were done there would be little incentive for any company to make the sort of piecemeal concessions that encourage consumers to think they can obtain lower prices by delaying purchases.

Public monopoly might be able to achieve a closer approach to the best relations between investment, output, and current demand at any given time than would result from any of the other kinds of market control. Presumably it would also have appreciably lower capital costs. In the long run, however, the disadvantages might well outweigh the gains in countries where most fields offering

[38] During depression the latter might improve on the results of simple oligopoly. See above, p. 351.

great opportunities for dynamic change are open to private enterprise, where the executives of private companies enjoy more freedom to carry out new policies than would be accorded to the officers of a public corporation, and where the traditions of free private enterprise are strong. In this kind of institutional setting government might find it difficult to obtain the services of men best fitted to promote progressiveness. The aluminum industry is one in which wide opportunity for new developments still remains. In countries like the United States of the present time private oligopoly seems to offer the best conditions for efficiency and rapid progress in this industry, while the competition of a public corporation would appear to be the most effective instrument for securing desirable relations between investment, output, price, and current demand.

No definite policy toward the aluminum industry is advocated here for any government. It has not been possible to determine whether undesirable consequences of existing types of control attain a magnitude which is worth bothering about. No study of the governmental aspects of the general problem of effective government control has been attempted. From the standpoint of the economic aspects alone the analysis contained in Part III affords an indication, however, of the sort of government activity best suited for improving market relations in the industry, if it is found that the existing relations leave a considerable margin for improvement. On economic grounds we are led to the conclusion that government competition with private producers — the number depending upon the relation of the effective scale of investment to the extent of the market — has the greatest potentialities for obtaining the best combination of efficiency, progressiveness, and desirable relations between investment, output, price, and demand. Wherever this policy was adopted rapid attainment of desirable market relations with minimum waste of resources would call for establishment of the public corporation first and encouragement of additional private firms as demand grew. Wherever the number of efficient firms could be increased by dissolution of the existing company such a measure might help to produce a condition of oligopoly sooner. Prohibition of all mergers not justifiable on grounds of greater efficiency and of competitive tactics which do not reflect relative efficiency would facilitate the continued existence of several efficient private producers.

Finally, it is plain, of course, that in an industry of this sort ideal results are not likely to be obtained by government action even with efficient, wise, and fair application of the proper principles. Although the government may be able to improve market relations, the actual results may fall far short of the ideal unless business men themselves understand concretely the nature of the best market relations and endeavor to carry out policies which will promote their attainment.

PART IV

RELATIONS BETWEEN MONOPOLY OF THE BASIC PRODUCT AND INDEPENDENT COMPETITORS AT LATER STAGES — SOME ASPECTS OF AMERICAN EXPERIENCE

CHAPTER XVI

THE NATURE OF COMPETITION IN SHEET PRODUCTION

1. Introductory

All the leading producers of ingot carry part of their output through some of the stages of fabrication and sell part to one-stage fabricators. In so far as the latter use virgin metal they are dependent for materials upon the integrated firms who also sell sheet, ware, castings, and other products in the same markets as the one-stage independents. These relations create various interesting problems which center about the control by the integrated firms of (1) the supply and price of material available for their own manufacturing divisions and the one-stage independents, and (2) the differentials between the price of ingot and the prices of various half-products and finished goods. In Germany where the capacity of semifabricating and finishing firms is adjusted to an export market which is served also by some of the leading European ingot producers, discussion has been concerned chiefly with the matter of rebates on the purchase price of ingot used to make goods for export. The antitrust laws of the United States have concentrated attention here upon questions of "unfair competition" between the integrated monopoly and its one-stage rivals. Some of the economic issues presented by the relations between the Aluminum Company of America and the independents in the rolling, utensil, and castings stages will be examined in Part IV.

Since 1910 there has been scarcely a year, except for the war period, during which at least one branch of the government has not been engaged in investigating the activities of the American aluminum monopoly, in particular its relations with the independent fabricators. Proceedings have been brought in only three instances. In 1912 the Aluminum Company consented to a decree enjoining it from the practice of various competitive practices in the utensil and castings branches. In 1919 the Federal Trade Commission attempted unsuccessfully to prevent acquisition by the Aluminum Company of a rolling mill which had formerly belonged to an independent utensil manufacturer. The third of the proceedings was initiated in 1925 with the issue of a complaint by the Federal Trade Commission charging unfair methods of competition in the utensil and castings fields. This complaint was dismissed in 1930 for lack of evidence to support the allegations. Government investigations seem to have been concerned chiefly with those activities of the Aluminum Company related to the utensil and castings branches of the industry, which have often contained some grumbling independents. Two firms operating rolling mills have recently engaged in private litigation with the Aluminum Company under the antitrust laws.[1]

The Aluminum Goods Manufacturing Company, a large utensil firm, has rolled much of the sheet used in its fabricating operations ever since its origin, and one other cooking-utensil concern has possessed a rolling mill since 1920; but neither of these companies has ordinarily sold sheet in the open market. It appears that the Cleveland Metal Products Company, which built a rolling mill in 1914, was the first independent to sell sheet in this country. The stimulus of high war prices brought into existence two small mills operated by the Bremer-Walz Corporation of St. Louis and the United Smelting and Aluminum Company of New Haven. In 1919 the St. Louis mill was sold to the Aluminum Goods Manufacturing Company. For several years the United Smelting and Aluminum Company remained the only independent producer of pure aluminum sheet, for in 1918 the Cleveland mill was taken over by the Aluminum Company under circumstances presently to be explained.

The total output of these three firms does not appear to have been large enough to affect price directly to any great extent, except perhaps in a few limited areas. Sales of the Cleveland mill amounted to about 1,600,000 pounds in 1916, and 1,500,000 pounds in 1917, while production of sheet by the Aluminum Company totaled 19,000,000 pounds in 1914, 36,000,000 pounds in 1917, and 48,000,000 pounds in 1919.[2] A large proportion of this was sold. The Bremer-Walz output was inconsiderable, and this company was not equipped to roll sheet over thirty inches in

[1] Below, pp. 384, 386.

[2] See pp. 12, 53, and 58 of Petitioner's Brief in *Aluminum Company of America* v. *Federal Trade Commission,* U.S.C.C.A., Third Circuit, no. 2721, October term 1921 (hereafter referred to as Petitioner's Brief).

width, which was required for a large part of the auto-body work. The United Smelting and Aluminum Company sold a small quantity of sheet in New England at prices higher than those charged by the Aluminum Company, depending on better and quicker service to retain its market.[3]

Since the war three other independent mills have entered the industry, two of which have produced chiefly alloy sheet.

2. The Rolling-Mill Case [4]

Organized in 1910, the Cleveland Metal Products Company engaged in the manufacture of enameled steel utensils and oil cooking stoves and heaters in which some aluminum was used. In 1913 it was decided to enter the aluminum-utensil field. It appears that assurance of a supply of European ingot was obtained. In 1915 a rolling mill was completed with an annual capacity of about 3,000,000 pounds of sheet, which was somewhat in excess of the immediate requirements of the utensil division. Owing to war conditions little European ingot was obtainable in the next few years. The company purchased most of its material from the Aluminum Company. The booming war demand for sheet permitted large profits through sale in the open market at prices considerably above those charged under the time contracts of the Aluminum Company. About three-quarters of the output of the Cleveland mill was sold in 1916. When the open-market price of sheet receded in the middle of 1917 in anticipation of government price fixing, the high profits also began to disappear. By the beginning of 1918 the Cleveland Company appears to have been losing $14,000 a month on its sheet business. As a consequence it was decided to restrict output to the needs of its utensil plant.

When informed of this intention by the president of the Cleveland firm, who visited Pittsburgh to request cancellation of part of its contract,[5] the Aluminum Company negotiated for the acquisi-

[3] FTC Docket 1335, Record, p. 3566.

[4] The information upon this incident is found in III Federal Trade Commission Decisions 302 (1921), and 284 Fed. Rep. 401 (1922). I have also used the briefs of FTC Docket 248.

[5] According to the opinion of the court this request was at first refused. The Aluminum Company maintained that cancellation was made as requested. (See,

tion of the mill, with the result that the following plan was carried out. A new corporation was formed called the Aluminum Rolling Mill Company. Of its $600,000 stock issue the Aluminum Company took two-thirds and the Cleveland Company one-third. The rolling mill and sheet business of the latter were sold to the new company. It appears that the mill was operated for a few weeks by two employees of the controlling parent before the new corporation was legally born. The reasons which actuated the Aluminum Company in obtaining this mill were doubtless several. Its rolling equipment was inadequate to meet the increasing war demands of the allied governments. Of importance for the future was the advantage of possessing a mill in the Cleveland district, which then consumed large amounts of aluminum sheet hauled by truck from Niagara Falls or New Kensington. An opportunity was afforded to assume control of a mill which might later become a substantial producer.

The new corporation was less than a year old when the Federal Trade Commission challenged its legitimacy by the issuance of a complaint against the parent company alleging violation of Section 7 of the Clayton Act, which prohibits the holding of stock of one corporation by another where the effect may be substantially to lessen competition between the two or tend to create a monopoly.[6] On March 9, 1921, the Commission ordered the Aluminum Company to divest itself of its holdings in the Rolling Mill Company.[7] Sale to any person, natural or corporate, connected with the Aluminum Company, or any of its subsidiaries or affiliated concerns, was prohibited. Under the Clayton Act the order was clearly called for by the circumstances. While war-time operation of this

for instance, Petitioner's Brief, pp. 62, 65.) The Commission's attorneys contended that the cancellation did not occur until after the preliminary negotiations for securing the mill had taken place. (See p. 5 of their brief opposing petition for rehearing and modification.)

[6] The Commission had already informally secured the retirement of one officer of the Aluminum Company from the board of directors of the Rolling Mill Company to avoid prosecution under Section 8 of the same act.

[7] The findings of fact upon which the order was based appear somewhat partisan. No mention is made of the war demand for sheet, and the profits of the Cleveland Company are shown without any suggestion of its losses. There is no intimation that this company intended to curtail its production. One would infer from the findings that the motive of the Aluminum Company was simply to stifle competition.

mill was imperative — and this the Commission should have recognized — the real question was the effect for the future.[8] Had the aim of the Commission been attained, there would have been an opportunity for the continuance of independent operation of the mill under normal peace-time conditions.

After the Court of Appeals had sustained its order, the more powerful arbiters, supply and demand, shaped somewhat by the new tariff, refused to support it. The Cleveland Company became insolvent and declared that it would never operate the mill.[9] The Rolling Mill Company, which had never earned profits, owed the Aluminum Company about $600,000 on ingot purchases. When informed of the probable outcome the Commission was apparently at a loss for the best policy to pursue. After public offering in 1923 of the Aluminum Company's stock in the Rolling Mill Company had failed to attract any purchasers, these shares were sold to the Cleveland Company for a nominal sum, and the erstwhile owner announced its intention to sue on the notes due from the Rolling Mill Company, and, if necessary for collection, to bid in the property at sheriff's sale. This is what happened, following an ineffectual effort by the Commission to secure modification of the court order so as to prevent the purchase of the physical property at an execution sale. The Commission grounded its application upon an allegation that the indebtedness of the Rolling Mill Company was entirely fictitious and created for the purpose of allowing the parent concern to acquire the plant at execution sale. This contention was based upon a comparison of the ingot prices charged the Rolling Mill Company with those charged the United States Aluminum Company, a 100 per cent subsidiary. In accordance with a long-standing practice of transferring materials to each of its subsidiaries at an arbitrary and unvarying figure the Aluminum Company had apparently transferred ingot upon the books to the United States Aluminum Company at 18½

[8] Cf. statement of the court concerning the defense that the transaction was justified by its motive to increase production for war needs. "With these matters we surmise, we have no present concern. They have to do with the motive for the transaction. We have to do only with the effect . . . its effect upon actual competition as well as in destroying potential competition in a way later to make actual competition impossible was substantially to lessen competition . . . and the stock acquisition did, in effect, tend to create a monopoly."

[9] It sold its aluminum-utensil business in 1923.

cents a pound.[10] The Rolling Mill Company had evidently paid the regular market price, which was higher. Its finished sheet was sold at market prices by the Aluminum Company on a commission basis. In denying the petition the court concluded:

> Grounding our decision solely on the inability of the Federal Trade Commission to establish fraud in the indebtedness on which the Aluminum Company proposes to seek recovery at law in another court, we are constrained to deny its petition to amend the decree previously entered.[11]

If the Rolling Mill Company had been a 100 per cent subsidiary of the Aluminum Company, there might have been no indebtedness, and the question of how to divest stock which no one will buy would have been squarely presented to the Commission. As it was, the indebtedness complicated the matter. The language of the concluding sentence of the court's opinion suggests that it might have considered favorably a scheme which would keep the mill on the market for a time, until it was certain that independent enterprise would not purchase it, and which also made some provision for the payment of the debt. Apparently the Commission presented no plan.

The enforcement of the Clayton Act in this instance resulted in giving potential competitors an opportunity to purchase a small mill at a time when business conditions were moderately favorable. No purchasers appeared, and the Aluminum Company was enabled through a peculiar set of circumstances to retain the mill. Unimportant in itself, this case perhaps provides an example of the operation of elements unfavorable to the development of many competitors which will be explained in the next two sections.

3. Independent Rolling Mills and Ingot Supply

Rolling aluminum sheet is, as officers of the Aluminum Company have often pointed out, a simple operation that any brass

[10] This is, of course, a general practice among holding companies. After the Federal Trade Commission had used these "sales" at arbitrary prices in an attempt to show fraudulent indebtedness, the practice was altered so that legal title at all times remained in the parent company. Before this, title had passed with physical transference of the material from one company to another. Cf. below, p. 463.

[11] There is no evidence to suggest that the Rolling Mill Company had purchased more material than it needed for its operations.

or copper mill could easily undertake. Explanation of the hesitancy of independent enterprise to enter this field must be found in a distaste for dependence upon monopoly for ingot supply and in the power of integrated monopoly over the ingot-sheet price differential, with both of which elements the tariff has been intimately connected. Three firms took up the aluminum-sheet business after the war. The Sheet Aluminum Corporation of Jackson, Michigan, was organized in August 1925 to roll coiled and flat sheet.[12] With a capacity of about 5,000,000 pounds per year it has produced sheet, rod, wire, stampings, and fabricated products. For five years or so it purchased virgin ingot chiefly from the Aluminum Company. Suffering from financial difficulties which included a growing debt to the Aluminum Company, this firm turned to secondary metal about 1931. It has smelting furnaces in which scrap is converted into ingot for rolling, die-casting, and other operations. Several years ago the Fairmont Manufacturing Company purchased a copper rolling mill in Fairmont, West Virginia, which had been built during the post-war inflation. The capacity of this mill, which has produced coiled and flat sheet and sheet circles, seems to have been at least double that of the Sheet Aluminum Corporation.[13] Recently, it has entered upon the production of high-strength alloys. A third firm, the Baush Machine Tool Company of Springfield, Massachusetts, took an enterprising part in the pioneering of duralumin alloys in this country. Production of duralumin sheet and forgings was begun in 1919 and rod, wire, and tubing were added in subsequent years. In 1922 this firm took up the production of pure aluminum sheet and manganese-alloy sheet. Baush aluminum-alloy products have been sold largely to the automobile and aircraft industries.

It has already been said that the United Smelting and Aluminum Company of New Haven sells a small amount of aluminum sheet in New England. The sheet capacity of this concern is, perhaps, smaller than that of the Jackson mill. As the name indicates, it smelts scrap. These four companies have sold pure alu-

[12] Originally called the Northern Manufacturing Company. Early in 1926 the present name was substituted without any change in organization.

[13] The Fairmont Company was described in the February 1930 issue of *Courtesy and Service,* published by the Monongahela West Penn Public Service Company.

minum sheet and alloy sheet and various fabricated products in markets also served by the Aluminum Company.[14] It is not likely that their combined sales of pure aluminum sheet have ever totaled one-quarter of those of their large rival.

The independent rolling mill may obtain its ingot supply from the domestic producer of virgin, from foreign producers, in the secondary market, or by remelting scrap itself. It may be doubted that a substantial venture would care to rely wholly upon secondary metal. Although certain grades of secondary aluminum can be rolled satisfactorily, it is questionable whether there is sufficient assurance of regularity in supply of a suitable grade or of ability to obtain additional amounts when demand for sheet is mounting.

An independent who elected to rely principally upon the domestic producer for virgin aluminum would face some rather disconcerting possibilities. Refusal to sell or delivery delays might result in serious financial embarrassment to him. Whether the cause of these difficulties were in the inelasticity of the reduction process, inefficiency, or monopolizing intent, they would be most likely to occur at times of brisk business when they would be most disastrous. Not being endowed with a "public interest" the Aluminum Company may, in the absence of any purpose to create or maintain a monopoly, legally refuse to sell to anyone. If the purpose of the refusal were not to promote monopolistic control this might not be readily apparent; or, on the other hand, increased requirements of the company's own fabricating divisions and larger demands for ingot from markets held less securely might be taken as a full explanation even though they were not. With either inefficiency or uncontrollable inadequacy of supply under conditions of booming demand, the effects upon independents might vary only in degree, or not at all, from those resulting from an intent to monopolize. What has been said of refusal to sell applies equally to the disturbing possibility of delivery delays, whether due to monopolistic intent or to the other conditions cited.[15] Until the late twenties past experience had led customers to expect delays of

[14] Some of them make and sell casting alloys and finished castings. Here we are interested only in sheet and the products made from sheet.

[15] Compare the conclusions of the Department of Justice after investigation of the company's treatment of utensil firms with respect to delivery delays. See below, p. 426.

some months at the height of the business cycle and occasionally at other times.

The consent decree of 1912 attempted to prevent refusal to sell and delivery delays when used to monopolize or favor the Aluminum Company's subsidiaries at the expense of competitors. The ineffective language of its injunctions and provisos goes far, however, toward nullifying everything except its spirit; while its applicability to an actual situation was probably removed several years ago when the Aluminum Company adopted a policy of retaining title to the metal at all stages as it goes through operations which are performed by its various subsidiaries on a toll basis.[16] It may also be noticed that the test which the decree set up — equality of treatment between the independents and their competitors — would not necessarily enable the independents to survive and prosper. A prolonged shortage of metal, due to factors over which the Aluminum Company had no control, or simply to lack of vision in foreseeing the movement of demand, might spell serious consequences for an independent while the former was incurring losses small in proportion to its aggregate financial resources. Furthermore, the Aluminum Company's best policy for its own sheet production at any given time is influenced by the demand and supply conditions in all other markets with which it deals. There are conceivably many conditions under which the company would find it most profitable to restrict the amount of ingot devoted to sheet production in its own mills.[17] A proportionate restriction of sales to independent rolling concerns, quite without intent to embarrass them, might result in substantial financial loss to them. And, of course, the conditions responsible for this

[16] Discussion of this decree is deferred until a later chapter because the charges of violation have centered about the activities of the company in the utensil and castings branches.

[17] Consider the effect of an appreciable increase in the demand for castings due to steel and copper price advances which are likely to be temporary. In order to secure this new market in the first instance and hold it later the Aluminum Company would need to divert substantial quantities of ingot from the rolling mill to the foundry, unless there was adequate capacity to meet the new demand without resort to this expedient. If filling the larger castings demand created a shortage with reference to the old demands of all kinds, the most secure of these would have to suffer during the time necessary to bring into existence new ingot capacity. Since the utensil market is relatively more secure than many others, sheet production might be relatively lessened.

situation would, at the same time, enhance the inducement to curtail sales to independents in greater proportion than the reduction of the company's own sheet production. What tests would a court use for determining legitimate reductions to outsiders in circumstances such as these? [18]

The dread of dependence upon the domestic monopoly for ingot is cogently testified to by the policies of the few independent mills. The Sheet Aluminum Corporation appears to be the only one of these which has purchased virgin metal almost altogether from the Aluminum Company. The Baush firm bought exclusively from foreign producers between 1923 and 1932.[19] The Fairmont mill has apparently obtained its requirements from Europe.[20] Some years ago virgin ingot used by the United Smelting and Aluminum Company came from abroad.[21] From time to time it has been possible to obtain foreign metal a little cheaper than domestic, although the Europeans do not seem to have undersold the Aluminum Company as a general rule.[22]

While these few rolling-mill companies have evidently been able to fill their requirements abroad, it is rather doubtful whether a larger number of firms would be able to do so satisfactorily. To assure certainty of continuous supply independent mills would need fairly definite guarantees from foreign producers to fill their requirements regularly. A policy of "shopping around" in this imperfectly competitive market leaves one at times of booming demand in the unenviable position of being no one's regular customer. Would the European firms be willing to assure definite shipments on a substantial scale? Past experience shows that they have not, in fact, adjusted their capacity for large exports to America, although they probably could have disposed of much greater amounts of metal here to the independent foundries and other users. A large part of the imports from European producers seems to be accounted for by depression, temporary shortages in

[18] It is plain that the considerations just discussed apply also to the case of companies which fabricate finished articles from sheet rolled in their own mills. There are additional complications, however. For instance, the Aluminum Company might prefer, for various reasons, to sell them sheet at certain times.

[19] BMTC appellant v. ACOA, Exhibit 38. No information on its purchases since 1931 is at hand.

[20] *Courtesy and Service,* February 1930, p. 4. Cf. BMTC appellant v. ACOA, fol. 2436. The president of this company was for several years an importer of foreign aluminum.

[21] BR, p. 147.

[22] See above, p. 159.

this country, or temporary surpluses abroad. As long as the United States maintains a substantial tariff and the Aluminum Company keeps its ingot price fairly close to the foreign price plus the duty, the Europeans are not likely to make investments for the purpose of export to America. The possibility of reductions in the prices of ingot and sheet here, whether due to monopolizing intent or diverse forces peculiar to the domestic market, would threaten the profits of the foreigners or the independent mills in this country. The latter would find it necessary to turn to the Aluminum Company unless the former were willing to shoulder the major part of the burden.

Dependence upon foreign producers by a large independent rolling-mill group here would, of course, be impossible if there were explicit or tacit understandings between the European and American ingot producers that home markets would not be aggressively invaded, or if either group refrained from vigorous incursion on account of the fear of retaliation by the other. The situation described in Chapter VII affords some reason to believe that the latter condition has obtained part of the time.

Finally, there is some question of the extent of the market open to independent rolling mills. Probably many fabricators would prefer to purchase from the Aluminum Company rather than from rolling mills dependent upon foreign sources of supply, and would prefer to deal directly with the Aluminum Company rather than with independents who obtained their material from the domestic monopoly. If one is to depend on monopoly for material it is doubtless more satisfactory to become a regular customer of the monopoly itself than to trust a middleman to cultivate its good will.

4. The Ingot-Sheet Price Differential

The power of integrated monopoly over the ingot-sheet price differential constitutes another strong deterrent to the existence of independent enterprise in rolling sheet. Under pure competition this differential would tend to equal the full cost of rolling sheet with the most effective known organization of the factors of production.[23] Returns to capital and enterprise would tend to be the

[23] If there were any marked advantages of integration of the ingot and sheet stages, independent firms engaged only in rolling could not exist under pure com-

same per unit at both the reduction and rolling stages of an integrated concern. However, when a company which has substantial monopoly power in the stages below the production of sheet also rolls sufficient sheet to affect its price, there is no reason to expect that the price differential will necessarily measure exactly the cost of rolling sheet or that the returns per unit of capital and enterprise will be equal in each operation. Some discussion of the various possibilities is appropriate.

The most profitable policy would be to adjust the price of ingot to the demands of those markets for half-products or finished goods which the integrated firm could not or did not wish to monopolize. If the Aluminum Company should sell ingot to outside mills instead of rolling sheet itself, it would need to consider the demand for sheet products in setting its ingot price. If, on the other hand, it produced all the sheet, it could set the price of ingot in complete disregard of the demand for sheet and sheet products, to which it might adjust its sheet quotations directly. Now if it should happen that the most profitable price at which to sell ingot destined for markets other than those of sheet products was much higher than the most profitable price at which ingot might be sold to rolling mills, the most lucrative policy would require a monopoly of sheet production. The sheet price could then be set so that it contained, in addition to the cost of rolling, a lower and more profitable price for ingot than would be charged for direct sales of the latter. Monopolization of rolling operations would be required for the most profitable discriminatory price structure,[24] and would be facilitated by the price policy appropriate thereto — that is, an ingot-sheet price differential which did not cover the full costs of rolling. A low "development" price on sheet would constitute a special instance of the general case just discussed. Should the demand for sheet products, on the other hand, be such that a higher price for ingot were appropriate for this market than could be charged in other markets without reducing profits there, the appropriate price policy would be expressed in an ingot-sheet differential which exceeded the full cost

petition. Little of the available evidence suggests any material advantages of integration of these two stages in the aluminum industry.

[24] Cf. the discussion of monopolistic discrimination, above, pp. 217 ff.

of rolling. If the extra profits were large enough to offset the deterrent elements explained earlier, new enterprise would enter the field. Faced with the possibility of enough newcomers to threaten competitive price making for rolled products, the best policy might be to raise the ingot price to the point required to keep the number of independents sufficiently small to permit oligopolistic price making; the solution would depend upon the relative intensities of demand for ingot in other markets and for sheet.

Another motive for maintenance of an ingot-sheet differential which did not cover the full outside costs of conversion might be to discourage indirectly attempts on the part of new enterprise to engage in the production of aluminum ingot. In general, the fewer the consumers of ingot the smaller would be the number of those who considered undertaking to produce the metal itself. Similarly, the fewer the semifabricators the fewer the opportunities for foreign aluminum to find a market here. The strength of the obstacles both to entry at the ingot stage and to steady importation of large amounts of European metal suggests that the ingot-sheet price differential could probably be regulated chiefly in accordance with the most profitable discriminatory price policy. Yet in so far as the ultimate effects of permitting several independent mills to thrive seemed quite unpredictable, some profits might be sacrificed to guard against the possibility of larger reductions in revenue in the future. Furthermore, it is clear that with our antitrust laws the existence of a substantial group of independent rolling mills dependent for material on the integrated firm would handicap pursuit of the most profitable price policy at times when that called for an ingot-sheet differential lower than the outside conversion cost. Financial losses on the part of the independents would doubtless be accompanied by much litigation. Finally, the policy of the company with respect to the ingot-sheet price spread might be affected by factors unconnected with, or indeed opposed to, the maximization of profits.

Table 28 shows the list prices of the Aluminum Company for 99 per cent ingot, which is ordinarily used for sheet, and certain classes of sheet, and the resulting spreads. It appears that the differentials declined markedly in the second part of the post-war decade.

TABLE 28

LIST PRICES OF ALUMINUM COMPANY OF AMERICA FOR 99 PER CENT INGOT AND CERTAIN CLASSES OF SHEET AND RESULTANT SPREADS, 1918–1931 *

(Cents per Pound)

Date of Change	(1) 99 Per Cent Ingot	(2) 2S † and 3S † Flat Sheet No. 20 B & S Gauge	(3) Spread (2)–(1)	(4) 2S † and 3S † Coiled Sheet No. 20 B & S Gauge	(5) Spread (4)–(1)	(6) 17S † Coiled Sheet No. 20 B & S Gauge	(7) Spread (6)–(1)
1918							
Mar. 6	32.2			37.2	5.0		
June 1	33.2			38.8	5.6		
1919							
Mar. 1	33				5.8		
1920							
Feb. 16				40.8	7.8		
Apr. 19	35				5.8		
Aug. 10				42.2	7.2		
Oct. 1	33				9.2		
Dec. 20	28.5			37.7	9.2		
1921							
July 15	25 ‡			34.2	9.2		
Oct. 20				30.2	— ‡		
Nov. 15	20				10.2		
1922							
Feb. 27						55.7 §	35.7
June 27						41	21
Sept. 26	22			31.2	9.2		19
Oct. 25						43	21
Nov. 1	23				8.2		20
Nov. 22	25			32.2	7.2	45	20
1923							
Feb. 7	26			33.2	7.2		19
Nov. 13						46	20
Nov. 23	27			34.2	7.2		19
1924							
Jan. 10	28				6.2		18
June 15				33.8	5.8		
July 1						74.8	46.8
Dec. 10						74.5	46.5

TABLE 28 — *Continued*

Date of Change	(1) 99 Per Cent Ingot	(2) 2S † and 3S † Flat Sheet No. 20 B & S Gauge	(3) Spread (2)–(1)	(4) 2S † and 3S † Coiled Sheet No. 20 B & S Gauge	(5) Spread (4)–(1)	(6) 17S † Coiled Sheet No. 20 B & S Gauge	(7) Spread (6)–(1)
1925							
Mar. 20						53.2	25.2
June 20		43	15	33.9	5.9		
Oct. 22	29		14		4.9		24.2
Dec. 17	28		15		5.9		25.2
1926							
July 1	27		16		6.9		26.2
Aug. 16		39.6	12.4	32.8	5.8	52.2	25.2
1927							
Jan. 15	26	40.7	14.7	31.8	5.8	51.2	25.2
Aug. 1		37	11	31.0	5	48.5	22.5
Sept. 1		36	10				
Oct. 20	25	33	8		6	47	22
Dec. 10		32	7	29.5	4.5		
Dec. 21	24.3		7.7		5.2		22.7
1928							
Jan. 3		31.5	7.2	29	4.7	46.5	22.2
Mar. 7		31.3	7.0	28.5	4.2	46	21.7
May 25						45	20.7
1929			No changes				
1930							
June 26	23.3	30.3	7	27.5	4.2	44	20.7
1931							
May 26						40	16.7

* Sources of price data:

2S and 3S flat sheet, 1925–1931, from BMTC appellant v. ACOA, Exhibit 133.

Ingot and 2S and 3S coiled sheet and spread, 1918–1931, from BMTC v. ACOA appellant, Exhibit 550.

17S coiled sheet, 1922–1924, from BMTC appellant v. ACOA, Exhibit 239; 1925–1931, *ibid.*, Exhibit 133.

All sheet prices, 1918–1931, are for 15-ton lots. A few minor changes in prices have been omitted from the table. No explanation is given of the striking increase in the price of 17S sheet in 1924 shown by the figures.

† 2S is commercially pure aluminum, 3S is a manganese alloy, 17S is an alloy of the duralumin class.

‡ From August 15 to November 15, 1921, scheduled prices of ingot were suspended, owing to unsettled conditions of the market.

§ 18 and 19 gauge.

Reductions in the prices of most other gauges of the varieties of sheet included in the table narrowed the differentials for those also. The same was evidently true of most gauges of the strong alloy sheet 25S and 51S, both coiled and flat, as well as of tubing.[25]

Two of the three independent firms which entered the rolling branch after the war do not seem to have prospered. No information about the experience of the third, the Fairmont Manufacturing Company, is at hand. It appears that the United Smelting and Aluminum Company, an older concern, has not expanded its rolling operations to sizable proportions.

Sales of duralumin sheet by the Baush Machine Tool Company increased rapidly from 1919 to 1925. Thereafter they declined markedly, while sales of similar alloy sheet (17S) of the Aluminum Company continued to increase. Table 29 shows the sales by both companies in the period 1925–1931 of similar alloy sheet in sizes for which the Baush Mill was equipped, and the declining proportion obtained by the Baush firm. The amount of duralumin forgings sold by the latter also fell off substantially after 1925. Its sales of duralumin rod, wire, and tubing grew steadily until the depression began, with the result that the average of the total pounds of duralumin products sold by this firm in 1928 and 1929 was about equal to its sales in 1924, although its total for 1925 was not subsequently reached.[26] Sales of pure aluminum sheet and alloy sheet not of the duralumin class by this firm never constituted a sizable portion of the market for these types of product.[27] It appears that the metals division of the Baush company incurred losses in every year during the period January 1, 1919–June 30, 1931.[28] In 1931 the Baush firm brought suit against the Aluminum Company for damages alleged to have resulted from the price policies of the latter. Testimony concerning the relative efficiency of the Baush firm was conflicting.[29]

[25] BMTC v. ACOA appellant, Exhibit 123.

[26] BMTC v. ACOA appellant, Exhibit 308. Of the total weight of duralumin and 17S products sold by the two firms the Baush proportion dropped from about 30 per cent in 1925 to a little over 7 per cent in 1929 and about 3 per cent in 1931 (*ibid.*, Exhibit 244). However, a substantial part of the increase in total sales evidently occurred in products or sizes which the Baush firm was not equipped to produce. Figures for competitive sizes and products were not separated except for sheet.

[27] *Ibid.*, Exhibit 308.

[28] *Ibid.*, Exhibit 195.

[29] It would appear that the output of pure aluminum sheet of the Baush firm was

TABLE 29

SALES OF SIMILAR DURALUMIN ALLOY SHEET IN COMPETITIVE SIZES BY THE BAUSH MACHINE TOOL COMPANY AND THE ALUMINUM COMPANY OF AMERICA, 1925–1931 *

(*In Pounds*)

	1925	1926	1927	1928	1929	1930	1931
Baush duralumin	272,584	256,118	178,595	167,741	187,332	50,524	52,256
17S of Aluminum Company †	136,326	233,048	329,864	458,518	837,959	427,684	660,517
Total	408,910	489,166	508,459	626,259	1,025,291	478,208	712,773
Baush percentage	66.66	52.36	35.12	26.78	18.27	10.56	7.33

* The data of this table appeared in Exhibit 522 in BMTC v. ACOA appellant.

† 18 inches width and under, not Alclad.

The Sheet Aluminum Corporation began to produce pure aluminum sheet in 1926 in a plant at Jackson, Michigan, advantageously located with respect to utensil concerns and the automobile trade. As the ingot-sheet spread declined, this firm turned its attention increasingly to the making of products from a new series of strong alloys called Hyblum which it introduced in 1928. It is said that the change was made in the belief that the more individual product would permit a greater price spread. The Sheet Aluminum Corporation appears to have experienced financial difficulties from the first year of the depression on. Late in 1934 this firm instituted proceedings against the Aluminum Company similar to the Baush suit. It seems questionable whether the output of pure aluminum sheet of the Sheet Aluminum Corporation ever reached a very efficient size.

Explanation of the narrowing of the ingot-sheet price differential in the second half of the twenties might be found in any one or more of the following factors. (1) Reduction in the cost of rolling. (2) Desire on the part of the integrated firm to weaken or drive out independent rolling mills. (3) A change in demand conditions making profitable a greater degree of discrimination between the prices returned per pound of the basic component in the form of sheet products and in the form of products of other important fabricating branches in which a substantial part of the business was done by independents who purchased ingot from the integrated firm.[30] (4) Simply a desire to extend the use of sheet products, particularly the strong wrought-alloy products, for the pleasure of seeing aluminum gain a position beside steel as a basic construction material, even though profits were sacrificed in the process. To distinguish the last point more precisely from the third, let it be assumed that the third point includes a policy of sacrificing immediate profits with the purpose of increasing profits some few years hence in all cases where the present value of the expected future gain exceeds the present sacrifice; while the last point signifies a policy of sacrificing more than the present value

far below the scale required for efficiency. Whether or not this was true of duralumin sheet does not seem clearly established.

[30] Price of the basic component is derived by subtracting the cost of converting it into the final product from the price received for the latter. The reader should recall the discussion of discrimination above, pp. 217 ff.

of the expected future gain.[31] (5) Overinvestment in sheet facilities due merely to mistakes in estimating demand.

A study of the available evidence of all sorts bearing upon the ingot-sheet price differential does not yield any clearly defined explanation of the declining spread. The Aluminum Company has installed larger and more efficient rolling mills in the past fifteen years. There seems to be no question that its rolling costs have been reduced in that period. Unfortunately, no thoroughly satisfactory data upon the relations between cost of rolling and the ingot-sheet price differential are obtainable. For the years prior to 1925 there are no data which enable comparison of these two quantities; and the fragmentary cost figures applying to the whole period since the war do not afford any adequate measure of the extent of cost reductions.[32] Although one of the chief allegations in the Baush suit was that the Aluminum Company had narrowed its differentials between the prices of ingot and various products until they did not cover its fabricating costs, the data introduced relevant to this charge were meager.

Table 30 shows annual composite averages of price and of fabricating cost of all aluminum and aluminum-alloy sheet sold by the Aluminum Company in the years 1925–1930, and compares both the average price differential with average fabricating cost and the derived price of 99 per cent ingot (average price of sheet less average fabricating cost) with the average list price of 99 per cent ingot sold in the market. Since average fabricating cost apparently covers only direct operating expenses, plant overhead, depreciation, and taxes, and does not include any expense on account of administrative overhead, use of plant, working capital, or selling, it would appear that the prices which the company re-

[31] Expenditures by those in control pursuant to the latter policy would really constitute consumption rather than investment, from their point of view.

[32] The Federal Trade Commission secured from the Aluminum Company figures showing total weight of aluminum sheet fabricated in each year 1921–1924 and the aggregate fabricating cost, evidently direct operating cost plus plant overhead (FTC Docket 1335, Commission Exhibit 735). These indicate that annual average cost varied in these years from about 8.5 cents to about 10 cents per pound. An officer of the company testified in 1928 that average cost was then 5.5 to 6.5 cents per pound (*ibid.*, Record, pp. 3265–3266). Since these figures evidently represent composite averages of different varieties, sizes, and gauges there is little reason to suppose that they provide an accurate measure of changes in efficiency. Comparable average price differentials for the years in question are not obtainable.

TABLE 30

ALUMINUM COMPANY OF AMERICA — PRICE AND COST DATA FOR ALUMINUM AND ALUMINUM-ALLOY SHEET OF ALL CLASSES AND 99 PER CENT INGOT, 1925–1930

(*Cents per Pound*)

Year	(1) Average Price Received for Sheet	(2) Average List Price of 99 Per Cent Ingot	(3) Average Ingot-Sheet Price Differential (1)–(2)	(4) Average Fabricating Cost of Sheet	(5) Average Price of Sheet Less Average Fabricating Cost (1)–(4)
1925	37.61	28.15	9.46	9.33	28.28
1926	36.30	27.50	8.80	8.50	27.80
1927	32.98	25.82	7.16	7.93	25.05
1928	30.25	24.30	5.95	7.72	22.53
1929	32.37	24.30	8.07	8.27	24.10
1930	31.04	23.78	7.26	6.99	24.05

The figures in columns 1, 4, and 5 appear in Exhibit 313, BMTC v. ACOA appellant; those in column 2 are computed from figures in Exhibit 344, *ibid.* The figures of column 1 evidently represent the average gross receipts per pound for all sheet sold by the Aluminum Company. It appears that the figure of 36.28 cents per pound for the year 1926 given in Exhibit 313 was the result of a slight mistake in computation. The figures of column 2 represent yearly averages of the list prices for 99 per cent ingot rather than the average gross receipts per pound, which are not available. The average fabricating costs in column 4 include all items of direct expense and plant overhead, depreciation, taxes. Administrative overhead, use of plant, cost of working capital, and selling expense are evidently not included (BMTC appellant v. ACOA, fols. 2685 ff.).

ceived for its sheet of all sorts did not return sums equal to its list price of ingot plus the average full cost of fabrication, including normal returns on fabricating investment, in some of these years at least. To the extent that ingot was sold to rolling mills at prices different from the list prices during these years the actual differentials with respect to such sales differed from those shown in column 3. Over a period of about twenty-three months in 1928, 1929, and 1930 the Aluminum Company evidently sold 99 per cent ingot to the Sheet Aluminum Corporation at 23.8 cents per pound when its list price was 24.3 cents.[33] Whether concessions were made to other mills by the Aluminum Company does not appear.

With one exception the data of Table 30 were not separated for the different classes of aluminum and alloy sheet nor for different gauges, so it does not appear whether very low prices on a few varieties of sheet pulled down the composite average. Average prices and average fabricating costs (similar in make-up to those of the above table) were presented for 17S sheet.[34] It appears that in this case the differential between the average sales price of sheet and the average list price of ingot barely covered average fabricating cost in 1925 but exceeded it thereafter by amounts ranging from about 2 cents to about 8.5 cents. Whether or not these margins covered all other costs in various years does not appear.[35] An officer of the Aluminum Company has testified that, although the company had worked up a substantial tonnage in duralumin alloys (which include other alloys as well as 17S) and their products between 1916 and 1929, it had incurred over the whole period a net loss upon the several million dollars of investment devoted to this part of the business.[36] It was not stated whether the company made any earnings in that department of the business in any single year during that period or in subsequent years. The available data upon the ingot-sheet cost and price differentials, although quite inadequate for thorough analysis, seem to indicate that in part of the period 1925–1930, at least, the price differential of the Aluminum Company was below its full conversion cost for some varieties or gauges of sheet.[37] Whether these were competitive

[33] BMTC v. ACOA appellant, Exhibit 262.

[34] *Ibid.*, Exhibit 335.

[35] Allocations of total administrative overhead and selling expense between various products were computed by attorneys for the plaintiff on the basis of relative dollar value of sales. For 17S sheet these two items amounted to about 3.5 cents in each year 1925–1929. It is highly questionable whether such a method of allocation yields even rough approximations to true cost.

[36] BMTC appellant v. ACOA, fols. 5108 ff. It would appear that this statement referred to the investment in fabricating facilities only, not the whole investment from ore up used for duralumin.

[37] Certain other data on this matter introduced in the Baush case do not serve to amplify this conclusion. Testimony on the fabricating cost of certain gauges of aluminum sheet in 1926 appears in BMTC v. ACOA appellant, fols. 5333 ff. Data on the marked reduction in cost of rolling hard alloys after installation of a new mill at Alcoa in 1931 are also given (*ibid.*, fols. 5313–5318). The Aluminum Com-

or noncompetitive varieties and gauges is not revealed by the information at hand.

If, as seems likely, the price differential did not always cover the full cost of rolling in the years in question, the explanation is not altogether clear. No intent to weaken or suppress competition seems to be proved by the somewhat conflicting testimony in the two trials of the Baush case.[38] The whole of the evidence seems to allow either the interpretation that the declining price differential was partly due to a desire to take business away from the independents or that it is explainable on other grounds entirely, but does not indicate that discouragement of competition was the chief motive.

Several considerations suggest that the most profitable policy for the Aluminum Company would have required narrowing the ingot-sheet price differential by substantial reductions in sheet prices in the absence of competitors in the rolling branch. In the early twenties aluminum sheet began to suffer from the competition of sheet steel and tubular steel in the automobile and aircraft markets, and stainless steel assumed competitive importance. Shortly thereafter, the new strong alloys reached the stage of development where wide industrial application was feasible. Marked price reductions on virgin ingot, alloy ingot, or alloy sheet were evidently required to stimulate demand and test out its elasticity in the potential markets for alloy sheet and its products — structural and decorative uses in building construction, furniture, bus and truck bodies, railroad cars, and so forth. Lowering the price of alloy sheet would enable maintenance of the most profitable price of virgin ingot sold in many other markets, old and new, where drastic price reductions might diminish profits. General sale of alloy ingot at low prices might have resulted in some transference of demand for virgin ingot to the former.[39] Furthermore,

---

pany followed a policy of retaining title to the metal and paying tolls to its 100 per cent subsidiary which produced sheet. Attorneys for the plaintiff computed that the difference between the prices received by the Aluminum Company for 17S sheet and the toll charges paid by it to its fabricating subsidiary was less in each year 1926 and 1928–1930 than the ingot price. The basis upon which the toll charges were set up does not appear, however.

[38] Some little testimony indicating such intent was presented by certain witnesses and denied by the chief officials of the Aluminum Company.

[39] Given a price advantage, this would occur in so far as the alloy in question

it is doubtful if independent enterprise would have been willing to assume the risks of attempting to develop the potential markets for the high-strength alloys as rapidly as the Aluminum Company evidently desired, especially in the face of the deterrents to the establishment of independent rolling mills which have been explained above. An executive has given it as his opinion that with existing conditions in the steel and other metal markets gross dollars received from the sale of strong-alloy products would have been less during the years 1925–1931 had their prices been several cents higher.[40]

A decrease in the prices of the new alloy sheet may have required cuts in the price of pure aluminum sheet and manganese-alloy sheet. If consumers of these two types of sheet began to purchase the new alloy sheet, which is more expensive to produce, the profits on the latter might be increased by much less than the reduction in the profits in the two other markets. Furthermore, in some cases different varieties of sheet could be used in complementary fashion on the same job, provided the prices of all were low enough. And, as mentioned above, sheet steel and tubular steel were becoming stronger competitors. It would appear that the demand for ordinary sheet, as well as the new alloy sheet, was becoming more elastic at lower prices in the twenties than had been the case earlier. Evidently the prices of several varieties and gauges of sheet were reduced to the point where they contained, after subtracting the full costs of conversion, lower prices for the basic component than it had formerly brought in the form of sheet or than it was then returning in the form of some other products. Apparently earnings per unit of investment were lower on that part of the investment from ore up devoted to the production and sale of high-strength alloys than in most other parts of the business. Losses may have been incurred on the former. No figures are given which would enable such a computation. During the twenties several million dollars were expended for additions and improvements in rolling equipment.[41]

---

could be substituted for virgin metal or other alloys or if the virgin metal could be easily extracted from the alloy.

[40] BMTC v. ACOA appellant, fol. 5324. By implication the indicated elasticity of demand seems to depend on conditions in other metal markets rather than on the existence of competing producers of hard alloys.

[41] *Ibid.*, Exhibit 473.

Capacity seems to have been expanded much ahead of demand. Unless this is to be explained entirely by mistakes in estimating demand, which is to be doubted, it appears that profits which might have been earned by investment in something else were sacrificed by a policy of overexpanding the investment devoted to the hard alloys to reduce direct cost, and lowering their prices drastically to build up consumption more rapidly than would otherwise have been accomplished.[42]

What is to be said as a matter of principle about the ingot-sheet price policy of an integrated firm possessing a monopoly of aluminum ingot? Two questions arise, one concerning the effect upon competition in the rolling branch, the other relating to discriminatory prices received for the basic component. Considerations of efficiency seem to afford no justification for single-firm monopoly at the rolling stage.[43] Obviously the existence of oligopoly, or of conditions approaching pure competition, if they become possible, depends upon the maintenance of an ingot-sheet price differential which ordinarily covers the average full cost of an efficient mill. This means that the price of sheet must tend on the average to equal the price at which ingot is sold in the market plus the full cost of conversion. If monopoly profit is received on the ingot sold it must be received on the ingot fabricated into sheet, or else competitors cannot survive.[44]

As explained in an earlier chapter, a discriminatory price pattern does not in general seem economically justifiable in the aluminum industry, since technological conditions do not necessitate the existence of substantial overcapacity for long periods. In so far as the maintenance of an inadequate ingot-sheet price differential facilitates discrimination in prices of the basic component it must be regarded as uneconomic. The same would be

[42] Additional evidence which seems to support this view is contained in the testimony of officers of the company and others. See particularly BMTC appellant v. ACOA, fols. 4310, 4362, 5108 ff., 5195 ff.

[43] It is not clear whether monopoly of production of alloy sheet would have been justified by efficiency for the early years of its development (which may not yet be over). If competition is likely to be the better form of market control it will probably be more effective when some competitors enter at an early stage.

[44] According to the testimony of one executive it has been the general policy of the company to set prices of fabricated products so as to cover the market price of ingot plus cost of fabrication, including profit (BMTC v. ACOA appellant, fol. 5713).

true of a differential which more than covered the full conversion cost. Consumers of sheet products would pay less or more for the metal contained than consumers of products whose prices more nearly equaled the sum of the market price of ingot and the full conversion cost; and the former would consume larger or smaller percentages of the amounts which they would take at prices just equal to the market price of ingot plus the cost of fabrication.

It may be urged, however, that discrimination would be justifiable when incident to temporarily unprofitable prices designed to develop rapidly the markets for new products, such as the strong alloys, or to widen the markets of old products, such as 2S and 3S sheet. In the first place, true overinvestment — a condition such that the present value of future gains does not offset present deficits measured from normal returns — is not justified, providing the capital would be alternatively invested where its social value would be greater.[45] Secondly, low development prices in some markets would not be justifiable if they would not be adopted unless the resulting deficits could be recouped by prices returning more than normal profits from other markets; under such circumstances one set of consumers would be forced to subsidize another group. Provided returns on all parts of the investment devoted to serving other markets tended to be equal and not above normal, special low prices to cultivate a new market should not be considered uneconomic discrimination as long as the temporary overinvestment is not so large as to constitute true overinvestment as defined above[46] — the discrimination is a necessary temporary prelude to a better allocation of economic resources. It is plain, however, that such a policy pursued over several years, though unobjectionable in itself, might, as a result of the accompanying inadequate ingot-sheet price differential, result in restricting the freedom of entry at the rolling stage and weakening or eliminating existing single-stage mills which were not provided with large financial reserves. Under certain circum-

[45] If the funds would not be so invested, and if the persons who decide what to do with them are also the owners of these sums, overinvestment for consumption (see n. 31, p. 387 above) is not objectionable unless it prevents the realization of some social benefits generally held superior.

[46] The content of the terms here involved must include allowance for reasonable mistakes in estimating future demand.

stances it is impossible to have both the existence of independent rolling mills and the advantages of a price and investment policy on the part of the integrated firm which results in rapid development of new markets. Obviously the same consequences might occur, in the absence of any intent to discourage competitors, from a policy involving unjustifiable discrimination. If effective freedom of competition is to be preserved, neither justifiable nor unjustifiable discrimination can be permitted to result in a differential which is insufficient to sustain independent mills. It follows that the antitrust laws must fail of their purpose unless the maintenance by an integrated firm of an inadequate price differential is considered unlawful, even though there is no intent to injure competitors. According to the criteria of unfair or uneconomic methods of competition given in the next chapter such a policy represents a practice which is uneconomic in effect, whatever the intent.

Inasmuch as marginal direct cost is often employed in this book as the criterion of correct pricing, it should, perhaps, be explained why the analysis of the last few pages has been put in terms of average full cost. As the trained economist will realize, the situation treated here concerns a policy which involves additional investment. When the additional investment is under consideration, marginal cost includes normal returns to capital and becomes very nearly equal to average full cost in the absence of a marked tendency to increasing or diminishing average cost. The dilemma presented by the fact that competitors may suffer if the stronger firm, after creating overinvestment, properly sets prices according to marginal direct cost will be examined in Chapter XX.

It is, perhaps, questionable whether freedom of competition for one-stage sheet firms should be preserved. Possibly the advantages accruing from more rapid development of new markets by one or several integrated firms would outweigh the social benefits incident to free competition. It should be recognized, however, that the maintenance of free competition in all fabricating stages would reduce the extent of uneconomic discrimination, although it would not prevent restriction of investment or output devoted to ingot, and hence restriction of output of all final products. In any event the choice between the two alternatives — free competition, or rapid widening of markets which requires deficits measured from normal earnings for some years — should be made by the government and announced in the form of a definite set of principles governing price policies, so that the integrated firm, existing independents, and those contemplating entry will know the rules of the game. As long as the choice is in favor of free competition in the fabricating stages it should be decreed that the integrated firm may make low development prices on fabricated products only if it lowers the price of ingot correspondingly.

It is obvious that the power to control the ingot-sheet price differential, the fear that it may be regulated for the benefit of the Aluminum Company, regardless of the effects upon others, and the knowledge that under certain circumstances independents may be adversely affected, are sufficient to prevent the growth of any substantial independent rolling-mill group unless there exists a conviction that government agencies will ensure that the differential ordinarily covers the full cost of an efficient rolling firm. The history of the relations between the government and the aluminum industry has not as yet afforded such assurance, as will be evident after reading the later chapters.

While competition in the production of virgin aluminum ingot has never existed in the United States, the less difficult obstacles to entry into sheet production have resulted in a small investment of capital and enterprise in independent rolling mills. But the development of a substantial group of independents has been deterred by elements among which the more important seem to be the dangers of dependence for ingot supply upon either the domestic monopoly or the foreign producers, and the fear of an ingot-sheet price differential which, in fluctuating for the benefit of the Aluminum Company, may injure others, whether or not it be manipulated to discourage competitors.

CHAPTER XVII

COMPETITIVE METHODS AND GOVERNMENT CONTROL

1. Consent Decree of 1912

We have seen that early in its career the Aluminum Company entered the utensil and castings branches of the industry. Just as its reduction patents were expiring it secured a prominent position, which has been held ever since, in these divisions, by participating in two consolidations. It is to be doubted that the company would have attempted to monopolize completely the utensil and castings branches, where the economical scale is small and entry is easy.[1] Complete monopolization would have required outright refusal to sell materials to fabricators or other drastic policies. Monopolization of these branches would not be required in order to maintain a price structure which discriminated between the two broad markets for fabricated goods. As long as the company controlled the sheet price the ingot quotation could be adjusted to the other market.[2]

In 1909 the Aluminum Company of America took 40 per cent of the stock of the Aluminum Castings Company, formed as a consolidation of the Allyne Brass Foundry Companies of Ohio, New York, and Michigan, the Syracuse Aluminum and Bronze Company, the Eclipse Foundry Company of Detroit, and the foundry department of the United States Aluminum Company. It appears that the new corporation, which manufactured aluminum, brass, and bronze castings in plants at Detroit, Buffalo, Syracuse, New

[1] Below, pp. 408 and 443.

[2] While complete monopolization of these broad markets might have enabled somewhat more discrimination between the sub-markets of each, it is probable that the most profitable margins of discrimination between the sub-markets within either one have been less than the most lucrative degree of discrimination between the broad markets. Until recently, at least, it would appear that the elasticities of demand from the sub-markets of one broad market did not differ much, because they were influenced by competition of much the same materials; whereas between the two broad markets the competing materials differed — e.g., brass, bronze, iron, and steel in the castings market, and enameled iron, tin, wood, and steel in the market for sheet products.

Kensington, Fairfield, Connecticut, Manitowoc, Wisconsin, and Cleveland, produced over half of the total output of aluminum castings in the country.[3]

The United States Aluminum Company had been an important factor in the utensil branch from its organization in 1901, but its business had constituted much less than half the output in that field. In 1908 three of the leading utensil companies decided to end a price war by consolidating. The Aluminum Company of America was called in to arbitrate and help finance the combination.[4] In this instance its interest was one of 25 per cent, and the utensil factory of the United States Aluminum Company was not included in the Aluminum Goods Manufacturing Company, as the new corporation was called. Since 1911 two executives of the Aluminum Company have held two of the six directorships of the Goods Company, and in 1914 an accountant of the Pittsburgh concern was made secretary and treasurer of the new corporation.[5] Shortly after the formation of the Goods Company internal dissension arose between Wisconsin and New Jersey members. When the dispute was settled, through the purchase of the eastern interests by the western, an agreement was entered into between the Aluminum Company and G. A. Kruttschnitt and J. C. Coleman, the New Jersey disputants, whereby the latter agreed not to enter any branch of the aluminum business in that part of the United States east of Denver for a period of twenty years.

[3] Tariff Hearings, 1912–1913, pp. 1527, 1532. In its petition the government averred that this combination was formed "with the declared purpose of 'closing the only door remaining open to our complete control of the aluminum industry in America.'" The government further alleged that "it was the purpose and intent of the defendant, in the formation of said castings combination, as then declared, to then take in only such plants as were needed to do the castings business, and thereafter to so discriminate in their favor and to give them such preferential prices on their ingot, beginning with 2 cents and increasing same to 5 cents per pound, if necessary, as would either destroy their competitors or compel them to come into said combination" and that "such declared intent has been and is being carried out, with the result that several competing castings companies have been compelled to sell their business at a loss to said castings company." In the absence of adjudication the truth of the allegations cannot be accepted.

[4] See KFR, p. 71. The government alleged in its 1912 petition that the Aluminum Company had formed the consolidation to further its control of the utensil and novelty field, and had discriminated in favor of its new offspring (Tariff Hearings, *op. cit.*, p. 1528).

[5] BR, p. 52.

According to the Department of Justice, the Aluminum Company had employed many unfair methods of competition in conjunction with these combinations and contracts, and those concerning bauxite and foreign competition, with the intent of destroying substantial competition in all branches of the industry. In line with the newer antitrust policy of that time, the government's petition for a decree against the company did not ask to have the combinations dissolved, but requested annulment of restrictive agreements and perpetual injunctions against the use of certain unfair methods of competition which the company was alleged to have employed. The decree which issued in 1912 conformed to these requests. Lack of adjudication makes it impossible to know the actual extent to which these methods had been used, or the consequences.

2. Government by Investigation

The post-war decade witnessed repeated charges that the Aluminum Company had violated the consent decree and the Clayton and Federal Trade Commission Acts. An endeavor to determine the extent of truth in these allegations will be prefaced by a brief review of the government investigations in order to provide some understanding of the political currents which have blown about the government agencies concerned with enforcement of the antitrust laws in this industry.

When Mr. Andrew Mellon became Secretary of the Treasury in 1921 and the Republican Congress raised the duty upon aluminum to 5 cents per pound, an invitation to political maneuvering was extended. While engaged in an inquisitorial orgy in 1924 the Senate appointed a select committee to investigate the Bureau of Internal Revenue. Under the chairmanship of Senator Couzens, whose enmity for Mr. Mellon had made itself known some time before, the committee scrutinized the tax refunds to the Mellon companies without finding any creditable evidence of the exercise of influence by the Secretary. During the presidential campaign of the following autumn a virulent Democratic attack was launched upon the administration for its tariff favors and tax refunds to the Aluminum Company. Just a month before the election the Federal Trade Commission transmitted to the Senate volume III of its report on the House Furnishings Industry, in which it stated that the methods of competition employed by the Aluminum Company in the cooking-utensil industry *appeared* to show repeated violations of the consent decree of 1912. It was extremely unfortunate that this report made its appearance during the height of the political attack. It did not actually demonstrate that violations of the decree had occurred. Further investigation by the Department of Justice was required to decide whether the evidence was conclusive. What had seemed sure-fire ammunition for the heavy artillery which the Democrats now swung into action merely raised a cloud of smoke which has obscured the truth. It would appear that the government agencies charged with impartial policing were to some extent drawn into this sham battle. Shortly after the submission of the report to the Senate, the Commission, as required by law, sent a copy of it to the Attorney General. A letter of acknowledgement to the Commission on January 30, 1925, written by Assistant to the Attorney General Seymour, to whom the report had been referred, and signed by Attorney General Stone, stated that it was apparent from the report that the Aluminum Company had violated several provisions of the decree in the period covered by the Commission's investigation (through the year 1922), and that the Department was undertaking an inquiry to bring the facts down to date.[6] Owing to charges by the officers of the company and others that the statements of the Trade Commission report were unreliable, a fairly comprehensive investigation extending over a year was carried out by the Department.[7]

On January 2, 1926, before the investigation was completed, the Department abandoned its customary policy, apparently to forestall an anticipated newspaper attack, and gave out a statement reviewing the history of its relations with the Aluminum Company and explaining that a careful inquiry of the complaints had

[6] *Hearings before the Committee on the Judiciary, United States Senate, 69th Congress, 1st Session, pursuant to Senate Resolution 109, directing an inquiry by the Committee on the Judiciary as to whether due expedition has been observed by the Department of Justice in prosecuting the inquiry in the matter of the Aluminum Company of America* (Washington, 1926), pp. 1 and 121.

[7] On March 18, 1925, Mr. John Garibaldi Sargent succeeded Mr. Stone as Attorney General, and shortly thereafter Col. William Donovan took Mr. Seymour's position.

not as yet revealed evidence to support the charge that the decree had been violated.[8] It was said that the investigation would be completed in about three weeks. This was the signal for a Senate resolution directing the Committee on the Judiciary to institute an inquiry to determine whether the Department of Justice had exercised due expedition in the prosecution of its investigations.[9] The committee hearings which followed were conducted by that experienced investigator, Senator Walsh of Montana. They disclosed a deplorable lack of coöperation between the Department of Justice and the Trade Commission, a disposition toward thoroughness combined with a rather casual attitude toward the need for celerity, and a somewhat inept lack of knowledge about the affair on the part of Attorney General Sargent. Inasmuch as the Department's inquiry had not been completed, the officials were unable to testify definitely whether or not there had occurred any violations of the decree. The majority report of the committee recommending a Senate investigation of the methods of competition practiced by the Aluminum Company was not accepted by that chamber.

The report of the Department's investigators was presented to the Senate when completed in February and printed as a Senate document.[10] It cannot be said that it altogether justified the pretensions to an exhaustive and intelligent thoroughness which had been claimed for the investigation.[11] In several places it appears to reveal a perhaps not unnatural desire to refute the criticisms of Senator Walsh and the allegations of the Federal Trade Commission, instead of the impartial attitude proper to such an agency. The report unequivocally exonerated the Aluminum Company of any violation of the decree. The appearance of the Kitchen Furnishings Report during the height of the presidential campaign

[8] Committee on Judiciary Hearings, p. 213.

[9] The resolution recited that the statute of limitations applicable to proceedings for criminal contempt would, by the end of January, have run against any prosecution based upon acts committed before the letter of Attorney General Stone written on January 30, 1925. Later both the Senate committee and representatives of the Department agreed that the one-year statute of limitations did not apply to such a decree as that entered against the Aluminum Company, and that the general statute of three years was the applicable law. See Senate Report no. 177, p. 9 (69 Cong., 1 Sess.).

[10] This was the Benham Report.

[11] Claimed in testimony of officials before the Committee on the Judiciary.

had been freely ascribed by many to a political purpose; and part of the Department's investigation was carried out in an atmosphere charged with allegations of influence in high places and counter endeavors to protect the administration from discredit. It is undoubtedly true that in this instance, as in most political attacks, much of what was charged by the various groups was without foundation. But it is impossible for an outsider to know whether all the charges and countercharges were groundless.

In the meantime the Federal Trade Commission had embodied several of the allegations of the Kitchen Furnishings Report in a formal complaint which also included charges of employment of unfair methods in the sand-castings industry and attempted monopolization of the scrap-aluminum market. During the course of the proceedings, which stretched over nearly five years, the attorneys for the Commission endeavored unsuccessfully to have the complaint amended to contain charges of exclusion of potential competition in ingot production by purchases of foreign bauxite and water-power sites, and unfair methods in the sale of alloy sheet. Finally in April 1930 the original complaint was dismissed by the Commission for lack of evidence to support the charges.

The Department of Justice also investigated the activities of the Aluminum Company in the sand-castings industry and the scrap market,[12] but has published no report on this matter. No proceedings were instituted. A few years earlier the attention of the Department had been directed to the projected lease by the Aluminum Company of the plants of its castings subsidiary, Aluminum Manufactures. After inquiry the Department stated that no evidence had been disclosed which would justify it in interfering.[13] During 1934 the Department of Justice again undertook an investigation of various aspects of the aluminum industry. No results have been made known as yet.[13a]

Finally, the National Recovery Administration has examined certain matters bearing upon the relations between the Aluminum Company and independent fabricators in connection with the operation of the aluminum code. A report prepared by Leon Hender-

[12] BR, p. 85.

[13] BR, pp. 8–11.

[13a] In April 1937, while this book was in press, the Department filed suit against the Aluminum Company alleging violation of the Sherman Act and asking for dissolution of the company.

son, director of the division of research and planning, was issued in 1935.

Careful study of the public reports and the testimony and exhibits of the hearings and the Trade Commission case described above yields disappointing results.[14] Although several thousand pages of information have been made public, it is impossible to reach definite conclusions upon many of the chief questions. This unfortunate result is due partly to misconception of the issues, and partly to the failure to obtain adequate data of the right sort.

3. Criteria of Unfair Methods of Competition

At this point it will be well to restate the criteria for uneconomic or "unfair" methods of competition.[15] (It should be understood that the concept of unfair methods here presented is one developed by economists and is not a definition of unfair methods according to law.) An unfair method of competition may be defined as any practice which gives to someone a differential financial advantage not resulting from greater effectiveness or purely fortuitous circumstances. This simple definition needs some elaboration. The central test of unfairness is interference with the tendency for the most effective producers to supply the demands of consumers. Methods which are instrumental in creating single-firm monopoly not based upon superior efficiency are clearly unfair. But the matter is rarely as simple as this, so there must be standards for judgment in instances where the granting of advantages which do not measure differential effectiveness falls far short of producing complete monopoly. In formulating such standards we may use the two tests of intent and effect, and it should be

[14] Both the Department of Justice and the Economic Division of the Federal Trade Commission appear to have accumulated extensive files of information on the aluminum industry which are not available for study by the author.

[15] General usage seems to render it desirable to keep the term "unfair," which was originally applied to practices involving fraud, misrepresentation, stealing trade secrets, or the like, that were condemned by the accepted ideas of right conduct. The use of the same term to refer to practices such as those under examination in these chapters, which are often not inconsistent with the generally accepted ethical code, is, perhaps, unfortunate. The confusion thus engendered, from which many never escape, may, however, be more than repaid by the possibility of gaining ethical sanctions for the best economic tests. The business world seems much more approachable on its ethical soundness than its economic sanity.

recognized that a practice may be unfair in either purpose or result with regard to the stage of industry in which he who uses it operates, or with respect to the stages in which the recipient of favors, or injury, is engaged. A method employed with the intent to secure for the user an advantage over his competitors which his relative effectiveness would not allow is unfair. And so also is a practice the purpose of which is to grant to a purchaser a differential advantage over his competitors, or inflict upon him a disadvantage which he would not otherwise suffer — that is, to accord different treatment to purchasers whose demands upon the seller are similar in nature (e.g., specifications, shipping instructions, and the like) and for whom the seller's conditions of supply are alike. Obvious illustrations of these two general cases are, respectively, selling below cost to steal some of a competitor's market, and special discounts to a purchaser to enable him to undersell his rivals. Any arbitrary changing of conditions to avoid discovery of such discrimination would be evidence of unfair intent.

Competitive practices which yield any of these results are unfair in effect, whatever the intent, unless their employment is due to abnormal conditions outside the control of the user. In the actual business world intent is usually so difficult to ascertain surely that attempts to control competitive practices must rely mainly on the test of effect. While it seems desirable to use the broad tests given above it should not be concluded therefrom that, in the absence of unfair intent, the government agencies should attempt to deal with all methods which are unfair in effect. In a world where monopolistic competition is typical of most markets there are many practices, unfair in effect according to these criteria, the eradication of which by government would cost more than it would be worth to the community. But since the very existence of monopolistic competition makes easier the getting and giving of advantages not based upon effectiveness, it seems desirable to use the broad tests and hope that government agencies will single out those instances of unfair methods, elimination of which will most benefit the community.[16] With regard to those methods which

[16] There are two alternative types of criteria for unfairness. One would assess each case on its own merits without the trouble of working out any general criteria and decide, in each instant situation, whether there was a net balance of social ill or welfare proceeding from employment of the method at issue. (The approach

affect the stage of the industry where the purchaser is engaged, the intensity of the effect may be greater the greater the degree of monopoly power at the seller's stage. Consider, for instance, the case of several manufacturers of special steels who sell to a large number of machine-tool companies. There will probably be many instances of price discrimination which, judged by these criteria, are unfair methods of competition, but it does not seem likely that the instances could be as numerous or the margins of discrimination as wide as might be so if there were only one or two producers of special steels. Thus a method may be unfair but not worth bothering about because its quantitative results are small, whereas the same method used under a different set of circumstances may demand remedying because its undesirable results are much greater. This is the justification for scrutinizing more carefully the activities of firms which possess a large degree of monopoly power.

4. Injunctions of the Consent Decree of 1912

We have seen that the consent decree of 1912 was ineffectual in creating conditions under which substantial competition in ingot

---

of the Trade Commission often seems to be closer to this type than any other.) This criterion would often bring the same results as the one given in the text, but it is really no more than the most general statement of the aims of social control and provides no tests for measuring the balance of social consequences. The other criterion is the one explained in the text, limited to instances where the results are different from those which would naturally obtain under oligopoly or monopolistic competition. In an earlier chapter I have argued that the true consequences of monopoly can only be judged by comparison with the probable consequences under the degree of oligopoly or monopolistic competition proper to the industry under examination. But the competitive policies of oligopoly and monopolistic competition do not provide a useful bench mark from which to begin measurement of unfairness. The true benefits of production on a great scale, which justify the existence of a small number of competitors, may accrue to consumers only if the few compete upon a basis of effectiveness. If public policy does not insist upon this, some of the few may employ the elements of monopoly power already possessed to acquire more, or to endow favorites at another stage of industry with undue advantages. General economic welfare is injured if resources are not allocated among industries and firms on the basis of equal marginal products. In a free capitalistic economy the state does not allocate resources directly. But it should set such standards for trade practices as will minimize departure from the best allocation of resources.

and sheet production would emerge. In order to control the activities of the company in the finishing stages of the industry it provided a set of injunctions against more or less specifically designated practices. The next two chapters will be devoted to an examination of the success of the decree in preserving free and beneficial competition in the utensil and castings branches, and a study of the activities of the Federal Trade Commission working in the same cause under the 1914 legislation.

In order to "prevent the unlawful acquisition by defendant of a monopoly in any branch of manufacturing from crude or semi-finished aluminum" the decree enjoins the company from: (*a*) any form of combination for the control of output or price of any product manufactured from aluminum; (*b*) arbitrary delivery delays to competitors, refusal to ship or continue shipments already on order without reasonable cause, purposely delaying bills of lading, shipping known defective materials; (*c*) discriminating in price, under like or similar conditions, between independents and any of the companies in which the Aluminum Company has a financial interest, or charging higher prices to any competitor with the purpose or effect of putting him at a discriminatory disadvantage in bidding on contracts against the Aluminum Company or any concern in which it possesses a financial interest; (*d*) refusal to sell metal to prospective competitors upon the same terms as the company sells under similar circumstances to any concern in which it has a financial interest; (*e*) requiring, as a condition precedent to sale to a competitor, the divulgence of the terms proposed to secure the work for which the material is desired; (*f*) requiring, as a condition precedent to sale, an agreement not to compete with the Aluminum Company or any concern in which it is financially interested; (*g*) intimating that unless a competitor buys metal from the Aluminum Company or its interests he will be unable to obtain a supply sufficient in amount or cheap enough in price to permit competition with them; (*h*) taking the position that a company manufacturing any kind of aluminum goods which expands or engages in enterprises competitive with the Aluminum Company or its interests will, for that reason, be unable to procure its supply from the Aluminum Company or firms in which it has an interest.

The term "competitor" as used in these sections was defined to include

all persons, firms, or corporations engaged in or who are actually desiring or about to engage in the manufacture of any kind of products or goods from crude or semifinished aluminum, whose business is not controlled or not subject to be controlled by defendant, its officers and agents, either by virtue of ownership of all or a part of the capital stock of such concerns, or through any other form or device of financial interest.

Three provisos at the end of the decree seem to be especially related to the injunctions just noted. One states that nothing in the decree shall prevent the Aluminum Company from making special prices or terms in order to enlarge the use of aluminum in new employments or in competition with other materials. The other two attempt to make it impossible for the company to achieve the results proscribed in the above injunctions while not actually violating a literal construction of them.

Provided further, that nothing herein contained shall obligate defendant to furnish crude aluminum to those who are not its regular customers, to the disadvantage of those who are, whenever the supply of crude aluminum is insufficient to enable defendant to furnish crude aluminum to all persons who desire to purchase from defendant, but this proviso shall not relieve defendant from its obligation to perform all of its contract obligations, and neither shall this proviso, under the conditions of insufficient supply of crude aluminum referred to, be or constitute a permission to defendant to supply such crude aluminum to its regular customers mentioned with the purpose and effect of enabling defendant or its regular customers, under such existing conditions, to take away the trade and contracts of competitors.

Probably this proviso reflected recent events in the industry. It appears that in 1906, 1907, and again in 1912 a condition of "insufficient supply" had existed.

It seems extremely doubtful whether this section, as worded, could have any force at all. Clearly the purpose is to prevent the company from using a condition of shortage as a valid excuse for supplying itself, its subsidiaries, and its regular customers outside, to the exclusion of those independents who are not "regular customers," or supplying the latter with much less of their needs than the former are able to secure. But the two parts of the proviso seem to result in a stalemate. Possible interpretations of "regular customers," "disadvantage," "insufficient supply," "take away *the* trade and contracts," would furnish several pages of speculation. The section is worded in absolute rather than relative terms. If the company elects to follow its permission to furnish aluminum (evidently any aluminum at all) only to its regular customers it is obvious that some of the trade and contracts of others will be taken away unless they can buy on the same general terms from abroad, which may be impossible at times of scarcity. On the other hand, if the company must order its dealings with buyers in such a way that occasional customers do not have *any* of their "trade and contracts" taken away by itself or its regular customers the permission granted in the first paragraph becomes meaningless. As worded, the whole proviso directly invites the courts to avoid the difficult questions of economic relationships involved and settle any issue by a legal choice of emphasis rather than an inquiry into the facts of the market place and a judicious assessment of their significance.[17] Finally, restriction of this particular proviso to dealings in crude aluminum (evidently ingot), which appears quite inexplicable, renders it inoperative with regard to relations with utensil manufacturers, nearly all of whom have always purchased sheet.

In the third proviso attention is clearly directed to the business facts of the situation.

> Provided, further, that the raising by defendant of prices on crude or semifinished aluminum to any company which it owns or controls or in which it has a financial interest, regardless of market conditions, and for the mere purpose of doing likewise to competitors while avoiding the appearance of discrimination, shall be a violation of the letter and spirit of this decree.

[17] The decree included no explicit injunction against artificial creation of a situation of shortage. Probably this would constitute "delaying shipments without reasonable cause," which is forbidden in paragraph (*b*).

CHAPTER XVIII

COMPETITIVE METHODS IN THE UTENSIL INDUSTRY

1. Introductory

At the time of the consent decree it was said that the Aluminum Company controlled the production and sale of more than 75 per cent of the aluminum cooking utensils made in the United States.[1] Undoubtedly this estimate was based on the output and sales of the Aluminum Goods Manufacturing Company and of the United States Aluminum Company, a 100 per cent subsidiary of the Aluminum Company which manufactured the ware sold by the Aluminum Cooking Utensil Company. About ten years later these two concerns apparently produced not less than 65 per cent of the total output.[2] Some aspects of the growth of the industry since 1914 are shown in the accompanying table. The persistence of a fairly large number of relatively small establishments suggests that there are few economies of large-scale production. Ten or a dozen companies have accounted for the larger part of utensil production, and among these size varies from the Goods Company down to concerns whose output is only from 15 to 20 per cent of its production.[3] The nature of the operations of stamping or spinning suggests that there is a wide variation in size of the production unit within which unit cost varies but little. Financial, marketing, and management problems would seem to be relatively simple. Evidently there is little opportunity for large-scale economies in these departments. This industry seems to present a specimen which slips easily into the box labeled "constant cost."

During the growth of the industry the Aluminum Company of America through its utensil subsidiary has taken a leading part in the development of high-quality ware and has accomplished much in educating consumers to know the true advantages and limitations of this product.[4] This attitude stands in contrast to that of

[1] Tariff Hearings, 1912–1913, p. 1532.
[2] KFR, p. 65.
[3] BR, *passim*.
[4] "Wearever" aluminum utensils (TACU product) have been a heavier gauge line than that made by most of the companies.

some of the independent firms, which have given consumers high claims coupled with poor quality. Many of the independent concerns have received much aid from the Aluminum Company in the form of liberal credit terms and technical advice;[5] but until recently, at least, it appears that the company often assumed a rather dictatorial attitude in dealings with its customers.

TABLE 31

STATISTICS OF GROWTH OF ALUMINUM-WARE INDUSTRY *

Year	Number of Establishments	Number of Wage Earners †	Capital	Value of Product	Value Added by Manufacture
			(Thousands of Dollars)		
1914	37	4,614	$11,088	$19,597	$ 5,176
1919	39	7,821	29,052	50,478	17,304
1921	46	9,328	—	37,212	16,117
1923	45	8,777	—	39,344	20,754
1925	—	—	—	32,052	—
1927	50 ‡	—	—	28,989	—
1929	55 §	—	—	35,100	—
1931	—	—	—	22,279	—
1933	—	—	—	16,400	—

* Largely cooking utensils. Except as noted the figures of this table are taken from the Census of Manufactures. After 1923 the census reported detailed figures only for the aggregate of establishments producing aluminum manufactures including sheet, shapes, cable, castings, etc., as well as ware.

† Average for year.

‡ Summary of Tariff Information, 1929, p. 727.

§ Fifteenth United States Census, Census of Distribution, *Products of Manufacturing Industries*, 1929, p. 100.

In October 1924 the Federal Trade Commission submitted to the Senate the results of its investigations pursuant to a Senate resolution directing it to study price conditions, combinations, and competitive methods in the principal branches of the house-furnishings industry. A section on aluminum described the organization of the aluminum-utensil industry, the position of the Aluminum Goods Manufacturing Company, and the origin and growth of the Aluminum Company of America and its relations to

[5] See BR, pp. 75–77, 262–286; and FTC Docket 1335, *passim*.

the utensil branch of the industry.[6] The Commission recited numerous complaints by utensil manufacturers against the Aluminum Company with respect to deliveries. These complaints were classified as follows:

(1) Cancellation of quotas.
(2) Refusal to promise shipment.
(3) Unreasonable delays in deliveries.
(4) In case of orders of metal for articles requiring two or more kinds of gauges, some were delivered and others held up.
(5) Large quantities of metal on which delivery had been unreasonably delayed were subsequently dumped on manufacturers in quantity.
(6) Deliveries of metal in quantity were made to manufacturers on unreasonably delayed orders shortly after their purchase of foreign metal.[7]

As illustrative evidence of the existence of these difficulties with deliveries the Commission printed excerpts from correspondence between the company and its customers, quotations of interviews with the latter and with officers of the company and statistical data concerning delays. Apparent instances of price discrimination, the shipment of defective metal, and the hindering of expansion were also adduced. A section on competitive conditions was closed with the following statement:

> A comparison of these provisions of the consent decree with the methods of competition employed by the Aluminum Company of America, described above, especially with respect to delaying shipments of material, furnishing known defective material, discriminating in prices of crude or semifinished aluminum and hindering competitors from enlarging their business operations appears to disclose repeated violations of the decree.[8]

The receipt of the report by the Attorney General [9] launched the investigation of the Department of Justice. As we have seen, the

[6] Report of the FTC on House Furnishings Industry, III, Kitchen Furnishings and Domestic Appliances, October 6, 1924. The aluminum section also contained a study of prices and profits of utensil manufacturers for the years 1920–1922.

[7] KFR, p. 98. Further investigation seemed to show that some of these complaints were groundless. See below, p. 416.

[8] *Ibid.*, p. 112.

[9] The whole of the volume on kitchen furnishings was sent to the Attorney General. In addition to the chapter on aluminum utensils this volume included the results of investigations with regard to vacuum cleaners, washing machines. refrigerators, sewing machines, etc., and the association activities of hardware dealers.

The Department of Justice had been investigating trade associations in the refrigerator industry since 1922. Subsequent to the report of the Federal Trade Commission the Department obtained the convictions by a plea of guilty of about

results of this inquiry, which were printed in the Benham Report, led to the conclusion that there was no evidence to support a prosecution for violation of any of the provisions of the consent decree.

The Federal Trade Commission, however, believing that it possessed sufficient evidence to start proceedings under Section 5 of its organic act, had issued a complaint on July 21, 1925, which charged the Aluminum Company with employing

a scheme the purpose and/or effect of which was and is to gain and maintain a monopoly of aluminum raw material, of aluminum ingots and sheets, of secondary aluminum, and of aluminum fabricated products and/or aluminum alloy products . . . and, in order to carry out such a scheme, . . . using the following practices:[10]

(*a*) Arbitrary neglect or refusal to supply ingot or sheet to manufacturers in competition with the company, or its subsidiaries.
(*b*) Arbitrary delays in deliveries of ingot to such competitors.
(*c*) Arbitrary delivery to such competitors of an insufficient quantity of aluminum or aluminum ingot, or a quantity less than ordered.
(*d*) Delivery of defective sheet and ingot to such competitors.[11]

A comparison of these charges with the much longer list of practices evinced in the Kitchen Furnishings Report reveals that many of the latter were eliminated in the transfer of the aluminum matter from the economic to the legal division of the Commission or by the commissioners themselves. Particularly noticeable is the omission of the charge of price discrimination. It will also be remarked that the charge of delivery delays, which was the most important in view of the actual facts, relates only to ingot. In so

---

twenty companies in this industry. Cases were also successfully brought against associations in the chair industry and in bedroom and dining-room furniture, on all of which the Commission had reported in earlier volumes of the house furnishings inquiry. In each instance the evidence was a combination of that secured by each of these government agencies. See pp. 124 and 229–230 of *Hearings before the Committee on the Judiciary, United States Senate, 69th Congress, 1st Session.*

[10] The broadness of this charge was undoubtedly designed to allow the Commission's attorneys to use any evidence of practices charged which might turn up in the hearings, whatever branch of the industry should be involved. Under the rule laid down by the Supreme Court in Federal Trade Commission v. Gratz (253 U. S. 421) no order may be issued which does not apply to a practice definitely specified in the complaint. The major portion of the complaint in the aluminum case had to do with the sand-castings branch of the industry. This will be discussed in the next chapter.

[11] See Appendix E for text of the complaint.

far as this accusation was designed to cover delays to utensil manufacturers it is practically meaningless, for most of these firms purchase only sheet.

Hearings in the case began just a week before the Benham Report was introduced into the Senate.[12] They were continued from time to time over more than three years until the middle of 1929. Oral argument before the Commission took place in the early spring of the following year, and finally, on April 7, 1930, that body ordered the complaint dismissed for lack of evidence to support the charges. Thus the Commission apparently arrived at the conclusion reached three years earlier by the Department of Justice — a conclusion directly opposite to that suggested by its earlier report. Yet neither the Kitchen Furnishings Report nor the Benham Report makes everything clear, while the Commission's order of dismissal was, according to regular procedure, published without any explanatory discussion of issues, facts, and conclusions. Even a study of all the evidence appearing in the two reports and in the record of testimony and exhibits of Docket 1335 is by no means entirely conclusive, although it makes some aspects of the matter much clearer.

During the subsequent analysis certain characteristics of the two reports should be borne in mind. The aluminum section of the Kitchen Furnishings Report, prepared by the economic division of the Commission, establishes certain facts but does not probe the surrounding circumstances. For the most part neither the reasons for the actions of the company nor the consequences are disclosed, although the general tone of the report suggests that both were undesirable. It is not definitely concluded that the practices reviewed constituted unfair methods in purpose or effect, but in places that seems to be the implication. With regard to the decree it is said that the practices "appear to disclose repeated violations." No exhaustive study of these matters was made for this report. Such was hardly contemplated by the resolution directing the Commission to inquire into conditions in a number of industries and report at the earliest possible time.

The Benham Report, on the other hand, purported to be a comprehensive study of one question — whether there existed any

[12] Federal Trade Commission, in the matter of the Aluminum Company of America, Docket 1335.

evidence of the nature and degree of proof required to support a prosecution of the Aluminum Company of America under the decree of 1912. To maintain a successful prosecution it would be necessary to prove beyond a reasonable doubt, since the proceeding would be one of criminal contempt, that acts were done with a deliberate purpose to injure a competitor.[13] With this in view the investigation of the Department was concerned not only with the practices of the company but also with the reasons for them. Since the reasons appeared to the investigators to be found in conditions beyond the control of the company, it was not necessary to inquire into the consequences of these acts. For what it attempts, the Benham Report is, on the whole, a fairly comprehensive and careful piece of work, although its general tone suggests a somewhat uncritical readiness to accept explanations favorable to the company.[14]

It would appear that the acts enjoined by the decree of 1912 would under certain conditions constitute "unfair methods" within the meaning of Section 5 of the Federal Trade Commission Act; and that the tests for considering them unfair methods would be both broader and less severe than the criteria for making them violations of the decree. Also it is obvious that the decree did not include all the unfair methods which might be used.

Some of the following analysis concerns the question of preferential treatment accorded the Aluminum Goods Manufacturing Company by the Aluminum Company of America. Hence it seems advisable to examine the administrative relationship existing between these two companies during the post-war decade. Since the early years of the Goods Company, somewhat less than one-third of its common stock has been owned by the Aluminum Company, which held two of the six directorships. The rest of the stock was held principally by officers and directors of the Goods Company. The Trade Commission alleged in the Kitchen Furnishings Report that "the Aluminum Company of America exercises a close control over the operations of the Aluminum Goods Manufacturing

[13] Memorandum of law prepared by A. F. Myers, special assistant to the Attorney General, BR, pp. vi–xi.

[14] In parts it sounds like argument designed to refute the allegations of the Trade Commission report rather than objective examination of the questions. See esp. pp. 52, 55, 66–71, 71–73, 84.

Company,"[15] but it offers in the report only three instances as supporting evidence, and these seem far from conclusive.[16] The Benham Report counters with a statement that "the allegation frequently made that the Aluminum Company of America controls or dominates the Aluminum Goods Manufacturing Company is wholly without merit,"[17] and prints forty letters and telegrams passing between officials of the two companies as supporting evidence.[18] The truth appears to lie somewhere between these opposing views. The evidence indicates the lack of any coöperation between the operating departments of the two companies. Competition between the "Mirro" brand of the Goods Company and "Wearever," produced by the Aluminum Cooking Utensil Company, appears to have been keen, although perhaps tempered by some exchange of price lists and comments about price policies.[19] Price, production, and investment programs were evidently decided by the Wisconsin officers with counsel from Pittsburgh at times.[20] The advice of the Aluminum Company representatives was followed on most matters of financial policy.[21] Banking connections of the Aluminum Company rendered financial services to the Wisconsin firm; and the latter benefited from the services of the Aluminum Company in the purchase of various sorts of capital equipment.[22] It appears that the Aluminum Company maintained a paternalistic attitude toward the little colossus of the utensil industry, and that the policies of the latter were shaped somewhat by the influence of the Pittsburgh directors in spite of their infrequent attendance at meetings.[23]

[15] Page xxiii.

[16] *Ibid.*, pp. 72–74. The report gives no intimation that the Commission was in possession of any more evidence, and in the subsequent case the Federal Trade Commission attorneys were unable to develop anything additional of much significance.

[17] BR, p. 53.

[18] BR, pp. 54, 161–172.

[19] FTC Docket 1335, Commission Exhibits 581–630; BR, pp. 163–168.

[20] The advice was not always taken. When the Wisconsin directors proposed the erection of an additional rolling mill in 1922 the Pittsburgh directors apparently advised strongly against it. Nevertheless the mill was built, in spite of the fact that this action seems to have been regarded by the latter as a breach of confidence between partners (KFR, pp. 73–74; BR, pp. 54, 167).

[21] FTC Docket 1335, Record, pp. 1726 ff., and Commission Exhibits 152, 156, 582, 598.

[22] *Ibid.*, Commission Exhibits 89, 138, 581, 597, 600, 601.

[23] *Ibid.*, Commission Exhibits 581–630.

2. Miscellaneous Practices

At the beginning of the analysis of competitive practices in the utensil field it should be recognized that there is little reason to believe that the Aluminum Company would have any incentive to attempt complete monopolization of the utensil industry. The process is simple, and potential competition would need but a few hundred thousand dollars to become active.[24] Monopolization would require maintenance of price spreads between ingot and sheet and utensils which were inadequate to support outside producers, or effective tactics to keep aluminum out of the hands of those who wished to manufacture ware. The latter policy would be difficult to enforce without the former as long as the company sold substantial amounts of ingot or sheet in the market. Even in the absence of antitrust laws the former policy might be undesirable because it was unprofitable. Preservation of a preventive price differential by charging high sheet prices would reduce profits in any markets for other sheet products where demand was more elastic. Were the deterrent differential put into effect by low utensil prices, profits in the utensil branch would suffer. However, assurance that the Aluminum Company has not, in all probability, attempted complete monopolization of the utensil branch does not make it unimportant to ask whether any of the company's practices have discriminated, in effect, between utensil firms. Is the fact that the Aluminum Cooking Utensil Company and the Goods Company continued to be much the largest units in an industry not subject to marked large-scale economies to be explained by the use of unfair methods or on other grounds? And did the Aluminum Company favor any among the independents? In the following analysis it should be understood that the criteria of unfair methods explained in Section 3 of the preceding chapter are employed. A conclusion that according to these criteria unfair methods have been used would not necessarily mean that these practices were unlawful. That would depend upon whether the Trade Commission and the courts adopted the same criteria. In this connection it will be recalled that the Trade Commission dismissed the complaint of 1925 without order.

[24] Several lines of manufacture can be adapted for production of aluminum ware at small expense. Some companies take up aluminum utensils as a side line when that is profitable and slip out at other times (BR, p. 131).

We may dismiss without extended discussion all the charges of unfair methods and violations of the consent decree with the exception of those concerning delivery delays, price discrimination, and discouragement of potential competition. Upon the other charges the evidence presented in the government reports and in the testimony and exhibits of the Trade Commission case is inconclusive where it does not indicate that the charges were unfounded. Both the Kitchen Furnishings Report and the Benham Report cite the existence of instances of refusal to promise shipment, delays in needed sizes while shipping sizes less needed, large shipments of back orders which looked like dumping, cancellation of contract quotas, and shipments of defective metal. Citation of instances of the first four are few. For reasons explained above one can conclude nothing from the Federal Trade Commission report. The Department of Justice states that interviews with a large number of utensil producers and the examination of correspondence files of these firms and of the Aluminum Company produced no evidence that these practices were the result of an intent to injure competitors but indicated, on the contrary, that they were due to abnormal conditions over which the company had no control and which it was striving to meet as well as possible. The Benham Report prints correspondence seeming to substantiate the conclusions reached. It will be recalled that these three charges were not incorporated into the formal complaint issued by the Commission in 1925. This seems to indicate either very little faith that the continued practice of these methods could be established or belief that they were not employed unfairly. Since the agents of the Commission and the Department of Justice were given free access to the files of the Aluminum Company, the almost complete lack of evidence to substantiate these charges may be taken to mean that they were groundless.

Shipments of defective metal were evidently due largely to the existence of unusual industrial conditions in 1920 and 1921 or to internal inefficiency. The Benham Report shows that the Goods Company received some poor metal. Very likely quantitative measurement of the relative shipments of poor metal to different companies would have shown that there was no marked discrimination in this respect. The Benham Report does not demonstrate this.

3. Price Discrimination

The questions of price discrimination, unreasonable delivery delays, and discouragement of potential competition require more analysis. The first raises several interesting questions. It was the custom of the Aluminum Company to quote a slightly lower schedule price on "50 ton lots" than that listed for orders of smaller amounts. Some utensil manufacturers contended that the "50 ton bracket" was designed to enable the Goods Company to buy more cheaply than its competitors, most of whom, it was said, did not purchase in such large quantities.[25] The Benham Report shows a long list of utensil firms "who during all or any of the years 1920, 1921, 1922, and 1923, got the benefit of 50 ton lot prices." [26] But it is not disclosed whether this benefit was gained regularly or sporadically by the majority. There is no attempt to ascertain cost differentials with a view to determining whether the brackets in the Aluminum Company's quantity discount were fixed according to considerations of cost or demand.

During 1921 and 1922 the Goods Company received an additional concession averaging slightly above 5 per cent on all aluminum purchased. It is explained that in earlier years the Aluminum Company had denied the former this differential but granted it in 1921 in order to utilize what otherwise would have been idle rolling capacity. At this time the company evidently made concessions to any customers whenever necessary to secure orders needed to keep the mills running.[27] It is concluded that the Aluminum Goods Manufacturing Company received a quantity price differential consistent with its consumption, although the only support offered for this conclusion is the fact that the annual purchases of the Goods Company amounted to at least three times those of any of the independents. There is no analysis to discover whether this discount actually measured the difference in the direct or variable costs of rolling and selling the larger tonnage of sheet which the Goods Company took as compared to the other utensil firms. Naturally, the most profitable policy for the Aluminum Company would involve discrimination in order to maximize net revenue over and above its variable costs. The data of Table 32 might suggest that such discrimination occurred. It is

[25] BR, pp. 68, 131–132. [26] *Ibid.*, p. 68. [27] *Ibid.*, p. 66.

conceivable, of course, that the differences in discounts reflected cost differences only, but hardly likely. If such a policy conferred upon the Goods Company, or for that matter any utensil firm, an advantage relative to its competitors which found no basis in a difference in the cost of serving them, it is to be regarded as uneconomic. Depression provides no justification on broad economic grounds for discrimination, no matter how salutary the results may be from the standpoint of stockholders.[28]

The investigators of the Department of Justice were satisfied that there had been no illegal discrimination when they ascertained that during 1921 and 1922 varying price concessions had been granted to many different customers in the utensil and other industries. The table shows the difference between published schedule prices and actual sale prices of raw and semifinished aluminum sold by the Aluminum Company to several utensil concerns and to companies in other industries. The table tells little. The utensil firms are not distinguished from the others. It is not certain that all the important utensil companies are included in the list. Nor are we told upon what basis the companies from other industries were selected. For all that the table reveals, every utensil company except the Goods Company may have received a discount of less than 1 per cent. Cumulative arrangement of the data does not support the conclusion that "this tabulation shows that . . . each of the companies considered received substantial discounts." [29] It seems quite probable that the large discounts were not secured by the utensil producers, but represented an attempt on the part of the company to preserve the employment of alu-

[28] The more general economic objection to discrimination has been explained above, p. 222.

[29] BR, p. 68. The statement may mean that each of the utensil companies received substantial discounts, but it does not say this definitely and the tabulation does not show this.

Discount	Number of Companies
Less than 1%	10
Less than 2%	16
Less than 3%	19
Less than 4%	19
Less than 5.26%	21
More than 5.26%	4

TABLE 32

DISCOUNTS FROM SCHEDULE PRICES OF RAW AND SEMIFABRICATED ALUMINUM OF THE ALUMINUM COMPANY OF AMERICA ACCORDED VARIOUS COOKING-UTENSIL FIRMS AND USERS OF ALUMINUM IN OTHER INDUSTRIES DURING THE PERIOD NOVEMBER 15, 1921–OCTOBER 3, 1922

Customer	Total Sales (*Thousands of Pounds*)	Percentage of Variation between Schedule and Actual Invoice Price
Aluminum Goods Manufacturing Company	16,343	5.26
A	2,818	14.06
B	2,703	1.07
C	2,097	2.50
D	1,847	5.10
E	1,173	9.10
F	758	0.35
G	714	0.00
H	702	1.08
I	620	0.76
J	554	0.26
K	353	6.90
L	286	1.33
M	280	4.42
N	256	0.61
O	231	0.70
P	112	1.70
Q	53	6.30
R	51	0.69
S	40	0.18
T	33	0.08
U	17	1.93
V	11	0.07
W	3	2.02
X	597	2.20
Y	530	1.14

The data of this table are taken from Exhibit 27 in the Benham Report, p. 228. The figures include those from the seven cooking-utensil concerns referred to in the Kitchen Furnishings Report.

minum in uses where the demand was more elastic because of keener competition with other metals.

Furthermore, the table yields no clue to the proportions in which ingot and sheet were sold to the concerns receiving discounts. A separation of these two is essential for a clear picture because the Goods Company, unlike the other utensil firms, possessed its own rolling equipment.[30] A discount upon ingot sales to this company would have the same effect upon the competitive situation as a discount upon sheet. As a matter of fact, during 1921 and 1922 this company rolled just about two-thirds of the sheet which it used, and in subsequent years it rolled a much larger proportion. A detailed inspection of the individual sales items from which the table in question was derived indicates that the sales to the twenty-five concerns other than the Goods Company were made up almost entirely of sheet.[31] Of the total 16,343,440 pounds purchased by the Goods Company, 9,000,000 pounds were definitely ingot, and the rest appear to have been sheet. Evidently the discount on ingot to this company was 4.17 per cent and that on sheet 6.15 per cent. Upon the *sheet which it purchased* it received more of an advantage, as compared with the majority of the concerns, than the table shows; but on the sheet which it rolled itself it received less of an advantage, for not only was the percentage discount lower than that shown in the table (4.17 per cent as compared to 5.26 per cent) but the discount of 4.17 per cent on ingot is actually equivalent to a lower rate upon the higher-priced sheet. The figure 5.26 per cent is not, therefore, strictly comparable to the percentage figures for the other firms who were buying only sheet. When a comparable figure is derived by re-pricing the ingot sales at an approximate sheet price, it is seen that the composite discount received by the Goods Company was equivalent to 4.3 per cent.[32]

In summary, the Benham Report is inconclusive in regard to price discrimination. There is no analysis to show whether the "50 ton lot" bracket for quantity discount was based upon cost considerations or something else. Although there is depicted a situation of varying price concessions beyond this quantity discount

[30] The Aluminum Products Company was the only other exception.

[31] These data are given in BR, pp. 229–235.

[32] Ingot was converted to sheet at the price of 30 cents, which was the Aluminum Company's price for coiled sheet during most of this period.

during 1921 and 1922, the report does not present an adequate comparison of the treatment accorded the Goods Company and that received by other utensil units. The discount to the former is computed, incorectly from the standpoint of the true competitive situation, but its relation to differences in cost is not disclosed.[33] Nothing is said of the price policy of the Aluminum Company in the years between 1922 and 1926, when the report was issued. This section of the report appears to be directed more largely toward a refutation of the specific charges of the Kitchen Furnishings Report than the discovery of the reasons for differences in price and their results.

The Department of Justice was concerned chiefly, of course, with determining whether the decree of 1912 had been violated. Since the section of that decree relevant to price discrimination contained the vague phrase "under like or similar conditions," the Department may have felt that the courts would consider that differences in either demand or supply factors constituted dissimilar conditions. What is said, as well as what is left unsaid, in the Benham Report leads one to infer that perhaps the Department itself took this position.[34] The reasons for the conclusions about price discrimination would have been much clearer if the report had shown the significance of the evidence in the light of a thorough discussion of the meaning of this section of the decree. Finally, it must be concluded that the essential economic issues regarding price discrimination were not clearly perceived. It is partly due to this that the evidence adduced on price discrimination does not seem to demonstrate whether or not the decree was violated or unfair methods employed.

4. Delivery Delays

The three sources of information concerning the question of unfair methods in the cooking-utensil branch of the industry all describe a situation of serious delays in deliveries by the Aluminum Company during the years 1919–1923.[35] A cursory glance at Tables 33 and 34 will reveal that in addition to the failure to

[33] The discounts appear to vary without reference to quantity or specifications.

[34] See conclusion at p. 84 of the report.

[35] KFR, pp. 99–104, 315; BR, pp. 58–63, 83, 156, 189–223; FTC Docket 1335, Record, *passim*, and Commission Exhibit 729.

deliver substantial proportions of metal within a sixty-day period there were considerable differences in the treatment accorded the various companies. No simple explanation of this discrimination appears. Degree of delay does not vary directly with size, nor do the Aluminum Company's interests invariably enjoy the quickest deliveries. Moreover, it is not always the same companies which receive the best or the poorest shipments.

When the Federal Trade Commission stated in the Kitchen Furnishings Report that the record of deliveries appeared to show a violation of the decree of 1912 it had not inquired into the causes of delay, nor had it secured data upon shipments to the Aluminum Goods Manufacturing Company and the Aluminum Cooking Utensil Company. The Department of Justice, being unable to find any specific complaints that delays were intentional on the part of the Aluminum Company, went to the records of the company. Figures for deliveries to the Goods Company and the Aluminum Cooking Utensil Company were compared with the data for four independents, two of which had the least favorable shipments as shown in the table of the Kitchen Furnishings Report, and two of which had the better deliveries. The Department was completely satisfied that there had been no violation because

> these tables of fiscal month shipments indicate that no unreasonable delays occurred in respect to the shipments to the four outside companies listed, practically all of their material having been shipped by the end of the second month. The tables also show that the Aluminum Cooking Utensil Company (the Aluminum Company's own subsidiary) received the least favorable treatment and that the Aluminum Goods Manufacturing Company (in which the Aluminum Company is a stockholder) received by no means the best treatment.[36]

Unless shipments of only 65 to 75 per cent may be regarded as "practically all their material" the first conclusion is not supported by the data.

Although apparently convinced that no unreasonable delays occurred, the Department inquired into the causes of delivery

[36] BR, p. 60. It was also explained that supplying the utensil producers is a custom business, and that some portions of an order "may fail to pass the inspection department, in consequence of which another batch has to be rolled later. It is for reasons of this character that there are frequently (as shown by the tables) trivial amounts of an order which are not shipped within what might be described as the schedule period, namely, the first 30 or 60 days after receipt of the order (*ibid.*, p. 59).

delays by requesting the company to "prepare a statement which might reflect any unusual conditions as to the company's labor supply covering the period from January 1920 to November 1925, or any other conditions which might have militated against deliveries." Correspondence was scrutinized and fifty-five letters and telegrams printed in the report. Like most exhibits of correspondence, this one was quite inconclusive. The statement of the company describing abnormal conditions is more enlightening.[37] Especially in 1920, and to a lesser degree in 1922 and 1923, the company was operating under unusual conditions over which it could exercise no control. Undoubtedly these conditions were responsible for a large part of the delays in deliveries. Owing to the transfer of the major portion of its bauxite-mining activities from Arkansas to South America, the unanticipated business revival of 1919 found the Aluminum Company with an uncompleted plant in the latter region and a much reduced scale of operations in the former. The resultant bauxite scarcity of 1920 was intensified by car shortages on the railroads, and inability to secure adequate supplies of efficient labor and of natural gas, which is employed in drying the ore. During 1920 only 64 per cent of the company's requirements of bauxite were shipped to the alumina plant, and the output of the latter was 64,000,000 pounds less than could have been produced had the bauxite been available. Strikes and lack of efficient labor reduced the output of the reduction plants by about 8,000,000 pounds, according to the company's estimates.

Concerning unusual conditions in 1922 and 1923 the company has less to say. The East St. Louis alumina plant was entirely closed down for a part of 1922 on account of a coal strike. Transportation was hampered by freight embargoes. Business revival in 1922 and 1923 required the recruiting of new labor which was of low efficiency. As a result, both production and quality in all of the company's plants were lowered. The sudden increase in demand for aluminum in 1923 found the company with low stocks, and this condition occasioned some delay in shipments of sheet during 1923.[38]

[37] BR, pp. 60–63.

[38] Production was cut back in 1921 to 38 per cent and in 1922 to 54 per cent of the 1920 output (BR, p. 63). Unfilled orders of sheet show a substantial increase in the latter part of 1922 and the first half of 1923 (*ibid.*, p. 194).

TABLE 33

PERCENTAGES OF OBLIGATIONS SHIPPED BY THE ALUMINUM COMPANY OF AMERICA IN CERTAIN PERIODS TO VARIOUS COOKING-UTENSIL COMPANIES *

Company	Obligations in Pounds	Percentages Shipped †					
		Within 60 Days ‡	Third Month	Fourth Month	Fifth Month	Sixth Month	Unshipped after Sixth Month
January 1920–December 1921 §							
Buckeye Aluminum Co.	1,423,129	61.44	11.97	10.41	4.64	3.16	8.38
Illinois Pure Aluminum Co.	1,181,311	64.04	10.51	6.84	2.83	2.29	12.05
Landers, Frary & Clark	780,158	52.36	7.95	8.61	6.20	4.28	20.13
Wheeling Stamping Co.	699,975	55.03	10.44	9.99	11.33	7.26	6.02
Porcelain Enameling Ass'n.	514,121	62.99	8.33	8.46	10.01	4.93	5.34
West Bend Aluminum Co.	1,704,405	42.75	11.70	6.00	4.76	6.69	27.71
Kewaskum Aluminum Co.	463,402	66.72	12.36	6.76	4.92	6.66	1.98
Aluminum Goods Mfg. Co.							
All	18,372,101	79.20	7.85	2.64	2.01	2.91	5.39
Ingot	14,044,231	91.92	7.41	0.71			
Other	4,327,870	37.91	9.28	8.88	8.52	12.35	23.06
January–December 1922 ‖							
Buckeye Aluminum Co.	1,452,411	90.01	9.20	0.79			
Illinois Pure Aluminum Co.	3,046,800	75.53	24.10	0.03	0.09	0.05	0.20
Landers, Frary & Clark	690,383	67.59	20.58	9.09	1.30	0.89	0.55
Wheeling Stamping Co.	490,842	95.27	3.77	0.96			
Aluminum Cooking Utensil Co.	7,744,722	64.31	16.67	8.55	3.44	2.47	4.56
Aluminum Goods Mfg. Co.							
All	22,023,245	92.12	7.35	0.23			
Ingot	14,007,500	100.00					
Other	8,015,745	78.33	20.20	0.63	0.01	0.83	

			January–June 1923			
Buckeye Aluminum Co.	795,485	87.45	9.01	0.76		2.78 °
Illinois Pure Aluminum Co.	420,000	77.44	21.61			0.95 °
Landers, Frary & Clark	465,912	63.81	23.79	10.51		1.89 °
Wheeling Stamping Co.	174,082	97.10	2.51	0.39		
Aluminum Cooking Utensil Co.	4,905,354	32.17	14.67	13.63		39.53 °
Aluminum Goods Mfg. Co.						
All	5,876,517	81.90	4.04	10.25		
Ingot	2,599,501	68.20	5.61	22.34	3.85	3.85 °
Other	3,277,016	92.77	2.79	0.67		3.77 °

* The economic division of the Trade Commission requested the Aluminum Company to submit delivery data for certain companies whose complaints had been noted. Data for the years 1920, 1921, and 1922 were requested, but the company sent only figures for the year 1922 and a part of 1923. Furthermore, the company failed to separate the amounts shipped during the month of maturity from those delivered during the next succeeding month, and apparently the figures represented calendar rather than fiscal months. (See pp. 100–102 of the Kitchen Furnishings Report.) Subsequently the officials of the company contended that the tables of this report did not fairly reflect the situation, since they were prepared on the basis of calendar months. Data were submitted to the Department of Justice showing delivery delays on the basis of fiscal months for 1922 and the first half of 1923. The Trade Commission attorneys obtained similar data for the period 1920–1921. In constructing these tables I have used fiscal months data except where otherwise noted.

† The percentages for each of the three periods included in the table represent averages of percentages on all individual orders in that period.

‡ In the Kitchen Furnishings and Benham reports this column is headed "percentages shipped month of maturity or first succeeding month." For spot orders this means within 60 days after receipt of order. Most of the metal is ordered on long-time contracts calling for the shipment of a certain number of pounds each month. We are not informed, in the source of delivery data, how the "order date" for such contracts is determined for purposes of computing delays.

§ Compiled from data presented in Commission Exhibit 729, FTC Docket 1335.

|| For 1922 and 1923 the data on deliveries of commodities other than ingot to all the companies were secured from the tables of Exhibit 18 of the Benham Report (p. 189), and the figures for ingot shipments to the Goods Company from Docket 1335, Commission Exhibit 729.

° Unshipped after fourth month.

Since there were no complaints that the delays were intentionally designed to injure the purchasers and the evidence "fully discloses that such delays as occurred . . . were caused by conditions beyond the control of the company," the Department decided that the decree had not been violated.[39] Here again the Department accepted the results of inadequate analysis.[40] The fact that many producers, including the Aluminum Company interests, had suffered was regarded as sufficient proof that none had been unfairly made to suffer. The existence of abnormal conditions which undoubtedly caused some of the delays was considered adequate proof that none of the delays were due wholly or in part to any other cause. There was no inquiry into results. Three important aspects of the situation the Benham Report neglected to analyze adequately. No explanation is offered of the apparent discrimination between the various companies. The investigators again did not perceive the significance of the large ingot purchases of the Goods Company. Thirdly, they failed to raise the question of artificial restriction of output.

The margins of discrimination suggested by the tables appear to be rather large. However, since the figures are in each case averages of the deliveries during a period of several months, it is more than probable that they do not measure accurately the actual margins of discrimination which occurred. Discriminatory effects as between companies would depend upon (1) the relative degree of initial delay at any time, and (2) the relative celerity with which the delayed orders were cleaned up; and the relation of these two to periods of better and poorer business. Obviously the actual discrimination would be that indicated by the figures if the actual delays in each short period — say a month — coincided with the average for the whole period. Had this condition obtained, the

[39] BR, p. 83.

[40] The lack of complaints of unfair treatment may have seemed convincing enough to render a searching analysis unnecessary. But the absence of complaints should not have been allowed to assume much significance, for the purchasers who depended so completely on the Aluminum Company for their metal would hardly have believed it good policy, at a time when deliveries had been satisfactory for two years, to complain about their previous experience. Furthermore, it is not likely that buyers would possess any information which would actually prove that they were securing a smaller proportion of their legitimate orders than others. It is only fair to the Department to notice that it did consider that a study of the records of the company and some attempt to ascertain the causes of delay were required.

disadvantage to the companies which received the poorer deliveries would not have been as great as it actually may have been. The differences in delay at the beginning of the period would have been as shown, and as soon as the delayed orders of the first few months were shipped all the companies would have been receiving in each month the full amount of their orders for that month.[41] In 1920–1921 there was a substantial variation in the average rates at which back orders were completed, but in 1922 and 1923 practically all the back orders were shipped by the end of the third month. Had actual deliveries in these two years coincided with the averages, fairly large inventories at the beginning of 1922 might have nullified any adverse effects of the initial delays.[42]

But, of course, actual deliveries did not correspond to the averages. And since the average of *aggregate* delays in deliveries *to all companies* varied considerably at different periods of each year, the range within which the actual discrimination may have diverged from the average is fairly wide.[43] Actual discriminatory effects were worse than average if any concern with a poorer average received, in the months of better demand, delays greater in initial amount or more protracted than its average, while, in the

[41] This may be simply illustrated. Assume that Company A receives 50 per cent in 60 days, and 25 per cent each in the third and fourth months, while Company B is shipped 70 per cent in 60 days and 30 per cent during the third month. Assume also that half the metal delivered in the first 60 days is received in each of these two months. Then after the third month each company is receiving each month the full amount of its orders.

Received in	Company A	Company B
January	25	35
February	50	70
March	75	100
April	100	100
May	100	100
/		
December	100	100

Of course, if orders were increasing steadily from month to month the "catching up" could not occur until such increase stopped or the percentage of deliveries became better.

[42] Finished product inventories of ten utensil concerns at the beginning of 1922 averaged an amount nearly equal to 16 weeks' average 1921 production. Only one company had a finished inventory equal to less than two months of the 1921 average production. See KFR, pp. 117–118.

[43] See Table 34. Of course, actual discrimination might have diverged from the average in the absence of seasonal fluctuations in demand for utensils and in aggregate deliveries.

same period, any firm with a better average was receiving treatment better than or equal to its average, or less than its average by a smaller amount. Conversely, actual discriminatory effects were not so great as the average figure wherever the opposite occurred. This reasoning applies to an initial period of good business when inventories at the start are in direct proportion to the relative scale of orders. In cases of greater discrimination, during a time of good business, than indicated by the average for the period shown in the table, opposite treatment during slack months would bring some compensation in the form of relatively larger inventories at the start of the next period of good business. This advantage then might continue or be whittled away month by month through the same relative treatment as occurred during the first season of better demand. Enough has been said to show that the patterns of the actual discrimination are not sketched in by the averages of the tables.

A table showing the monthly distribution of shipments to seven companies at different periods was included in the Kitchen Furnishings Report.[44] Since it is figured upon the basis of calendar

[44] Appendix Table 24, at p. 315. Two companies were shipped during the period January 1922–June 1923 averages of 69 and 64 per cent respectively in the month of maturity or first succeeding month (calendar month basis). The average shipments in percentages at four periods of this time were as follows:

Period	Company	Maturity Month or Next	Second Succeeding Month	Third Succeeding Month
January–March	69	60.22	39.78	—
1922	64	29.32	44.79	25.89
April–August	69	64.79	34.64	.57
1922	64	94.07	5.69	.24
September 1922–	69	62.26	31.20	6.54
March 1923	64	51.41	32.94	14.52
April–August	69	90.61	.05	9.34 *
1923	64	97.62	1.39	.99 *

The following figures for two other companies are also interesting.

Period	Company	Maturity Month or Next	Second Succeeding Month	Third Succeeding Month
September 1922–	A	30.28	37.07	24.40
March 1923	B	90.79	7.33	1.65
April–August	A	86.16	11.79	2.05 *
1923	B	77.13	20.89	1.98 *

* Unshipped at close of second succeeding month.

TABLE 34

PERCENTAGES OF OBLIGATIONS AT DIFFERENT SEASONS SHIPPED BY THE ALUMINUM COMPANY OF AMERICA TO FOUR MANUFACTURERS AS COMPARED TO SHIPMENTS TO THE ALUMINUM GOODS MANUFACTURING COMPANY AND THE ALUMINUM COOKING UTENSIL COMPANY BASED ON A FISCAL RATHER THAN A CALENDAR MONTH *

Period	Obligations in Pounds	Percentage Shipped in 30 Days	Percentage Shipped after Maturity						Unshipped after Seventh Month
			Second Month	Third Month	Fourth Month	Fifth Month	Sixth Month	Seventh Month	
Independents									
Jan.–Mar. 1922	881,147	27.46	51.81	18.10	2.63				
Apr.–Aug. 1922	2,344,862	54.68	41.35	3.29	0.48	0.08	0.12		
Sept. 1922–Mar. 1923	3,636,644	29.05	39.14	27.62	2.61	0.35	0.30	0.16	0.77
Apr.–Aug. 1923	915,658	50.58	39.09	9.23	1.10†				
Aluminum Company of America Interests									
Jan.–Mar. 1922	4,903,789	18.14	56.43	20.59	0.56	1.05	1.04		0.219
Apr.–Aug. 1922	4,372,226	18.59	44.21	24.47	9.56	1.25	1.00	0.48	0.44
Sept. 1922–Mar. 1923	13,222,162	16.58	40.85	13.98	9.70	5.90	3.52	4.50	4.97
Apr.–Aug. 1923	789,453	46.40	33.69	11.97	4.55	0.19			3.20

* From Benham Report, Exhibit 22, p. 192. Headings and footnotes have been altered somewhat in form. The figures for the Aluminum Company interests are averages of the aggregate shipments to the Aluminum Goods Manufacturing Company and the Aluminum Cooking Utensil Company except those for the period April-August 1923, which refer to the Aluminum Goods Manufacturing Company only.

† Unshipped, end of third month.

months rather than fiscal months inferences must be tentative; but it suggests that some firms with poorer averages than certain of their competitors suffered more discrimination in the two periods of 1922 and the first half of 1923 than is indicated by the average figures of Table 33. Although the available data do not measure accurately the discrimination in delivery delays, they seem to leave no doubt that it existed in an appreciable degree. The Benham Report offers no explanation.

According to the data presented, the Aluminum Cooking Utensil Company, the wholly owned subsidiary, received the poorest deliveries of all. It would appear that there was no endeavor to build up the business of this concern at the expense of other utensil manufacturers. It is not explained in the Benham Report, however, whether this 100 per cent subsidiary orders in the same way as the independents. If that is not the case, the delivery figures may not be strictly comparable. Nor are the relative inventories of this subsidiary and of the independents presented. Furthermore, for the most part the independents were making a lighter gauge ware which did not compete directly with the Aluminum Cooking Utensil Company.[45]

The apparent discrimination between independents in the matter of delays may have been due to differences in specifications, or to the fact that, in times when deliveries are poor, some buyers order more than they really need in the hope of securing a larger allotment than their competitors,[46] or to other reasonable causes. It might have been to the advantage of the Aluminum Company, however, to use the general situation of delays as a cloak to discriminate against those whom it wished to force into a more subservient attitude, or whose business it wished to hamper for one reason or another, or whose bargaining power was weaker. Unfortunately, there is little or nothing in the Benham Report or in

[45] See FTC Docket 1335, Record, pp. 1285–1286; and BR, p. 135. The competition is really between substitute goods of different quality and price.

[46] Inter-office correspondence of the Aluminum Company intimates that the Illinois Pure Aluminum Company and the West Bend Aluminum Company employed this device in 1920 (BR, pp. 209–212 and 217–219). It may be noticed that the delivery figures for the latter are the worst for 1920–1921 and that the former appears from statistics to have suffered much more than some others in 1922 and 1923. However, the fact that the Illinois company *took* practically all of the metal it ordered in 1922 suggests that its orders in that year were legitimate.

the evidence in the subsequent case before the Federal Trade Commission which demonstrates conclusively whether or not this sort of thing occurred. If there were any reason to expect that the Department or the Federal Trade Commission had investigated this aspect of the matter the lack of evidence would be significant. But its consideration was apparently omitted. If the proviso in the 1912 decree designed to prevent the use of a condition of shortage as an excuse for discrimination in deliveries had been worded more incisively and made applicable to sheet as well as "crude aluminum," a thorough inquiry into the whole question would have been required.[47] Actually there was no attempt to explain fully either the causes or the comparative results to the companies. One would like to know whether the comparative changes in net earnings of the utensil concerns were in direct proportion to the relative delivery delays.[48]

In the light of several indications in the Benham Report that the Goods Company rolled a large proportion of its sheet one finds it surprising that the Department investigators neglected to ascertain the facts concerning deliveries of ingot to this company.[49] The Kitchen Furnishings Report had pointed out that the Goods Company possessed a substantial advantage over competitors

[47] It may be objected that government agencies have neither the time nor the money to inquire so carefully into all the economic issues and facts involved, but this does not, of course, deny the necessity of such detailed investigation if we are to be assured that no unfair methods have been used. When put in terms of social cost and social utility the objection has much force. But it is also important to recognize that the effectiveness of government control could be improved materially without increased expenditure if the government agencies perceived the issues more correctly and obtained more of the useful and less of the unimportant information.

[48] Gross sales figures for the various companies during 1922, 1923, and 1924 were included in the Benham Report, not, however, for this purpose. Gross sales data might be expected to reflect any serious losses of market. For the three years given there is no discernible relation of a causal nature between sales changes and deliveries. Analysis of this sort would require data for years before, during, and after the period of delays, and some study of any complicating factors. In its defense to the Federal Trade Commission the Aluminum Company evidenced the rapid growth in net worth of the Illinois Pure Aluminum Company, which had made more than half of the complaints found in the correspondence. This evidence is useless without similar data for other firms and knowledge of the extent to which factors other than earnings accounted for growth in net worth.

[49] This advantage was mentioned by a field agent in his report (BR, p. 133), but apparently no data were secured and the body of the report contains no discussion of it.

through its ability to get adequate supplies of ingot while the latter were suffering delays in shipments of sheet.[50] Equally serious delays of sheet to the Goods Company would mean little for the competitive situation if that concern were able to secure without delay sufficient ingot to keep its rolling mills operating at the ideal schedule at any given time. Figures obtained by the Federal Trade Commission's attorneys during the case indicate that this was what happened in 1920 and, to a somewhat lesser extent, in 1922 and 1923.[51] It is indicated in Table 33 that while the Goods Company received very poor deliveries of sheet in 1920–1921, it obtained practically 92 per cent of its ingot requirements within sixty days and all the remainder within the next two months. The ingot orders amounted to more than three times the sheet orders. At this time the Aluminum Company was shipping to the independents on the average only 45 to 65 per cent of their orders in the first two months, and the remainder was delayed to such an extent that substantial amounts were still unshipped at the end of six months.

In order to compare the treatment of the Goods Company with that accorded the independents a figure has been computed showing the percentage of total pounds ordered of ingot and all other commodities which was shipped to the Goods Company in each of the periods. This simple calculation might suggest that the latter gained an appreciable competitive advantage through the receipt of almost 80 per cent of its raw material in the first two months, while the independents could obtain but 45 to 65 per cent.[52] Inspection of the table reveals that unfilled orders of the independents at the end of sixty days were not worked off at a much faster rate than those of the favored company. It appears that the Aluminum

[50] KFR, p. 74. For simplicity I shall use the term "sheet" to mean all commodities other than ingot. The utensil producers use mostly coiled sheet and sheet circles.

[51] The attorneys did not, however, use the data to make the point definitely that the Goods Company was receiving preferential treatment.

[52] This analysis is subject to the qualification that actual discrimination may have diverged largely from that indicated by the averages. The tendency of the independents to order more material than they needed at such times makes the averages still less conclusive. The extent to which this practice occurred cannot be determined. Testimony in the Trade Commission case is conflicting. It is to be doubted that such overstatement of delays to the independents would account for the major part of the difference indicated by the average figures.

Company did not actually possess sufficient rolling capacity in 1920 to take care of the demand for sheet, and that this inadequacy was caused by the abnormal post-war conditions.[53] Clearly, this discrimination could hardly be considered a violation of the decree as worded. And, at first sight, this seems to be an instance of a competitive method which would be unfair in effect were its practice not due to abnormal conditions. However, the use of the limited rolling capacity to meet demands other than those of utensil manufacturers introduces complications. It seems to be true that sheet delays to the utensil firms were greater than those to other consumers whose profitable custom would be more easily lost. In the absence of quantitative evidence upon this point no definite conclusion is possible. If it were shown that the independent utensil companies suffered substantially because a part of their usual share of the Aluminum Company's rolling capacity was temporarily diverted to fill more profitable demands, while the Goods Company received materials much more promptly, it would be clear that the whole policy constituted an unfair method.

In 1922 the Aluminum Company possessed adequate rolling capacity.[54] The sales manager of the company stated that after completion of the new sheet mill at Alcoa, Tennessee, in August 1920, the company had ample rolling capacity to take care of whatever demand emerged.[55] By 1922 the difficulties incident to getting a new capital unit into satisfactory operation would have been overcome. The shortage of ingot in 1922 was created by voluntary curtailment of output at the reduction plants during 1921, lasting until a sudden pickup of demand sometime in 1922.[56] Since the shortage proceeded from the ingot stage or below, the slack filling of sheet orders for the independents while shipping the Goods Company's requirements of ingot 100 per cent complete

[53] See correspondence, BR, pp. 197, 201, 202, 217–219.

[54] Testimony of its president was reported as follows: "In the first place, unless you get clearly into your head the difference between a shortage of ingot and a lack of rolling capacity, you do not comprehend the situation at all. There never has been a shortage of rolling-mill capacity on our part . . . whatever shortage there has been in the sheet business is a reflection of the shortage in the ingot business" (KFR, p. 100).

In view of the evidence referred to in the preceding footnote the statement must have been wrong as far as it applied to 1920, or incorrectly quoted.

[55] *Ibid.*

[56] BR, pp. 62–63.

within two months is to be considered an unfair method in effect, whatever the motive. Whether it constituted an infringement of the 1912 decree would depend upon the interpretation of the wording of that decree. The amount of discrimination shown by the averages in favor of the Goods Company in 1922 is not large, however, and it does not persist after the second month. The situation in 1923 is very interesting, exhibiting as it does somewhat of a *volte-face*. In the first six months (the figures carry us no further) the Goods Company ordered more sheet than ingot and received about 93 per cent of its sheet within sixty days after the order matured. Ingot shipments, which had always been very prompt, as we have seen, were seriously delayed. It appears that the Goods Company was persuaded to buy sheet, with the result that the Aluminum Company could meet pressing demands for ingot elsewhere and also keep its rolling mills in operation at an ideal schedule.[57] During the years in question the Goods Company

[57] This is suggested by the following excerpt from inter-office correspondence of the Aluminum Company.

"At the present time, we are still very hard pushed for ingot deliveries, even though we have not delivered anything to Aluminum Goods for some weeks. In order for us to deliver, say 1,000,000 pounds to Aluminum Goods Manufacturing Company, in addition to keeping up our regular deliveries to Werra, Oberdorfer, United States Steel Corporation, and dozens of others users, will mean [*sic*] that we will practically have to shut down another sheet mill, something we can not undertake to do just now. You know that our Cleveland mill is shut down, and Niagara, New Kensington, and Edgewater are all crimped back for lack of ingot and also for lack of orders.

"Mr. Vits told me if they could not get delivery of as much metal as they could use they would bring in their salesmen; in other words, they would stop selling goods for the balance of the year. Such a step would not be necessary at all just because we could not deliver ingot to them and shut down our own sheet mills. The Aluminum Goods Manufacturing Company could very nicely give us sheet orders which would keep our sheet mills going and would allow them to have some raw material for fabrication and shipment on their salesmen's orders. We are of the opinion that Mr. Vits was largely bluffing when he threatened to stop selling aluminum goods, because he can always buy sheet from us and at the present time we can give him very good deliveries.

"The job cut out for you is to go to Manitowoc, see Mr. Vits and persuade him to give us sheet orders which will allow the Aluminum Goods Company to run full tilt and also keep our own sheet mills going, rather than have us decide to shut down our own sheet mills and give them ingot so they can run their new sheet mill. Of course, if Mr. Vits decides he must buy 2,500,000 pounds, I suppose we will sell it to him at today's price of 26 cents per pound, but at the present time I can not see how we could start delivery on such a contract, and if he buys it we will expect him to take every pound, even if the price goes down" (BR, p. 169).

seems to have obtained material of one sort or another somewhat more quickly than some of its competitors.

The Department of Justice and the Federal Trade Commission did not analyze the question of artificial restriction of output by the Aluminum Company.[58] The facts available are as follows. At the end of 1921 the company reported its ingot capacity to the Bureau of Internal Revenue as approximately 160,000,000 pounds.[59] Production for 1921 was about 55,000,000 pounds, for 1922 about 75,000,000 pounds, for 1923 nearly 130,000,000 pounds. The small output during 1921–1922 represented a voluntary curtailment of production due to the business depression. Sometime in 1922 capacity operations were again resumed,[60] but during 1922 and 1923 the company's output was restricted by labor and transportation difficulties. The wisdom of the directors in curtailing operations so greatly may, perhaps, be questioned, but it is evident that the shortage of ingot which made it impossible to meet promptly all the demands for both sheet and ingot in the latter part of 1922 and during 1923 was not artificially induced to throw a screen over harassing tactics toward utensil firms. The chief explanation of delays in deliveries to the latter is probably found in the fact that other markets for the metal were less secure. A large amount of testimony upon deliveries to customers other than utensil makers was taken in the Trade Commission case. Although it is somewhat conflicting, one receives the impression that many of these customers enjoyed much quicker deliveries. This is particularly true of the automobile-body manufacturers.[61] When

[58] This question was raised by Senator Walsh in the inquiry by the Senate Judiciary Committee into the expedition of the Department of Justice in the Aluminum Company matter. See Hearings, pp. 172–175 and 193–202.

[59] *Hearings of the Select Committee on the investigation of the Bureau of Internal Revenue,* 68 Cong., 2 Sess., p. 1815.

[60] FTC Docket 1335, Record, p. 696. Cf. production figures, p. 572 below.

[61] See the testimony of Charles T. Fisher (Record, pp. 2007–2018) and of the purchasing agent of the Budd Manufacturing Company (Commission Exhibit no. 860, answers to 93rd and 94th interrogatories). Mr. Fisher stated that "we had no trouble at all with deliveries." For testimony of various purchasers see Record, pp. 3870–4750. An inter-office letter of the Aluminum Company referring to a request by the Fisher people for an extra discount is relevant.

"Now we would like to make it very clear to him: That we consider the Fisher Body Company one of the very best customers we have and that we always intend that our negotiations with them shall reflect this appreciation; that this appreciation in times of easy delivery when there is plenty of aluminum for all, is shown by the

faced with the inability to meet all demands promptly it would be only natural to supply first those large consumers whose custom would be most easily lost. In automobile bodies and castings and, to a lesser extent perhaps, in electrical equipment the use of aluminum in competition with steel, iron, and copper depended upon an advantage so slender as to disappear with delivery delays. It seems likely that the utensil fraternity suffered the worst treatment because they could not turn so easily to the use of some other material.

It cannot be concluded that there was no practice of methods unfair according to the criteria here applied; and the conclusion that there was no violation of the decree does not seem grounded on thorough analysis. When delivery delays are analyzed upon a comparable basis the evidence suggests (in so far as the averages are significant) that the Aluminum Goods Manufacturing Company received more prompt shipments than several others in 1920 and in 1922. The evidence in insufficient to determine whether or not this constituted an unfair method in 1920. If the discrimination suggested by the averages for 1922 occurred, it constituted an unfair method according to the criteria given above, although the consequences may not have been very great in that particular instance. Since neither the Department of Justice nor the Federal Trade Commission seems to have penetrated far enough into the economic intricacies involved in the questions of discrimination between the various independents and restriction of sheet production for the utensil market, it is impossible to reach certain conclusions upon these two matters. One might infer that the main cause of delayed deliveries to the cook-ware industry lay in the relative security of the market. Evidently discrimination between companies did not result in forcing any from the field, but we do not know the extent of financial loss suffered by the poorly treated firms.

---

very favorable price, numerous instances of which you can readily point out; that in times like these with the open market above 40 cents we expect our full schedule price and that our regard for them is shown by taking care of their requirements in such a manner as not to interrupt their production" (Commission Exhibit no. 115). Cf. also excerpt from letter printed in footnote on p. 434 above.

5. Attitude toward Potential Competition

It has been shown that the Aluminum Goods Manufacturing Company apparently received much more rapid delivery of ingot than sheet during most of the time when the independents were suffering from delayed shipments. One of the latter, the Aluminum Products Company of La Grange, Illinois, equipped itself with a rolling mill in 1920 when delays were especially annoying.[62] Interviewed by an agent of the Department of Justice, the president of this company said that he had always been able to secure adequate supplies of metal with reasonable promptness. In his report to the Department the agent noted that the president of this company "was one of the few independent manufacturers who were particularly friendly toward the Aluminum Company of America at the time the representatives of the Trade Commission conducted their investigation," and also that the correspondence files of this company did not indicate the same difficulties with deliveries as were experienced by others.[63] He drew the conclusion that the reason was possession of a rolling mill. Was the failure of other utensil firms to erect rolling mills due to coercion by the Aluminum Company? The Kitchen Furnishings Report raised this question by showing excerpts from correspondence and interviews which indicated that late in 1921, when the ingot-sheet price differential was higher than usual, three of the independents had considered the erection of rolling mills but had later abandoned the idea.[64] No evidence was found, however, by the Commission or the Department to prove that any coercion had been employed, although it would appear that officials of the Aluminum Company adopted a strenuously discouraging attitude, expressed by letter and conference, toward the Illinois Pure and the Kewaskum firms.

The section of the Benham Report dealing with the discourage-

[62] The Aluminum Products Company had only a small mill, which did not entirely meet its own sheet requirements. No sheet had ever been sold. It was testified that an officer of the Aluminum Company told the president of the Aluminum Products Company, when the latter consulted him before building the mill, that there was not a banker's profit in rolling sheet (*ibid.*, Record, pp. 1538–1539). This was later denied by that officer of the Aluminum Company (BMTC appellant, v. ACOA, fol. 5434).

[63] BR, p. 126.

[64] KFR, pp. 106–108. See Table 28 on p. 382 for changes in the price differential.

ment of potential competition exhibits an inadequate grasp of the economic issues involved, which reflects the failure to recognize the significance of prompt delivery of ingot and delayed delivery of sheet, and lack of understanding of the forces determining the ingot-sheet price differential. As far as the decree is concerned direct attention is confined to paragraph (*f*) of Section 7, which prohibits agreements not to compete in any line, and no violation is found, of course. It is surprising that the Department did not also give its opinion as to whether the attitude of the Aluminum Company violated paragraph (*h*), which enjoins:

> taking the position with persons . . . engaged in the manufacture of any kind of aluminum goods that if they attempt to enlarge or increase any of their industries or engage in enterprises . . . competitive with defendant . . . [they] will for that reason be unable to procure their supply of material from defendant . . . .

When asked whether he had told Mr. Walker, president of the Illinois Pure Aluminum Company, "who was talking rolling mill," that the latter might be unable to secure ingot, an officer of the Aluminum Company is said to have replied,

> Yes, I told Walker that, but not exactly that way. I said that it had been our policy in the past to supply customers first and let our own mills suffer, but that in the future, if there was a shortage, it would be our policy to supply our own requirements first and that, if there was any surplus, that is what we would sell.[65]

This attitude seems to come close to the sort of thing paragraph (*h*) attempted to enjoin. However, the ineffectual wording of the proviso relative to use of a general shortage as an excuse for discrimination in deliveries would be the deciding element if application of paragraph (*h*) to the given circumstances were held subject to that proviso.

The Department of Justice raised the question of the exercise of threat or influence — without relating it to paragraph (*h*) — evidently in order to reply to the charges of the Federal Trade Commission, for consideration of this question was not required to establish lack of violation of paragraph (*f*). The conclusion that

[65] KFR, p. 108. Mr. Walker said that it was also pointed out to him by the same officer that no one in this country had ever been able to make any money in rolling aluminum sheet.

the Aluminum Company had not used any undue pressure or influence seems to depend chiefly upon acceptance of statements of officers of the Aluminum Company, and interpretation of the correspondence about Mr. Walker's proposal to build a mill to mean that he was bluffing for the purpose of obtaining better deliveries and price concessions.[66] This interpretation may not be true. Perhaps Mr. Walker was partly bluffing and partly serious. The reasons for both would be found in the same factors: poor deliveries of sheet, and an ingot-sheet differential regarded as greater than cost plus ordinary profit. Undoubtedly Mr. Walker had no intention of building a mill if he was satisfied on these points. As a matter of fact, officers of the Aluminum Company were sufficiently impressed to attempt dissuasion and make an arrangement late in 1922 regarding deliveries of sheet which would satisfy him.[67] The most reasonable explanation of Walker's failure to build the mill, although he continued for some time to express dissatisfaction with the price differential, is to be found partly in the cessation of delivery delays, partly in the forbidding considerations discussed in Chapter XVI — dependence upon tariff-protected monopoly, and the fluctuating ingot-sheet differential — and partly in the difficulties of a new manufacturing and marketing problem. Into these aspects of the matter the Department did not go. The ingot-sheet price differential is mentioned, but in place of analysis we are solemnly informed that

> no information or evidence is at hand to support any contention that the prices of ingot and sheet are arbitrarily fixed by the Aluminum Company of America without regard to fundamental principle of economics; that is, the law of supply and demand [*sic*].[68]

An official of the Kewaskum Aluminum Company was quoted in the Kitchen Furnishings Report as saying that the Aluminum Company "did everything they could to discourage us from going into the rolling-mill game." [69] Yet apparently this company was not visited by agents of the Department of Justice.

The matter of discouraging the erection of rolling mills was omitted from the complaint of the Trade Commission case. The evidence available indicates that the Aluminum Company engaged

[66] BR, pp. 71–73, 84, 124, 157.
[67] *Ibid.*, pp. 236 ff.
[68] *Ibid.*, p. 73.
[69] Page 108.

merely in dissuasion. The principal deciding elements, described in Chapter XVI, were, of course, self-evident. The result is not undesirable, provided delivery delays occur but rarely, and movements in the price differential do not handicap the one-stage firms. The absence of rolling mills attached to the smaller utensil firms probably represents a social economy as well as a saving to the utensil manufacturers themselves.

6. Summary

The results of study of the available evidence upon matters considered in this chapter may be summarized as follows. The unmeasureable constructive activities of the Aluminum Company in the utensil industry may have overbalanced the unmeasured results of any uneconomic conduct. To its educational efforts and its rigid stand for high quality a substantial part of the growth of the utensil industry is due. To a large extent its dealings with the independents seem to have been fair and helpful, if somewhat arbitrary and dictatorial. However, except during the last several years, there appeared at times of business boom numerous complaints, particularly with regard to delays in deliveries and the shipment of defective metal. From time to time there have been complaints of undue price discrimination and discouragement of the expansion of independents.

Since the consent decree of 1912 involved no adjudication it is not known to what exent the activities enjoined were previously practiced. From 1922 through 1929 the Federal Trade Commission was engaged upon investigation of the company's activities, and during part of this time an independent inquiry was made by the Department of Justice. Unfortunately neither of these agencies perceived all the important economic issues clearly; and both evidently encountered a considerable reluctance to talk on the part of persons dependent in a business way upon the Aluminum Company.[70] As a result, although analysis of the evidence and opinions

[70] See esp. BR, pp. 140–141; FTC Docket 1335, Record, pp. 3602 ff., and Hearings before Committee on Judiciary, pp. 295 and 356. A reluctance to give the same testimony on the stand (with officials of the Aluminum Company present) which was apparently given to agents in private appears throughout the Trade Commission case. Part of the modification and denial of statements must be ascribed to another element, which also probably explains partially why the Department found

of both bodies sheds some light upon the problems, many of the important questions remain undecided. We do not actually know the extent to which delivery delays were due to abnormal post-war conditions, the inelastic capacity of the reduction process, inefficiency, burdening the markets of most security and least bargaining power. The evidence available shows no proof of intent to injure, but for the reasons already given, it is hardly conclusive upon discrimination between companies. When analyzed correctly the figures obtained by the two government agencies suggest that the Aluminum Goods Company had prompt deliveries of ingot while the independents were suffering rather long delays on shipments of sheet, although failure to break down the data into statistics for shorter periods precludes a definite conclusion that this was so. If this indeed occurred, it constituted, in 1922 at least, an unfair method of competition, according to my criteria, whatever the intent.

Shipments of defective metal seem to have been caused by uncontrollable conditions and some inefficiency. Since no analysis was made of the relations of cost to discounts, or of the relative price concessions between utensil concerns, no conclusion at all can be reached upon price discrimination. It seems reasonable to believe that, if coercion to prevent expansion had existed, evidence would have been forthcoming. Voluble attempts to dissuade two independents from building rolling mills were evidently successful, largely because all the other factors were so strongly opposed to any long-run advantage. Unfortunately the public documents do not exhibit a thorough analysis of the causes and nature of the practices complained of, and they provide little data for ascertaining the results. It was shown definitely that no companies were forced out of business by the Aluminum Company.[71] In the Kitchen Furnishings Report a study of rates of return upon investment of each of eleven utensil concerns during 1920–1921 demonstrates that the rates earned by the Aluminum Goods Manufacturing Company were higher than those of all but one of the

less complaint than the Commission — the fact that the original inquiry by the latter was carried out during and just after the occurrence of the delays and other incidents, while a few years, during which relations had improved, intervened before the Department's investigation.

[71] BR, pp. 77–79. Testimony in the Trade Commission case was to the same effect.

companies and appreciably above the average returns.[72] There is no analysis to determine the extent to which this was due to large-scale economies, more capable management, better deliveries, or to other factors. For the most part the allegations of the Kitchen Furnishings Report were not well substantiated in subsequent examination. In any event there was little reason to suppose that the Aluminum Company was trying to monopolize the utensil field. Whatever uneconomic methods may have been employed represented merely some relative advantage or disadvantage for the firms affected.

The hearings of the Trade Commission case disclosed no indication of unfair methods in the utensil branch after about 1923, and no further complaints concerning the activities of the company in this division have come to public notice since the close of the case in 1930.[73] It is the opinion of one official of the Commission that the work of the Commission in that case contributed to an improvement in the relations between the Aluminum Company and the utensil manufacturers.

[72] KFR, p. 115.

[73] An NRA code proposed in 1933 for the cooking-utensil industry, but never adopted, contained clauses treating the matter of price differentials. See below, p. 476.

CHAPTER XIX

SCRAP AND SAND CASTINGS

1. Complaint of Unfair Methods

In the decade 1920–1930 the number of foundries pouring aluminum sand castings in the United States ranged between 2,000 and 2,700.[1] Nearly all of these foundries were very small units. The bulk of the output has come from fairly large plants, some of which are departments of automobile companies and other manufacturing enterprises. In 1921 it was estimated that about fifteen foundries turned out 35 or 40 per cent of the total production, one-third was produced by several sand foundries subsidiary to the Aluminum Company, and the remainder was scattered through hundreds of small units.[2] The number of larger foundries probably did not change appreciably in the ensuing ten years.[3] This industry, also, fits the assumptions of constant cost fairly well. Mechanization cannot extend far in sand castings, and although there are substantial economies in technical foundry control, their enjoyment is not confined to the largest companies.

Both virgin and secondary aluminum are used in making sand castings, but the greater proportion of metal going into this form is secondary. The metal cost accounts for more than half the expense of manufacturing sand castings. The large foundries have always melted scrap themselves, while the smaller ones buy secondary ingot from large remelters. Secondary ingot sells below virgin except when there exists a shortage of the latter.

The formation of the Aluminum Castings Company in which the Aluminum Company came to own a 50 per cent interest has been described. At its organization in 1909 this concern was said to produce over one-half of the castings output of the country. A decade later its successor (which is about to be introduced) was

[1] MI, xxxiii, 29 (1924); *Metal Industry,* XXVIII (October 1930).
[2] United States Tariff Commission, *Digest of Tariff Hearings* (1922), p. 230; and MI, xxx, 16 (1921).
[3] Cf. testimony in FTC Docket 1335, *passim.*

reported to have only a third, or less, of the total sand castings business.[4] In 1919, when the condition of the Aluminum Castings Company was not prosperous, a new company, Aluminum Manufactures, was incorporated in Delaware to acquire all the assets of the former concern and extend its activities.[5] Aluminum Manufactures raised several million dollars of new capital. Part of this was contributed by the Aluminum Company, which assumed a controlling interest. During 1920 the new corporation incurred a loss of several hundred thousand dollars exclusive of depreciation charges, which were not figured in. Deficits continued through 1921 at a lesser rate, and the company did not quite break even in the first half of the following year.[6] At this juncture the Aluminum Company of America leased all the assets of Aluminum Manufactures for twenty-five years, assigning the lease to its fabricating subsidiary, the United States Aluminum Company. As rental the Aluminum Company agreed to pay dividends on the preferred and common stock of Aluminum Manufactures. The company averred that its reasons for making the lease were to preserve its own financial reputation and to protect the public investors who had bought the stock of Aluminum Manufactures upon the supposition that the Aluminum Company of America was behind it.[7] In the latter part of 1922 and in 1923 the Aluminum Company purchased large amounts of scrap aluminum, and the price of scrap rose in greater proportion than the price of virgin.

The Federal Trade Commission was advised that the Aluminum Company was practicing unfair methods of competition in the castings industry. After investigation a formal complaint was

[4] *Digest of Tariff Hearings, loc. cit.* The Castings Company also made brass and bronze castings, as did most aluminum founders, and permanent mold pistons.

[5] *Wall Street Journal,* November 24, 1919, p. 12, and June 22, 1922, p. 12. Prior to the formation of this new company certain officials of the Castings Company had withdrawn to start an independent enterprise, the Charles B. Bohn Foundry Company. Rapid growth and merger has brought this concern to a position in the industry second only to that of the Aluminum Company.

[6] *Ibid.* It is not clear whether the financial ills of the new corporation were largely inherited or were chiefly due to the increase in capitalization and plant just a year before the beginning of business depression. The *Wall Street Journal* (June 22, 1922) quoted the president of Aluminum Manufactures as saying that soon after organization it was found that the condition of the Castings Company had not been so prosperous as supposed. It was not explained how it happened that the true condition of the Aluminum Castings Company had been misapprehended.

[7] Statement to Department of Justice (BR, pp. 8 ff.).

issued against the company on July 21, 1925.[8] The first count charged violation of Section 2 of the Clayton Act by price discrimination in the sale of sheet to foundries. Since foundries do not as a rule purchase sheet, this charge was not sustained by the evidence.[9] The second count charged employment of several unfair methods in violation of Section 5 of the Federal Trade Commission Act in carrying out "a scheme the effect of which was and is to gain a monopoly of the aluminum sand-castings industry in the United States." [10] It is very unfortunate that in this instance the Commission attempted to meet the rule laid down by the Supreme Court in the Gratz case by drawing the complaint in such sweeping terms.[11] It is quite unlikely that the Aluminum Company would attempt to monopolize the sand-casting branch. In the absence of artificial control of the differential between the prices of materials and the prices of castings there are no substantial barriers to entry into this industry. Hence artificial control would have to be stringent in order to gain and hold even that portion of the business done by the large foundries, unless the company were prepared to buy out substantial competitors as fast as they sprang up. All but a small fringe of ignorant or lucky competitors could be effectively shut out by keeping the selling price of castings at a level unprofitable to independent enterprise; or by preventing would-be competitors from obtaining adequate supplies of scrap or virgin ingot at prices enabling them to compete. Maintenance of price differentials between castings and their materials, virgin and scrap aluminum, which were inadequate to support independent foundries would seem the more feasible method. If virgin ingot was sold for other uses, however, this method would be profitable only when the demand for the basic component was more elastic in the castings market than in other uses for which ingot was sold.[12] At times when demand in the castings market

[8] This complaint, which also included the charges of unfair methods in the utensil branch, was dismissed in 1930 for lack of evidence to support the charges.

[9] Attorneys for the Commission abandoned it for that reason (FTC Docket 1335; FTC brief, p. 2).

[10] The full text of the complaint and answer of the company is printed in Appendix E.

[11] The rule of the Gratz case permits the orders of the Commission to deal only with allegations definitely charged in the complaint.

[12] Cf. above, pp. 217 ff., and 386 ff.

was less elastic, differentials exceeding the cost of conversion could not long be maintained without attracting independent enterprise. Preservation of a destructive scrap-castings price differential by purchasing continuously substantial amounts of scrap at relatively high prices would doubtless be unprofitable. Attempts to discriminate in price between foundries and other buyers of the company's ingot would probably be ineffective. Finally, it is doubtful whether the company could prevent independents from securing adequate supplies of domestic, or perhaps foreign, virgin ingot at little more than the ruling price.

Although it seems certain that the company was not trying to monopolize this branch of the industry completely, it cannot be so easily concluded that there was no employment of unfair methods with the purpose or result of advancing the trade of the foundries leased from Aluminum Manufactures. It must be reiterated that although the use of unfair methods may rarely achieve complete monopoly, it may result in substantial advantages not based upon efficiency. To deal intelligently with such situations requires formulation of concepts in terms of degree rather than in absolutes. In the instant case it may well be that the executives of the company, feeling that they were under a charge to which considerable stigma attached — and which may have seemed ridiculous — were quite sincere in their vehement denials of everything and anything connected with the charge; whereas if the complaint had been worded in less sweeping language, the way might have been opened for a more objective discussion of the issue whether natural business policy had in this instance brought results which were undesirable on economic grounds.

The particular practices with which the Aluminum Company was charged may be summarized as follows: [13]

(*a*) Arbitrarily fixing a differential between its price of virgin ingot and the open market price of scrap by an extensive campaign of buying scrap

[13] The complaint alleged that these practices had been and were still being employed. In another paragraph the charges of arbitrary neglect or refusal to supply metal to competitors of the company or its subsidiaries, of arbitrary delivery delays, and so on, were made applicable to the activities of the Aluminum Company in the castings branch as well as the utensil industry. (These are summarized above, p. 411.) Comparatively little attention was devoted to these matters in the consideration of competitive methods in the castings branch, so there is no basis for a discussion of them here. The complaint is reprinted in Appendix E.

for which it sometimes paid more than its cost of manufacturing virgin ingot.

(*b*) Making concessions in the price of sheet to customers engaged in the fabrication of aluminum on the understanding that they sell back to the company at high prices their total supply of scrap.

(*c*) Transferring virgin ingot to its subsidiaries at arbitrary prices below cost of production or below the price at which ingot was sold to competitors of the subsidiaries in the manufacture and sale of sand castings.

(*d*) Selling sand castings to the automobile industry at prices approximating or less than the cost of manufacture and at prices less than those at which competing foundries could sell at a profit, taking into consideration the prices they had to pay for virgin and scrap.

After enumeration of the practices set out above in (*a*) through (*d*) the complaint implies in the next subparagraph that the chief concern of those who drew it was with the scrap-purchasing activities of the company. Verbiage eliminated, this section reads:

The practices of respondent as set out [above] . . . have been made and are being made with the purpose and/or effect of curtailing the supply of raw material used by the independent . . . foundries . . . and of compelling [them] . . . to purchase virgin aluminum . . . from respondent and with the purpose and/or effect of eliminating as a source of supply . . . the scrap aluminum theretofore available.

It will be noticed that the practices connected with price cutting do not seem to be regarded as main components of the scheme. Emphasis is laid on curtailment of the supply of materials and forcing the independents to pay arbitrary prices for them, with the result that the true economic issues are much confused. This unhappy consequence is, perhaps, to be explained by the concern of the Commission's attorneys with the following letter.

ALUMINUM COMPANY OF AMERICA [14]

Internal correspondence

Sept. 9, 1922.

FROM —— ——
CLEVELAND OFFICE

TO —— ——
PITTSBURGH OFFICE

I was in Detroit last Friday and spent most of the day talking to —— and —— about the feasibility of our controlling the market on aluminum scrap and the advantages to be gained by us, principally to our sand-castings

[14] Commission Exhibit 525. A copy of this letter had been taken by a field agent of the Commission who had been granted access to the files of the company during the kitchen-furnishings inquiry.

business, by boosting the price of scrap as close to the price of new metal as possible. I described a scheme to you when I was talking to you in Pittsburgh, and it involves nothing more than deciding for ourselves upon an arbitrary differential between the price of new ingot and the price of reclaimed scrap, and in buying enough scrap ourselves for use in the castings plant to put the price of scrap to that level and to hold it there.

The effect will be to put all jobbing foundries, including our own, on the same metal level; to permit us to take full advantage of the products of the recovery plant at Niagara Falls and at Cleveland and to permit us also, by means of the products of these recovery plants to offset, where necessary, any peculiar advantages in manufacturing conditions that some of our competitors may enjoy.

I outlined the scheme to ——— and to ———, and for half a day we tried to pick flaws in it, and the only possible flaw that any of us could see in the scheme rested in the fact that none of us were quite certain as to the relation between the total tonnage of scrap offered for sale and the tonnage of castings business offered by the trade.

I talked this feature of it over with Mr. ———, who was of the opinion that scrap prices could be held up to an arbitrary level by the purchase of perhaps considerably less than half of that which is offered.

I would like to sit in a meeting one of these times, called for the purpose of throwing stones at this idea, and then if nobody can smash it, I would like to see the management proceed with it.

——— — ———

This letter was written to the sales manager of the Aluminum Company by a district sales manager. The persons mentioned in it included another district sales manager and two officials connected with the leased foundries. Executives and officials testified that the company had not adopted any such policy as the one advocated. Attorneys for the company introduced a large number of letters and memoranda as evidence of the innocent purpose behind its scrap policy; but the whole file of correspondence dealing with this matter was not introduced, nor made available to the examiner and the Commission's counsel. Over objection of the latter the examiner allowed the company to submit only what it cared to.

Unfortunately, in the brief for the Commission there is no adequate analysis of the conditions under which an arbitrary narrowing of the price differential between scrap and virgin would grant to the Aluminum Company any margin of competitive advantage. It may be asked whether the alleged scrap policy could gain for the company any differential advantage over its competitors, other than that of helping to deprive them of adequate supplies of metal, *without resort to price cutting in the sale of castings.* The importance of the alleged less-than-cost sales of castings is recognized in that one-half of the argument of the brief is devoted to this matter. But the relation between the scrap policy and the sale of castings below cost is not clearly set forth. Nor is there any satisfactory analysis of the standards of measuring cost at one stage of the operations of an integrated concern, or of the proper computation of cost in the short run.[15]

2. Preliminary Analysis

Before presenting the facts of the case it will be well to discuss some of the possibilities of uneconomic employment of practices included in the complaint and to examine the possible relations between scrap purchases and price cutting in the sale of castings. (The reader should recall the concept of unfair methods explained above, pp. 402 ff.) In dealing with these matters it will be convenient, as a device of exposition, to suppose first that before initiation of the large scrap purchases, the Aluminum Company and other founders were able to sell castings for prices which fully covered cost, including ordinary profit in this branch, that investment was well utilized, and that the total demand for castings did not recede during the period covered by the scrap purchases. In other words, we start from a condition of equilibrium in which demand and supply are nearly perfectly related. Marginal cost and full average cost per unit are equal;[16] those to whom the marginal analysis is not familiar will encounter no difficulties in following the argument. Later the discussion can be broadened to include conditions of demand under which the full product of the investment in the castings industry could not command prices equal to full average cost.

Clearly the company could expand the sales of its own foundries

[15] On the whole the economics of this case were handled by the attorneys for the Commission in no less capable fashion than many economists display in dealing with legal intricacies. The puzzling issues of the case called for economists as well as lawyers.

[16] This does not mean, of course, that castings which differ in size, components, time required for production, and the like, will have the same average cost. In perfect equilibrium the full average cost for each type of casting would cover an amount of overhead (including profit) which was equivalent to the proportionate utilization of investment incident to the production of a unit of that sort of casting.

at the expense of its weaker competitors by selling castings at less than cost — in other words, by narrowing the differential between the prices of the materials (scrap and virgin ingot) and the prices of the finished products. In the absence of such price cutting, the company might gain a competitive advantage by putting its rivals at a disadvantage relative to its own foundries with respect to amounts of materials obtainable and the celerity with which they could be secured. A determined scrap-purchasing campaign might make it difficult for the independents to buy their usual amounts of secondary metal at any price. And the company could undoubtedly hinder them in obtaining virgin ingot, although the broad ingot market would probably render ineffectual any attempts to cut them off altogether from the latter.

At first sight it appears that the Aluminum Company could, within the limits of the practices charged in the complaint, win no advantage over its competitors, other than hindering their supplies of materials, without resort to the sale of castings below cost. For it can easily be shown that the *price movements* incident to an extensive program of scrap absorption would not give the company any *margin* of advantage over its competitors — the effect upon the costs of all would be the same. The relative amounts of scrap, or "remelt" ingot made from scrap, and of 98–99 per cent virgin entering the mix for sand castings depend upon a number of factors. In general, there seems to be an advantage in using a large amount of secondary metal because the resulting product is sounder and machines more easily.[17] But it is obvious that price changes, if wide enough, might become the determining element. Ordinarily scrap sells for a few cents a pound less than virgin, which is an additional reason for its extensive use by founders. Any unusually large demand for scrap, such as that incident to the alleged campaign of the Aluminum Company, would induce competitive bidding on the part of the independents. The price of scrap could go at least as high as that of its substitute, 98–99 per cent virgin.[18] But narrowing or wiping out the scrap-virgin differ-

[17] There was much testimony to this effect in the record, and some contradictory opinion.

[18] Scarcity of scrap might result in a price somewhat below or somewhat above that of 98–99 per cent primary. On the one hand, the waste in scrap would act to keep it below; on the other, the better machineability of castings made with some

ential would increase the metal costs of all founders by the same amount. As long as the relative prices favored the large use of scrap, both the company and the independents would be affected alike as far as the increased price of the secondary metal is concerned. If its price were forced to a point where it was cheaper to make castings chiefly from virgin, the independents would suffer no differential disadvantage as long as they had free access to the latter at the ruling price.[19] As has been remarked, the broad ingot market offers so many possibilities of resale that price discrimination between founders and other buyers would hardly be practicable even if imports were limited to insignificant proportions. Hence it appears that, except for artificial restrictions upon the free flow of scrap and virgin, whatever their prices, an extensive scrap-purchasing campaign which resulted in diminishing or eliminating the scrap-virgin price differential could confer no competitive advantage upon the Aluminum Company, unless its sale price of castings were not raised by an amount equal to the increase in cost.[20]

Further reflection shows, however, that under certain conditions the Aluminum Company (or indeed any unintegrated foundry company with sufficient financial resources) might, without selling castings below cost, gain some advantage for itself by forcing the price of scrap up substantially. The metal costs of all foundries would, as we have seen, be increased alike. If average costs of production varied appreciably between foundries, some high-cost firms would suffer unless demand were growing at the same time to a position where the same total production of castings could be

---

secondary metal, and the uncertainty of obtaining virgin, might explain a premium. Of course, the price of scrap might rise substantially above that of virgin if the Aluminum Company kept its price of virgin down during a scarcity of primary metal.

[19] If they always had free access to virgin the price of scrap could not go above the point of equal relative utility, unless some powerful interest, or a demand from other markets, kept it above.

[20] It is obvious that if the foundries leased by the Aluminum Company had previously used virgin ingot nearly altogether and purchased it at the market price, these foundries would gain some competitive advantage from a scrap policy which resulted in an increased use of scrap in making their castings and a narrowing of the virgin-scrap differential. No direct evidence was taken on this point, but it appears unlikely that these foundries had formerly used virgin ingot nearly altogether. In any case the resort to cheaper material would not in itself constitute unfair competition.

sold at prices covering the costs of these firms.[21] Provided the foundries of the Aluminum Company were relatively low-cost units, it could, after raising the costs of all foundries by forcing up the price of scrap, obtain part of the erstwhile business of the high-cost firms which abandoned the field,[22] or at least became weaker competitors than previously, and then allow the price of secondary metal to drop to its usual level again.[23] During this whole process the Aluminum Company would, of course, have no greater margin of cost advantage (or disadvantage) over any of its competitors than before; it could gain trade only from those whom the uniform increase in cost pushed beyond the margin of profitable production. Furthermore, even if demand were increasing rapidly enough to allow all the concerns formerly making ordinary profits to continue to do so, the company might be able to expand more than the independents. For it could extend its own foundries in anticipation of the increased sale of castings when it allowed the price of scrap to fall.[24] Inasmuch as the Aluminum Company held the whip hand, its competitors might not care to do this.[25]

It may be urged that such a scheme, when employed without the sale of castings below cost, even if technically an unfair method

[21] Since the evidence available upon cost variations between existing foundries, differing scales of production, and different combinations of the factors of production, is very meager, it does not seem worth while to carry theoretical analysis of this point into further refinements dealing with submarginal firms, different returns to different units of capital and enterprise within the same organization, the effects of expansion on cost, the possibilities connected with financial reorganization of the weaker concerns, and so on.

[22] In this industry investment appears to be neither so large nor so specialized as in those fields where financial reorganization is nearly always preferred to abandonment or shift to another industry.

[23] The better independents would probably secure some of the business lost by the less efficient ones.

[24] If the demand for castings were very inelastic, the sale of castings would increase but slightly with a fall in their cost of production. At times when demand is not a function of price in the ordinary way sales might be unaffected or decrease with a drop in price.

[25] It is conceivable, of course, that the Aluminum Company would attempt to increase the costs of its foundry competitors merely by raising its price of primary aluminum. This could be accomplished provided there were sufficient transference of demand from primary to secondary metal to force up the price of the latter. But it is doubtful if the company would raise its price of virgin ingot above the figure which was most profitable, considering all of the many demands for this metal, merely to enlarge its foundry business.

of competition, would not be socially injurious, since high-cost firms would be eliminated, thus enabling expansion of more effective business units. Expansion of the latter might bring higher costs, however, and, in the actual dynamic world, at any given time some of the high-cost firms may be young concerns which, if unhampered, will later achieve low costs through growing up to the best scale, acquisition of experience by executives, and the like.

Upon the assumption of uniform average costs between foundries (no matter how the costs of all change with variations in total output), it is obvious that, after the price of scrap had been forced up, the Aluminum Company would have to sell castings below cost unless demand had at the same time moved forward to a position where the increased metal costs could be included in price without lessening the number of units sold. (Otherwise all foundries would suffer some underutilization and less than normal returns.) If demand advanced beyond this, the company might be able, as in the case of varying costs, to expand while competitors were afraid to, or could not obtain capital owing to their poor financial record resulting from a period of sales below cost. Assuming differing average costs between foundries and stationary demand, the Aluminum Company would have to sell castings below cost if its foundries were among the higher cost units.

It is clear that the sale of castings below cost, when demand and supply conditions permitted cost to be fully covered with normal capacity utilized, would constitute an unfair method of competition, according to the criteria set forth above, without any attempts to increase the cost of materials or to hinder the supplies of competitors. It has been pointed out that substantial additions to metal costs would make it impossible, under rather keen competition between firms having about the same cost conditions, for any of the concerns to obtain normal returns unless demand increased enough. In the event that the demand for castings did not grow sufficiently, the higher metal costs resulting from the Aluminum Company's scrap purchases could not be fully covered. This would constitute an instance of unfair less-than-cost selling unless the whole increase in cost of materials were explainable by elements outside the control of the company.

The discussion must now be generalized to cover all possible relations between investment and demand in the castings industry as a whole; in other words, to include situations where marginal cost and average cost are not equal and where prices (in the absence of discrimination) bring in less or more than ordinary profits. Under these conditions the justifiable minimum for price is marginal cost. Any price for a unit of business which fails to return the full amount added to aggregate expense on account of that unit — in other words, the expense which would not be incurred if that unit were not produced and sold — is unjustifiable. In application to the conditions of demand now assumed, the argument of the last few pages requires no change in substance. This will be apparent if it is understood that full average cost was there regarded as the justifiable minimum for price only because average cost was equal to marginal cost. Under all circumstances the test of relative efficiency in meeting society's demands is marginal cost. It must be understood, of course, that the items of expense properly included in marginal cost are determined by the size of the increment of output in question. Since the evidence presented in the case treated in this chapter was confined to individual orders taken over a short period of time and since no data on changes in capacity were given, it is not necessary at this point to examine the problems presented by continuous expansion which maintains overcapacity for a long time. The difficult problems incident to the application of the test of marginal cost to such a situation will be considered in the concluding chapter of this section of the book. At times when demand will not take at prices covering average cost the full output of the normally efficient capacity in an industry, sales below average cost are not to be considered unfair competition, provided they return marginal cost. When demand yields prices above average cost, and marginal cost is higher than average cost due to straining existing capacity, prices above average cost but below marginal cost would constitute unfair pricing. In the absence of reasonable mistakes there can be little motive for sales below marginal cost under any conditions except that of gaining an advantage not based upon efficiency or conferring a benefit which finds no explanation in efficiency.

It must be inquired, then, in reviewing the evidence presented in this case, whether any of the following competitive practices, or any combination of them, was used unfairly by the Aluminum Company.

(1) Hindering competitors from obtaining with reasonable celerity adequate supplies of scrap and/or virgin metal.

(2) Forcibly raising the price of scrap, and hence the costs of all foundries, with any of the following results:

(*a*) Pushing high-cost firms (if cost varied between foundries) beyond the profitable margin of production.

(*b*) Reducing the profits of all foundries (if costs were uniform between foundries) because of the impossibility of selling the same amount of castings at prices which fully covered the increased costs. (This case does not, however, involve prices below marginal cost of the smaller amount of castings sold.)

(*c*) Enabling the Aluminum Company to expand the capacity and output of its own foundries ahead of those of its competitors.

In addition to these possible consequences the situation so created might be used as a cloak for price cutting.

(3) Selling castings at prices below marginal costs. Under conditions such that price could be made equal to full average cost with equipment fully utilized, average cost, being equal to marginal cost, would represent the minimum justifiable price. When demand was less intense the lowest fair price would be below average cost. Sales of castings below cost might be entirely unrelated to the other practices mentioned above, or they might take the form of failure to advance the price of castings enough to make up the increased metal cost incident to extensive purchase of scrap. (This case is distinguished from 2*b* above by the fact that price does not cover marginal cost here.)

It is not enough, however, to ask whether these practices have been employed at all, for most of them might result from circumstances over which the Aluminum Company possessed no control. In fact, this was the principal defense of the company. It will be advantageous to examine this defense early in the presentation of facts and evidence in order that the tests for unfairness may be framed as specifically as possible.

3. Scrap

In the years 1922–1925, which were covered specifically by the complaint, demand for aluminum grew rapidly in most markets. The demand for castings was no exception. It was only toward the end of this period that the shift to iron crankcases began. The forward-moving demand reflected business recovery, increasing prices of other metals, and a higher price of aluminum abroad. It would have been strange if the domestic price of aluminum had not gone up under these conditions, particularly since the tariff of 1922 raised the duty from 2 to 5 cents.

The few undisputed facts about the scrap purchases seem to be substantially as follows.[26] A few months subsequent to leasing the plants of Aluminum Manufactures, the Aluminum Company began to buy large amounts of scrap. Contracts were made with a great number of customers to sell their scrap back to the company. In several instances the contracts called for the return of all the scrap.[27] Total amounts purchased were as follows:

	Pounds
1922	
July 1–Sept. 30	327,063
Oct. 1–Dec. 31	2,033,683
Total for last six months	2,360,746
1923	
Total for the year	10,063,356
1924	
Total for the year	3,651,748
1925	
Total for first six months	2,463,839

[26] Unless otherwise stated, the information upon this case was secured from the record of testimony, exhibits, and briefs of Docket 1335. Footnote references will be given only for special points.

[27] In at least one instance the company evidently made a special price for sheet contingent on the return of all scrap at a fixed price. It will be recalled that the complaint charged unfair use of this practice. It does not appear from the testimony that this practice was extensively employed; and in the one instance dealt with at length, the testimony seems to show that the practice was not used unfairly. Several million pounds of sheet were sold to the Budd Manufacturing Company at a special price for use in making Ford sedan bodies. The contract called for return of all resultant scrap at a set price. When the Budd company subsequently sold a substantial portion of the scrap to the Bohn Foundry and various utensil makers through a New York broker, there arose a dispute with the Aluminum Company over the interpretation of the term "scrap" as used in the contract

It is impossible to ascertain the exact proportions which these amounts made of the total scrap appearing on the market. The figures [28] of the Bureau of Mines showing the secondary-aluminum recovery during these years were incomplete, owing to failure to report on the part of some firms engaged in melting scrap. Furthermore, since these figures referred to *recovered* metal, allowance must be made for the shrinkage in melting, which appears to have averaged somewhere between 10 and 20 per cent of the weight of the scrap.[29] On the other hand, apparently some part of the scrap, from which the secondary metal represented in the Bureau's figures was obtained, did not go through the market. The United States Aluminum Company probably reported secondary metal recovered from sheet clippings, foundry borings, and dross, and so on, originating in the plants of the Aluminum Company and its subsidiaries, as well as that recovered from purchased scrap. The same may be true of other firms. The proportion of market scrap absorbed by the Aluminum Company in 1923 was probably somewhere between 10 and 25 per cent. For the subsequent years it must have been less than 10 per cent.[30]

According to testimony the price of scrap went up during 1922 and 1923 in greater proportion than that of virgin. In the last quarter of 1922 sheet scrap rose from 14 to 20 cents, while virgin

(Officials of the Budd Company maintained that, according to their understanding, the contracts called for return of only those pieces from which no product could be directly made — i.e., which were useless without remelting and conversion into ingot.) The correspondence over this matter seems to establish the fact that the chief concern of the Aluminum Company was to make a special arrangement with the Budd Company which would enable the cost of the Ford bodies to remain unchanged on an advancing aluminum market. Calculation of the contract prices of sheet and scrap may have been based upon an expected rise in the price of the latter. In any event, when the market quotations on scrap rose above the contract figure the Aluminum Company naturally preferred to retain this advantage for itself. (This matter is dealt with in Commission Exhibit 860 and Respondent's Exhibit 153, and much space is devoted to it in the briefs.)

[28] See Table 39.

[29] Testimony indicated an average of about 20 per cent shrinkage, but estimated recovery on the total scrap purchased by the Aluminum Company in these years was about 90 per cent (Commission Exhibit 854).

[30] Attorneys for the company estimated the proportions, after allowing for shrinkage and including the recovery of one large purchaser of scrap who had not reported to the Bureau, as follows: 18 per cent for 1923, 5 per cent for 1924, and 4 per cent for 1925 (Respondent's brief, p. 73). These estimates do not include the amount of metal recovered from "own scrap." No inquiry at all was made about this matter.

ingot went up only 3 cents. During part of 1923 sheet scrap sold for as much as the Aluminum Company's primary ingot.[31] The latter was then cheaper, of course, since it costs something to convert scrap into "remelted" ingot. Apparently the purchases of scrap by the Aluminum Company in 1922–1923 were extraordinary as compared with the experience of the foundries when operated by Aluminum Manufactures.[32] Evidently the company's purchases of scrap in the years following 1923 were also much smaller.

From what has been said in earlier chapters it is clear that the Aluminum Company was hard put to it to supply the demand for aluminum in various forms during the latter part of 1922 and all of 1923. All the data submitted to the Commission upon inventories, production, orders, and the like bear out this fact. The company's principal defense was that, on account of its shortage of primary metal relative to the rapidly increasing demands, it needed the scrap which it bought, not only for its newly acquired foundries, but also to fill orders for aluminum in forms other than castings. Evidently much less than one-half of the metal scrap purchased during 1922–1923 was used in the foundries.[33] The rest apparently went into regular ingot, sheet, and other products, or inventory. The examiner of the Commission, before whom the testimony was taken, was so impressed by the condition of shortage that his findings of fact were considered by attorneys for the company to be an adequate statement of the facts of the situation for their purpose.[34]

It is quite possible that the scrap purchases of the Aluminum Company and their effect upon price were entirely due to the impact of rapidly growing demand upon the company's small and rather inelastic supply of primary metal — i.e., to forces which,

[31] Inter-office correspondence of Aluminum Company, FTC Docket 1335, Respondent's Exhibit 110; also Record, p. 930.

[32] There is no direct evidence on this point. No figures were called for or submitted. The fact that the respondent presented no evidence to the contrary, as well as inference from the testimony and exhibits in general, seems to substantiate the statement.

[33] Inferred from figures given by an officer of the company (Record, p. 5279).

[34] On the scrap aspect of the case they present his findings verbatim, with scarcely any addition, as their statement of the facts. Indeed, the company's attorneys filed no exceptions to any of the findings of the examiner in this case. Careful study of the evidence leads to the conclusion that the examiner's findings do not present a comprehensively realistic picture of the whole situation.

for the purpose of deciding upon the use of unfair methods, must be regarded as outside the control of the company, or fortuitous.[35] A marked shortage of virgin would result in transference of some demands from primary to secondary metal. Provided the outlets ordinarily using virgin would pay prices at least as high as the lowest prices at which the castings industry would take all the scrap supply, bidding between the new and old consumers of the somewhat inelastic supply of scrap would force its price up, and give to the former a portion of it. With the industry organized as it is, the Aluminum Company would be the instrument through which scrap flowed into many products (e.g., sheet and all sheet products).[36]

It is clear that the actual shortage of virgin metal would justify the employment *in some degree* of the practices listed under (1) and (2) on page 455 and that no condition of scarcity could make the methods listed under (3) fair competitive practices. Obviously measurement of the extent to which use of the practices under (1) and (2) was required by the general situation would present difficult problems. Neither the Commission nor the company attempted it. It is impossible to conclude definitely from the available evidence whether the company purchased much more scrap than was necessary to meet the market situation in the most economical manner or bid the price up higher than this required. Certain figures suggest, however, that the company bought more scrap than it needed to meet the emergency.[37] If this was so, the over-

[35] It is the unpredictability of demand, of course, which requires that the situation be considered outside the control of the company. The wisdom of the directors in curtailing output so much in 1921 and 1922 may be questioned, but it would be unreasonable to judge the use of unfair methods by a debatable lack of foresight.

[36] Ingot could be bought from producers of "remelted" ingot. But regular ingot customers of the Aluminum Company would probably prefer to buy from it. Apparently producers of sheet products were not at that time accustomed to the use of secondary metal (R. J. Anderson, *Secondary Aluminum*, Cleveland, 1931, pp. 453 and 470).

[37] Commission Exhibit 736 includes the number of pounds of "purchased secondary aluminum" used by the Aluminum Company in each year, 1920–1925. The amount used during the three years of scarcity, 1922–1924, appears to fall short of the sum of scrap purchased during these years (including the inventory taken over from Aluminum Manufactures, July 1, 1922) by nearly 3,000,000 pounds. For the shorter time, 1922–1923 (which excludes the business recession of 1924), the total available purchased scrap exceeds indicated use by almost exactly 3,000,000 pounds. How-

buying might have represented either a substantial overestimate of the amount needed or a purpose to disadvantage competitors.

In fact, in spite of several thousand pages of testimony and hundreds of exhibits, it is impossible to reach very definite conclusions upon any of the more important issues presented by this case. For judgment upon the intent behind the scrap purchases and their results — we have seen that unfairness cannot be determined by results alone in this case — it would have been exceedingly useful to have the actual facts about relative costs of the important foundries and some conclusion about the relations between capacity and demand in the castings industry during the period in question.

The nature of the sand-castings business and the available facts concerning its organization seem to permit the inference that neither unit costs nor marginal costs would vary markedly, even from year to year, as between substantial foundry companies managed by men of average ability, unless there was a failure to keep abreast of improvements, or some companies had overexpanded, new firms had just entered the field, or other factors of temporary importance caused a difference. The metal cost is a large part of the whole cost, and the opportunities for variation in administrative, capital, or operating technique do not appear to be extensive. There was some suggestion in the evidence that the foundries of the Aluminum Company enjoyed somewhat lower recovery costs due to more effective apparatus. Integration may possibly result in small economies at this stage. If we may assume that costs were nearly uniform between foundries, it follows that the scrap policy could have gained no unfair advantage through weakening or eliminating higher cost firms. There was, in fact, no testimony that this had happened.

With regard to the other practices under (2) — using the scrap

---

ever, it is not clear whether the figures for "purchased secondary aluminum used" represented scrap or metal recovered from scrap. (The terminology is confusing, for the Aluminum Company did not buy secondary ingot; it bought scrap.) When these figures are reduced by 10 per cent (the estimated average "shrinkage" of all scrap purchased July 1922–June 1925) the amount of scrap used in 1922–1923 falls short of the total purchased scrap available for use by more than 1,500,000 pounds. For 1922–1924 there is an indicated carry-over of 1,000,000 pounds. Furthermore, the amount of "purchased secondary" used in 1925 was roughly 5,190,000 pounds, while only 3,925,000 pounds were used in 1924.

policy to force a condition such that profits of all foundries suffered, thus weakening them relative to the Aluminum Company and perhaps enabling the latter to expand faster — no definite conclusion can be reached. No records of yearly capacity, output, and earnings of the important foundry concerns were obtained. The rapid increase in consumption of aluminum by the automobile industry [38] and the marked general business revival suggest that demand for castings may well have moved forward to such an extent that there was no need for foundry profits to suffer in spite of the higher metal costs, but this cannot be known for certain.[39] If this was so, then the scrap purchases could have gained no unfair advantage for the Aluminum Company except in so far as they facilitated the sale of castings below cost, or the hindering of competitors in obtaining materials, or created uncertainty about the future margin between the prices of scrap and virgin. There was no testimony that the independents were unable to purchase all the scrap they could use at the high prices ruling. Unfortunately, the question of the degree to which independent founders were supplied by the Aluminum Company with virgin, when the price relationship made it more economical to use large amounts of the primary metal in the mix, was not thoroughly examined. The scarce supply of virgin should have been allocated between industries on the basis of relative intensity of demand, and between firms in the same industry according to some equitable plan. Probably this was done to a large extent; but it also seems likely that the foundries may not have been very well supplied because their demand for primary aluminum must have ordinarily been somewhat irregular and often small. Depriving the foundries as a whole of some of the virgin which they desired would not be unfair to the extent that this was necessary to fill higher paying demands. But the implication of the evidence is that the foundries of the Aluminum Company were quite well supplied with virgin all the time. Although there is no definite proof of this, it appears

[38] See Table 5.

[39] An officer of one foundry company testified that during 1923 and 1924 the price of scrap was too high to permit a fair profit in making castings. There was no general testimony bearing on this point, although many foundrymen appeared as witnesses in this case. Had profits in general suffered severely, it seems probable that the record would have shown this.

that the total supply of virgin made available to the foundries as a whole was divided in such a way that the Aluminum Company gained some advantage over its competitors when the price of scrap was so high as to make the use of large amounts of primary ingot desirable. Because of the general reluctance to take the stand against the monopoly at a time when their relations with it were fairly satisfactory, neither the failure of founders to complain of this at the hearings nor the many testimonials from founders that they received good treatment demonstrate conclusively that they had not suffered earlier. If the case had been heard in 1923 and 1924 instead of later, the testimony would probably have had greater evidentiary value.

Unfortunately, no evidence of the relative expansion of the foundries of the Aluminum Company and those of other concerns was presented.[40]

The evidence presented in this case does not establish any unfair intent or results with respect to the scrap policy; but for the reasons given it is impossible to reach definite conclusions on some issues. The Aluminum Company may have bought more scrap than it needed, though neither the fact itself nor the purpose is demonstrated. It is probable, but not certain, that independent foundries were not as well supplied with virgin as those of the Aluminum Company. It is possible, though it does not seem very likely, that demand conditions were such that the scrap policy itself had the effect of impairing profits all round. With the exception of the possible qualifications just noted, which cannot be resolved one way or the other, it seems probable that the scrap policy by itself — without relation to the sale of castings below marginal cost — was not used with unfair purpose or result.

4. Pricing of Castings

One of the two charges connected with unfair pricing was that the Aluminum Company transferred metal to its fabricating sub-

[40] Figures in MI, xxx, 16 (1921), and those given by Anderson in *The Metallurgy of Aluminium and Aluminium Alloys,* pp. 483 and 490, indicate that the annual capacity of the foundries of the Aluminum Company remained at 30,000,000 pounds between 1921 and 1924, while the capacity of other aluminum foundries increased from 60,000,000 pounds to 120,000,000 pounds. Neither the basis for these estimates nor the meaning of annual capacity is explained. Moreover, it is not altogether clear that the figures represent capacity for sand castings only.

sidiaries at a lower price than it charged the outsiders. As a matter of fact, the company did not transfer title to the metal to its subsidiaries at all. The United States Aluminum Company, which operated the foundries, charged its parent concern a toll which was supposed to include the operating and overhead costs of making castings. After deducting these items, and the value of the virgin metal used, from the sales invoice of the Aluminum Company the profit went to the subsidiary also; but the parent company retained title to the castings until they were sold.[41] To this legal arrangement the charge just mentioned did not, of course, apply.[42] However, the charge of selling castings below the cost of manufacture enabled examination of the whole matter.[43]

To deal with this satisfactorily something must be said of the standards for measuring less-than-cost selling at one stage of an integrated firm which has a monopoly at lower stages. Unless the costs of the integrated concern are calculated in such a way as to ensure economic competition between its own plants and independent one-stage firms, the proscription of unfair methods of price making will be meaningless. In determining the cost of production of castings made by the Aluminum Company it must be decided how to figure the cost of its virgin ingot and its own scrap. It has already been suggested that the company arranges its price schedule so that the largest earnings come at the ingot stage.[44] Should the cost of virgin used in making castings be based on the market price, which may often include high profits, or should its virgin go in at a figure which represents normal profits at the ingot stage? This question was not discussed in the case at hand. It may appear that the latter basis should be chosen, upon the ground that the total price policy of the monopoly limits the expansion of independent foundries, and also of the company's foundries, and of other finishing plants. Presumably a lower ingot price would permit larger sales of all aluminum products except

[41] Commission Exhibit 657.

[42] Transfer of title was superseded by the present arrangement sometime after the Rolling Mill Case. See above, p. 373.

[43] It is obvious that the transfer of title is of legal importance only. Whatever legal arrangements govern ownership of the product and participation in expenses and profits, the economic problem, whether costs are fully covered, is essentially the same.

[44] Above, p. 223.

those produced only by the Aluminum Company. However, as has been shown, it is unlikely that the company could charge a different ingot price to founders than to other consumers, or that it would fix its ingot price with reference to the castings demand only. Hence it may be concluded that, in all probability, the company does not ordinarily use its price for ingot as a tactical weapon in the competitive struggle in the castings industry. The desirability of allowing restriction of output of all aluminum products through a high price of ingot is a different question from that of competitive methods at a given stage of the industry. If the price policy of the company is permitted to include a high charge for ingot, this type of price pattern must be accepted in evaluating tests for distinguishing methods of price making in the finishing stages which would prevent true measurement of the relative effectiveness of the independents and the company. According to the criteria used here, economic or fair competition in the stages above ingot requires that the company be compelled to make the same profit upon the ingot used in its own finishing operations as on that sold outside.[45] Otherwise, to the extent that its competitors use virgin, their investment and earnings may be changed by the Aluminum Company at will, simply by pricing castings so as to earn less than the sum of the profit accruing on ingot and the ordinary competitive profit at the castings stage. Hence the calculation of the company's cost of producing castings should include virgin ingot "at the market." Scrap from the plants of the Aluminum Company which is used in making castings should also go in at the price for which it could be sold in the market. Or the remelt ingot made from the scrap should be reckoned at the price which the scrap would command plus the cost of converting it into remelt ingot.[46]

[45] The company's attorneys took the following position: "Respondent does not concede that if it so desired it could not use pig at *cost,* for there is certainly no requirement of any kind that we have been able to discover that compels this respondent to make a profit on every stage of its manufacturing operations" (Respondent's brief, p. 118).

[46] In general, the value of the company's own scrap used in making castings is, of course, the same as that of similar scrap actually sold in the market. Were the company to sell and buy back those amounts of the different kinds of its own scrap which it uses in the operations, the market prices of these types of scrap would remain unchanged. This is subject to various qualifications, none of which seem important for the issues at hand. The wording of the complaint and the argu-

The general presumption in favor of efficiency would call for recognition, when calculating cost, of any gains from integrating the production of virgin ingot and castings. However, if these advantages were slight compared to the social benefits to be expected from competition in the castings industry, they should be neglected in reckoning the minimum justifiable price. While no conclusive evidence on this point is obtainable, it would appear that efficiency is not markedly altered by integration of the manufacture of virgin ingot and castings.[47]

It appears that in many instances the foundries of the Aluminum Company used virgin pig aluminum instead of ingot when making low-grade castings. Pig is the first product of the reduction cell and contains impurities of bath material which are eliminated by remelting before the aluminum is run into ingot for sale. The company valued its pig at one cent below ingot, a difference which was supposed to represent the cost of turning it into ingot. Pig was used in making sand castings only at the most efficient of the company's foundries. When so used the metal cost was figured at one cent per pound below the price of ingot. It appeared that the company had never sold any pig to the outside market. Its representatives testified that the independent founders would not be able to make castings of good quality out of pig. It is not clear whether this "saving" of one cent per pound (it would actually be a little less than one cent, because the foundry costs would be slightly higher with pig) represents a real advantage of integration. Some of the independent foundries appear to be large and efficiently run. It would seem that they might acquire the knowledge, experience, and equipment to use pig successfully. If they could not, the economic penalties would soon make them aware of the fact that they should use ingot. It is true, of course, that the penalties might fall partially on other makers of castings.

---

ment of the brief show that the Commission took the position that virgin should be "costed" at the market price. The company's procedure will be explained in a moment. The question of how to figure the cost of the company's own scrap was not raised.

[47] In so far as this is so it would follow that the value of own scrap coming from departments other than the foundries should be reckoned at the price at which it could be bought rather than the price at which it could be sold. For large amounts of scrap of known origin, such as that of the Aluminum Company, the difference between these two prices would doubtless be very small.

In so far as the use of pig by founders in general was nearly certain to result in poor castings, with resultant damage to the reputation of aluminum castings in many uses, the Aluminum Company would be quite justified in refusing to sell pig. But it seems doubtful that the better independent founders could not use pig satisfactorily; and purchasers of castings could specify the use of ingot whenever they cared to. The competitive advantage to the Aluminum Company of using pig instead of ingot on a million-pound order for which the mix included 30 per cent virgin would amount to nearly $3,000.

Specific instances of alleged sales below cost are now to be considered. From July 1, 1922, through the year 1924 the Aluminum Company filled 24,097 orders for sand castings, resulting in a total sale of about 24,900,000 pounds of metal in this form. Counsel for the Commission selected five cases in which there seemed reason to believe that the Aluminum Company had bid or sold below cost. In two instances the Aluminum Company secured no order. Of the remaining three, two were argued to be instances of sales below cost. Together these two contracts, made in the fall of 1922, resulted in twenty-four orders for a total sale of 2,520,000 pounds of castings to the Hudson and Chandler companies.[48] It is evident that the attorneys singled out for attention a few of the large contracts. No others were specifically mentioned, which seems to indicate that questionable bids and sales were not general practice of the company. It cannot be concluded, however, that these three contracts were necessarily the only ones which presented any issue of sales below cost. Considerations of time required for the acquisition of knowledge about casting technology and for careful analysis of the complicated issues concerning the costs of "mixes" composed of varied content would have prevented examination of more than a few cases. Also, we are again confronted by the sort of imponderables which hampered the inquiries in other matters — reluctance to testify and the very evident inability of the independents to be

[48] The third contract resulted in a sale of 880,000 pounds of castings to the Buick Company, which had apparently made its requirements in its own foundry since 1915. Testimony suggested that the Buick Company found that it could buy castings from the Aluminum Company in 1924 more cheaply than they could be produced in its own foundry. Little evidence was taken upon this matter, and its meaning is not clear.

certain whether the Aluminum Company was selling below cost or not.

Turning to the contracts with the Hudson Motor Company and the Chandler Motor Car Company, which counsel for the Commission considered were gained by sales below cost, one is confronted by complicated evidence which does not result in definite proof that unfair pricing was, or was not, employed. The record of testimony presented the cost estimates used by the Aluminum Company as a basis for the bids which secured these contracts. The estimates were composed of two items: (1) metal cost, and (2) conversion cost plus general expenses and profit. Metal cost was divided into detailed expense items of the proposed metal mixes — e.g., 30 per cent virgin pig, 20 per cent borings ingot, and so on. Virgin was "costed" at the company's sale price of 98–99 ingot less one cent, inasmuch as pig was used directly. It appears that the company did not have in its possession at the time of bidding nearly enough scrap to make up the mixes it proposed to use in the castings.[49] Hence the real question with respect to the other metal costs would seem to be whether the figures employed in the bids represented a reasonable estimate of the prices at which the company could buy the necessary scrap at that time, or in the future as it made up the castings which were to be delivered over a period of months. The prices at which the company actually bought during the months immediately following October and November 1922, when the bids were submitted, might suggest that some of the estimates, at least, were too low to represent reasonable errors in judgment of the scrap market.[50] Some testimony implied that the bids of competitors reflected the rising market more than those of the Aluminum Company. The item of borings ingot was included in the proposed mix for the Hudson bid at 13.5 cents per pound, while borings ingot was figured in the Chandler bid at 19.4 cents. Although both bids were calculated at about the same time no satisfactory explanation of this discrepancy was given. Internal correspondence of the company seems to show that the higher figure was considered to represent the real cost at that time.[51] A differ-

[49] See Commission brief, exceptions to examiner's findings, pp. 26 ff.; Respondent's brief, p. 117; and Commission Exhibit 856.

[50] See references in preceding footnote.

[51] Commission Exhibits 656, 657, 658. The discrepancy is hardly to be explained

ence of only 5 cents a pound on the borings ingot, which constituted 20 per cent of the mix in this instance, would mean a difference of one cent on a pound of castings.[52]

The Chandler contract was obtained at a figure of 27 cents a pound. In a letter discussing this sale, a district manager wrote to the sales manager, "At the outset let me say to you that I simply cannot justify this price in cold figures." The cost estimate which the district manager had prepared came to 27.6 cents a pound, and this included 40 per cent of ingots from skimmings and dross at 12 cents a pound. Of the latter figure he said, "This price is quite arbitrary and agreed upon for this particular job by telephone conversation with you."[53] The ingot from dross and skimmings was evidently the company's own scrap and should have been figured at the market price of such scrap plus the cost of making it into remelt ingot. Unless the latter expense were very much less at the foundries of the Aluminum Company than at those of other large producers of remelt ingot there should not be any marked divergence between the cost of such ingot to the company (reckoned as above) and the market price of equivalent remelt ingot. In this case the price of 12 cents appears to have been at least 4 cents less than the market value of that sort of ingot, a difference equal to about 1.5 cents per pound of castings. As has already been remarked, there is some reason to believe that the other estimates of metal cost were somewhat low.[54]

---

by any change in the market price of this kind of scrap, for the bids were evidently made within a few days of each other.

[52] It was testified that this Hudson bid resulted in depriving the Bohn foundry of business which it had enjoyed fairly regularly. The evidence is not, however, conclusive. Testimony concerning the relative figures of the bids submitted by the Bohn Company and the Aluminum Company was conflicting. Other foundries obtained a part of the total Hudson contracts closed at the time, partly, at least, because the Hudson Company desired to divide the business between several sources.

[53] Commission Exhibit 656.

[54] Furthermore, it seemed probable to counsel for the Commission that, after employing scrap prices as a basis for bidding on the Chandler contract, the scrap was largely replaced by virgin pig in the actual mix from which the castings were made. (See Commission brief, pp. 82 ff.) With the strong demand for ingot from many markets it does not seem likely that this would have been done unless the cost of ingots made from purchased scrap became greater than the value of virgin pig. At the time of the bid (October 1922) the latter exceeded the cost of the secondary ingots (from purchased scrap) specified in the estimate by one-half cent to one cent. During the period within which the orders were filled the price of 98–99 primary ingot (and so the value of virgin pig) rose 4 cents. Hence, if the company

While the evidence is not altogether conclusive, it would appear that in the case of the Hudson and Chandler bids the estimates of that part of marginal cost made up of metal expense were too low to cover adequately the required expenditure. This does not necessarily mean, however, that the total cost figures were below true marginal costs in either case. The Commission's attorneys did not examine the make-up of the second expense item for which figures were obtained, conversion cost (labor and foundry overhead) plus general expenses and profit. Evidently they had in mind a sort of average cost concept. With this conception of cost, demonstration that the estimates of metal cost were too low would be sufficient in itself to show sales below cost.[55] Prices below average full cost would have been justified, however, if marginal cost were below average cost because some capacity was unused. Since the lump-sum item for labor, foundry overhead, general expenses, and profit was not broken down we have no way of computing how much or how little of this item would be properly reckoned as a part of marginal cost. Consequently it is impossible to know whether the Aluminum Company was practicing uneconomic pricing of castings or was simply selling below average full cost, but not below marginal cost, in order to use capacity which would have remained idle at higher prices.

In addition to the lack of adequate data for computing marginal cost there is an absence of other sorts of data which might have been illuminating. Statistics of the changes in capacity, out-

---

did replace scrap with virgin in making castings, one would infer that the price of scrap rose more than 4 cents. In that event, unless the estimates of cost of secondary ingot were stupidly made without regard for the prevailing scarcity condition, it would appear that they were purposely set low. If the company had bought enough scrap shortly after the bids this would not follow. The evidence appears to indicate that, as a matter of fact, it did not, during 1922, purchase sufficient scrap borings to use in filling the Hudson and Chandler contracts alone, to say nothing of other contracts.

Additional evidence upon the question of low estimates of metal cost concerns the marked differentials between the estimates by the company of the cost of different types of secondary ingot. It seems probable that some of these remelted ingots, though made from different materials, were essentially the same in chemical analysis and utility, and would have sold for about the same price in the market. To the extent that this was true, the differing cost estimates would have been justified only if the company were able to pick up bargains in scrap. It is doubtful if such large amounts of scrap could have been obtained at bargain prices.

[55] Unless there were reason to think that the overhead and profit items were too large.

put, sales, and earnings of the chief foundry companies would have afforded a general picture of the relations between investment, price, and demand, as well as providing a check upon cost calculations.[56] The extensive testimony in this case gave no indication that the castings industry as a whole possessed overcapacity in the years 1922–1924, which may, perhaps, be interpreted to mean that excess capacity was not in any case substantial. The vague sketch given of the condition of Aluminum Manufactures in the years prior to the lease suggests, however, that this corporation possessed foundry capacity exceeding the demands of the period of good business before the depression of 1921. One might infer that to become fully utilized in 1922 and 1923 these foundries needed a larger proportion of the total sales of castings than they had previously enjoyed.

TABLE 35

SALES OF ALUMINUM CASTINGS BY THE ALUMINUM COMPANY OF AMERICA AND TWO INDEPENDENTS IN CERTAIN YEARS

(Thousands of Pounds)

Year	C. B. Bohn Foundry Co.	General Aluminum & Brass	Total for the Two Companies	Aluminum Company of America
1920	6,265	2,432	8,697	
1921	4,433	2,355	6,788	
1922	7,752	4,182	11,934	4,310 (last half)
1923	5,381	5,752	11,133	10,886
1924	7,999	3,985	11,984	9,696
1925	—	—	17,214	7,275 (first half)

In the fall of 1924 the General Aluminum and Brass Company was merged with the C. B. Bohn Foundry Company to form the Bohn Aluminum and Brass Corporation. The figure shown for the C. B. Bohn Foundry Company for 1924 includes the sales of the consolidated company during November and December of that year. Hence the figure for General Aluminum and Brass for 1924 covers the first ten months only. The figures for the Aluminum Company cover the last six months only of 1922 and the first six months only of 1925. The figures of the table appear on pp. 1193 and 1194 of the Record and in Commission Exhibit 731.

[56] It is, of course, doubtful whether the Commission possesses legal authority to obtain such data from firms which are not charged with the violation of law. This simply means that the present law is inadequate for effective government control of business practices.

The accompanying tables present certain data collected by the Commission. Table 35 shows the weights of sand castings sold in certain years by the Aluminum Company, the Bohn firm, and a concern absorbed by Bohn in 1924. Officers of the Bohn company were outright in asserting that their business had been harmed by the sales practices of the Aluminum Company.

It is plain that the sales of the Bohn Foundry Company suffered appreciably in 1923, which was a much better business year in the aluminum industry (as in general), according to the testimony and exhibits, than either 1922 or 1924. Furthermore, the dollar value of the sales of aluminum castings by the Bohn Company dropped from 71 per cent of the value of its brass and bronze business in both 1921 and 1922 to 58 per cent in 1923, rising again to 82 per cent in 1924.[57]

The earnings record of the sand-castings department of the Aluminum Company during the period in question appears in the following table. Unless the large increase in gross profits of the

TABLE 36

EARNINGS OF THE ALUMINUM COMPANY OF AMERICA ON SAND CASTINGS *

Year	Weight of Castings Sold (*Thousands of Pounds*)	Gross Profit	Administrative Expenses and Loss on Returned Goods	Net Profit	Gross Profit per Pound
1922					
July–Dec.	4,310	$9,041	$56,452	$47,411†	$0.002
1923					
Jan.–Dec.	10,886	169,509	139,130	30,378	0.016
1924					
Jan.–Dec.	9,696	433,951	139,746	294,205	0.045
1925					
Jan.–June	7,275	375,408	64,523	310,884	0.052
Total	32,167	$987,910	$399,853	$588,057	

* Both gross and net profits are calculated with virgin aluminum at the company's list price. All the figures except those of gross profit per pound are given in Commission Exhibit 731.

† Deficit.

[57] Record, p. 1195.

Aluminum Company in 1924 was to be explained by savings in operating expenses it was symptomatic of much higher prices than those of the preceding years. Since the earnings record of the Bohn Company for separate years was not given, nothing can be inferred about its prices.

While it would appear that the Aluminum Company promoted its sand-castings business in 1922 and 1923 (perhaps at the partial expense of the Bohn foundry) through lower prices than it charged in subsequent years, it is manifest that the information in these tables does not show whether or not prices were set below marginal cost. Testimony of a large number of witnesses, among whom were competitors, customers, and writers for metal-trade journals, was largely to the effect that independent foundries had not been harmed by the Aluminum Company during the years covered by the inquiry, and that the competitive methods of the company had been fair. How much reliance to place in this testimony cannot be known, for reasons already explained. An investigator of the legal division of the Trade Commission, who made an inquiry before the issuance of the formal complaint, has stated that "complaints regarding treatment being received from the Aluminum Company were quite general — not universal, however." [58] Conclusions by competitors, purchasers, or trade-journal writers that fair or unfair methods have been used are of questionable worth. It is doubtful if many of these persons understand sufficiently the real issues or correct tests, or possess enough relevant information to form valuable opinions.

5. Summary

A study of the testimony, exhibits, and briefs treating the policies of the Aluminum Company in purchases of scrap and the sale of sand castings yields disappointing results. Some of the charges were shown to have no foundation; but several of the more important issues are not resolved by the evidence. This unfortunate outcome must be partially ascribed to the failure of the Commission to frame the issues correctly and to obtain the

[58] Senate Committee on Judiciary Hearings, p. 414. The statement probably referred to all sorts of complaints from all classes of aluminum consumers, competitors or otherwise.

right sort of data. No doubt the Commission's work was also hampered somewhat by a reluctance to testify, at a time when conditions were much better from the standpoint of the independents, upon matters which were past history.

It seems certain that one chief purpose of the scrap purchases was to add to the company's scarce supply of virgin a substantial amount of secondary metal; but it is not certain that a program appropriate to this end was not exaggerated somewhat with the object of helping to restore the earning power and former dominant position of the failing foundries by hindering competitors from obtaining adequate supplies of materials, or weakening their financial condition, or creating uncertainty about the future movement of the scrap-virgin differential. Little evidence appeared on these questions, upon which attention was not directly and clearly focused. It may be inferred that these consequences would have been brought to the attention of the Commission had they existed in large numbers or with manifestly dire results in individual instances. Nevertheless, the situation represented such a complicated concatenation of elements that individual founders could hardly have assessed correctly the significance of the limited facts within their range of knowledge. No evidence of a general and continuing policy of sales below cost was presented. It would appear that the Aluminum Company obtained two large contracts at prices below average full cost. But it cannot be determined whether these prices covered marginal costs, which represented the significant test for distinguishing unfair pricing. Unquestionably the evidence which appeared in this case afforded no basis for anything but dismissal of the complaint charging unfair methods in connection with scrap purchases and the sale of sand castings.

CHAPTER XX

THE EFFECTIVENESS OF GOVERNMENT CONTROL

1. The NRA Code

Before summarizing the results of government activities concerned with the aluminum industry a brief description should be given of the fair-practice provisions of proposed codes submitted to the National Recovery Administration and of the code which was approved. Although these fair-practice clauses bore little relation to the avowed objects of the National Industrial Recovery Act, they are interesting in connection with the problems raised in this section of the book. In the summer of 1933 a code for all stages of the industry from ore through finished products was submitted by the newly organized Association of Manufacturers in the Aluminum Industry, whose membership represented over 95 per cent of the known producers of virgin aluminum and its products in this country and about 97 per cent of the volume of production.[1] After some revision this code was finally approved on June 26, 1934. In the form originally submitted it contained a provision that if any member of any branch of the industry so requested, the code for that branch should provide that an integrated firm must conduct its business in such a way that the total profit or loss resulting from the production and sale of any given article would be "equitably distributed between the several branches of the industry employed in producing said product," and that no branch would obtain "excessive or disproportionate gain or profit therefrom to the exclusion of any other branch so employed."[2] This measure was evidently regarded as ineffective by a small group of independent fabricators who submitted a proposed code for the aluminum-fabricating industry which was defined to mean the manufacture and sale of semifinished goods

[1] *NRA Code of Fair Competition for the Aluminum Industry,* p. 115. (This pamphlet is paged 113–129.)

[2] *NRA Code of Fair Competition for the Aluminum Industry,* as submitted on August 18, 1933, by the Association of Manufacturers in the Aluminum Industry.

made from aluminum and its alloys.[3] Their proposal included the following provision:

> No member of the industry shall, in the ordinary course of business, sell any product fabricated by him from aluminum for less than the sum of the market value of the aluminum used in the fabricating process and the cost of such process.[4]

Sales at prices below such an amount, occasioned by extraordinary conditions, were to be reported to the supervisory agency. Whereas the code proposed by the Association designated as administrative authority the Association or such agency as it selected, this group of fabricators contemplated a supervisory agency consisting of a person appointed by the administrator of the NIRA. Their code also provided for the collection of detailed statistics of basic information.

The Aluminum Company declined to accept the fair-practice clause submitted by these independents, or any other which would restrain sale of fabricated or semifabricated articles at prices below the sums of the ingot prices and the costs of conversion. Three reasons were advanced to show that such a provision would be unfair to the company.[5] It would impede the development of new markets through low prices, hamper the competition of aluminum with other materials, and obstruct price reductions for the purpose of disposal of accumulating stocks during depression.

Doubtless the term "cost" as used in the clause proposed by these independents meant average full cost. Insistence that average cost, including all overhead, be covered by price when demand is slack during depression would be likely to operate against the interest of the economy as a whole. A code designed to promote the public interest would require all firms to reduce prices to marginal direct cost in order to ensure the best possible utilization of facilities and avoid diminishing the number of employed. Such a code would include a provision that the market price of ingot must be equal to marginal cost and a clause such as the one under discussion in which the term "cost" would signify marginal direct cost. The fair practice provision (Article IX) of the code for the alu-

[3] Apparently the members of this group had not joined the Association (HR, *passim*).

[4] *NRA Code of Fair Competition for the Aluminum Fabricating Industry,* as submitted on August 29, 1933.

[5] BMTC appellant v. ACOA, fols. 5195 ff.

minum industry which was finally adopted and approved by the NRA read as follows:

> No member of the Aluminum Industry who produces aluminum ingot from virgin aluminum alone or from virgin aluminum in combination with scrap which has not left his possession shall discriminate under like conditions in the prices charged for such ingot (whether charged to himself in case of fabrication prior to sale, or charged to others, in case of sale for fabrication) either between himself and controlled companies on the one hand and other purchasers on the other hand; or between such other purchasers. Such ingot to the extent it is available shall be sold to anyone whose credit warrants. Provided, that nothing herein contained shall be construed to prohibit such member from meeting prices quoted for imported aluminum in any specific instance which shall be reported to the Secretary of the Code Authority within ten (10) days after a sale or quotation by such member at such a price, together with a statement that such price was made for the purpose of meeting such competition. Provided, further, that nothing in this Article shall impose any liability on such member by reason of quotations or sales made without intent to discriminate or made to meet foreign prices believed in good faith to have been quoted.

Article IX also stipulated that these provisions were not to be evaded by any subterfuge, and provided for a study of their operation by the code authority. It is plain that the phrases "under like conditions" and "charged to himself" would require exact definition before the economic efficacy of the provisions of Article IX could be determined. It will be noted that the measure adopted did not deal with the problem of differentials between the prices of ingot and fabricated goods, although both the original code proposed by the Association and the one submitted by the independent fabricators had treated this matter. The approved code provided for administration by a code authority elected by the members of the several branches of the industry. It included no provisions for filing of prices or collection of detailed statistics of basic economic data.

Before this code was finally approved, a code for the cooking-utensil branch was proposed.[6] This contained a clause stipulating

[6] NRA Proposed Supplemental Code of Fair Competition for the Aluminum Cooking Utensils Manufacturing Industry. For code purposes the aluminum cooking-utensil industry was regarded not as a part of the aluminum industry but as a subdivision of the fabricated metal products manufacturing and metal-finishing and metal-coating industry. A code for the secondary aluminum industry was approved earlier. Its provisions are of no interest in connection with the problems discussed here.

that any utensil manufacturer who also produced aluminum ingot or sheet must keep separate books of account for each branch and use as its aluminum-sheet or ingot costs in the utensil division the open-market prices of such products. In parenthesis it was remarked that the substance of this measure was a part of the consent decree of 1912, and two clauses were reprinted from that decree. Apparently this proposed code never became law.

The general aluminum code met with such strong objection from some independent fabricators that it was approved for a trial period only, during which an investigation of the practices of the industry was made by the research and planning division of the NRA. A report embodying the results of this study was submitted about the end of 1934 and made public a few months later.[7] From time to time the code was extended without alteration — except for the removal of Article IX as explained below — until it died with the Recovery Act. In the course of the investigation all known members of the industry were invited to submit complaints regarding the scope and operation of the code. No complaints were received indicating that the code had occasioned any increased inequity between the members of the industry. It appeared that there had been a tendency toward lower and more uniform prices charged for virgin ingot to competing fabricators. It was stated that the different groups represented on the code authority had expressed a general feeling that benefits were being obtained from the existing code. Members of the code authority had no suggestions for additions to the fair practice provisions. The code authority had made no interpretation of the meaning of Article IX, with the result that interpretation of the ambiguous phrases relating to ingot prices was left entirely to the Aluminum Company. It appeared that in accordance with its interpretation of no discrimination in prices charged "under like conditions" the company was charging different prices to separate groups of buyers, classified according to the degree of competition of substitute materials.

The only complaints received by the research and planning division came from three firms which had made the same objections at the original hearings. Two of these were engaged in private liti-

[7] This is the report which has been cited earlier under the name of Leon Henderson, director of the research and planning division. The following information has been taken from this report, except as otherwise noted.

gation with the Aluminum Company involving the issues raised by their complaints to the NRA. These complaints were summarized as follows.

(*a*) An artificial price of aluminum maintained above cost of production by monopoly control and world agreements.
(*b*) Domination of the Association of Manufacturers in the Aluminum Industry by the Aluminum Company.
(*c*) Sale of products by the Aluminum Company at prices less than the sum of the market value of the metal contained and the costs of fabrication.

The first complaint concerned matters which were for the most part outside the scope of the code. The research and planning division concluded that the membership of the Association of Manufacturers in the Aluminum Industry and the membership of the code authority included too many diverse interests for domination by a single firm. In connection with the third point the complaining firms reiterated their original proposal for a provision prohibiting the sale of any fabricated article at a price less than the market value of the metal used, plus a reasonable allowance for cost of fabrication. While the report of the research and planning division fully recognized the importance of the problem of differentials, no change was made in the code to further its solution. The report advocated as a minimum an exact interpretation of Article IX and the reporting to the NRA of the prices at which virgin ingot was sold to others and the ingot charges set up by the integrated producer in each of its own fabricating branches. It would appear that the report also favored the filing of list prices which would be made public.

It is said that during the early months of 1935 attempts were made by the National Industrial Relations Board to modify Article IX of the code in such a way as to increase the protection for small enterprises.[8] Amendments designed to accomplish that aim were apparently rejected as unsatisfactory by a majority of the members of the industry. On March 25 Article IX was suspended. It was said that the board had reached the conclusion that the code had not operated to protect small enterprises from oppression or discrimination or helped to effectuate the policy of Title I of the NIRA.[9] While this conclusion was not necessarily inconsistent

[8] *Metal and Minerals Market,* March 28, 1935, p. 3.

[9] *Ibid.*

with the findings of the research and planning division, the reasons for the suspension of Article IX have not been made clear.

2. Results of Attempts at Government Supervision

It is questionable whether the net results of the activities of government agencies concerned with the relations of the integrated firm and the single-stage independents have been very beneficial to consumers, to independent enterprise, or to the Aluminum Company. In spite of continuous investigation the chief problems concerned with relations between an integrated firm possessed of substantial monopoly power at earlier stages and its competitors at later stages cannot be said to have been settled satisfactorily. This is due in considerable measure to the failure of government agencies to perceive the true issues and in part to their lack of adequate legal authority. The consent decree of 1912 was not drawn with sufficient precision or breadth to establish an effective code. The inquiry of the economic division of the Federal Trade Commission, pursuant to a Senate resolution calling for an early report upon matters covering a wide range, was necessarily superficial. The results of the investigation of the Department of Justice, which was carried on in an atmosphere charged with political recrimination, left several of the more important questions unanswered. The Trade Commission case raised several interesting and rather perplexing problems, but the issues were not clearly presented, and the evidence was insufficient to resolve many of them.

The unfortunate aspects of the omission of opinions by the federal trade commissioners explaining their orders and dismissals are well demonstrated in a case such as this, involving a large corporation whose monopoly power was frequently alleged to have been strengthened by political favors, and upon the rectitude and economic statesmanship of whose officers the success of the competitive process in the semicompetitive stages of the industry must be in large measure dependent. If the commissioners believed that the Aluminum Company, which had for years been the target of various charges, was fully vindicated by the evidence, the publication of this opinion and the full reasons therefor was certainly due the company. If, on the other hand, the commissioners found the evidence conflicting and the important issues unresolved, some assurance that the Commission would keep the industry under careful scrutiny should have been afforded the one-stage manufacturers of semifabricated and fabricated products, who, in venturing to compete with the integrated company, make their fortunes partially depend upon the judgment, efficiency, policy, and caprice of its management.

It is not apparent that the National Recovery Administration contributed in substantial measure to the solution of the problems here at issue. Its chief objects, of course, concerned other matters.

It is difficult to determine the effects of government investigations upon conditions in this industry. Doubtless the constant attention of government agencies has forced upon the Aluminum Company a certain degree of caution in the formulation of its policies; but few if any guiding principles for distinguishing economic from uneconomic, lawful from unlawful, conduct have been afforded the company. It is to be doubted that independent enterprise has drawn much encouragement from the results of government investigations. That government agencies have not, in the course of their long-continued investigations involving considerable expense to the government and to the Aluminum Company, developed a body of economic principles capable of salutary application to the difficult problems presented by the relations between the integrated firm and the single-stage rivals is particularly regrettable. The difficulties of analysis and of measurement involved in these problems are indeed formidable; it is not intended to minimize the efforts of those who have struggled with them. The disappointing character of the results must be ascribed in important measure to the fact that the investigations and proceedings have been carried on for the most part by persons whose training has been chiefly legal, and to the restricted scope of their activities occasioned by existing law.

3. Private Litigation

The case of Haskell against the executors of the Duke estate and the litigation against the Aluminum Company by the Baush Machine Tool Company and the Sheet Aluminum Corporation appear to be the only instances in the history of this industry in which private individuals have invoked the provisions of the anti-trust laws for suits to obtain damages for injuries resulting from alleged violations of these laws.

In its suit against the Aluminum Company the Baush firm alleged that monopolization of the production and sale of virgin aluminum in the United States, as a consequence of the suppression of independent attempts to enter the industry, agreements with foreign producers or the exercise of coercive influence upon the foreigners, and the fact that control of Aluminium Limited and the Aluminum Company rested in the same hands, had resulted in artificially high prices in this country for virgin ingot of both domestic and foreign origin. Secondly, it was charged that the Aluminum Company had attempted to monopolize the production and sale of sheet and certain other semifabricated articles made from aluminum and its alloys by reducing the differentials between the prices of ingot and the semifabricated products below the costs of fabrication and below the spreads required to yield profits to an independent mill. The plaintiff alleged that it had suffered damage from the conditions in both the markets for aluminum ingot and for semifabricated products. The suit was filed in the summer of 1931. During the trial, which consumed ten weeks in the fall of 1933, about 2,500 printed pages of testimony and exhibits were introduced. A verdict for the defendant was returned.

Upon appeal the Circuit Court of Appeals for the Second Circuit reversed the judgment and ordered a new trial, apparently on the following grounds.[10] Since the plaintiff's claim for damages was based upon the allegation of injuries resulting from monopolization within the meaning of the statute, it should have been afforded adequate opportunity to demonstrate such monopolization. Certain evidence offered, consisting of transactions and contracts with domestic and foreign corporations, which purported to show monopolization, was erroneously excluded. The charge of the trial court restricted the jury too narrowly in their consideration of the evidence bearing upon the question of control of imports of virgin aluminum. Furthermore

[10] 72 Fed. Rep. (2) 236 (1934)

the instructions to the jury which directed that the jury would have to find for the appellee [defendant] if they found no agreement in effect between appellee and the foreign producers or between appellee and Aluminium Limited to control the price of aluminum or to restrain competition in the United States was [*sic*] erroneous.[11]

The significance of this statement does not seem clear. It appears to mean that in order to establish violation of the antitrust laws it was not necessary under the circumstances for the plaintiff to prove control of importation if it demonstrated unlawful monopolization of the domestic market for virgin aluminum.[12] Finally, the trial court had instructed the jury that they must find an agreement with the foreign producers or with Aluminium Limited resulting in control of the price of ingot in the United States in order to consider the question of injury from an inadequate price differential between ingot and semifabricated products. The Circuit Court stated that the jury should be instructed in the second trial that the plaintiff might recover damages caused by unfair price fixing which rendered impossible profitable operation of the plaintiff's business, if such price fixing was employed with the purpose of monopolizing; and that the plaintiff should be given opportunity to show that the price differentials were inadequate to permit it to operate without loss. It does not seem entirely clear whether this was considered as a separate issue from the question of control of the ingot price. The trial court in the second trial apparently regarded these as separate issues, as will be explained in a moment.

The second trial, which took place early in 1935, lasted about eight weeks. Inclusion of additional evidence brought the full record to more than 3,200 printed pages. The judge charged the jury that it should find for the plaintiff if the evidence established damage as a result of any one or more of the following circumstances. (1) Monopolization of the domestic market in virgin aluminum by combination or price agreement with foreign importers. (2) Monopolization of the domestic market in virgin aluminum (subsequent to organization of Aluminium Limited and distribution of its shares to stockholders of the Aluminum Company) by combination or agreement with Aluminium Limited that

[11] 72 Fed. Rep. (2) 241.

[12] The district judge sitting in the second trial evidently adopted the interpretation given in the text. See his charge, BMTC v. ACOA appellant, fols. 6502 ff.

the latter would not compete in the United States. (3) Monopolization of the domestic market in virgin aluminum by the Aluminum Company itself or by its subsidiaries. (4) Intentional monopolization or attempt to monopolize a substantial part of the domestic market in aluminum products and alloys by putting into effect differentials between the prices of these articles and the prices of ingot which were inadequate to enable the plaintiff, if efficient, to operate at a profit.

The jury brought in a verdict awarding $956,300 damages to the plaintiff. In accordance with the law this sum was trebled to $2,868,900. The sum of $300,000 was assessed against the defendant for attorney's fees.

The appeal of the Aluminum Company from this verdict was based to a large extent upon technical questions of procedure, but raised two interesting questions concerning interpretation of the antitrust laws. While denying that any evidence in the case established uniformity of prices charged by the Aluminum Company and importers, or constancy of price over long periods, or prices unreasonably high, the attorneys for the Aluminum Company maintained in exceptions to the charge that if such conditions should come into existence as a result of independent action by each of several sellers there would be no violation of the antitrust laws.[13] In other words, it was contended that oligopolistic policies would not infringe the antitrust laws. Secondly, the attorneys for the defendant, while insisting that it had made profits at all stages, took the position that maintenance by an integrated firm of price spreads between different stages which do not cover conversion costs is not in itself unlawful.[14]

In September 1935 the Circuit Court of Appeals again reversed the verdict arrived at in the lower court and directed a retrial.[15] The chief grounds for reversal seem to have concerned matters of procedure.[16] The suit has since been settled out of court. From

[13] *Ibid.*, fols. 6578 ff.

[14] *Ibid.*, fols. 5692 and 6633 ff.

[15] 79 Fed. Rep. (2) 217 (1935).

[16] In particular the following. During the second trial plaintiff's attorneys read to the judge, in the hearing of the jury, portions of the opinion of the Circuit Court on the first appeal in which the court expressed its opinion upon the conclusions which might be drawn from certain sorts of evidence. A majority of the court believed that this procedure "had the effect of a peremptory command to that jury to draw the conclusions of fact we had said might be drawn" (p. 224) and hence impaired impartial assessment of the meaning of the evidence.

the standpoint of development of the interpretation of the anti-trust laws it is unfortunate that the courts did not have an opportunity to pass upon the interesting issues mentioned above.

One more point remains. The archaic character of our present laws concerning the problems of monopoly could find no better demonstration than that afforded by the litigation of the Baush firm and the Sheet Aluminum Corporation with the Aluminum Company. To expect an ordinary jury, untrained in any but the simplest elements of economics, to resolve successfully and fairly the complicated and perplexing issues raised in cases like these calls for either a belief in divine guidance or an exalted faith in the salubrious qualities of ordinary common sense.[17] Few of the economic issues can be dealt with by the higher courts. Private litigation can contribute but little to the formulation of a satisfactory code of principles to govern the relations of the integrated firm and the single-stage independents. The enormous expenditures of effort in litigation of this sort would have been rendered superfluous had government agencies built up an efficacious set of principles. In the course of the Federal Trade Commission case, attorneys for the Commission had endeavored to have the complaint amended to include charges of unfair pricing in the sale of duralumin products. It is particularly regrettable that the Commission neglected its function in this instance by refusing to amend the complaint and make a thorough examination of this matter. However, as has been said earlier, it is to be doubted that the Commission can with its present organization, practice, and legal powers deal satisfactorily with this sort of question. The problems involved in this kind of controversy should be submitted to a court or commission whose membership comprises persons with legal training and persons with economic training, and also, serving in an advisory capacity at least, scientific experts and men of business experience who have a broad economic outlook. Such a group might be expected to work out a body of effective rules and apply them with judicious effectiveness. Some of the possibilities for creating conditions under which such controversies would not be likely to arise will be examined in the concluding section.

[17] This statement is not intended to imply anything at all as to the respective merits of the opposing contentions in these particular cases.

4. Possibilities for Economic Reform

The question treated in this section is how to secure desirable relations between investment, output, price, and demand in the various fabricating branches of the aluminum industry. The analysis in earlier chapters suggests that the existing antitrust laws are not likely to produce in all fabricating branches the sort of competition which tends to bring the desirable market relations. This objective might be obtainable by establishing conditions under which competitive forces would operate with greater effectiveness, or alternatively through direct government regulation of investment, price, or earnings in these branches, government competition in fabrication, or consumers' coöperation. In each case the appropriate instrument would depend partly upon the relation between the most effective scale of investment and the market demand. In any branch in which single-firm monopoly was necessary for maximum efficiency consumers' buying coöperatives or direct government regulation would be the only measures that might not impair efficiency. In any such instance private oligopoly or government competition might, however, bring better total results, particularly if they made for greater progressiveness.

It appears that maximum efficiency in most fabricating branches does not require single-firm monopoly. Wherever considerations of efficiency dictate few rather than many producers the problem differs in accordance with the presence or absence of single-stage firms. In a few instances the advantages of integration in productive operations may exclude the latter. The case of oligopoly of *integrated* producers alone in a fabricating branch has, in effect, been examined in Chapter XV, where it was concluded that if more public control was desirable the competition of an integrated government enterprise seemed potentially the best device.

In several branches it would appear that the presence of single-stage firms alongside one or more integrated companies would not diminish efficiency and might tend to bring better market relations as a result of stronger competitive forces. In some instances oligopoly would be appropriate for efficiency; in others there could be enough sellers to approximate, as far as numbers are concerned, the conditions of pure competition. But in either case competitive forces are not likely to produce the most desirable results of which they are capable until the problems of supply of materials to single-stage firms and of price differentials are satisfactorily settled. Under certain circumstances the most advantageous policies for an integrated firm possessing a considerable degree of monopoly power in the sale of materials used in fabrication may operate to impair the effectiveness of competitive forces in fabricating stages. The analysis of the preceding chapters in Part IV has shown some of the possibilities; a few moments' reflection will suggest others. Changes in the discriminatory price pattern which involve alterations in the differentials between prices of ingot and later products may have the effect of weakening independent competitors in some branches. Shifting allocation of the supplies of ingot or semi-fabricated products may unduly advantage subsidiaries or some independents at the expense of others. The same result may be produced by discriminatory treatment with respect to price, credit, quality, terms of delivery, coöperative technical work, and the like. The nature of competition may range from a fairly smooth and effective process to one which is halting, hit-and-miss, wasteful, or anarchic. There is no guarantee that under most circumstances a maximum of private advantage will coincide with a maximum of public benefit. We have seen that the attempts at control by government agencies acting under the antitrust laws do not seem to have advanced the solution of these problems very far. Reliance upon competitive forces in private industry to bring desirable market relations wherever that is consistent with a high degree of efficiency seems generally superior to other methods of obtaining this objective. Accordingly, attention will be given chiefly to alternative methods of creating conditions such that competitive forces will be as effective as possible in promoting desirable market relations in all branches where single-stage firms operate.

At this point it may be noted that government competition in *fabricating stages only* could not deal effectively with the problems of supply of materials, of discrimination in quality, credit, or service, or of inadequate price differentials. That would require government competition or regulation at the ingot stage.[18] Consumers' producing coöperatives which did not integrate back through the making of ingot would be as dependent upon the integrated com-

[18] Some indirect control at the ingot stage might, of course, occur if the possibility of government competition or regulation at that stage were stronger because of the existence of a government producer of fabricated products.

pany as ordinary single-stage fabricators. Opportunity for improvement in market relations through government competition in fabrication alone or through consumers' coöperation in fabrication alone would be limited principally to reduction of price differentials in any branches in which existence of oligopoly made possible unnecessarily high differentials. It is doubtful whether consumers' producing coöperatives would be established even in these circumstances unless there were definite assurance of continuous supplies of materials and of maintenance of adequate price differentials.

One other point should be noted here. In all branches where secondary aluminum is used in large measure, the problems of availability of materials and of discrimination with respect to price of materials, quality, service, and so on, have much less significance. This is not, however, generally true with regard to the problem of price differentials. The price of secondary ingot does not change by the same amount whenever the integrated firm changes its prices of fabricated goods. It seems to move more closely in step with the price of virgin ingot. Hence use of secondary ingot does not free fabricators from the possibility of varying differentials between prices of materials and prices of fabricated goods; the differentials may be altered by the policies of the integrated firm.

An endeavor to establish circumstances favorable to effective operation of competitive forces in fabricating branches might follow one of two courses: (1) creation of conditions under which the extent of government control in these branches could be restricted to a small minimum; or (2) formulation and application of a rather elaborate code of principles to govern the relations between the integrated firm and the single-stage enterprises. Although it might not solve all problems, exclusion of ingot producers from later stages would constitute the most effective device for creating conditions requiring only a relatively small amount of government control. But since it is probable that marked social benefits ensue when integrated ingot producers engage in research upon new variations and adaptations,[19] this solution seems undesirable.[20]

[19] Cf. Chap. XV.

[20] It is possible, however, that if only one integrated firm existed, advantages of this sort would not be large enough to outweigh the ease and lack of cost involved in this method of securing effective competition in all stages above ingot.

The existence of several integrated ingot producers might solve all problems of the supply of materials to one-stage firms and minimize discriminatory treatment of the latter with respect to price, quality, credit, and so forth. It is quite possible that the several integrated companies would compete to an extent sufficient to prevent the maintenance of any inadequate price differentials. But there is no certainty, of course, that individual oligopolists would act in this way instead of using their monopoly power to maintain a discriminatory price structure.

Removal or appropriate reduction of the import duty on ingot could be regarded as complementary to either of the above measures if it had the effect of increasing the satisfactory sources of supply for fabricators. Reduction of the duties on semifabricated and finished aluminum products might contribute to the maintenance of more effective competition here.[21] Tariff policies of this sort might also be considered complementary to a code regulating relations between integrated and nonintegrated producers, which constitutes the other broad alternative for securing effective competition.

From the standpoint of economic considerations alone, the production and sale of ingot by a government corporation seems to offer the greatest promise for maintenance of conditions favorable to effective operation of competitive forces in fabricating branches. The government corporation could keep the price of ingot low enough so that no integrated firm would find it profitable to set inadequate price differentials. The possibilty of discriminatory treatment of fabricators with respect to prices, quality, service, credit, and so forth could be eliminated in considerable measure by the existence of a government corporation. Government competition in the sale of ingot could not, it is true, directly prevent the narrowing of price differentials with intent to weaken or drive out single-stage fabricators; but it would tend indirectly to render such tactics unprofitable in the long run. In any such instance the detailed information about conditions in the industry which the government corporation would have could be placed at the dis-

[21] No data are available to estimate relative costs of producing aluminum or any of its products in Europe and in the United States. We are not concerned here with the question whether domestic ingot production or domestic fabrication is economically justifiable.

posal of government agencies enforcing the general law of unfair competitive methods. More important, the government corporation could follow a policy of entering any branch in which such tactics were used. Presumably no integrated firm would consider it worth while to incur losses for a time merely in order to exchange the competition of private one-stage firms for that of the government enterprise. Although government competition could not directly control inadequate differentials proceeding from nonprofit motives, the sums available for policies which were not expected to yield good profits would be small if government competition kept profits in the industry close to the normal costs of private capital and management.[22]

We now turn to the other alternative for creating conditions favorable to effective operation of competitive elements, the development of a code of principles to govern the relations between integrated and single-stage producers. The scope and provisions of such a code would depend somewhat on the number of integrated firms producing ingot and later products. The following principles are developed to fit the case of one integrated producer only. With several integrated companies the problems might be fewer, but also more complicated.

The chief problems to be treated include conditions of supply of virgin ingot and half-products, price discrimination in sales to producers of the same article, and price differentials. Unless removal or reduction of the ingot tariff would result in making available to American fabricators ample supplies of foreign metal on favorable terms, the domestic integrated producer should be obligated to sell metal in any desired form to all comers who satisfy reasonable credit requirements.[23] Rules against discrimination in respect to amounts sold, quality of metal, deliveries, and other aspects of service would need to be more incisively formulated than those of the consent decree of 1912. They should apply to discrimination between independents and between independents and competing divisions of the integrated firm or companies in

[22] In cases where management holds most of the shares it may, of course, prefer to forego some part of ordinary returns in order to carry out some other desired policy. This sort of thing could be controlled, if at all, only by an elaborate code.

[23] Article IX of the NRA code provided that virgin "ingot to the extent it is available shall be sold to anyone whose credit warrants." The phrase "to the extent it is available" would require precise definition.

which an interest is owned directly or indirectly by the integrated firm or any of its officials or leading stockholders. Discriminatory treatment of competitors who use different materials, such as utensil manufacturers who buy both ingot and sheet, would need to be prohibited. The code should provide that differences in the prices of materials furnished to competitors (defined as above) must not diverge from the actual differences in marginal cost.[24]

The problem of price differentials is not easily settled. It is clear that the minimum justifiable price for any article made from ingot must be the sum of the market price of ingot or semifabricated metal, as the case may be, and the marginal cost of conversion. (In the following discussion let marginal cost signify the sum of these two amounts.) It is also obvious that effective competition cannot exist continuously unless on the average the differential ordinarily covers the full average conversion cost of an efficient single-stage company. As long as there is no considerable overcapacity with respect to a price including the market value of material and the average full cost of conversion, the latter will not diverge appreciably in any branch from the marginal fabricating costs (as just defined) exclusive of the value of metal contained. The use of marginal cost, in general the proper test of efficiency, presents a dilemma, however, in that single-stage competitors would be weakened or eliminated without sales below marginal cost on the part of the integrated firm if the latter kept in existence continuously over a period of years substantial overcapacity in any fabricating branch.

Maximum economical production and consumption would occur, of course, only with the sale of ingot at its marginal cost. A requirement that marginal cost should constitute a maximum for the ingot price is not, however, a necessary part of a program to ensure effective competition in the fabricating branches. This objective calls merely for measures which place all competitors on a par with respect to metal cost and maintain adequate differentials on the average.

It does not seem probable that long-run overcapacity in any fabricating branch is unavoidable. If the collection and publica-

[24] If the costs of the integrated firm and its subsidiaries were properly computed it would be immaterial whether metal was sold to subsidiaries or furnished on some other basis.

tion of basic statistics were required there would be little reason to expect that independents would seriously overinvest the industry. The chief problem would appear to be the prevention of uneconomic expansion in any fabricating branch by the integrated firm. In other words, the independents can probably be relied upon not to plan an investment and price policy which is likely to involve the sale of goods at prices below their marginal cost of increasing output — the market value of metal plus the average full cost of fabrication with the added investment; but, as we have seen, the integrated firm may for one reason or another find it desirable to do that. If it is permitted to expand investment with the expectancy of receiving something less than the full marginal cost of the additional output — that is, any amount between that and the marginal direct cost after the additional investment has once been made — effective competition cannot survive. The use of marginal cost as the test for the minimum justifiable price at any given time is inconsistent with the existence of effective competition in the fabricating branches unless overcapacity is limited to that which is due only to mistakes in estimating future conditions of demand and supply; and unless overcapacity, except for severe business depression, is of infrequent appearance and short duration. In other words, under certain circumstances, it is impossible to have effective free competition and the maximum economical utilization of capacity at all times.

Several methods of dealing with this problem suggest themselves. The investment of the integrated firm in the fabricating branches could be subjected to government control. Requirement of government approval for every addition to capacity does not, however, seem desirable. Both the burden placed upon government administrators and the possibilities of friction between government and business would be greater with this provision than with others. A different measure would provide that whenever overcapacity existed in any branch and the investment of the integrated firm in that branch had been increased in greater proportion than that of the independents as a whole, the integrated firm should be required to retire from operation the disproportionate part of its investment immediately. Exemption would be appropriate when it could be shown that production with new equipment was markedly superior in efficiency to production with older facilities which had been regarded as a part of the normal operating capacity. Administration of this provision would call for determination of a measure of overcapacity and specification of the length of the period adopted for comparing the relative increase in facilities of the integrated firm and its single-stage rivals. These problems, which are difficult but by no means insuperable, would need to be left to the administrative authorities. The principles which should form the basis for a measure of overcapacity need no elaboration here. The length of the period within which comparison of relative expansion is to be made should probably be not less than a year. Clearly the provision would not work well if it extended over many years. The only general principles which can be laid down are the following. A new period should always begin after a condition of equilibrium with best utilization of facilities has been in existence for some months. Otherwise the advantages of competition in investment would be lost, since either the integrated firm or its rivals could later be penalized for disproportionate additions which had resulted merely in carrying total investment up to but not beyond the ideal amount of capacity. Secondly, a new period should always begin immediately after the retirement provision had been invoked. It is manifest that this measure would produce salutary results only if administered with care, economic understanding, and fairness. Its real worth should proceed chiefly from its effectiveness in preventing the development of a situation in which it would become operative.

Other possible devices to secure effective competition have to do with prices. The integrated firm might be prohibited from ever charging a price less than the sum of the market price of ingot (or material in other form) and the average full cost of fabrication at best utilization of fabricating facilities, which would be nearly equivalent to the long-run marginal cost. While such a provision would doubtless tend strongly to restrict the capacity of the integrated firm to economical proportions, it is open to serious objections. Rivals who retained price freedom could deprive the integrated firm of much of its business whenever overcapacity existed. Such a prohibition laid upon the integrated company alone might encourage one-stage firms to create overcapacity in order to capture business from the former in the belief that they could retain it when price rose again with growing demand and the integrated company was able to reënter the market. If the provision were made applicable to all competitors it would prevent price from falling, as it should, to marginal direct cost at times of slack demand. A requirement that the prices of the integrated company, or of all fabricators, must never be reduced below the sum of the market price of material and the average full cost of an efficient outside firm would entail the same sort of undesirable consequences. Although prohibitions of this sort put in terms of marginal cost would promote better utilization of investment at all times, they would not tend to prevent uneconomic expansion unless it was stipulated that, for a specified period following the introduction of additional capacity, marginal cost must be considered to include the resultant added overhead, except when demand fell off. Administration of these provisos would, of course, be attended by great difficulties in formulating and applying standards of quantitative measurement. Prohibition of the sale of any product at any time at a price which failed to cover marginal direct cost (including the market price of the material) at the time of sale would, however, be desirable on various grounds.[25]

A different scheme would require that whenever the integrated firm reduced the price of any article made from ingot or half-products it must lower the price of the material by the same amount unless any smaller reduction or no reduction at all would maintain a differential at least equal to average full cost of conversion. If the differential exceeded average full cost it could be narrowed until it equaled that without disturbing the price of the material. If it were equal to average full cost any cut in the price of the article would require an equivalent drop in the price of the material, unless it was wholly or partially offset by a reduction in the cost of conversion. In the latter case the price of the material should be lowered by the difference between the

[25] The time of sale should ordinarily be considered as the time at which the order was entered. If it should turn out that this did not work well the time of delivery could be used. The former test would, however, seem to contribute more to better utilization of capacity at all times. Exceptions could be made for inventories unsalable at such prices owing to changes in style or quality or other abnormal circumstances.

fall in the price of the article and the decrease in its conversion cost.[26]

If the integrated firm followed every reduction in the prices of fabricated products made by its rivals, such a provision in unqualified form would allow the single-stage firms to drive the integrated company out of fabrication by the simple expedient of cutting prices to the point where the latter found it unprofitable to compete at all. If the integrated firm did not meet reductions in the prices of fabricated goods, and hence did not need to reduce its prices of ingot and other materials, the outcome would be different. With higher prices than its rivals the integrated firm would lose much of its business and have much idle fabricating capacity. As long as the others had overcapacity they could continue to charge prices lower than those quoted by the integrated firm,[27] but they would not cover all their costs. If demand increased rapidly they would soon emerge with a larger share of the business than they had formerly had. But if the condition of overcapacity persisted for long, then there might appear the curious spectacle of a powerful firm "driving out" the weaker by maintaining higher prices! Perhaps fear of losses would ordinarily prevent fabricating companies from reducing their prices below those of the integrated firm, but the temptation to do this would be strong. In order to avoid the undesirable consequences, the measure requiring the integrated firm to maintain a differential equal to average full cost of conversion should be applied only when it had expanded its capacity in any division in greater proportion than the independents as a whole. Differentials covering average full cost of conversion would then be maintained except when overcapacity existed and the fabricators had expanded in greater proportion than the integrated firm. Periods for comparison of relative expansion could be determined in the way discussed above in connection with the provision for retirement of capacity. Measurement of average full cost of conversion

[26] Initial application of this measure would need to occur at a time when the differentials covered average full cost.

[27] If they did not have overcapacity, the provision that prices must not go below marginal cost would prevent them from charging prices below those of the integrated firm except when the latter maintained differentials exceeding average full cost of conversion with best utilization (which is equal to marginal cost of conversion).

would present difficulties, particularly with respect to the allocation of overhead on account of factors used in more than one branch or stage; but it is probable that base figures which were sufficiently accurate to give fair and desirable results could be obtained for most ordinary products.

The scheme just discussed really represents an alternative to the expedient of requiring the integrated concern to retire some equipment whenever overcapacity has developed and it has enlarged its capacity in greater proportion than the independents as a whole. This is so because the integrated firm would not lower prices by an amount exceeding cost reductions except when overcapacity existed or when it expected to bring in new investment in the near future to replace facilities abandoned by others. The prohibition of sales below marginal direct cost would in fact prevent a policy of cutting prices to remove outside investment before the new capacity destined to take its place had been introduced.

A provision to deal with narrowing differentials resulting from increase in the prices of materials unaccompanied by a corresponding rise in the prices of later products would also be necessary to ensure effective competition in several different branches.[28] The integrated firm could be required to raise the prices of all products by the same amount whenever it increased the prices of their materials. This would not of itself be sufficient in cases where consumption was reduced on account of higher prices. The integrated firm would also have to restrict its output of these fabricated goods enough to permit its rivals to produce at the same rate as before.

A requirement that the integrated firm always charge the same price for each fabricated article as its one-stage rivals would resemble in some ways the scheme for keeping the integrated firm's price differentials equal to its average conversion costs, but there are several drawbacks which seem to render it inferior. Obvious advantages in simplicity of administration would be offset partly, at least, by the difficulties inherent in differentiation of

[28] If demand became less elastic for the products of some branches the most profitable adjustment might be to raise the prices of ingot and of those products by the same amount. For other branches where demand for the final products had not changed the price differentials would be narrowed.

product as between firms. In finished goods, and to some extent in semifinished products, there are some differences in design, composition, or quality which may be accompanied by differences in price. The price for a differentiated product of the integrated firm which would be equivalent to the prices of similar articles sold by other firms might be very difficult to establish, particularly if these latter prices were different. Secondly, this provision would prevent the integrated enterprise from lowering price in step with its cost reductions unless its rivals had also cut their costs by the same amount. Finally, whenever the integrated company had created overcapacity in a fabricating branch, although it could not narrow the price differential it could increase its selling expenditures, with the result of forcing its rivals to do the same or to lower their prices. Whichever course they chose, their profits might fall below normal. This result would not occur with the other scheme, by which the price differentials of the integrated firm must cover its own average costs of conversion; because, as its selling expenses grew, its average conversion costs would increase, necessitating a wider differential.

With but one integrated firm in the industry a code designed to maintain effective competition in all branches of the industry succeeding the production of ingot would need to be composed of something like the following provisions.

(1) The integrated firm must supply materials (ingot, sheet, etc.) in any form desired to all would-be purchasers who can satisfy reasonable credit requirements.

(2) There must be no discrimination with respect to amounts sold, quality of metal, expedition in delivery, or other aspects of service as between independent companies or between independents and competing departments of the integrated firm or companies in which an interest is owned directly or indirectly by the integrated firm or any of its officials or leading stockholders. This provision must apply where competitors purchase materials of different form as well as where they buy metal in the same form.

(3) No article made from ingot or half-products may be sold at any time at a price below marginal direct cost, including the market price of the material, at the time of the sale.

(4) Differences in prices of materials furnished to competitors must not diverge from the actual differences in marginal cost.

(5) Whenever overcapacity, with respect to a price equal to the market price of materials plus the average full cost of conversion of the integrated firm, develops in any branch and the capacity of the integrated firm in that branch has been enlarged during a certain preceding period in greater proportion than that of the independents as a whole, the integrated firm must immediately retire from operation the disproportionate increment of its investment, unless it chooses instead to follow the procedure laid down in the next paragraph. To the extent that the added investment enables more efficient production than any older facilities which have been considered a part of the normal operating capacity of the branch in question the retirement provision need not apply.

(6) Whenever the integrated firm reduces the price of any product made from ingot or half-products after it has expanded its capacity in that branch during a certain period in greater proportion than the single-stage firms as a whole, it must lower the price of the material by the same amount, unless any smaller reduction or no reduction will result in maintaining a differential which covers its average full cost of conversion.

(7) Whenever the integrated firm raises the price of ingot or of any half-product, it must raise the prices of all products made from this material by an amount sufficient to keep its differential equal to its full average cost of conversion; and it must restrict its own sales of the fabricated article enough so that the sales of others do not decline.[29]

Administration of these measures would require the frequent reporting of basic statistics of capacity, output, sales, inventories, unfilled orders, costs, and the like. Publication of total or average figures for each branch would be desirable. The code measures already mentioned might be supplemented by a provision restricting the sales of the integrated company in any branch to a certain proportion of the total. It is to be doubted that any material social advantages would be sacrificed if the integrated firm were limited to 30 or 40 per cent of the business in many branches. Finally, it would be desirable to include in the code a general prohibition of all uneconomic methods of competition by

[29] No attempt is made here to suggest the actual wording which would make an effective code of the seven provisions.

the independents or the integrated firm, in order to permit the commission to work out new principles to meet new situations.

Discussion of the governmental measures which would be required in order to adopt either of these two alternative programs for the maintenance of effective competition in fabricating branches of the aluminum industry is beyond the scope of this book. Several of the principles explained above could be formulated into working rules by the Federal Trade Commission and the courts acting under Section 5 of the Federal Trade Commission Act. Congressional action would probably be required for others. It is questionable whether control of investment of the integrated firm could be accomplished without constitutional amendment.

It is plain that an elaborate code of the sort developed here would probably fall far short of achieving its purpose even when administered with vigilance and judicious care. The task of making the measurements necessary for its administration would involve perplexing difficulties, and numerous mistakes would be inevitable. Administration of such a code might be nearly as difficult as outright regulation of all investment and prices, which would seem to be the least desirable of the various alternatives. It should be quite as evident, however, that with intricate relationships like those inherent in the present organization of the aluminum industry the application of simple rules provides no real assurance that desirable market relations will prevail. Where the problems are complex the rules must be complex. The analysis of this chapter, which has been confined chiefly to the economic aspects of the matter, suggests that in this instance it would be better for government to give up its traditional role of rule maker and umpire, and enter the game itself.

Government production of ingot has several advantages as compared with other methods of achieving desirable market relations in the fabricating branches. The problems of availability of materials and discrimination in price, quality, or service could be made very largely to disappear with existence of a government corporation producing ingot; and the problems of price differentials could be handled with more ease, better knowledge, and greater efficacy. Competition between private fabricators could then be relied upon to bring desirable results in many branches, while the government corporation could enter those fabricating branches in which competitive forces were not strong enough to produce an approximation to ideal market relations. Or, with government production of ingot, consumers' coöperation in fabrication might prove to be a feasible substitute for government fabrication in some of those branches.

As long as the present organization of the industry continues, the success of the competitive process in fabricating branches will depend partly on the economic statesmanship of the chief executives of the integrated company, for no code can remove all opportunity for uneconomic exercise of power. In the small segment of industry examined in Part IV of this book it appears that conditions might be improved by a major operation upon industrial organization which removed some economic power, or by very elaborate government regulation. But it does not seem likely that a maximum of desirable results could be obtained in either of these ways. That goal might be attained only if government entered the production of ingot, and perhaps some fabricated products; or if the executives of the integrated firm or firms understood thoroughly how to promote the public welfare and placed it ahead of private advantage whenever the two conflicted.

APPENDICES

APPENDIX A

ALUMINUM AND THE ELECTROCHEMICAL REVOLUTION

SCIENTIFIC study of the general principles governing the relations between electric currents and chemical changes originated with the science of electricity. The successful economic adaptation of these principles to particular problems, which proceeded rapidly in the last quarter of the nineteenth century, altered technical operations in some industries, and introduced several new products which have assumed considerable economic significance. The role played by aluminum in this electrochemical revolution was twofold. It was among the most important of the new products; and the search for a successful method of reducing it from refractory ores advanced the knowledge of electrochemical technique and contributed to the development of several other new products. The following survey describes the development of some of the chief principles of electrochemistry and the results of the work of several inventors who attempted to apply them to the reduction of aluminum.

Shortly after the invention of the voltaic pile, which first provided electric currents of effective magnitude, Sir Humphry Davy succeeded in isolating potassium and sodium by electrolysis. Then, turning his battery of one thousand plates upon aluminum oxide, he attempted a number of experiments which, though failing to yield pure aluminum, had, by 1809, established the fact that alumina can be decomposed while fluid in the electric arc, and its metal reduced as an alloy of iron. Davy also decomposed other oxides and hydrates, but his work did not extend far into the realm of analysis and explanation. His pupil, Michael Faraday, eclipsed the master by his comprehension and formulation of the fundamental principle that there exists a definite quantitative relation between the amount of current passing in any electrolyte and the chemical effect produced. The chemical effect upon each substance was explained by Faraday as being directly proportional to its equivalent weight and to the time during which the current passed. Elucidation of these laws led Faraday to believe that the dissolved molecules of an electrolyte consisted of oppositely charged atoms which he called ions. Under the influence of the electric current the positively charged atom (cation) was attracted to the negatively charged pole (cathode) of the circuit and the negatively charged atom (anion) migrated to the positively charged pole (anode). At the pole each gave up its charge. Thus there collected at each pole groups of neutralized atoms, each of which constituted a molecule of the substance or element now existing in a free state. When metallic compounds were electrolyzed the atoms of the metals became cations, and their migration to the cathode resulted in the free existence of metal at that pole, where it could be collected.[1]

Less than a decade after Faraday's communications on electrolysis Joule formulated the principle that there is a definite mathematical relation between the quantity of electricity passing in a conductor and its heating effect. "By these two laws (Faraday's and Joule's) nearly all of modern practice of electrochemistry and electrometallurgy is governed." [2]

In 1819 Hans Christian Oersted, professor at the University of Copenhagen, found the relationship between electricity and magnetism which philosophers had long suspected. This epochal discovery at once attracted the attention of Ampère of France, Professor Joseph Henry of the United States, and Faraday in England. Ampère and Henry worked out the relationship in detail. It was the English genius who, aided by their researches, discovered that an electric current could be induced in a coil of wire by revolving it in a magnetic field, and thus revealed the fundamental principle of the dynamo. Several magneto machines were constructed upon this principle during the thirties and forties. The typical machine consisted of a permanent steel magnet between or near the poles of which rotated an armature composed of wire coils wound about iron cores. By multiplying the number of coils and magnets, and utilizing steam power, currents of sufficient strength to produce illumination for lighthouses were generated, but it was found impossible to obtain steel magnets of

[1] Of course, theories of electricity and electrolytic action have changed and expanded since Faraday's work, but the fundamental principles which he formulated remain valid.

[2] E. A. Ashcroft, *A Study of Electrothermal and Electrolytic Industries* (New York, 1909), Pt. I, p. 6.

strength enough to induce powerful currents. The true dynamo did not make its appearance for several decades.

Thus, before the century was half over, some of the fundamental principles of electrochemistry were thoroughly understood, and the magneto machine had been demonstrated as a practical success. But the scientific achievements of the laboratory could not emerge into the economic world until the dynamo made its momentous appearance. Investigations of the effects of electric current upon various chemical compounds continued, however, with some important results. Professor Bunsen of Heidelberg succeeded in obtaining barium, chromium, and manganese solutions with the battery. He also decomposed fused magnesium chloride by the electric current.

Aluminum had been first isolated in 1825 by the Danish chemist, Oersted, who reduced aluminum chloride with potassium amalgam. Two years later Friedrich Wöhler, professor of chemistry at the University of Göttingen, secured the white metal by a slightly different chemical process. Henri St. Claire Deville of the École Normale in Paris is, however, entitled to the honor of first obtaining fairly large amounts of aluminum in a state of almost perfect purity and determining its true properties. Deville employed potassium as the reducing agent. Encouraged to continue his researches by a financial grant from the Academy, he directed his attention to electrolysis, since potassium was both dangerous to handle and very expensive. Bunsen's success with electrolysis of magnesium chloride stimulated Deville to try the same experiment with aluminum chloride. After a few weeks of experiment he obtained aluminum in March 1854. Scientific discoveries often have a habit of descending upon two men in different places at nearly the same time. Bunsen had published in Poggendorf's *Annalen* an electrolytic process quite similar to that of Deville just a week before the latter, in complete ignorance of Bunsen's success, read a paper to the Academy describing his own results. The reduction of aluminum by electrolysis was clearly the stimultaneous discovery of both of these men. Neither of them, however, attempted to apply the discovery to industrial production. Deville, who was seriously interested in promoting an aluminum industry, realized that the large consumption of zinc in the battery would entail a prohibitive cost. He reverted to chemical methods and invented the sodium process, which gave birth to an aluminum industry of small proportions. Professor Richards, writing in 1896, when electrolytic aluminum had been on the market for several years, observed that

> . . . the great advances made in dynamo-electric machinery in the last decade have led to the revival of the old methods of electrolysis discovered by Deville and Bunsen, and to the invention of new methods of decomposing aluminum compounds electrolytically. It will be recalled that the first small pencils of aluminum made by Deville were obtained by electrolysis and that he turned back to the use of alkaline metals solely because the use of the battery to effect the decomposition was far too costly to be followed industrially. This fact still holds true, and we cannot help supposing that if Deville had had dynamos at his command such as we have at present, the time of his death might have seen the aluminum industry far ahead of where it now is.[3]

Just three years previous to the successful experiments of the French savant and Bunsen upon aluminum Charles Watt in England had taken out what one authority refers to as "the master patent of the electrochemical and electrometallurgical industries in the United Kingdom."[4]

> In this patent, Watt described in some detail how the electric current might be employed for producing alkali hydrates and chlorine, hypochlorites, or chlorides, and how it might be utilized for refining copper, silver, or other metals, or for separating these from their ores. . . .
>
> Watt's ideas, as put forth in this patent of 1851, could not, however, receive practical trial until the dynamo was developed and improved, and it was not until 1869, when Elkington erected the first electrolytic copper refinery at Pembrey, in South Wales, that the industrial development of the facts and ideas gathered during the previous years of the century commenced.[5]

Just why there was such a long time lag between the understanding of scientific principle and the adaptation to industry in the instance of the dynamo is not altogether clear. It would appear, however, that the possibilities of utilizing powerful electric currents in industry must have been manifest; so one would

[3] J. W. Richards, *Aluminium,* p. 24. See also R. J. Anderson, *The Metallurgy of Aluminium and Aluminium Alloys,* p. 3; MI, I, 12 (1892); *Engineering News,* LXXIII, 177 (1915).

[4] J. B. C. Kershaw, *Electro-metallurgy,* p. 6.

[5] Apparently this was one of the first electrochemical patents. It is not without significance that until at least the middle of the century most of the discoveries appeared in the transactions of the learned societies rather than in the patent office. The number of patents for electrochemical processes and equipment appears to increase through the sixties and seventies to a veritable outbreak in the eighties.

infer that the chief resistance was found in the difficulties of overcoming serious technical problems. The magneto machine maundered along through the fifties and sixties with no startling change. The Siemens armature designed in 1856 increased the maximum current strength, but this could not remove the main obstacle, which was the weakness of the magnetic sphere of influence. It was in the latter part of 1866 that Siemens of Berlin and Wheatstone in London independently made a change in the construction of the machine which provided the necessary improvement. In place of the permanent steel magnet they substituted an electromagnet consisting of a core of soft iron wound with insulated wire which was connected to the revolving coils of the armature. Although soft iron possesses but a trace of magnetism, this trace is sufficient to induce a feeble current in the coils of the armature. A portion of this current passes through the wire wound around the iron, thus magnetizing it. This increased sphere of influence enhances the strength of the current in the armature, with the reciprocal result of a further access of magnetism in the iron, and so on. By the cumulative action of this process the dynamo, as the machine was called after this change, generates currents of far greater strength than the magneto machine was able to produce. The perfecting of the dynamo in the following years was due largely to the work of Gramme.

As is indicated in the passage from Kershaw quoted above, the first important development of industrial electrochemistry followed close on the heels of the appearance of the dynamo. But the electrolytic copper refinery mentioned was for some years the only important electrochemical works. Although electrolytic refining of copper was inaugurated twenty years earlier it did not expand substantially until it had caught the contagion which spread from the electrochemical outbreak in the eighties.

Progress in dynamo construction in the seventies, by cheapening cost as well as increasing the electric horsepower available for industrial application, aroused more interest in practical experiments aiming at the reduction of highly refractory metallic compounds. The results of this experimentation were twofold. In the first place the electric furnace was developed and adapted to many particular industrial applications. Secondly, several electrochemical processes were discovered and perfected which could be carried out successfully in the electric furnace. The natural consequences of the two related achievements appeared in the birth of several new industries, of which the aluminum industry was the first and most robust infant. The experimental and developmental work done upon aluminum compounds in the eighties possessed also a wider significance. In 1891 a writer describing the early achievements in electrochemistry referred to the development of electrometallurgy [6] as "largely due to the attempts to produce aluminum economically." [7]

During the eighties a large number of inventors were at work upon the problem of electrolytic or electrothermal reduction of aluminum. The first proposal to use the dynamo for electrolytic reduction of aluminum was made in Berthaut's patent in 1879. His process was similar to that of Deville. Grätzel in 1883 and Kleiner in 1886 each proposed an electrolytic process using a dynamo. Although none of these processes proved successful in industrial application, they were significant in showing that attempts were being made to utilize the possibilities of the new power generators. The first successful industrial work was accomplished by the brothers Alfred and Eugene Cowles of Cleveland. In 1883 they purchased with their father a zinc mine in New Mexico. The extreme refractoriness of the zinc ores led them to an investigation of electric-furnace reduction which resulted in the designing of a successful electric furnace. For this initial task and the long series of inventions, experiments, and practical achievements in electrometallurgy which followed, the brothers possessed a rare combination of qualities. Alfred brought an alert mind and a training in science which were evidenced by a distinguished career of several years' study at Cornell University. Eugene, also resourceful and ingenious, had acquired much practical experience in metallurgy and electrical engineering, in addition to executive experience as manager of an electric-lighting plant.

Encouraged by the success in smelting refractory zinc ores in their electric furnace, the brothers turned their energetic en-

[6] Electrometallurgy is, of course, one branch of electrochemistry. The early development was largely metallurgical.

[7] R. L. Packard, MR, 1891, p. 147. Cf. E. E. Slosson, *Creative Chemistry* (New York, 1919), p. 245: "The industrial development of the electric furnace centered about the search for a cheap method of preparing aluminum."

thusiasm upon aluminum at about the time when Charles M. Hall was graduating from Oberlin College after a classical course which had left him with a passion for chemical experiment. The Cowleses appeared to have the prize within their grasp when they developed a commercially successful method of obtaining copper-aluminum alloys in the electric furnace; but it was wrested away overnight when Hall discovered an electrolytic process which yielded pure aluminum.

The Cowleses' attack upon the problem of adapting their electric furnace to aluminum smelting and finding a suitable reagent had consumed only a few months when their success was announced by Professor Charles F. Mabery, one of their associates, in a paper before the American Association for the Advancement of Science.[8] The process was simple. In a furnace of the so-called resistance type a high temperature was obtained by introducing coarsely pulverized carbon, which presented a great resistance to the electric current. Carbon was, at the same time, the most easily available substance for the reduction of oxides. The attempt to secure pure aluminum by this method was unsuccessful because the aluminum combined with the carbon to form a useless carbide. However, it was found that by introducing another metal into the furnace a useful alloy of aluminum would be yielded at a cost per pound of aluminum equivalent to a third of the price at which the pure metal was then selling. For the production of aluminum bronze there was placed between the electrodes of the furnace a charge of alumina in the form of granulated corundum, mixed with charcoal and granulated copper. The most successful Cowles furnace was, of course, adapted for continuous working. Aluminum-alloy production was undertaken in a plant at Lockport, New York, where cheap water power was available from a tailrace of the Niagara overflow. The Cowleses also built and put into operation a plant at Milton, England. At one of these plants (reports seem to divide the honor) there was installed a dynamo of almost 400 h.p. capacity which aroused so much interest, by virtue of its novel magnitude, that it became famous under the sobriquet of "the Colossus." The alloy business flourished for a few years until the advent of pure aluminum announced its doom. Metal mixers preferred to make their

[8] August 28, 1885.

own alloys when the pure metal became available at a favorable price.

While aluminum was for some years their primary interest and their first industrial venture with the electric furnace, the Cowles brothers did not confine their early experiments or their later commercial activities to this one metal. A committee of the Franklin Institute of Philadelphia in recommending the award of two medals [9] to the Cowleses in 1886 reported in part:

> The essential and valuable novelty of the process is the ingenious application of the intense heat obtained by the passage of a powerful current of electricity through a conductor of great resistance, to the reduction, in the presence of carbon, of the most refractory ores, some of which have hitherto resisted all similar attempts at reduction. . . .
>
> This process is applicable to the reduction of all kinds of ores, but particularly to those unreducible by other means . . .; already aluminum alloys of iron, silver, tin, cobalt, and nickel have been prepared; silicon, boron, potassium, sodium, magnesium, calcium, chromium, and titanium as well as aluminum have been obtained in a free state.[10]

Albert W. Smith in a brief biographical sketch of Professor Mabery credits the Cowleses with even more discoveries.

> In this work, the development of the electric furnace, they were the first to produce all the many electric-furnace products which have since become such important items in industrial chemistry — calcium carbide, carborundum, fused alumina, and artificially-made graphite — although they did not interest themselves in the commercial development of these products.[11]

Perhaps the claims made in these statements are somewhat exaggerated. A few years later Moissan in his researches on the electric furnace showed that many of the first conclusions regarding electrothermal carbon smelting to secure pure metals required important modification, because useless carbides were the typical yield. Very likely the Cowleses and their employees did produce

[9] The John Scott Legacy premium and medal given by the City of Philadelphia to encourage science in the arts, and awarded on recommendation of the Franklin Institute; and the Elliott Cresson gold medal, the highest honor given by the Franklin Institute.

[10] Report of the Committee on Science and the Arts on the process and furnace for reduction of refractory ores, and the production of metals, alloys, and compounds invented by Eugene D. and Alfred H. Cowles (*Journal of the Franklin Institute,* CXXII, 51, July 1886).

[11] *Journal of Industrial and Engineering Chemistry,* XV, 314 (March 1923).

in one form or another most of the products which Smith credits to them. For the most part, however, they did not produce these products with commercial success until others had demonstrated their economic usefulness.

The part played by these innovators in the electrochemical revolution may be summed up as follows. They built and operated several electric furnaces which were not only successful in industrial application but also provided an arresting demonstration of the possibility of utilizing the electric furnace in a wide range of industrial activities. In their electric furnace they decomposed for the first time many metallic oxides which had previously resisted the most determined efforts of the fuel furnace. They were responsible for the first commercial production of aluminum in the form of alloys on a fairly large scale at a cheap cost. Their work possessed great significance for the development of the dynamo. "In the early days of the Cowles Company their experiments and improvements did important pioneer work for the United States and other countries in the use of large dynamos." [12] The success of the Cowles furnaces was followed by a broadcast of their results in the leading scientific journals of Europe and America. Literature concerning aluminum was widely disseminated by the Cowles company. In attracting world-wide attention to the potentialities of electric smelting by furnishing other workers with both knowledge and stimulus, the Cowles brothers played a leading part in the industrial development of electrochemistry.[13] As far as aluminum itself is concerned, although they did not obtain the metal in pure form, their attainments with the electric furnace were doubtless of aid to the inventors of the electrolytic process, while their prosperous business in aluminum alloys presented a stimulus to those who were still at work upon the more difficult problem. It should be added that, while the Cowles company found it unprofitable to operate the electrothermal reduction of aluminum after 1892, Alfred Cowles continued his

[12] *Electro-chemical Industry,* I, 56 (October 1902). Another authority declares that "the Cowles process attracted much attention, as it was the pioneer of dynamo-electrometallurgy in the United States" (R. L. Packard, in MR, 1894, p. 359, being *Annual Report of the United States Geological Survey,* XVI, Part III).

[13] Cf. Dr. Leonard Waldo's supplementary note in his translation of Adolphe Minet, *The Production of Aluminium and Its Industrial Use,* p. 255.

interest in aluminum in particular and electrometallurgy in general for many years.[14]

The progress of the Cowleses in electrometallurgy was closely paralleled on the other side of the Atlantic by a brilliant Parisian inventor, M. Paul L. T. Héroult. In 1886 M. Héroult began a long career of experiment and industrial application characterized by achievements in nearly every department of electrometallurgy. His attention, also, was first claimed by aluminum. In 1886 he independently discovered and patented in Europe an electrolytic process yielding pure aluminum which was in all essentials precisely similar to the Hall process. At the time he did not realize the possibilities attaching to the metal in its pure state. Speaking at the Metallurgical Congress in Paris during the World's Fair in 1900, he explained how it came about that he did not immediately attempt to perfect his process for industrial production.

> My practical knowledge of chemistry was at the time [1886] that of a student of twenty-three; of special knowledge I had as good as none at all. Under these circumstances, it is needless to say that after I had taken out my first patent I sought the counsel and encouragement of those men who were then considered authorities on this subject. Péchiney (Salindres), whom I first approached, explained to me that aluminum was a metal of restricted usefulness; at most it might be used for opera-glasses; and whether I wanted to sell the kilogram for 10 or 100 francs, I would not be able to dispose of one kilogram more. It was otherwise in the case of aluminum bronze, of which considerable quantities were handled commercially, if I could produce it cheaply; I would then, beyond a doubt, come out even in my reckoning.
>
> I had then already in this connection undertaken some successful experiments; and I therefore laid aside for the time being the production of pure aluminum and turned to a series of new researches which in the year 1887 led to a second patent.
>
> In this additional patent a system of electric furnaces and a process were described which made possible a continuous production of alloys of aluminum, and particularly of all metals difficult to melt and reduce.[15]

The furnace and process of the 1887 patent were essentially like the Cowles apparatus for aluminum alloys. Héroult's European patents for production of aluminum and its alloys were purchased by the Schweizerische Metallurgische Gesellschaft, formed in October 1887 by Swiss industrial interests.[16] During 1888 and

[14] Eugene Cowles died in 1893.

[15] Quoted in Minet, *op. cit.*, pp. 115–116.

[16] Schulthess, *op. cit.*, p. 4.

a part of the following year aluminum alloys were produced with the aid of Héroult himself in a small plant at Neuhausen, Switzerland. In 1888 this company and the Deutsche Edison Gesellschaft (later the Allgemeine Elektrizitätsgesellschaft) organized the Aluminium Industrie A. G., which thenceforth operated the aluminum patents in an enlarged plant at Neuhausen. Dr. Martin Kiliani, who had experimented upon the production of pure aluminum for the AEG was largely responsible for the successful industrial adoption of the Héroult process for producing pure aluminum to which attention was turned in 1889. Héroult returned to France, where he aided in perfecting electrolytic production of aluminum by the Société Électrométallurgique Française, founded in 1888 with the aid of the Neuhausen company.

Héroult next turned his attention to the electrothermal production of calcium carbide, which had been discovered in 1892 and 1893 by Willson and Moissan. The furnace which had served for aluminum alloys was used for calcium-carbide production at La Praz and perhaps at other plants of the Société. The later and more important accomplishments of this versatile inventor were concerned with ferroalloys and steel refining. A passage in *The Electric Furnace* by J. N. Pring is particularly apt in that it shows, by tracing the repercussions of Héroult's early work in connection with aluminum upon other fields in electrometallurgy, a trend which was characteristic of the activities of several aluminum experimenters.

> The work of Héroult on the production of steel in the electric furnace followed as an outcome of the aluminum process which had been established in 1887. The possibility of producing various ferroalloys was shown in a similar type of furnace consisting of a metallic casing of crucible form, the bottom of which is carbon lined to form one pole, whilst the movable carbon electrode making contact with the surface of the charge forms the second electrode. . . .
> The production of low-carbon ferrochromium led to work on the production of steel and to the establishment of a furnace in which, by the use of special slags, high-grade steel can be obtained directly from highly impure iron. Furnaces for this purpose were brought into operation at La Praz and Froges in France at the aluminium works of P. Héroult, and in 1906 the process was applied at Remscheid, in Germany.[17]

[17] J. N. Pring, *The Electric Furnace* (London: Longmans, Green & Co., 1921), p. 210. Héroult also developed the electric-furnace smelting of iron ores.

If it be said that the Cowles brothers introduced the electric furnace to the industrial world, it must be said that Héroult assisted at the introduction and took a leading part in its subsequent adaptation for several important uses. The popularity of the Héroult steel furnace in the early years of electric steel was testified to by Wright in 1910.[18] Figures on electric furnaces in the British steel industry in 1919 show that the Héroult installations outnumbered those of every other type by a substantial margin.[19]

Another American inventor who first studied aluminum reduction, but made his real contributions to the development of electrochemistry through his work on other materials, was Thomas L. Willson. In 1885 Mr. Willson was an employee of the Brush Company of Cleveland, manufacturers of dynamo machines for the Cowles brothers. Willson's interest was caught by the experiments of the latter with their electric furnace. He left the Brush Company and began experiments upon aluminum reduction in an electric furnace at Spray, North Carolina, which was situated near deposits of corundum. These experiments resulted in a patent for the reduction of aluminum alloys in 1890, just as they were beginning to be supplanted by Hall's electrolytic aluminum. Willson discontinued further work on aluminum soon after he had accidentally discovered calcium carbide while attempting to obtain metallic calcium in the electric furnace for the purpose of trying this metal as a reducing agent upon aluminum.[20] Willson and Moissan, who obtained calcium carbide at about the same time in France, were the first to perceive the economic usefulness of this compound for the production of acetylene gas. Industrial

[18] J. Wright, *Electric Furnaces and Their Industrial Application* (London, 1910), p. 118.

[19] Pring, *op. cit.*, p. 272. Héroult furnaces numbered 49 of the 144 electric furnaces installed. Total capacity of these 49 was 195 tons compared with 79 tons capacity of the 34 furnaces of the next most common type.

[20] The honor of this discovery must be shared with Moissan in France, who announced his results at about the same time as Willson (the end of 1892). Willson appears to have been the first to manufacture calcium carbide, beginning as early as 1891 in Spray. Moissan's researches, which continued for several years, were the more scientific and the more valuable for the economic utilization of this product. The discovery of calcium carbide by these two men was in fact a rediscovery, for Robert Davy, cousin of Sir Humphry, had discovered this compound originally in 1855, and Wöhler had repeated Davy's experiment in 1862 and engaged in further study of the compound. This is another example of the time lag between laboratory discovery and industrial development.

production of calcium carbide and acetylene gas was inaugurated on a fairly large scale in the United States and England under the Willson patents, and on the Continent under patents taken out in the name of Bullier, an assistant of Moissan.

Calcium carbide, which had appeared as an indirect result of experiments upon aluminum, was also destined to lead into further developments in electrometallurgy. In the last three years of the century calcium carbide and acetylene developed an inflationary boom complex in England and upon the Continent. Capital and enterprise poured into these new industries with cautionless optimism, and acute overcapacity was rapidly created. In 1900 the collapse occurred with even more swiftness. The difficulties of the new companies were increased by court actions in which the patents of Bullier and Willson were declared invalid in most of the European countries. Under the influence of manifold troubles the owners of many carbide plants with cheap water power turned their interest to the production of ferroalloys or sold their plants to persons who had been experimenting in this field.[21] In the United States Willson saw his second venture with the electric furnace end without great financial success to himself when, as a result of the patent litigation, the control of carbide manufacture was awarded to the owners of the Bradley patents.[22] Willson likewise turned to the production of ferroalloys in plants at Holcomb Rock and Kanawha Falls, Virginia, and Ste. Catherine in Canada. In 1900 *The Mineral Industry* reported that the Virginia plants were devoted exclusively to the manufacture of ferrochromium. Subsequently other ferroalloys were produced, but ferrochromium continued to be the chief product. A few years later the Electro-Metallurgical Company was formed to take over the Willson patents and Virginia furnaces. A new plant was built at Niagara Falls by this company. One authority on the develop-

[21] France was the leader in the development of electric-furnace ferroalloy production.

[22] It is interesting to note that this owner was none other than the Cowles Electric Smelting and Aluminum Company, which henceforth owned an interest in the Union Carbide Company. A part of the fortune which Willson had failed to win from his early experiments upon aluminum came to him later through Saguenay water power, which is now used for aluminum production. J. B. Duke acquired a large part of the Saguenay water rights from Willson, who, it is said, had bought them from the Quebec government in 1912 for one thousand dollars.

ment of the electric furnace summarizes the evolution of the ferro-alloy industry as follows:

> It is in the manufacture of ferroalloys that the electrochemical industry met with its greatest success and most rapid development. The main incentive which led originally to the progress of this work was the decline in the calcium-carbide industry which followed its early extension. Thus in 1900, experiments were made in France on the production of ferrochromium, ferrosilicon, and other ferroalloys. Carbide furnaces were found applicable for this manufacture, and the success obtained has finally led, in the case of many of these alloys, to the complete replacement of the older processes by electric-furnace manufacture. Large works manufacturing ferroalloys are now in operation mainly in Savoy, and Isère in the South of France and in the United States, Switzerland, and Scandinavia.[23]

With the exception of Héroult's electrolytic process for aluminum reduction the achievements so far described involved the use of the electric furnace in electrothermal processes — that is, the electric current was used to produce heat of a sufficient intensity to allow chemical reactions which would not occur at a lower temperature. During the eighties a few men perceived that successful reduction of the alkali and alkaline earth metals and aluminum would be attained by electrolysis rather than electrothermal methods. Charles S. Bradley became convinced by experiments that many of the highly refractory metallic ores or compounds which were nonconductors in an unfused state could be reduced by electrolysis when the electric current was at the same time used to fuse the electrolyte and maintain it in a state of fusion.[24] The peculiar feature of Bradley's work was the use of electric current to fuse and maintain fusion, thus making it possible to dispense with external heat. Bradley seems to have been the first, in America, at least, to recognize the significance of internal heating of the electrolyte.

Almost contemporaneous with Bradley's work were the experiments of Charles M. Hall upon the electrolysis of aluminum. Before his graduation from Oberlin College in 1885 this brilliant young man became vitally interested in the possibilities of wresting the ubiquitous metal aluminum from its useless compounds. While

[23] Pring, *op. cit.*, p. 14. This book was published in 1921.

[24] Three-quarters of a century earlier Davy had reduced potassium and sodium by employing the electric current for simultaneous fusion and decomposition.

still an undergraduate studying the classics, he found time to experiment with the effects of the electric current upon aluminum compounds, and after graduation his whole interest became absorbed in this problem. He was shortly convinced that the chief requirement for an electrolytic reduction process capable of commercial development was an effective solvent for alumina, the cheapest aluminum compound.

As he has several times explained to the writer, he had in mind the analogy to dissolving a salt, such as copper sulphate, in water and obtaining the ingredients of the salt at the two electrodes without decomposition of the water.[25]

Hall's next experiments were directed toward the discovery of a substance which would dissolve alumina as water dissolved the copper sulphate in the illustration given.[26] He worked with many salts to find one which would (1) dissolve alumina fully, (2) conduct electricity, (3) yield only aluminum and oxygen from electrolysis, and (4) not volatilize or deteriorate on continued use.

A few months of elimination work sufficed to arrive at the discovery which was the essential part of Hall's invention. He found that the double fluoride of aluminum and the more electropositive metals possessed the qualifications for an effective solvent of alumina. It was on February 10, 1886, that he was delighted to remark that cryolite, the double fluoride of aluminum and sodium, readily dissolved considerable amounts of alumina. Using a gasoline burner to heat his crucible, he applied the electric current from a galvanic battery to a bath of alumina dissolved in cryolite. The result was not at first successful. A suspicion that the difficulty lay not in the bath but in the presence of silica in the lining of his clay crucible was proved to be correct when he lined the crucible with a mixture of ground carbon and tar. On February 23, 1886, he obtained his first button of pure

[25] J. W. Richards, in *Electro-chemical Industry*, I, 159 (January 1903).

[26] Aqueous solutions are unsatisfactory for the electrolysis of aluminum salts because the aluminum possesses such a great affinity for oxygen that the hydroxide of aluminum is yielded rather than the metal itself. See Minet, *op. cit.*, p. 57. Richards states (MI, XIV, 13, 1905) that Hall's earlier attempts at Oberlin convinced him that electrolysis from aqueous solutions was out of the question because the nascent aluminum at the cathode was immediately oxidized. The critical pressure for decomposition of water is lower than that of aluminum.

aluminum. Patents were applied for in July, and during the next two years Hall was engaged in the work of adapting his process for industrial production and securing financial backing. Throughout the experimental period, and apparently for a few months after industrial production had actually begun at the plant of the Pittsburgh Reduction Company, Hall relied upon external heating. It is not clear whether he really intended to continue external heat when the process was conducted upon a commercial scale, or whether he believed that as a natural consequence of larger-scale operations the electric current would maintain the bath in a fused state. It does not appear that Hall appreciated the principle of fusion by the electric current simultaneous with electrolysis until after some experience with industrial operation of his process. His patent claims did not specify this; on the contrary, they definitely mentioned external heating. In the absence of Bradley's work Hall would, of course, have adopted internal heating when operation demonstrated its advantage. It should be recognized, however, that Bradley was the first of the many investigators of this decade to appreciate the significance of the principle of simultaneous fusion and electrolysis by the electric current. It was Hall who overtopped his many competitors and carried electrolysis from the laboratory to its first successful application in the industrial world.

It may be asked why Hall and other inventors worked with alumina, which exists in nature (as corundum) in a very limited amount, rather than with some of the common aluminous ores such as bauxite, gibbsite, or kaolin. Any reduction of its ores which will yield aluminum will also reduce the metals associated with it in the ores and thus destroy all the valuable qualities of the aluminum. The refining of impure aluminum presented great obstacles. Alumina seemed the most promising point of attack. The alumina used in the Hall process has until recently been produced only by an expensive chemical treatment from the ore bauxite.

Hamilton Y. Castner, a chemist from Columbia University, also turned his attention to aluminum at the same time as the others. His attack, however, was upon the chemical method which reduced aluminum at a high cost by the chemical action of sodium, which was itself obtained only at great expense. It was stated in

1883 that 57 per cent of the cost of aluminum was attributable to the sodium. Castner invented a process of reducing sodium which lowered its cost from one dollar a pound to about twenty or twenty-five cents. A new aluminum company was set up in England by Castner and an English associate in 1887. Four years later the competition of electrolytic aluminum forced this company to discontinue aluminum manufacture and write its capital down from £400,000 to £80,000. The firm continued to produce sodium, and Castner, perhaps disgruntled by his failure in trying to patch up an obsolete process, began to experiment with electrolysis. His ability was both recognized and rewarded a few years later when he invented an electrolytic method for reducing sodium from common salt, which soon replaced the chemical methods.[27] Castner also contributed to the industrial application of electricity to the production of caustic soda and chlorine and cyanides.

The development of the electric furnace and the rapidly increasing knowledge concerning its application in the industrial arts had other repercussions before the turn of the century. In 1891 E. G. Acheson, while experimenting on artificial production of diamonds in an electric furnace, accidentally secured an extremely hard crystalline material which he supposed to be a compound of carbon and alumina (corundum). It was only after he had given it the name carborundum that he identified it definitely as the simplest compound of carbon and silicon — CSi or silicon carbide. After investigation had shown the usefulness of this material, which is next to the diamond in hardness, Acheson proceeded to produce and sell it under the name of carborundum. A small factory in Pennsylvania was superseded in 1895 by larger works at Niagara Falls, where furnaces absorbing 1,000 h.p. were set up. This represented a significant development, for until this time only small furnaces had been used in applied electrometallurgy. A few years later Acheson founded successful industrial processes for the manufacture of artificial graphite and refractories (siloxicon and aloxite) in the electric furnace.

Up to this point we have surveyed the early course of the electrochemical revolution, which received its stimulus from the search for a practical process of reducing pure aluminum on a

[27] This was similar to the method used by Davy in 1807.

large scale. Laboratory discovery and the formulation of the fundamental principles of electrochemistry in the first half of the century brought no immediate industrial consequences owing to the absence of cheap electric power in large amounts. The introduction of effective dynamos in the late sixties and seventies precipitated an outbreak of experimentation aimed at the use of the electric power now made available for electrochemical processes. Electrolytic copper refining, although first upon the scene, did not reach sizable proportions until after further advances in industrial electrochemistry, which resulted in the birth and growth of several new industries. Most of the experimentation of the eighties was concerned with aluminum reduction and somewhat incidentally with the reduction of other refractory oxides. The invention and development of the electric furnace by the Cowles brothers in this country, and by Héroult and Moissan among others abroad, marked the first important step. Alloys of both aluminum and silicon were produced industrially by the Cowleses in the latter eighties. After Hall and Héroult had seized the aluminum prize with their successful electrolytic process, Willson, Moissan, the Cowleses, and others, who had been working upon electrothermal processes in the electric furnace, developed the manufacture of calcium carbide and acetylene and ferroalloys. During the first decade of the present century these industries became firmly established. Under the influence of ferroalloy production a beginning had been made in the production of electric furnace steel. This decade also witnessed the growth of additional industries which seemed to appear naturally once the applicability of the electric furnace had been established — e.g., carborundum, artificial graphite, and refractories. While the existence of an effective dynamo was a necessary condition for the birth of electrochemical industries, their growth in turn demanded the further cheapening of power which was obtained by hydroelectric developments. The early promotion of hydraulic electricity at Niagara Falls in the nineties found about half its market in electrochemical plants which were built there to use this cheap power. The rapid progress of industrial electrochemistry in this decade and an aroused interest in its development are testified to by the simultaneous establishment in 1902 of the *Electro-chemical Industry,*[28] a scientific

[28] Now *Chemical and Metallurgical Engineering.*

and trade journal, and of the American Electrochemical Society, whose membership embraced both industrial and academic men.

It is interesting to note that in addition to his earlier achievements already described, Alfred Cowles and his associates had by this time come to hold the most prominent position in electric smelting in the United States. By patent litigation or purchase the following companies had become more or less subsidiary to the Electric Smelting and Aluminum Company: the Union Carbide Company, the Willson Aluminum Company, the Electric Gas Company, the Acetylene Illuminating Company, the Acetylene Company.[29] In 1913 the Cowles company won an infringement suit against the Carborundum Company over the Cowles electric-furnace patents. Although these had already expired, the decision legally established the Cowles brothers as pioneers in electric smelting.

While it would be beyond the purposes of this essay to describe the further development of electrochemistry, it may be of interest to indicate briefly the economic significance connected with the later development of the industries whose birth has been noticed, and to mention the more recent electrochemical applications.

Since the early nineties the electrolytic copper-refining industry has shown an enormous expansion, principally in America. The inadequacy of the supply of "lake" copper to fill the rapidly growing demand for metal of high purity for electrical uses necessitated the exploitation of the great deposits of the western states. Ordinary metallurgical methods could not secure the metal from these ores in pure enough form for the electrical industry. As a consequence, electric refining of the raw copper obtained by smelting these ores developed rapidly with the expansion of the electric industry. This development was facilitated by the presence of silver and gold in American raw copper, the recovery of which materially reduces the cost which must be borne by copper. The absence of these valuable constituents in the European copper ores has been adduced as the main reason why the growth of electrolytic refining in Europe was so slow by contrast.[30]

Calcium carbide, which was earlier employed chiefly as a base for acetylene manufacture, has not lost its importance with the

[29] MR, 1903, p. 267.

[30] Kershaw, *op. cit.*, p. 109.

waning of acetylene illumination. Oxyacetylene cutting and welding has exerted a marked influence upon the development of the metal industries. Furthermore, acetylene, C_2H_2, is useful as a starting point in building up higher compounds of carbon and oxygen by synthetic chemistry. Some of the products thus derived are alcohol, acetic acid, acetone, and methyl. The largest use of calcium carbide in recent years has been in the manufacture of calcium cyanide, now employed extensively as a fertilizer.

It was pointed out above that the production of ferroalloys in the electric furnace received part of its initial stimulus from the overcapacity of calcium-carbide furnaces. Since then the furnaces used for ferroalloy production have been modified in details, and the ferroalloy industry has enjoyed a tremendous expansion under the influence of an increasing interest in the development of ternary and quaternary steel for special purposes. Ferroalloys serve two purposes in the manufacture of steel. First, they act as purifiers and deoxidizers by combining with elements which would lower the quality of the steel unless removed. Ferrosilicon and ferromanganese are the principal alloys used for this purpose. For many years these two alloys were prepared by ordinary blast-furnace methods. The electric furnace, which yields alloys of much higher manganese and silicon content, has largely superseded the older methods in the preparation of ferromanganese and to a lesser extent in the making of ferrosilicon. Secondly, ferroalloys are employed to introduce into the steel a certain proportion of the alloyed metals, with the purpose of increasing the quality of the steel for special uses. Of the several alloys used in this way ferrochromium was the first to be produced in large amounts by the electric furnace, which has almost completely replaced the crucible process for this alloy. By virtue of its extreme hardness, steel with a small percentage of chromium has been instrumental in the development of gear machinery and cutting tools. Ten to twelve per cent chromium has given us "stainless steel." Ferrotungsten has had a remarkable growth as an electric-furnace product. It is utilized in making high-speed steel for cutting tools which, by retaining their edge even at the red heat induced by rapid machining, have revolutionized machine-tool performance. Ferromolybdenum added to steel imparts qualities similar to those of tungsten steel. Ferrovanadium is made largely in the electric furnace. Steel possessing very small amounts of vanadium resists shock and vibration more satisfactorily, and hence is used for axles, cranks, connecting rods, and so on. Ferrotitanium, ferrouranium, and ferrophosphorus are alloys of lesser importance which are produced to some extent in the electric furnace.

It has already been explained that the use of the electric furnace for steel refining was suggested by the success with the production of ferroalloys. About 1900 Héroult in France and Kjellin in Sweden successfully applied many of the carbide plants to the production of high-quality steel as well as ferroalloys. Since then, because it is more economical and gives larger masses of metals of uniform composition than the crucible process, the electric furnace has almost entirely replaced the crucible process for the production of "fine steels" [31] — i.e., high-grade carbon steels and the highly complex alloy steels, such as tungsten steel. The electric furnace is also constantly encroaching upon the domain of the open-hearth fuel furnace in the production of structural alloy steels for automobile and airplane parts. There is a growing tendency toward the adoption of the "cold melt." Further, the electric furnace has also been used somewhat to replace or supplement the open-hearth and converter processes for producing "tonnage" steel. In general, the advantage of the electric furnace is that it allows the use of less pure materials, while producing higher quality steel than the fuel furnace. But for "tonnage" production the electric furnace is economical as an alternative to fuel heating methods only in situations where fuel is quite dear and electric power relatively cheap. Actual replacement of fuel furnaces has been rare. However, there has come into increasing use a "duplex" process whereby steel is produced in bulk by subjecting it first to a preliminary refining in open-hearth furnaces or Bessemer converters and then transferring it in liquid form to the electric furnace for further refining.[32] The electric energy is usually generated by

[31] Professor S. S. Stratton is responsible for the term "fine steel." He has supplied much of my information on the electric furnace in various branches of the steel industry. See his unpublished doctoral dissertation, "Some Chapters on the Development of the Fine Steels Industry in the United States," Harvard University, 1930.

[32] In 1908 there was but one electric furnace in the United States. Pring reported 287 in operation in January 1919 (*The Electric Furnace,* p. 272).

gas engines driven by blast furnace gas. Most of the steel rails made in this country are now manufactured by this "duplex" process. It has enabled the Bessemer process to hold its own as a competitor of the open-hearth in the face of more exacting requirements in the quality of steels for rails and other products. Stassano, Héroult, Keller, and others built successful furnaces for electric smelting of iron ores, but industrial application has remained small, owing to the high efficiency of the fuel process already in existence. Large-scale operations have been carried on for several years in Norway, Sweden, and California, however, and electric smelting of iron ores appears to be gaining slowly.

Metal grinding and polishing has been revolutionized by carborundum, which has also enjoyed wide application as a refractory for lining various types of furnaces in which high temperatures are developed. Its high thermal conductivity renders it useful in the construction of furnace muffles which are required to transmit heat. Fused alumina or artificial corundum (sold under such names as alundum and aloxite) has proved a more satisfactory abrasive than the natural compound. Though less hard than carborundum, it is also less brittle, and therefore more efficient in grinding steel and malleable iron. It is also receiving increasing employment as a refractory in the form of crucibles and tubes. The Carborundum Company has lately reduced silicon in the electric furnace and now manufactures it for use in steel production, in the chemical industries, where its high resistance to acids is advantageous, and in the making of hydrogen. "Pyrex" dishes are electric-furnace products containing 80 per cent silica. Artificial graphite, because it is infusible and incombustible except at extremely high temperatures, is used extensively for crucibles and electrodes. It can be employed in the form of electrodes in fused alkali and aqueous solutions, and possesses electrical conductivity four times that of the best carbon electrodes. Furthermore, it can be readily machined with accurate threads, so that it can be connected up for structural purposes. Colloidal graphite in a medium of water or oil presents a useful lubricant for bearings.

Electrolytic reduction of sodium from fused sodium hydroxide drove out the chemical process for sodium products soon after Castner's invention. Sodium, however, has only restricted uses, and it is in the development of the production of chlorine and caustic soda that electrolysis of the alkalis has gone farthest. Electrolytic decomposition of common salt (sodium chloride) in aqueous solution yields chlorine and caustic soda. Chlorine is used largely in making bleaching powders. Many paper mills and other large users of bleaching powders have installed electrochemical plants to produce chlorine. Electrolytic caustic soda has largely replaced the products of the older methods. Electrolysis of salt also produces sodium hypochlorite, which is employed as a bleaching agent and for disinfecting purposes. The decomposition of potassium chlorate, used extensively where a strong oxidizing agent is needed, is obtained by electrolysis.

The most important of the more recent developments of electrochemistry seem to be the various electric furnaces for the fixation of atmospheric nitrogen into cyanides, cyanimides, or nitrides. Oxidation of ammonia to produce nitric acid is another electrochemical industry of growing significance. The electric current has a wide use to produce ozone from oxygen for use in water-purification plants. The attempts to apply electrometallurgy to metals other than those already mentioned have not met with success until quite recently. Electrolytic extraction of zinc from aqueous solution has lately begun to achieve a considerable degree of success, especially with complex ores containing lead and silver. An extensive development has also occurred in the electrolytic reduction of copper from aqueous solutions of copper ore. Electric smelting of copper and tin ores has been carried on to a slight extent only. Electrolytic gold refining now finds a wide range of employment. Electroplating of several metals has developed into an important industry. Magnesium is produced solely by electrolysis. This light metal is useful in alloys (e.g., magnalium, an aluminum-magnesium alloy, and Elektron) and will probably have an increased demand as aviation develops. The preparation of phosphorus and carbon bisulphide is now conducted almost wholly by electrothermal processes.

An interesting by-product of electrochemistry was described by Professor Richards in his presidential address to the American Electrochemical Society in 1903.

Such organizations as research companies, formed explicitly to combine research with practical application, are novelties in the industrial world which have originated with, and are almost peculiar to, electrochemistry. They invent, investigate, and develop electrochemical processes, and furnish facilities to would-be experimenters whose ideas might otherwise remain stillborn.

Industrial electrochemistry has thus branched far in many directions since the beginnings which were occasioned by interest in aluminum reduction. Some of the more important products and their applications have been noticed here, but it would require several volumes to describe all the ramifications and the repercussions upon the industrial structure. The importance of this new group of industries for the development of the most spectacular child of the twentieth century is well explained by Mr. Tone of the Carborundum Company.

The mechanical perfection of the automobile and the interchangeability of its parts have been made possible by the modern grinding wheel. Practically every part of the automobile must be ground with artificial abrasives at some stage of its manufacture. Take away from the automobile industry artificial abrasives and other products which the chemist has made available to it by the electric furnace, such as aluminum, alloy steel, and high-speed steel, and the labor cost of building a car would become prohibitive. The industry would cease to exist on its present lines.

APPENDIX B

PATENT LITIGATION IN THE UNITED STATES

THE first fifteen years of the electrolytic aluminum industry witnessed continuous litigation in the United States to settle conflicting claims to participation in the profits of the new industry. This litigation is of interest in connection with the question of the participation of several persons in the same invention and the distribution of rewards. It also bears upon the relation of patents to monopoly.

Charles M. Hall succeeded in reducing pure aluminum by electrolysis in February 1886. He filed his first patent application in the following July. In the course of correspondence between Hall and the Patent Office several apparatus claims were disallowed as being mere aggregations of well-known apparatus. S. C. Mastick, lecturing on chemical patents, states:[1]

> We have seen that the various parts of his apparatus were all old and that the essence of his invention consisted in fusing a compound composed of the fluorides of aluminum and of a metal more electropositive than aluminum, dissolving alumina therein and passing an electric current through the fused mass. It may be that at this point of time Hall himself did not appreciate that the process, regardless of the form of apparatus used, was the broad and valuable invention.[2]

During the correspondence with the Patent Office an interference was declared between the application of Hall and one filed by Héroult, who had independently discovered the same process in France. It was settled in favor of Hall because the date of actual success with his process (February 23, 1886) preceded the filing of Héroult's application. Patent number 400,766 was issued to Hall on April 2, 1889.[3] It contained three claims, all process

[1] For this discussion of the patent litigation I have drawn heavily upon a series of lectures given in 1915 by Seabury C. Mastick, special lecturer on chemical patents, Department of Chemical Engineering, Columbia University. The aluminum litigation was treated quite fully in these lectures, which are reprinted in *Industrial and Engineering Chemistry*, VII, 789, 879, 984, 1071 (September to December 1915).

[2] *Ibid.*, VII, 881 (October 1915).

[3] In the interim Hall had amended his original application in some respects. The only change of importance seems to have been specifying carbonaceous anodes

claims, viz.: (1) a bath of "fluorides of aluminum and a metal more electropositive than aluminum," and passing an electric current through the fused mass; (2) fluoride of sodium as the metal more electropositive than aluminum, and the use of a carbonaceous anode; (3) fluoride of lithium as an additional or alternative ingredient of the bath.

In the meantime Hall, who was without resources, had made several attempts to secure the financial backing necessary to perfect his invention for industrial operation. In the summer of 1886 he worked in Boston, where his brother had been able to raise a little money to defray expenses. After four rather discouraging months the inventor found his backing withdrawn and returned to Oberlin, where he used a large bichromate-sulphuric acid battery constructed by himself. The results of his work there were so encouraging that he went to Cleveland in December and attempted to raise funds for work on a larger scale. A decided lack of interest upon the part of Cleveland capitalists combined with the interference declared about this time by the Patent Office, and not immediately settled in his favor, led the disheartened inventor to enter into an optional agreement to sell his patent to the Cowles brothers, whom he had met during the preceding summer. Hall was to have current and facilities for experiment at the Cowles works at Lockport, and was to receive one-eighth interest in the Cowles company in the event that they decided to purchase. He worked at Lockport from the summer of 1887 until July 1888, endeavoring to perfect his process. It was during these months that there emerged a difference which may at first have been merely a difference of opinion, but which later developed into an acrimonious dispute between Hall and the Cowles brothers over the right to fundamental parts of the reduction process as it was finally developed. Previous to his Lockport work Hall had employed external heat to fuse the bath in his crucible and keep it in fusion. His patent applications specified an externally heated crucible, although he had included a statement similar to the following quotation from a letter written to his sister in August 1886. "Also it is evident from the experiments that the waste heat of electricity, which must be used anyway, will be nearly, if not quite, enough

rather than copper, which had not proved satisfactory. Four other patents covering minor details of the process were issued at the same time.

to keep the solvent melted."[4] At Lockport he continued to use external heat. Furthermore, at this time he was employing copper anodes, as he had done from the beginning. According to Mr. Mastick, the Cowleses believed that internally heated crucibles with carbon anodes were necessary, and internal heating was alleged to be within the scope of certain prior Cowles patents and applications.[5]

Testimony concerning the results of Hall's work at Lockport is contradictory. He himself relates that he experienced difficulties with his process for a time, but "after finally overcoming the difficulties which I have mentioned, I made several pounds of aluminum in small crucibles which I showed to Mr. Alfred Cowles and gave him all the facts in relation to the same, but he was not interested."[6] Mr. Cowles is reported as stating that the results, as far as they saw, were not sufficiently encouraging;[7] while it is said that Hall alleged that the current at his command was so small that he could not show the results which would come with larger-scale processing.[8] However this may be, the Cowleses allowed the option to lapse in July 1888.

One of their metallurgists, Romaine Cole, was interested, however. Before Hall left Lockport, Cole resigned and went to Pittsburgh, where he was able to gain the support of Captain Alfred E. Hunt of the Pittsburgh Testing Laboratory. Hall's arrival in Pittsburgh at the close of July was followed within a few weeks by the organization of the Pittsburgh Reduction Company, with Captain Hunt as its president. A capital of $20,000 was subscribed by Hunt and his associate, Mr. Clapp of the Pittsburgh Testing Laboratory, and four other Pittsburgh men. On the following Thanksgiving Day production was begun in a small plant in Pittsburgh. Apparently, external heating was employed for a short time, during which no startling success attended, and then was abandoned in favor of internal heating by electric current, which

[4] 111 Fed. Rep. 754.

[5] *Industrial and Engineering Chemistry,* VII, 986 (November 1915).

[6] Remarks of Mr. Charles M. Hall in acknowledgement of the Perkin Medal (*ibid.,* III, 148, March 1911). Cf. also Hall's report to the Cowles company, printed in *Aluminum Industry,* I, 21; and statement in an anonymous biography of Hall that he was able to produce aluminum at Lockport "in nearly as large a quantity, in proportion to the power employed, as had ever been done since" (*Aluminum World,* I, 66, January 1895).

[7] *Electrochemical Industry,* I, 10 (September 1902).

[8] *Ibid.* Cf. also report of Hall cited above.

immediately proved more effective.[9] Carbon anodes were substituted for copper by Hall at some time during the early months of his work with the Pittsburgh Reduction Company. Progress during 1889 and 1890 was rapid.

Probably the future success of electrolytic aluminum was perceptible to the Cowleses, experienced as they were in electrometallurgy. As soon as Hall departed they had begun experiments with cryolite and alumina, keeping the bath fused by electric current. For a while attempts were made to merge the two companies. When this failed, the Cowles company threw down the gauntlet with a determined gesture. First they brought suit against the Pittsburgh company, alleging infringement of certain Cowles patents. They claimed prior invention and application of the use of electric energy to fuse ores preparatory to reduction.[10] Their next move is described by Judge Taft.

> The evidence leaves no doubt that the defendant company [the Cowles company] began their manufacture of pure aluminum in January 1891, with the aid of one Hobbs, who had been the foreman of the complainant company, and engaged for it in superintending the manufacture of aluminum by the Hall process.[11]

Hitherto the Cowleses had made and sold only aluminum alloys. Now they began to advertise pure aluminum at prices which undercut the Pittsburgh company's charge of $1.50 a pound. A short price war ensued, bringing the price down to $1 by the middle of March. At this juncture the Pittsburgh Reduction Company entered the legal arena with a countersuit alleging infringement of the Hall patent and praying for a preliminary injunction. Judge Ricks, of the Circuit Court of the Northern District of Ohio, denied a complete injunction but issued an order restraining the Cowles company from increasing its output during the trial of the suit and from selling below a price to be named by the complainant.[12]

[9] Mastick, in *Industrial and Engineering Chemistry,* VII, 986 (November 1915). Cf. *Aluminum Industry,* I, 24.

[10] E. P. Allen, "The Production of Aluminum," *Cassier's Magazine,* I, 419 (February 1892). This suit never went beyond the filing of bills in the Circuit Court at Pittsburgh.

[11] Opinion in Pittsburgh Reduction Company v. Cowles Electric Smelting and Aluminum Company; quoted from Mastick, *op. cit.,* p. 989.

[12] The Pittsburgh company named $1.50 as the price, but this was lowered to 50¢ in August to meet foreign competition.

The outcome of the suit was favorable to the Hall interests. Judges Taft and Ricks handed down an opinion on January 11, 1893, holding the Cowles company to be infringers and ordering the payment of $292,000 damages to the Pittsburgh Reduction Company. This sum was never paid because attempts to secure a rehearing lasted until another suit had turned the pecuniary tables. The results of this first case established the Hall company as the sole producers of aluminum in the United States by the electrolytic method, the only method then in commercial use for producing pure aluminum.

Nevertheless, the Cowleses were not willing to regard this defeat as final. Rather they pushed the struggle in the courts for another decade. The litigation concerned the ownership of the Bradley patents, which were believed to dominate the Hall patent, and the infringement of the Bradley patents by the Pittsburgh Reduction Company. Apparently the Cowleses had concluded after the adverse decree of 1893 that their own patents could not be used successfully in fighting the Pittsburgh company. At that time they had certain claims to two patents which had just been issued to Charles S. Bradley. The brothers had come into contact with Bradley in 1895 when the Patent Office declared an interference between some of their respective patent applications. Bradley sold out to the Cowleses. At this time there stood rejected at the Patent Office an earlier application of Bradley's, filed February 23, 1883, relating to the separation of metals from highly refractory ores which were nonconductors in an unfused state, by using the electric current to fuse, maintain fusion, and decompose by electrolysis. The use of this process for aluminum reduction was specifically claimed. The rejected application was brought to the attention of the Cowleses, and was the subject of a discussion between them and Bradley before a contract was finally signed which, in quite broad language, conveyed to the Cowleses "all interest in any and all discoveries and inventions relating to electric smelting processes and furnaces, and all patents they [Bradley and an associate] have obtained therefor and all applications now pending, and caveats on file, in the United States Patent Office, relating to electric smelting processes and furnaces, which do or may interfere with any application for patents made by Eugene H. and

Alfred H. Cowles of Cleveland, Ohio, now pending in the United States Patent Office." [13]

Bradley's application of 1883 lingered on in the Patent Office, with no interest shown in it by the Cowleses, until 1892, when the Board of Examiners-in-Chief allowed the issuance of two patents which Bradley promptly assigned to G. P. Lowrey.[14] The latter brought suit to restrain the Cowles company from claiming title to these patents, to which the company replied with a cross-bill praying that Lowrey be enjoined from claiming the title. Judge Taft held in the Circuit Court decision, rendered April 23, 1895, that the Bradley patents were not intended to be conveyed by the assignment of May 18, 1885.[15] The Circuit Court of Appeals, Judge Severenz delivering the opinion, reversed the lower court, holding that the inventions were intended to be included in the terms of the contract of 1885.[16] Immediately upon receiving title, the Cowles interests again took up the legal cudgels against the old rival in Pittsburgh, alleging infringement of the Bradley patents.[17] The Circuit Court finally dismissed the bill in October 1901, holding that the Hall process did not infringe.[18] Having become used to defeat in the lower courts, the Cowleses at once appealed the suit, and were rewarded two years later with a verdict that one of the Bradley patents had been infringed.

The judges of both courts were in apparent agreement that the novelty of Hall's process or the essence of his invention consisted in the discovery that alumina would dissolve freely in cryolite. The process actually operated by the Pittsburgh Reduction Company

[13] Quoted by Mastick, *op. cit.*, p. 1072.

[14] Patents no. 464,933, issued December 8, 1891, and no. 468,148, issued February 2, 1892. Lowrey, a shrewd patent attorney, had evidently hunted out this Bradley application and pushed it through. When the patents issued, Lowrey immediately notified both the Pittsburgh and the Cowles companies that they were infringers. An interesting side issue in the struggle between this attorney and the Cowleses over the Bradley patents was a shift from the Cowleses to Héroult, engineered by Lowrey, on the part of a group of Berlin capitalists whom the Cowleses had got together for the erection of an aluminum works in Switzerland which was to be operated under Cowles patents.

[15] 68 Fed. Rep. 354 (1895).

[16] 79 Fed. Rep. 331 (1897).

[17] Electric Smelting and Aluminum Company v. Pittsburgh Reduction Company, 111 Fed. Rep. 742. The Electric Smelting and Aluminum Company had been recently formed by the Cowles interests. The old Cowles Electric Smelting and Aluminum Company continued its existence as a subsidiary of the new company.

[18] 111 Fed. Rep. 742 (1901).

involved also simultaneous fusion and electrolysis by the electric current. Did this constitute infringement of the Bradley patent? The chief issues stressed in both opinions were two: (1) whether the construction to be placed on Bradley's patent should be broad or narrow; (2) whether the Hall process as operated employed electric energy in excess of the amount necessary for electrolysis, the excess being used for heating to fuse and maintain fusion. The judges of neither court were altogether successful in avoiding confusion with respect to the electrochemical relations which continually intruded upon the legal domain. In a muddled opinion which occasionally confused the two issues, Judge Hazel of the lower court concluded that the proper construction to be placed on the Bradley patent was too narrow to cover Hall's process, and that "the heat required to maintain fusion is obtained by the heat radiation, and from such sources as are incidental to the use of the process, and not from any independent process of electric heating."

Judge Coxe of the Circuit Court of Appeals, with his two associates concurring, delivered a clearer opinion, well ordered, more careful in logic.[19] After a survey of the prior art, which seems to be more penetrating if no more exhaustive, he concluded that the Bradley patent should have a liberal construction. Starting with the undisputed fact that before Bradley's work no one had ever been able to separate aluminum from its compounds solely by the use of electricity — i.e., without the employment of external heat — Judge Coxe went on to show that although it was previously known that metals contained in ores which were conductors could be separated therefrom by electricity, the problem of dissociating metals from nonconductors by this method had not been solved. The ores of aluminum are nonconductors at ordinary temperatures.

The principal expert for the defendant, Dr. Chandler, whose reputation for learning and ability is well known to the courts, although of the opinion that slight modifications of the previous methods would produce the Bradley process, nevertheless admits frankly, "I do not recall any one process which, when applied to the ore of aluminium, would without any modification whatever have produced aluminium, in which process both the fusion and the electrolysis would have been accomplished by the electric current."[20]

[19] Electric Smelting and Aluminum Co. v. Pittsburgh Reduction Co. on appeal (125 Fed. Rep. 926), decided October 20, 1903.

[20] 125 Fed. Rep. 932.

From Davy to Bradley no one had been able to produce any aluminum by electricity alone. The efforts of inventors were directed to the perfection of external heating processes, even after the introduction of dynamos; and, indeed, these efforts continued for several years after the Bradley invention. Hall himself employed external heat until 1889. The Court disposed of thc first issue by concluding:

We are unable to discover anything in the prior art describing this process or anything closely approximating thereto. The patent is, therefore, not anticipated, and its claims are entitled to a liberal construction.

The Judge of the Circuit Court, after a careful and painstaking research, reached the conclusion that Bradley had made a valuable invention, but failed to grant relief to the complainant upon the theory that the process which the defendant uses was an entirely separate invention, neither dependent upon nor subsidiary to the invention of Bradley. In this we think there was error. Hall's achievement should be considered in the light of an improvement upon Bradley's fundamental discovery.[21]

In taking up the second issue Judge Coxe stated that consideration of the Hall patent to which the court below had devoted much time was irrelevant, since the patent was not issued until 1889 and did not disclose the process which the Pittsburgh company used and of which the complainant complained. The material fact was that Hall's discovery that he could dispense with external heat came at least three years after Bradley's invention. Upon the question of excess energy beyond that necessary for electrolysis the Court believed that a current which fuses, maintains fusion, and electrolyzes must be of greater power than one which electrolyzes alone. The fact that the Bradley process was actually operated commercially in the United States and abroad is adduced.[22] In sum, when the proper construction is placed upon the Bradley patent, it was seen to be infringed because some of the electric energy was used to fuse and maintain fusion.

It is, of course, an indisputable fact that a substantial portion of the electric energy is converted into heat which results in constant fusion.

There are, however, certain electrochemical processes in which electrical energy is used both for heating and for effecting electrolytic resolution; the

[21] 125 Fed. Rep. 932 (quoted by Mastick, *op. cit.*).

[22] Probably the reference was to operation by the Cowles firm before it had acquired legal title to the Bradley patent.

most noteworthy instance is in the manufacture of aluminium by electrolysis of alumina dissolved in a double fluoride of aluminium and sodium. The bath is not only decomposed electrolytically, but is also kept fused by heat obtained at the expense of electrical energy passing between the electrodes.[23]

Indeed the conversion of electric energy into heat is inevitable in this electrolytic process. Professor Richards has explained that:

. . . it is impossible to pass any current whatever through an electrolyte without generating some internal heat in it, and therefore the question as to whether the heat thus generated internally shall be sufficient alone to keep the bath melted, at the proper temperature, is merely a question of increasing the size of the pot and the scale of the operation.[24]

It is clear that Hall obtained internal heat by this expedient. Either Hall's invention was not a complete one in the first instance, because he failed to realize that the electric current would fuse and maintain fusion as well as decompose,[25] and hence continued for some time to employ external heating, the use of which hindered him from increasing the size of the pot and the scale of the operation to the point where this principle would be made manifest; [26] or, recognizing the principle, with or without knowledge of Bradley's work, Hall was unable at first to apply it satisfactorily, and hence did not specify it in any of the patents which were issued to him in 1889; or else he did not consider that it could be or should be patented.[27] Whatever was true of Hall, Bradley — and perhaps the Cowleses — had recognized the importance of this principle earlier.

The question whether Hall had benefited from Bradley was

[23] Bertram Blount, *Practical Electro-chemistry* (New York, 1903), pp. 24, 167 ff. See also A. J. Allmand and H. J. T. Ellingham, *The Principles of Applied Electrochemistry* (New York, 1924), pp. 521 ff.; J. W. Mellor, *A Comprehensive Treatise on Inorganic and Theoretical Chemistry* (London, 1924), I, 166; Richards, *Aluminium,* p. 383; Minet, *The Production of Aluminium,* pp. 19 ff.

[24] *Aluminum World,* VIII, 132 (April 1902).

[25] This opinion was expressed by Professor F. Haber in reporting to the Bunsen Society of Germany upon the industrial development of electrochemistry in the United States after a visit to this country sponsored by the Society.

[26] Dr. F. Regelsberger says that of the four inventors Hall, Minet, Kiliani, and Héroult, the latter two alone recognized early that external heating could be replaced by means of a stronger current. Héroult specified this in his British patent. See *Aluminium,* VII, Heft 9, p. 1 (May 16, 1925).

[27] Upon one occasion Hall characterized this principle as resulting "from a law of nature and not from any invention." See his remarks in acknowledgement of the Perkin Medal, reported in *Industrial and Engineering Chemistry,* III, 148 (March 1911).

immaterial to a decision upon infringement. The same may be said of the question whether, at the scale upon which the Hall process was being operated, just the amount of energy required for electrolysis would necessarily generate just the amount of heat necessary to maintain fusion properly. The only possible construction for the Bradley patent, except a construction which would have nullified it, would appear to be that it covered any use of the electric current to secure internal heat for fusion simultaneous with electrolysis of aluminum and other substances specified, whether such generation of heat was unavoidable or not. The fundamental question is whether Bradley should ever have been granted a patent for his process.[28] As issued, the patent, if construed to mean anything, must necessarily have been infringed by the Hall process.

The final decision in this suit involved a judgment of nearly $3,000,000 against the Pittsburgh Reduction Company as infringers since 1892. A few months prior to this holding the Cowles group had succeeded in having the old case, in which Judge Taft had enjoined the Cowles company from manufacturing pure aluminum, reopened for the introduction of new testimony and reargument. The sum decreed against them by Judge Taft had never been paid because of litigation. Now a final agreement between the two companies settled the whole controversy. The Pittsburgh company paid a sum of money somewhat less than the damages awarded. It was agreed that this company should have the monopoly of aluminum manufacture until the expiration of the Bradley patent in February 1909, and should work under a license, paying royalties. The Cowles companies agreed not to manufacture pure aluminum but could buy and sell all grades of aluminum.[29] As far as the inventors were concerned the outcome was not, perhaps, far removed from the attainable optimum of human justice. Hall and his associates, who had actually made a commercial success of electrolytic alu-

[28] This is a question upon which I am not competent to pass judgment. In the view of this controversy given by J. D. Edwards in *Aluminum Industry,* I, chap. II, it is stated that Bradley never operated his process and it is implied that in the light of this fact and the state of the prior art this "paper patent" should not have been issued.

[29] The Cowles companies were, of course, free to continue the manufacture of aluminum alloys.

minum, remained alone in the field. The Cowleses, who had played a conspicuous part in the early development of electrochemistry, and who may have had some influence upon Hall's success, received a cash reward. And Bradley, who at least seems to have been the first in this country to grasp the importance of the idea of simultaneous fusion and electrolysis of aluminum, received recognition.

The patent controversy and its outcome were a typical instance of the industrial development of inventions under a system of patent law. As the *Engineering News* remarked:

> The situation is simply one which constantly recurs in the history of inventions, in which an inventor whose work reaches commercial success finds that he must settle with the owner of some earlier pioneer patent, whose claims are entitled to a broad construction.[30]

In its relation to monopoly the result of this litigation was of considerable significance. What may have seemed to the Pittsburgh Reduction Company a severe blow was transformed some time later into an undisguised blessing. The license to operate the Bradley patent extended the period of legal monopoly past the business boom of 1906–1907 into the middle of the succeeding depression, and really gave the company three extra years in which to become so well fortified against competition that none developed. Furthermore, the decision of the victors to refrain from the manufacture of pure aluminum removed the most logical competitor. Mr. Alfred Cowles and his associates were probably better fitted by experience to enter this new industry than any other group of men in the country, except those already operating the Pittsburgh enterprise. A study of the patent struggle also makes it clear that the electrolytic process was so simple in its elements as to permit no possibility of patenting modifications or variations upon the basis of which a competing enterprise might operate.

[30] *Engineering News*, L, 390 (1903).

APPENDIX C

INVESTMENT AND EARNINGS OF THE ALUMINUM COMPANY OF AMERICA, 1909–1935

ANNUAL financial reports have been published by the Aluminum Company of America only since the year 1926. Materials from which estimates of investment and earnings may be derived for the years 1909–1926 are composed of the following sorts of information: (1) approximate figures for particular years given to government bodies by the company; (2) general statements in security advertisements that earnings, before or after interest as indicated, exceeded, equaled, or did not fall below certain sums in certain periods or particular years; (3) balance sheets for the years 1920–1924 submitted to the Department of Justice and printed in the Benham Report, p. 92; (4) records of interest and dividend payments, and fragmentary records of capital expenditures; (5) study of market conditions. The object of the study is to ascertain approximately the average ratios of earnings to investment in certain periods. Earnings signify net earnings of capital *after* operating expenses, including depreciation, depletion, taxes, and interest on current debt, and *before* interest on funded debt. Investment is equivalent to total assets less depreciation and depletion.[1] In order to obtain average investment during each year, investment at the beginning and end of each year has been averaged, except in the case of years during which assets or securities were bought or sold. In such instances averages of investment during appropriate periods of each year have been computed.

With regard to the period prior to 1927 there have appeared figures of investment at the end of 1908, 1910, 1911, 1912, 1916, each year 1920–1924, and 1926, and data on earnings for the years 1909–1912, 1916, 1921, and 1924–1926.[2] Other information about earnings is as follows:

[1] No good-will item or other questionable asset account appears on the balance sheets of the Aluminum Company. Assets of "non-consolidated" subsidiaries are represented in the investments account.

[2] Sources are given in the notes to Table 37.

(1) 1915–1924 — Average annual earnings after interest were $9,843,133.33.[3]

(2) 1917–1926 — Average annual earnings before interest exceeded $12,000,000.[4]

(3) 1911–1920 — Average annual earnings after interest exceeded $10,000,000.[5]

(4) 1916–1919 — Earnings after interest were in no year less than $10,000,000.[6]

(5) 1915–1918 — Earnings were in no year less than $8,000,000.[7]

The general method used is shown in the following summary of figures.

	Total earnings after interest, 1915–1924		$98,431,333 [8]
	Total earnings paid as interest on funded debt, 1915–1924		$9,297,500 [9]
	Total earnings, 1915–1924		$107,728,833
Less	Earnings, 1915	$9,000,000 (estimated)	
	Earnings, 1916	$20,000,000 (official)	
	Total earnings, 1915-1916	$29,000,000	$29,000,000
	Total earnings, 1917–1924		$78,728,833
Less	Total earnings, 1921–1924 (figures for 1921–1923 derived from comparing balance sheets; figure for 1924 given by the company)		$30,425,300
	Total earnings, 1917–1920		$48,303,533
	Total earnings after interest, 1911–1920, equal at least to		$100,000,000
	Total interest paid on funded debt, 1911–1920		$1,500,000 [9]
	Total earnings, 1911–1920		$101,500,000
Less	Total earnings, 1917–1920	$48,303,533 (above)	
	Total earnings, 1915–1916	$29,000,000 (above)	
	Total earnings, 1911–1912	$9,560,000 (official)	
		$86,863,533	$86,863,533
	Total earnings, 1913–1914		$14,636,467

[3] *New York Times,* October 29, 1925, p. 38. [4] *Ibid.,* February 7, 1927, p. 28.

[5] *Wall Street Journal,* October 3, 1921, p. 3.

[6] *Commercial and Financial Chronicle,* CXI, 1853 (November 6, 1920).

[7] *Ibid.,* CVIII, 880 (March 1, 1919).

[8] This figure represents ten times the annual earnings after interest officially reported for that period. It does not appear whether the operating loss of about

[9] (See page 540 for note 9.)

The indicated total net earnings for each of the several periods were distributed over the individual years in accordance with considerations suggested by changing conditions. Growth in investment was estimated by calculation of reinvested earnings and additions to assets from sales of new securities. Recent testimony implies that there has been no substantial revaluation of any assets upward in the period covered by this study.[10] Some discussion of estimates of investment and earnings in certain periods is appropriate.

1913–1920

Figures of investment at the end of the years 1912, 1916, and 1920, and of earnings in the year 1916 have been given by the company. Figures for other years have been estimated according to the method just described. Investment grew from about $30,000,000 at the end of 1912 to about $80,000,000 at the end of 1916. The sum of the indicated reinvested earnings in these years is only $39,000,000. Apparently no new securities were sold. At the end of 1920 investment had increased to approximately $158,000,000. During the four years 1917–1920 reinvestment totaled $38,000,000, according to my estimates, and the sale of notes added about $24,000,000. The indicated increase in investment is only $62,000,000, while the actual increase was $78,000,000. During the whole period 1913–1920 investment increased about $128,000,000, according to the company's figures. Only $101,000,000 of this increase is accounted for by my estimates of reinvested earnings plus the proceeds of security sales.[11] The balance sheet for December 31, 1920, shows current payable items equal to $16,000,000. If current payables stood at zero at the beginning of 1913 the part of the growth in investment

---

$5,000,000 in 1921 was taken into consideration in computation of the average. If it was, the total earnings after interest during the period 1915–1924 were $5,000,000 larger.

[9] Calculated from data published in financial manuals.

[10] BMTC v. ACOA appellant, fol. 5707. A small write-up in 1925 is referred to on the next page.

[11] According to figures supplied by the company the total capital expenditures by the Aluminum Company and its subsidiaries in this period amounted to about $102,000,000. See *Hearings before Senate Committee on Investigation of Bureau of Internal Revenue,* 68 Cong., 2 Sess., Part 10, p. 1852.

which is unaccounted for in this study would be reduced to $11,-000,000. Evidently the estimates of earnings presented here for the period 1913–1920 are too low by at least this amount.

1921–1935

Figures of investment at the end of each year (except 1925) have been given by the company. An operating deficit of about $5,000,000 is reported for 1921. To this has been added an inventory loss of $5,000,000 suggested by examination of balance sheets, price data, and other information, on the chance that this loss may not have been included in the operating deficit. Earnings for 1922 and 1923 have been estimated after comparing balance sheets and studying other relevant material. Comparison of balance sheets was rendered difficult by lack of information to explain changes in inventory items and reduction in plant account. If the reduction in plant account of about $9,000,000 in 1922 represented a simple write-down, the estimated earnings for the period 1921–1924 are too low by about that amount. Figures for earnings before interest in the years 1924–1926 were given by the company. Net earnings for each year thereafter have been computed by adding to the published figure of earnings after interest and taxes the annual interest payments required for the average amount of bonds outstanding in the year. Total investment increased approximately $75,000,000 during the period 1921–1929. Of this, $70,000,000 can be accounted for by the algebraic sum of indicated reinvestment of earnings, net proceeds from security sales, and the diminution in current liabilities and sundry other items. This small discrepancy is about equal to an increase in the book value of marketable securities made at the time of the merger with the Canadian Manufacturing and Development Company in 1925.[12] Examination of financial reports for the years since 1929 does not indicate any substantial write-down of assets or recapitalization other than bond retirements.

In addition to the fact that the full growth of investment during the period 1909–1920, at least, does not seem to be accounted for with the estimates of earnings given in Table 37,

[12] BMTC v. ACOA appellant, fol. 5707.

there are other indications that the annual earnings and rates of return on investment were actually somewhat greater, during part of the time, at least, than the figures shown in the table. (1) The method used in making estimates of earnings in those years for which no official earnings figures appear has probably tended to understatement because lower rather than higher figures have been used in all cases of doubt. (2) The figures of Table 37 do not include interest paid on current liabilities which is part of true earnings of total capital investment. Since payments on this item could be computed for a few years only, they were uniformly excluded. (3) The investment and income of several partially or wholly owned subsidiaries were evidently not included in the consolidated balance sheets and income accounts of the Aluminum Company.[13] Only such part of the earnings of these subsidiaries as was paid to the parent in dividends would appear in the income account of the latter. At the end of 1924 (the latest date for which we have information) the non-consolidated group included several important subsidiaries or affiliates, such as foreign bauxite companies, the Norwegian aluminum firms, Aluminum Manufactures, and the Aluminum Goods Manufacturing Company. (4) It appears that during the years 1912–1925 the Aluminum Company spent several millions of dollars in the acquisition and development of foreign ore properties, particularly in South America. At the beginning of 1925 the aggregate capitalization of these subsidiaries seems to have been not much more than $1,000,000.[14] It is possible that some true earnings were used for these bauxite properties without entering the income or capital accounts of the parent or subsidiaries. (5) At the end of 1927 the reserve for amortization, depletion, and depreciation was equivalent to about 32 per cent of the undepreciated book value of land, plants, and facilities. Three years later the corresponding figure was about 35 per cent, and at the end of 1934 it was about 37 per cent. More than three-quarters of the present capacity represented by dams and powerhouses seems to have been added since 1912, and at least half of the present capacity since 1925. Other facilities have been greatly enlarged in the last ten or fifteen years. True annual depreciation on hydro-

[13] BR, pp. 92–96.
[14] *Ibid.*

electric dams, powerhouses, and equipment is, of course, very small.[15] Furthermore, the company has accumulated large reserves of bauxite. Although the annual charges to depletion and depreciation may not have exceeded true charges since 1926 — they averaged slightly less than 3 per cent of undepreciated book value of land, plants, and facilities during the years 1927–1934 — it is quite possible that the large reserve for depletion and depreciation contains a substantial amount of reinvested earnings which have not been included in the income figures.

Sales and cost data appearing in the Benham Report[16] enable rough computation of earnings at the ingot stage — i.e., earnings upon all ingot, which is made up of ingot sold in that form and ingot sold in the form of later products. When the average price received for the metal sold as ingot is considered as the average price received for all ingot, the difference between this figure and the average cost of producing ingot (exclusive of any profit) constitutes the average profit per pound of ingot. Net earnings at the ingot stage may then be estimated by multiplying the total amount of metal sold in all forms by the average profit per pound of ingot. This computation indicates that in 1923 earnings at the ingot stage represented about 45 per cent of the total net earnings estimated for that year. In 1924 earnings at the ingot stage appear to be almost equal to the amount which the company reported as its net earnings on all operations, while in 1925 earnings at the ingot stage during the first half year apparently represent more than 40 per cent of the total net earnings reported for the whole year, or perhaps 80 per cent of the net profit in the half year.[17] It is not specified whether the cost figures, which have been used in the computation here, include selling cost and general expense or not. If they do not, profit at the ingot would, of course, be less. If one quarter of the total selling cost and general expense for 1925[18] is included in cost of ingot

[15] J. D. Justin and W. G. Mervine estimate that the annual depreciation for typical hydroelectric plants will vary from 0.7 per cent to 1.5 per cent (*Power Supply Economics*, p. 150).

[16] Pages 47 and 118.

[17] The company may, of course, keep its books in such manner that a higher rate of earnings is shown on the investment in fabricating facilities.

[18] These expenses for 1925 appear in Exhibits 58 and 258 of BMTC v. ACOA appellant.

TABLE 37

INVESTMENT, EARNINGS, AND RATE OF RETURN OF THE ALUMINUM COMPANY OF AMERICA, 1909–1935

Year	Investment at End of Year [a] (*$1,000*)	Average Investment during Year [b] (*$1,000*)	Net Earnings [c] (*$1,000*)	Rate of Return (*per cent*)
1908	$24,000 [d]			
1909	27,000 [e]	$25,500	$3,600 [f]	14.1
1910	30,000 [d]	28,500	4,590 [f]	16.1
1911	26,300 [d]	28,150	5,100 [f]	18.1
1912	30,000 [d]	28,150	4,463 [g]	15.9
1913	36,750 [e]	33,380	7,500 [e]	22.5
1914	43,130 [e]	39,990	7,500 [e]	18.8
1915	50,900 [e]	47,015	9,000 [e]	19.1
1916	80,000 [h]	65,450	20,000 [h]	30.6
1917	91,750 [e]	85,880	14,000 [e]	16.3
1918	100,650 [e]	101,200	11,230 [e]	11.1
1919	120,880 [i]	114,830 [j]	10,500 [e]	9.1
1920	157,723 [k]	127,500 [j]	12,500 [e]	9.8
1921	145,331 [k]	150,220 [u]	def. 10,000 [l]	–6.7
1922	134,188 [k]	139,760	3,000 [e]	2.1
1923	145,016 [k]	139,600	14,000 [e]	10.0
1924	155,515 [k]	150,510	13,425 [m]	8.9
1925	190,000 [n]	170,000 [o]	22,892 [m]	13.5
1926	209,716 [n]	203,260 [j]	19,747 [m]	9.7
1927	250,170 [p]	248,100 [j]	18,160 [q]	7.3
1928	215,320 [p]	235,000 [r]	23,390 [q]	10.0
1929	232,517 [p]	223,220 [s]	27,330 [q]	12.2
1930	240,778 [p]	235,940 [s]	13,630 [q]	5.8
1931	245,133 [p]	242,240 [s]	6,495 [q]	2.7
1932	237,438 [p]	240,500 [s]	def. 510 [q]	–0.2
1933	233,452 [p]	234,660 [s]	3,400 [q]	1.4
1934	228,317 [p]	229,610 [t]	8,100 [q]	3.5
1935	221,703 [p]	221,000 [v]	10,820 [q]	4.9

[a] Total assets, except as otherwise noted.

[b] Average of investment at beginning and end of year; or, in the case of years in which assets were sold, securities sold or retired, or the like, average of investment at different periods of the year.

[c] Net earnings before interest on funded debt but after interest on current debt and taxes.

[d] Approximate figures given by an officer of the company (Tariff Hearings, 1912–1913, House Document no. 1447, II, 1491 ff.).

[e] Estimate by author.

[f] Testimony of an officer of the company that earnings were approximately 15 to 17 per cent of investment in these years (Tariff Hearings, *loc. cit.*). It has been assumed that this meant 15 to 17 per cent of investment at the beginning of each year.

[g] Figure given in bond advertisement, *Wall Street Journal*, Oct. 3, 1921, p. 3.

[h] Approximate figure given by an officer of the company. See *Congressional Record*, LV, 4592.

[i] Includes $12,000,000 increase in assets from sale of notes in 1919.

[j] Average of investment at beginning of year and that amount plus reinvested earnings, adjusted for increase or decrease in assets due to sale or retirement of securities during year.

[k] Figure from balance sheet in Benham Report, p. 92.

[l] This deficit consists of an operating loss of about $5,000,000 (FTC Docket 1335, Record, p. 5240) and an estimated inventory loss of about the same amount.

[m] Figure given in bond advertisement, *New York Times*, Feb. 7, 1927, p. 28.

[n] No official figure for Dec. 31, 1925, is obtainable. Figures for Sept. 30, 1925, and Dec. 31, 1926, appear in Exhibits 249 and 250, BMTC v. ACOA appellant. The figure for Dec. 31, 1925, has been computed by deducting from the investment at the end of 1926 the sum of the cash obtained by sale of notes in 1926 and the indicated reinvestment of earnings in 1926.

[o] The figure for 1925 represents an approximate average of investment during several periods of the year separated to reflect changes in investment occasioned by the merger of the Aluminum Company and the Canadian Manufacturing and Development Company on July 29 and retirement of $12,000,000 of notes on Nov. 1.

[p] Figure taken from annual report with deduction of preferred dividend payable next day.

[q] Net earnings after all expenses incident to operations and reserves for depreciation, depletion, income and franchise taxes from annual report, plus interest paid on funded debt computed from data in financial manuals.

[r] The figure for 1928 represents the average of average investment in the periods before and after exchange of certain assets for stock of Aluminium Limited and distribution of that stock to shareholders of the Aluminum Company.

[s] Average of investment at beginning and end of year adjusted for retirement of bonds on March 1.

[t] Average of investment at beginning and end of year adjusted for retirement of bonds on March 1 and purchase of $924,000 of bonds during 1934 for retirement in January 1935.

[u] Average of investment at beginning of year and that amount less deficit of $10,000,000, adjusted for retirement of notes.

[v] Average of investment at beginning and end of year adjusted for retirement of $6,000,000 of bonds, September 1, 1935.

for the half year, net profit at the ingot would approximate 35 per cent of the net for the whole year, or perhaps 70 per cent of the half year's earnings. Similar calculations have been made for the years 1926 and 1928.[19] When about half of the total sales and general expenses are added to the plant cost of ingot — probably a substantial overestimate — net earnings at the ingot stage appear to represent about 40 per cent of net on all operations in 1926 and about 60 per cent in 1928. It is questionable whether the plant cost figures, taken from an exhibit of the Aluminum Company do not include some items which are not a part of true cost from the standpoint of the question raised here.

[19] From data in Exhibits 58, 106, 117, 258, 291, 293, *ibid.*

APPENDIX D

DECREE

In the United States District Court, Western District of Pennsylvania
Session of 1912

UNITED STATES OF AMERICA, Petitioner,
v.
ALUMINUM COMPANY OF AMERICA, Defendant

DECREE

This cause coming on to be heard on this 7th day of June, 1912, before the Hon. James M. Young, District Judge, and the petitioner having appeared by its district attorney, John H. Jordan, and by Wm. T. Chantland, its special assistant to the Attorney General, and having moved the court for an injunction in accordance with the prayer of its petition, and it appearing to the court that the allegations of the petition state a cause of action against the defendant under the provisions of the act of July 2, 1890, known as the Anti-trust Act, that it has jurisdiction of the subject matter, and that the defendant has been regularly served with proper process, and has appeared in open court, by George B. Gordon, its counsel, and has given its consent to the entering and rendition of the following decree:

Now, therefore, it is ordered, adjudged, and decreed:

1. That sections 2, 4, and 5 of the agreement entered into as of date September 25, 1908, between the Société Anonyme pour l'Industrie de l'Aluminum of Neuhausen and the Northern Aluminum Co. (Ltd.), acting on behalf of the defendant corporation, as follows, to wit:

§ 2. The N. A. Co. agree not to knowingly sell aluminum directly or indirectly in the European market.

§ 3. The A. J. A. G. agree not to knowingly sell aluminum directly or indirectly in the American market.

§ 4. The total deliveries to be made by the two companies shall be divided as follows:

European market, 75 per cent to A. J. A. G., 25 per cent to N. A. Co.

American market, 25 per cent to A. J. A. G., 75 per cent to N. A. Co.

Common market, 50 per cent to A. J. A. G., 50 per cent to N. A. Co.

The Government sales to Switzerland, Germany, and Austria-Hungary are understood to be reserved to the A. J. A. G.

The sales in the U. S. A. are understood to be reserved to the Aluminum Co. of America.

Accordingly the A. J. A. G. will not knowingly sell aluminum directly or indirectly to the U. S. A. and the N. A. Co. will not knowingly sell directly or indirectly to the Swiss, German, and Austria-Hungarian Governments.

§ 5. The N. A. Co. engages that the Aluminum Co. of America will respect the prohibitions hereby laid upon the N. A. Co.

be, and the same are hereby, declared null and void, and that the defendant Aluminum Co. of America, and all its agents and representatives in whatever capacity, are hereby perpetually enjoined from directly or indirectly requiring the parties to said contract to abide by its terms, and defendant is further enjoined from either directly or indirectly entering into, through said Northern Aluminum Co., or any other person or corporation, and from making or aiding in making any agreement containing provisions of the nature of those hereinbefore set out, in so far as they relate to the sale of aluminum in the United States, or its importation into or exportation from the United States, or any contract or agreement, either verbal or written, the purpose and effect of which would be to restrain the importation into the United States, from any part of the world, of aluminum, or alumina, or bauxite, or any other material from which aluminum can be manufactured, or to fix or illegally affect the prices of such aluminum, alumina, bauxite, or other material, when imported.

2. That the fourth and eighth paragraphs of the agreement en-

tered into, under date of July 5, 1905, between the defendant Aluminum Co. of America, under its former name, Pittsburgh Reduction Co., and the General Chemical Co., a corporation, which paragraphs read as follows:

"Fourth. Said Chemical Co. further expressly covenants and agrees that it will not use or knowingly sell any of the bauxite sold to it by the said Bauxite Co. hereunder, or any other bauxite, or the products thereof for the purpose of conversion into the metal aluminum, and that upon proof that any of said bauxite or products thereof have been put to any such use it will not make any further sales or deliveries to the purchaser thereof.

"Eighth. It is understood and agreed that the bauxite sold hereunder by the said Bauxite Co. to the said Chemical Co. shall be used by the said Chemical Co. and by companies under its control or whose stock is largely held by it, and by no other person or party, and only for the manufacture of alum, alum salts, alumina sulphate or alumina hydrate for alum and its compounds, and for no other purpose whatsoever —"

be, and they are, hereby declared null and void, and are stricken out of said contract; and that the fifth section of said contract which reads as follows:

"Fifth. The said Reduction Co. agrees to use its good offices in the interest of said Chemical Co. so far as relates to promoting the trade of the latter in alum and alum products in the United States and in foreign countries; and said Chemical Co. reciprocally undertakes and agrees to use its good offices in the interest of said Reduction Co. so far as relates to promoting the metal business of the latter in the United States and in foreign countries —" in so far as it may be considered as an agreement upon the part of the General Chemical Co. to antagonize the interests of the competitors of the defendant company, be and it is hereby declared to be null and void, and that defendant and all its agents and representatives be, and they are, hereby perpetually enjoined from in any manner, and to any extent, requiring an enforcement of said provisions, and from entering into or acting in pursuance of any contract or agreement the purpose and effect of which would be to place any restraint upon the General Chemical Co. with reference to the right of said company to acquire and sell, or the quantity which it may acquire and sell, or the price at which it may acquire and sell any bauxite, alumina or aluminum of which it may become the owner by purchase, manufacture, or otherwise.

3. That the tenth and eighteenth sections of the contract entered into under date of April 20, 1909, between the defendant Aluminum Co. of America and the Norton Co., which sections read as follows to wit:

"Tenth. Norton Co. may mine and use bauxite from the said 40-acre tract of bauxite land referred to in paragraph D above, which shall be used for the purpose of manufacturing alundum, and may mine and sell from the said property bauxite or other mineral taken therefrom for any purpose except for the manufacture of aluminum, and Norton Co. shall not sell or otherwise dispose of said 40-acre tract except subject to the above restrictions.

"Eighteenth. Norton Co. shall not at any time during the continuance of this agreement use or sell any of the bauxite contained on the said 40-acre tract described in paragraph D above, or any other bauxite, or the products thereof, hereafter acquired by Norton Co., in the United States of America or the Dominion of Canada for the purpose of conversion into aluminum —" and all other parts of said contract, in so far as they restrain or seek to restrain the Norton Co. from exercising its free and independent will in using and disposing of the bauxite which it may receive under the provisions of said contract, or any other bauxite which it may obtain, be, and the same are hereby, declared null and void and are abrogated; and that the defendant, and its officers and agents, be perpetually enjoined from in any manner or to any extent enforcing or requiring recognition by the Norton Co. of such provisions, and from hereafter entering into any contract with said Norton Co., the purpose and effect of which would be to restrain said Norton Co. in the disposition of any bauxite which may be obtained from any source, or of any alumina or aluminum which it may manufacture from such bauxite, or may otherwise obtain.

4. That the following clause in a contract between defendant and the Pennsylvania Salt Manufacturing Co., to wit:

"The Pennsylvania Salt Manufacturing Co. agrees not to enter into the manufacture of aluminum as long as this agreement is in force — "

and the ratification and extension of said clause contained in a letter from the Pennsylvania Salt Manufacturing Co. to defendant, dated January 1, 1907, be, and the same are hereby, declared null and void; and that defendant Aluminum Co. of America and its officers and agents be, and they are hereby, perpetually enjoined hereafter from in any manner or to any extent enforcing or relying upon said clause and its ratification, and from entering into any contract with said Pennsylvania Salt Manufacturing Co., the purpose and effect of which would be to restrain said Pennsylvania Salt Manufacturing Co. from freely making any disposition that it may see proper, and at any price it may deem proper, of any bauxite, alumina, or aluminum the ownership of which it may acquire from any source.

5. That that part of the agreement entered into as of date November 16, 1910, by defendant Aluminum Co. of America and Gustave A. Kruttschnitt, of Newark, N. J., and James C. Coleman, of Newark, N. J., which provides that —

"As part consideration for the execution of this agreement by Aluminum Co., Kruttschnitt and Coleman hereby severally agree that for the period of 20 years from the date hereof, in that part of the United States east of a north and south line through Denver, Colo., neither Kruttschnitt nor Coleman will directly or indirectly engage or become interested in the manufacture or fabrication or sale of aluminum, or any article made substantially of aluminum, provided that either or both the said Kruttschnitt and Coleman may be employed by or become interested in the Aluminum Co. or said Aluminum Goods Manufacturing Co., without committing a breach of this contract — "

in so far as it constitutes a restraint upon said Kruttschnitt and Coleman from freely engaging in any part or branch of the aluminum business, be, and the same is hereby, declared to be null and void, and that the defendant and its officers, agents, and representatives be, and they are hereby, perpetually enjoined from entering into a contract with said Kruttschnitt or Coleman or with any other individual, firm, or corporation of a like or similar character to the above-quoted provisions of said contract, except as the same may be a lawful incident to the purchase of good will.

6. That the defendant and its officers, agents, and representatives be, and they are hereby, perpetually enjoined from entering into a contract with any other individual, firm, or corporation of a like or similar character to the above-quoted provisions in the contracts between the Aluminum Co. of America and the General Chemical Co., between said Aluminum Co. and the Norton Co., between said Aluminum Co. and the Pennsylvania Salt Manufacturing Co., and between said Aluminum Co. and Kruttschnitt and Coleman, or either of them, and from entering into or participating in any combination or agreement the purpose or effect of which is to restrict or control the output or the prices of aluminum or any material from which aluminum is directly or indirectly manufactured, and from making any contract or agreement for the purpose of or the effect of which would be to restrain commerce in bauxite, alumina, or aluminum, or to prevent any other person, firm, or corporation from or to hinder him or it in obtaining a supply of either bauxite, alumina, or aluminum of a good quality in the open market in free and fair and open competition, and from themselves entering into, or compelling or inducing, under any pretext, or in any manner whatsoever, the making of any contract between any persons, firms, or corporations engaged in any branch of the business of manufacturing aluminum goods the purpose or effect of which would be to fix or regulate the prices of any of their raw or manufactured products in sale or resale.

7. To prevent all undue discriminations upon the part of defendant and its officers and agents, or upon the part of any firm or corporation in whose business defendant owns or hereafter acquires a financial interest by stock ownership or otherwise, against any competitor of defendant and thus to prevent the unlawful acquisition by defendant of a monopoly in any branch of manufacturing from crude or semifinished aluminum, defendant and its officers, agents, and representatives, are hereby perpetually enjoined from committing the following acts, to wit:

(*a*) Combining either by stock ownership or otherwise with any one or more manufacturers for the purpose or with the effect of controlling or restraining the output of any product manufactured from aluminum, or fixing or controlling the price thereof.

(*b*) Delaying shipments of material to any competitor without reasonable notice and cause, or refusing to ship or ceasing to continue shipments of crude or semifinished aluminum to a competitor on contracts or orders placed, and particularly on partially filled orders, without any reasonable cause and without giving notice of same, or purposely delaying bills of lading on material shipped to any competitor, or in any other manner making it impossible or difficult for such competitor promptly to obtain the material upon its arrival, or from furnishing known defective material.

(*c*) Charging higher prices for crude or semifinished aluminum from any competitor than are charged at the same time under like or similar conditions from any of the companies in which defendant is financially interested, or charging or demanding higher prices for any kind of crude or semifinished aluminum from any competitor for the purpose or which under like or similar conditions will have the effect of discriminating against such manufacturers in bidding on proposals or contracts to the advantage of said defendant or any company in which it is financially interested.

(*d*) Refusing to sell crude or semifinished aluminum to prospective competitors in any branch of the manufacturing aluminum goods industry on like terms and conditions of sale, under like or similar circumstances, as defendant sells such crude or semifinished aluminum to any firm or corporation engaged in similar business in which defendant is financially interested.

(*e*) Requiring, as a condition precedent to selling crude or semifinished aluminum to a competitor, that such competitor divulge to defendant the terms which such competitor proposes to make in order to secure the work in which the desired aluminum is to be used, and from imparting to any one the purpose or purposes for which said competitor is intending to use said metal.

(*f*) Requiring or compelling the making of agreements by competitors not to engage in any line of business nor to supply any special order in competition with defendant or with any company in which it is financially interested as a condition precedent to the procurement of aluminum metal.

(*g*) Representing or intimating to competitors that unless they dealt with defendant or with companies in which defendant has a financial interest for their supply of metal such competitor will not be able to obtain a sufficient supply of metal or obtain it at a price that will permit them to engage in competition with defendant or with companies in which defendant is financially interested; or in like manner representing or intimating to consumers of aluminum in any stage of manufacture that unless they deal with defendant or with a company in which it is financially interested, their supply of material or manufactured products will be cut off for that reason.

(*h*) Taking the position with persons, firms, or corporations engaged in the manufacture of any kind of aluminum goods that if they attempt to enlarge or increase any of their industries or engage in enterprises that are or will be competitive with defendant or with the business of any firm or corporation in which defendant is financially interested, such persons, firms, or corporations will for that reason be unable to procure their supply of material from defendant or any of the companies in which it is financially interested.

The term "competitor," as used above, shall be construed to mean all persons, firms, or corporations engaged in or who are actually desiring or about to engage in the manufacture of any kind of products or goods from crude or semifinished aluminum, whose business is not controlled or not subject to be controlled by defendant, its officers and agents, either by virtue of ownership of all or a part of the capital stock of such concerns, or through any other form or device of financial interest.

Provided, however, That nothing contained in this decree shall be construed to prevent or restrain the lawful promotion of the aluminum industry in the United States.

Provided, further, that nothing herein contained shall obligate defendants to furnish crude aluminum to those who are not its regular customers, to the disadvantage of those who are, whenever the supply of crude aluminum is insufficient to enable defendant to furnish crude aluminum to all persons who desire to purchase from defendant, but this proviso shall not relieve defendant from its obligation to perform all its contract obligations, and neither shall this proviso, under the conditions of insufficient supply of crude aluminum referred to be or constitute a permission to defendant to supply such crude aluminum to its regular customers mentioned with the purpose and effect of enabling defendant or its regular customers, under such existing conditions, to take away the trade and contracts of competitors.

Provided, further, that nothing in this decree shall prevent defendant from making special prices and terms for the purpose of inducing the larger use of aluminum, either in a new use or as a substitute for other metals or materials.

Provided, further, that nothing in this decree shall prevent the acquisition by defendant of any monopoly lawfully included in any grant of patent right.

Provided, further, that the raising by defendant of prices on crude or semifinished aluminum to any company which it owns or controls or in which it has a financial interest, regardless of market conditions, and for the mere purpose of doing likewise to competitors while avoiding the appearance of discrimination, shall be a violation of the letter and spirit of this decree.

This decree having been agreed to and entered upon the assumption that the defendant, Aluminum Co. of America, has a substantial monopoly of the production and sale of aluminum in the United States, it is further provided that whenever it shall appear to the court that substantial competition has arisen, either in the production or sale of aluminum in the United States, and that this decree in any part thereof works substantial injustice to defendant, this decree may be modified upon petition to the court after notice and hearing on the merits, provided that such applications shall not be made oftener than once every three years.

It is further ordered that the defendants pay the cost of suit to be taxed.

JAMES M. YOUNG,
Judge.

APPENDIX E

UNITED STATES OF AMERICA
BEFORE FEDERAL TRADE COMMISSION

IN THE MATTER OF Aluminum Company of America, a corporation.	DOCKET NO. 1335

COMPLAINT

I

Acting in the public interest, pursuant to the provisions of an Act of Congress approved October 15, 1914 (the Clayton Act) entitled "An Act to supplement existing laws against unlawful restraints and monopolies, and for other purposes," the Federal Trade Commission charges that the Aluminum Company of America, a corporation, hereinafter referred to as respondent, has been and is violating the provisions of Section 2 of said Act, issues this complaint and states its charges in that respect, as follows:

PARAGRAPH ONE: Respondent, Aluminum Company of America, is a corporation organized, existing and doing business under and by virtue of the laws of the State of Pennsylvania, with its principal or executive offices in the City of Pittsburgh, in said State. The said respondent owns extensive bauxite deposits from which the aluminum ore is secured in Saline County, Arkansas, and in British and Dutch Guiana, South America, and also owns or has a controlling interest in bauxite deposits in France and Jugo-Slavia. Said respondent owns and operates crushing and drying apparatus in Saline County, Arkansas, a refining plant in East St. Louis, Illinois, reduction works where aluminum is made at Niagara Falls and Massena, New York, Maryville (Alcoa), Tennessee, and Badin, North Carolina; it owns and operates a wire and cable mill at Massena, N. Y.; a general fabricating plant at New Kensington, Penna.; a plant for the manufacture of aluminum bronze powder and aluminum foil at New Kensington,

Penna.; and rolling mills at Niagara Falls, N. Y., at Maryville (Alcoa), Tennessee, and at Edgewater, New Jersey. Respondent is the sole producer of virgin aluminum ingots in the United States and, since March, 1923, has produced over 95 per centum of the virgin sheet aluminum manufactured in the United States, the present sole competitor in this branch of the industry, the United States Smelting & Refining Company, of New Haven, Connecticut, producing not more than one per centum of said virgin sheet aluminum at higher prices for spot delivery. Respondent owns 36 per centum of the stock of the Aluminum Goods Manufacturing Company, the largest manufacturer of aluminum cooking utensils in the United States, and 100 per centum of the stock of the United States Aluminum Company, the second largest manufacturer of cooking utensils in the United States, these two companies producing not less than 65 per centum of the total output of said cooking utensils in the United States; 75 per centum of the stock in the American Body Company, which manufactures aluminum bodies for automobiles; 64 per centum of the stock of the Aluminum Manufactures, Inc., which company makes sand castings for automobile parts; 89 per centum of the stock in the Aluminum Die Castings Corporation; and 80 per centum of the stock of the Aluminum Screw Machine Products Company. Respondent owns a 50 per centum stock interest in the Norsk Aluminum Company of Norway, a one-third interest in Norske-Nitrid Company of Norway, and also is the sole owner of the Northern Aluminum Company, Ltd., of Canada, the only other manufacturer of virgin aluminum ingots in North America. The total holdings of respondent on December 1, 1922, comprised a 100 per centum stock ownership in 34 corporations, a greater than 50 per centum stock ownership in nine corporations, and a less than 50 per centum stock ownership in 17 corporations, engaged in various enterprises.

The said respondent, Aluminum Company of America, is now and has been for more than two years last past, engaged in the manufacture and sale in interstate commerce of pig aluminum ingots and aluminum ingots, aluminum sheet, tubing, moulding, wire, cable, foil and powder and, through affiliated subsidiaries and/or leased companies, is engaged in the manufacture and the sale in interstate commerce of fabricated aluminum products, and aluminum alloy products, in particular cooking utensils, aluminum sand castings, permanent mould castings and die castings, causing its aforesaid products, when so sold, to be transported from the place of manufacture in one State to purchasers thereof located in other States of the United States.

The sole sources of supply of aluminum metal required by foundries and/or manufactories engaged in the manufacture and the sale in interstate commerce in the United States of fabricated aluminum products, and/or products manufactured from aluminum alloy, in particular aluminum cooking utensils, aluminum automobile bodies, aluminum sand castings and permanent mould and die castings are (1) respondent, Aluminum Company of America, and its subsidiary, the Northern Aluminum Company, Ltd., of Canada, the estimated capacity for production of said companies annually, being about 175,000,000 pounds; (2) foreign companies engaged in the production of aluminum ingots and/or aluminum sheets, importations from which during the calendar year 1923 for companies other than respondent and its subsidiaries being about 28,000,000 pounds; and (3) domestic manufacturers of fabricated aluminum products who have for disposition scrap aluminum resulting from the aforesaid manufacturing. In recent years the supply of scrap aluminum available to foundries and manufactories in competition with respondent for use in remelting into secondary ingots and in the production of aluminum cooking utensils and aluminum castings has been very extensive. In the year 1922 the recovery of secondary aluminum as pig aluminum or in alloys amounted to slightly more than 32,000,000 pounds. Since that time and as a result of competitive practices of respondent of which complaint is made herein, practically all of this secondary aluminum has been removed from the market by respondent, for the purpose and/or effect of preventing its competitors from securing this secondary or scrap aluminum and in order to make respondent's monopoly of the aluminum raw material more certain and complete.

In the course and conduct of its said business the respondent was at all times hereinafter mentioned, and still is, in competition with other individuals, firms, partnerships and corporations likewise engaged in interstate commerce.

PARAGRAPH TWO: Respondent, Aluminum Company of America,

for more than two years last past, in the course and conduct of its business —

(a) Has adopted and maintained the practice of entering into contracts or agreements for the sale and is now selling and making contracts for the sale in interstate commerce of virgin sheet aluminum to manufacturing foundries at prices less than they have been and are selling said virgin sheet aluminum to jobbing foundries; and/or

(b) Has adopted and maintained the practice of entering into contracts or agreements for the sale, and are now selling and making contracts of sale with certain manufacturers of automobile bodies, of cooking utensils, and/or of other fabricated aluminum products, for the sale in interstate commerce to said manufacturers of virgin sheet aluminum at prices less than they have been and are selling said virgin sheet aluminum to other manufacturers of automobile bodies, of cooking utensils, and/or of other fabricated aluminum products, on the condition, agreement, understanding or contract that the said manufacturers to whom the lower selling price is made shall sell all the aluminum scrap resulting from their manufacturing operations to the Aluminum Company of America; and said discrimination in price between purchasers of virgin sheet aluminum engaged in the manufacture of automobile bodies, of cooking utensils and/or of other fabricated aluminum products by respondent, Aluminum Company of America, was not made on account of differences in the grade, quality or quantity of the commodities sold, nor did it make only due allowance for difference in the cost of sale or transportation, nor was it made in good faith to meet competition nor in the selection of customers in bona fide transactions.

The effect of such sales and/or contracts for sale, and agreements, conditions and understandings may be and is to substantially lessen competition and tends to create a monopoly.

PARAGRAPH THREE: The above alleged acts and things done by respondent, Aluminum Company of America, are all in violation of Section 2 of the Act of Congress entitled "An Act to supplement existing laws against unlawful restraints and monopolies, and for other purposes," approved October 15, 1914.

II

Acting further in the public interest, pursuant to the provisions of an Act of Congress approved September 26, 1914, entitled "An Act to create a Federal Trade Commission, to define its powers and duties, and for other purposes," the Federal Trade Commission charges that the Aluminum Company of America, a corporation, hereinafter referred to as respondent, has been and is using unfair methods of competition in commerce in violation of the provisions of Section 5 of the said Act, issues this complaint and states its charges in that respect as follows:

PARAGRAPH ONE: Proceeding in the public interest and as a further cause of action in violation of Section 5 of the above Act, the Commission charges and relies upon the matters and things set forth in Paragraphs One and Two under the First Count of this complaint to the same extent as though the allegations thereof were set out at length herein and the said Paragraphs One and Two of the First Count are incorporated herein by reference, and adopted as a part of the allegations of this Count.

PARAGRAPH TWO: Respondent, Aluminum Company of America, for more than two years last past has employed, and still employs, a scheme the effect of which was and is to gain a monopoly of the aluminum sand castings industry of the United States, and, in order to carry out said scheme, respondent has adopted, and used, and is now using the following practices, to-wit:

(a) It arbitrarily fixes a differential between the selling price of virgin aluminum ingots and the purchase price of scrap aluminum;
(b) It pays higher prices for scrap aluminum than it costs the respondent to manufacture virgin aluminum ingots;
(c) It makes concessions to automobile body manufacturers and/or to manufacturers of other fabricated aluminum products in the price of virgin sheet aluminum to said manufacturers upon the agreement, understanding, or contract that said manufacturers sell respondent their total

available supply of scrap aluminum at prices fixed by respondent approximating the actual cost of manufacture or at prices higher than it cost the respondent to manufacture the virgin aluminum ingots; and/or at prices higher than competing foundries engaged in manufacturing and jobbing fabricated aluminum products or aluminum castings or aluminum alloy products could pay for such metal and more than its intrinsic value when compared with virgin aluminum metal.

(d) It transfers virgin aluminum metal to its agents and/or its subsidiaries, at an arbitrary price below the cost of production and below the selling price of said metal to competitors of its said agents or subsidiaries engaged in the manufacture and the sale of aluminum sand castings;

(e) It makes sales of aluminum sand castings to manufacturers of automobiles or automobile products at prices approximating the actual cost of manufacture or at prices less than it cost the respondent to manufacture the aforesaid sand castings; and/or at prices less than competing foundries can sell aluminum sand castings at a profit taking into consideration the cost to the said competing foundries of virgin aluminum and scrap aluminum.

(f) The practices of respondent as set out in subparagraphs "(a)" to "(e)" of this paragraph, both inclusive, have been made and are being made with the purpose and/or effect of curtailing the supply of raw material used by independent and/or competing jobbing foundries or manufactories and of compelling said independent and/or jobbing foundries or manufactories to purchase virgin aluminum ingots and aluminum sheets from respondent at prices arbitrarily fixed by respondent and with the purpose and/or effect of eliminating as a source of supply for independent and/or competing jobbing foundries or manufactories the scrap aluminum theretofore available; in that the domestic source of supply of aluminum metal of the aforesaid independent and/or competing jobbing foundries or manufactories, with the exception of the aforesaid scrap aluminum, is limited to and dependent upon the supply obtainable from respondent; and that the effect of the aforesaid practices of respondent as herein set out has been and is to suppress competition and to tend to create a monopoly.

PARAGRAPH THREE: Respondent, Aluminum Company of America, for more than two years last past, in the course and conduct of its business as described in Paragraph One hereof, has employed and is still employing, a scheme the purpose and/or effect of which was and is to gain and maintain a monopoly of aluminum raw material, of aluminum ingots and sheets, of secondary aluminum, and of aluminum fabricated products and/or aluminum alloy products, throughout the United States, and, in order to carry out such scheme, respondent has adopted and used and is now using, the following practices, to-wit:

(a) It arbitrarily neglects or refuses to supply to manufacturers of aluminum goods and/or aluminum fabricated products, and/or aluminum castings the aluminum sheet metal or ingots required by said manufacturers, who are in competition with respondent or its subsidiaries.

(b) It arbitrarily fails to make shipment of aluminum ingots to its competitors or to the competitors of its subsidiaries at the time said products are ordered, and/or at the time specified for shipment;

(c) It arbitrarily makes deliveries of aluminum or aluminum ingots to its competitors or to the competitors of its subsidiaries in insufficient quantity and in amounts or quantities less than ordered;

(d) It makes deliveries of aluminum sheets, and/or aluminum ingots to its competitors or to the competitors of its subsidiaries of quality inferior to that required.

(e) The practices of respondent as set out in subparagraphs "(a)" to "(d)" of this paragraph, both inclusive, have been made and are being made for the purpose and/or effect of unfairly harassing the competitors of respondent or of respondent's subsidiaries and with the effect of suppressing competition between respondent and its competitors and of creating or tending to create or maintain a monopoly.

PARAGRAPH FOUR: The above alleged acts and things done by respondent, Aluminum Company of America, are all to the prejudice of the public and of respondent's competitors and constitute unfair methods of competition in commerce within the intent and meaning of Section 5 of an Act of Congress entitled "An Act to create a Federal Trade Commission, to define its powers and duties and for other purposes," approved September 26, 1914, and/or with the effect of suppressing competition and tending to create a monopoly.

WHEREFORE, THE PREMISES CONSIDERED, The Federal Trade Commission, on this 21st day of July, 1925, now here issues this its complaint against said respondent.

NOTICE

Notice is hereby given you, Aluminum Company of America, respondent herein, that the 9th day of September, 1925, at 10:30 o'clock in the forenoon, is hereby fixed as the time, and the offices of the Federal Trade Commission, in the City of Washington, D. C., as the place, when and where a hearing will be had on the charges set forth in this complaint, at which time and place you shall have the right, under said Act, to appear and show why an order should not be entered by said Commission requiring you to Cease and Desist from the violation of the law charged in this complaint.

IN WITNESS WHEREOF, The Federal Trade Commission has caused this complaint to be signed by its Secretary, and its official seal to be hereto affixed at Washington, D. C., this 21st day of July, 1925.

By the Commission:

Otis B. Johnson,
Secretary.

(SEAL)

L. A. R.

UNITED STATES OF AMERICA
BEFORE THE FEDERAL TRADE COMMISSION

IN THE MATTER OF
Aluminum Company of America, a corporation. } DOCKET NO. 1335

ANSWER OF THE ALUMINUM COMPANY OF AMERICA
TO THE COMPLAINT OF THE COMMISSION,
DATED THE 21ST DAY OF JULY, 1925.

And now, to wit, September 21st, 1925, comes the Aluminum Company of America, the respondent in this case, and makes the following answer to the complaint filed against it by the Federal Trade Commission, dated the 21st day of July, 1925.

FIRST. The respondent denies that any or all of the averments set forth in the complaint disclose any violation of law, or that the same, if true, would justify the making and issuing of any decree by the Commission against the respondent, and therefore prays that the complaint be dismissed.

SECOND. In answer to the averments and allegations contained in COMPLAINT I respondent avers:

1. In so far as PARAGRAPH ONE sets up the corporate organization of the respondent and the location of its principal office, the same is admitted. In so far as the complaint avers ownership of bauxite deposits, the same is admitted. Bauxite is the principal raw or natural material used in the production of the metal aluminum, and the bauxite owned by respondent is necessary in the reasonable conduct of respondent's business. It is true that the respondent, either directly or through subsidiary corporations, is the owner of bauxite deposits, refining plants, reduction works and fabricating plants as set forth in Paragraph One of Complaint I. It is also true that respondent is the sole manufacturer in the United States of aluminum ingots made from the ore (bauxite). It is not true that respondent has since March, 1923, produced over 95% of the virgin sheet manufactured in the United States, neither is the United States Smelting

& Refining Company at present respondent's sole competitor in the manufacture of such sheet aluminum.

The statements contained in said paragraph as to the ownership of stock by respondent in certain manufacturing companies whose names are given therein, are substantially correct. The United States Aluminum Company is and always has been a hundred per cent. subsidiary of the respondent. The Aluminum Manufactures, Inc. is not engaged in business; all its plants are leased to and operated by the United States Aluminum Company. In regard to the companies referred to whose names are not mentioned, those in which respondent owns one hundred per cent. stock are purely subsidiaries engaged in the holding of real estate or production of power or raw materials necessary and useful for the respondent in the conduct of its business or in the fabrication of the materials manufactured by respondent or in the transportation or sale of said raw materials or finished products. The other unnamed companies referred to in the complaint are engaged in business which has a direct relationship to the business conducted by the respondent and respondent's ownership in the stock thereof is lawful and useful and to a great degree consists of investments made by respondent in corporations engaged in the fabrication of materials manufactured by respondent entered into for the purpose of promoting the aluminum industry in the United States and to some degree represent temporary investments of surplus funds.

It is admitted that the respondent is engaged in interstate commerce.

It is admitted that the supply of aluminum metal in the United States consists of the primary aluminum manufactured by the respondent in the United States, the primary aluminum manufactured by the respondent and others outside of the United States, and scrap of various kinds, both from within and without the United States.

As to the annual production of ingots, sheets and scraps available in the United States, respondent has no precise knowledge. It varies from year to year, and if said facts are material to the present controversy, respondent demands proof of same.

It is not true that as a result of any practices, competitive or otherwise, practically all the secondary aluminum referred to in the said paragraph has been removed from the market by respondent, nor is it true that any purchases of secondary aluminum that the respondent may have made were made either for the purpose or had the effect of creating a monopoly or preventing any of its competitors from securing secondary aluminum, or to make respondent's alleged monopoly of aluminum raw materials more certain or complete.

It is admitted that in the course and conduct of its business the respondent was at all times and now is in competition with other individuals, firms, partnerships and corporations engaged in interstate commerce.

2. In answer to PARAGRAPH TWO of COMPLAINT I, the averments of subparagraph (a) are denied. The averments of subparagraphs (b) are denied. It is particularly denied that if it had been true (which it is not) that any such sales or practices as those referred to in this paragraph either had been in the past or were now in existence, such practices or sales, could, would or do have any effect upon competition or tend in any way to create a monopoly.

3. PARAGRAPH THREE of COMPLAINT I is denied.

THIRD. In answer to the averments and allegations contained in COMPLAINT II, respondent avers:

1. The same answer is made to PARAGRAPH ONE of COMPLAINT II as is made to Paragraphs One and Two of Complaint I, and with the same force and effect as though said answers were set forth herein at length.

2. The averments of PARAGRAPH TWO of COMPLAINT II as therein stated are denied; subject, however, to the following explanations:

Respondent admits that it sometimes pays higher prices for certain qualities of scrap aluminum than it costs the respondent to manufacture primary aluminum; in so far as respondent may transfer primary aluminum to subsidiaries at arbitrary prices, said conduct is purely one of bookkeeping and is a customary and convenient way of handling such transactions on the books of a parent company and its subsidiaries.

3. The allegations of PARAGRAPH THREE of COMPLAINT II are denied.

4. The allegations of PARAGRAPH FOUR of COMPLAINT II are denied.

FOURTH. In further answer to the complaint filed in this case, respondent avers that it engaged in the manufacture of aluminum in the year 1888 in accordance with the methods set forth in certain letters patent of the United States, which respondent lawfully acquired and respondent had, during the life of said patents, a lawful monopoly in the manufacture of aluminum by the methods set forth in said letters patent. The process set forth in said letters patent and so used by the respondent is the only method by which aluminum could be manufactured at a cost which made it possible to use the metal commercially. Although the last of the patents used by the respondent expired in the year 1907 the same processes are the methods still in use by the respondent and all other manufacturers of aluminum in the world. At the time when the respondent began the manufacture of aluminum there was no commercial market for aluminum, and it became necessary for the respondent, in the development of its business, to fabricate the metals into shapes in which it could be used and to induce the public to use it as a substitute for other metals. The respondent explicitly avers that since the expiration of said letters patent there has been nothing to prevent any person who so desired from engaging in the manufacture or fabrication of aluminum in the United States, and respondent never has done, neither has it attempted to do, anything which in any way prevented or embarrassed others from engaging in said business, but, on the contrary, has devoted its time to producing aluminum in quantities which the public needed and of the best quality that could be produced, and has encouraged and aided others to embark in the fabrication and use of articles in which aluminum is the sole or a constituent part. It has been its policy to give to consumers of aluminum the lowest possible prices, and respondent avers and charges that whatever complaints have been made as to respondent's prices and methods are largely, if not wholly, confined to complaints of middlemen who naturally have found their profits more or less interfered with by said policy of respondent.

WHEREFORE respondent prays that the complaint be dismissed.

ALUMINUM COMPANY OF AMERICA,

By Gordon, Smith, Buchanan & Scott

Its Attorneys.

UNITED STATES OF AMERICA
BEFORE FEDERAL TRADE COMMISSION

At a regular session of the Federal Trade Commission,
held at its office in the City of Washington, D. C.
on the 7th day of April, A. D., 1930

COMMISSIONERS:

Garland S. Ferguson, Jr., Chairman
C. W. Hunt,
William E. Humphrey,
Charles H. March,
Edgar A. McCulloch.

IN THE MATTER OF

Aluminum Company of America,
a corporation.

DOCKET NO. 1335

ORDER OF DISMISSAL

The above-entitled proceeding coming on for consideration by the Commission upon the complaint of the Commission, the answer of respondent, the record, briefs and oral argument of counsel for the Commission and for the respondent, and the Commission having duly considered same and being fully advised in the premises,

IT IS ORDERED that the complaint herein be and the same hereby is dismissed for the reason that the charges of the complaint are not sustained by the testimony and evidence.

By the Commission.

Otis B. Johnson
Secretary

(SEAL)

Sources of Data of Table 38 *

United States —1883–1903, *Mineral Resources;* 1904–1909, *Mineral Industry;* 1910–1912, given by A. W. Mellon, *Senate Judiciary Committee, Hearings on Aluminum Company of America,* p. 410; 1913–1919, given by Aluminum Company, *Hearings of Senate Committee investigating the Bureau of Internal Revenue,* Exhibit H; 1920–1924 given by Aluminum Company, *Federal Trade Commission,* Docket 1335; 1925–1931, given by Aluminum Company, Baush v. Aluminum Company, appellant, Exhibit 126; 1932–1935, *American Bureau of Metal Statistics.*

Canada — Canadian production is included in the figures for the United States until 1909. The figures for 1909–1913 represent Canadian exports; for 1914–1919, estimates of the *Mineral Industry.* These figures are probably overestimates, for this yearbook overestimated output in the United States in these years. For 1920–1924 and 1933–1935, *American Bureau of Metal Statistics.* For 1925–1932, estimates from a private source.

Europe — There seem to be no reliable government statistics of aluminum production except in Italy. Statistics have been given in government publications in England, France, and Norway, but there is reason to believe that they have often been incorrect. After some consideration of the statistics published by private institutions it was decided to use those of the *Metallgesellschaft* for all the countries of Europe. Until 1920 the production of Germany, Switzerland, and Austria appeared as one figure. Estimates of the output of Germany, 1916–1919, were taken from German press reports. The figures for Swiss output in these four years were corrected by subtracting the estimates of German production. Until 1920 the figures for Switzerland include the small production of Austria. Estimates of European output in 1935 are taken from the *American Bureau of Metal Statistics,* which reports the estimates of the *Metallgesellschaft* for Germany, Great Britain, Switzerland, and Austria.

* Pages 570–571.

APPENDIX F

TABLE 38

ESTIMATED WORLD PRODUCTION OF PRIMARY ALUMINUM BY COUNTRIES, 1890–1935 *

(Thousands of Metric Tons)

Year	United States	Canada	Total America	Total Europe	France	Germany	Switz-erland	Austria	Great Britain	Norway	Italy	Russia	Other Countries	Total World
1890	0.03		0.03	0.15	0.04		0.04		0.07					0.18
1891	0.07		0.07	0.26	0.04		0.17		0.05					0.33
1892	0.12		0.12	0.35	0.07		0.24		0.04					0.47
1893	0.15		0.15	0.58	0.14		0.44							0.73
1894	0.25		0.25	0.87	0.27		0.6							1.12
1895	0.42		0.42	1.0	0.36		0.65							1.42
1896	0.59		0.59	1.2	0.37		0.7		0.13					1.79
1897	2.0		2.0	1.58	0.47		0.8		0.31					3.58
1898	2.4		2.4	1.68	0.57		0.8		0.31					4.08
1899	2.9		2.9	3.15	1.0		1.6		0.55					6.05
1900	3.2		3.2	4.1	1.0		2.5		0.6					7.3
1901	3.2		3.2	4.3	1.2		2.5		0.6					7.5
1902	3.3		3.3	4.5	1.4		2.5		0.6					7.8
1903	3.4		3.4	4.8	1.6		2.5		0.7					8.2
1904	3.5		3.5	5.4	1.7		3.0		0.7					8.9
1905	5.1		5.1	7.0	3.0		3.0		1.0					12.1
1906	6.5		6.5	8.5	4.0		3.5		1.0					15.0
1907	11.8		11.8	11.8	6.0		4.0		1.8					23.6
1908	5.9		5.9	12.6	6.0		4.0		2.0		0.6			18.5
1909	6.8	2.8	9.6	15.2	6.0		5.0		2.8	0.6	0.8			24.8
1910	15.4	3.5	18.9	24.2	9.5		8.0		5.0	0.9	0.8			43.1

TABLE 38 — *Continued*

Year	United States	Canada	Total America	Total Europe	France	Germany	Switzerland	Austria	Great Britain	Norway	Italy	Russia	Other Countries	Total World
1911	16.8	2.3	19.1	24.7	10.0		8.0		5.0	0.9	0.8			43.8
1912	18.1	8.3	26.4	34.8	13.0		12.0		7.5	1.5	0.8			61.2
1913	21.5	5.9	27.4	36.4	14.5		12.0		7.6	1.5	0.8			63.8
1914	26.3	6.8	33.1	35.9	10.0		15.0		7.5	2.5	0.9			69.0
1915	41.1	8.5	49.6	28.3	6.0		12.0		7.1	2.3	0.9			77.9
1916	52.2	8.5	60.7	43.5	9.6	5.0	15.8		7.7	4.3	1.1			104.2
1917	58.9	11.8	70.7	53.5	11.1	10.3	15.7		7.1	7.6	1.7			124.2
1918	56.6	15.0	71.6	62.9	12.0	14.1	19.9		8.3	6.9	1.7			134.5
1919	58.3	15.0	73.3	59.4	15.0	11.2	20.3		8.1	3.1	1.7			132.7
1920	62.9	12.0	74.9	53.1	12.3	12.0	12.0	2.0	8.0	5.6	1.2			128.0
1921	24.7	8.0	32.7	43.1	8.4	11.0	12.0	2.0	5.0	4.0	0.7			75.8
1922	33.4	10.0	43.4	49.2	7.5	16.0	13.0	2.0	5.0	4.9	0.8			92.6
1923	58.4	10.0	68.4	70.6	14.3	17.0	15.0	1.5	8.0	13.3	1.5			139.0
1924	68.3	12.5	80.8	88.8	18.5	20.0	19.0	2.2	7.0	20.0	2.1			169.6
1925	63.5	15.0	78.5	104.1	20.0	27.2	21.0	3.0	9.7	21.3	1.9			182.6
1926	66.9	18.0	84.9	112.2	24.0	30.6	21.0	3.0	7.3	24.4	1.9			197.1
1927	74.2	36.0	110.2	108.6	25.0	28.4	20.0	4.0	7.9	20.8	2.5			218.8
1928	95.5	36.0	131.5	120.7	27.0	31.7	19.9	4.0	10.7	22.8	3.6		1.0	252.2
1929	103.4	31.0	134.4	132.7	29.0	32.7	20.7	4.0	13.9	24.4	7.0		1.0	267.1
1930	103.9	34.0	137.9	128.2	26.0	30.2	20.5	3.5	14.0	24.7	8.0		1.3	266.1
1931	80.5	31.0	111.5	107.5	18.0	26.9	11.4	3.3	14.2	21.4	11.1		1.2	219.0
1932	47.6	18.0	65.6	88.3	15.0	19.0	8.5	2.1	10.3	17.8	13.4	1.0	1.2	153.9
1933	38.6	16.2	54.8	86.3	14.3	18.3	7.5	2.0	11.0	15.5	12.1	4.4	1.2	141.1
1934	33.6	15.5	49.1	119.6	16.0	37.2	8.1	2.1	12.5	15.5	12.4	14.4	1.4	168.7
1935	54.1	20.6	74.7	178.9	21.8	70.7	11.7	2.1	15.1	16.0	14.0	24.5	7.0 †	257.6 †

* For sources of data, see p. 569.

† Includes 4,000 tons produced in Japan. No information is available concerning output in Hungary.

TABLE 39

ESTIMATED PRODUCTION AND FOREIGN TRADE IN ALUMINUM OF THE UNITED STATES, 1900–1935 *

(*Metric Tons*)

Year	Production of Primary Aluminum Ingot	Production of Secondary Aluminum Ingot	Total Production of Aluminum Ingot	Imports of Aluminum Ingot	Exports of Aluminum Ingot	Exports of Fabricated Aluminum
1900	3,240			116		
1901	3,240			256		
1902	3,310			338		
1903	3,400			226		
1904	3,490			234		
1905	5,150			241		
1906	6,510			350		
1907	11,790			396		
1908	5,900			211		
1909	6,800			2,318		
1910	15,420			5,566		
1911	16,780			1,893 †		
1912	18,140			10,324		
1913	21,450			10,517		
1914	26,300	4,110	30,410	7,367		
1915	41,050	7,730	48,780	3,871		
1916	52,210	17,550	69,760	3,015		
1917	58,890	14,640	73,530	27		
1918	56,580	13,680	70,260	766	9,141	1,806 ‡
1919	58,270	17,000	75,270	8,003	2,022	255 ‡
1920	62,890	14,090	76,980	18,178	3,348	919 ‡
1921	24,740	8,090	32,830	13,870	477	519 ‡
1922	33,400	14,820	48,220	18,122	698	3,342 §
1923	58,360	19,360	77,720	19,534	531	4,433 §
1924	68,280	24,550	92,830	13,333	1,523	4,432 §
1925	63,550	40,000	103,550	19,690	3,688	4,978 §
1926	66,850	40,180	107,030	33,965	266	4,156 ‖
1927	74,200	42,000	116,200	32,744	1,599	5,682 ‖
1928	95,500	43,450	138,950	17,189	1,084	6,505 ‖
1929	103,400	44,000	147,400	21,961	278	8,566 ‖
1930	103,900	35,090	138,990	11,112	276	8,431 ‖
1931	80,530	27,550	108,080	6,261	688	2,033 ‖
1932	47,600	21,820	69,420	3,631	1,771	622 ‖
1933	38,600	30,460	69,060	7,580	2,501	294 ‖
1934	33,646	42,180	75,826	8,333	3,653	375 ‖
1935	54,113	46,730	100,843	9,560	1,525	

* Sources of statistics of production of primary aluminum are given in the note to Table 38. Estimates of the recovery of secondary aluminum have been published in *Mineral Resources* (now *Minerals Yearbook*) since 1914. The figures underestimate the actual recovery because some firms do not report. It is believed that coverage has become increasingly broader in the last decade. Import statistics are general imports of aluminum in crude form and alloys, including scrap, for calendar years. (Figures for calendar years during the period 1900–1917 are given in the *Mineral Industry*.) The figures for 1934 and 1935 are imports for consumption. Before 1918 exports were reported by value only. The total quantity of fabricated aluminum exported was reported only for the four years 1922–1925. The classes covered in other years are indicated in the footnotes. The figures do not include small amounts of foreign metal exported from the United States. Imports of fabricated aluminum have rarely exceeded 500 tons.

† July–December only.

‡ Plates, sheet, bar, etc. Does not include tubes, castings, utensils, and other manufactures.

§ Total exports of semifinished and manufactured aluminum products.

|| Plates, sheet, bar, etc. and tubes, molding, and castings. Does not include utensils and other manufactures.

BIBLIOGRAPHY

BIBLIOGRAPHY

THIS bibliography includes literature on the aluminum industry and a few books and articles on economic theory or its application. The larger part of it consists of those sources of information about the aluminum industry which have been useful for this study. The best known government publications, such as annual statistical abstracts and statistics of foreign trade, have been omitted. Only a small part of the vast literature on the technology of the industry has been included. No attempt has been made to present an exhaustive list of books, articles, and government publications treating the industrial history and economics of this industry. I believe, however, that I have discovered most of the substantial works of this sort in English and German. Owing to the inadequate indexing of French economic and business literature, which is particularly manifest in the lack of an index of periodical material, I may have failed to discover some sources which would have been helpful. The paucity of cross references in French articles and books suggests, however, that the number of works of this sort in French is quite limited.

The short list of books and articles on economic theory or its application is included with a twofold purpose. It will afford the economist familiar with this literature an indication of some of the principal influences which have led me to the particular formulation of the problems and the type of analysis appearing in this book. It may also serve as a useful reference list for any lay reader whose interest may be attracted because of or in spite of inability to understand those portions of this book where the technical apparatus of economic theory is most obtrusive.

I. THE ALUMINUM INDUSTRY

BOOKS AND ARTICLES

Technology

Allmand, A. J., *The Principles of Applied Electrochemistry* (New York, 1912).

Allmand, A. J., and Ellingham, H. J. T., *The Principles of Applied Electrochemistry* (revised and enlarged ed.; New York, 1924).

"Aluminum Alloy Progress," *Iron Age,* CXXVI, 1455 (1930).
Aluminum Company of America, *Strong Aluminum Alloys* (Pittsburgh, 1928).
Anderson, R. J., *The Metallurgy of Aluminium and Aluminium Alloys* (New York, 1925).
——, *Secondary Aluminum* (Cleveland, 1931).
Ashcroft, E. A., *A Study of Electrothermal and Electrolytic Industries* (New York, 1909).
Barrows, H. K., *Water Power Engineering* (New York. 1927).
Berg, Hans, *Aluminium und Aluminiumlegierungen* (Frankfurt a.M., 1924).
Blount, Bertram, *Practical Electro-Chemistry* (New York, 1903).
Edwards, J. D., Frary, F. C., and Jeffries, Zay, *The Aluminum Industry,* 2 vols. (New York, 1930).
Fox, Cyril S., *Bauxite and Aluminous Laterite* (2nd ed.; London, 1932).
Gregory, R. A., *Discovery, or the Spirit and Service of Science* (London, 1916).
Higgs, Paget, ed., *Magneto- and Dynamo-Electric Machines* (London, 1884).
Howe, H. E., ed., *Chemistry in Industry* (New York, 1924).
Justin, J. D., and Mervine, W. G., *Power Supply Economics* (New York, 1934).
Kershaw, J. B. C., *Electro-Metallurgy* (London, 1908).
Mastick, S. C., "Lectures on Patents," *Industrial and Engineering Chemistry,* VII, 789, 874, 984, 1071 (1915).
Melchior, Paul, *Aluminium — die Leichtmetalle und ihre Legierungen* (Berlin, 1929).
Mellor, J. W., *A Comprehensive Treatise on Inorganic and Theoretical Chemistry,* vol. I (London, 1924), chap. V.
Pannell, E. V., *High Tension Line Practice* (London, 1925).
Perrine, F. A. C., *Conductors for Electrical Distribution* (2nd ed.; New York, 1907).
Pring, J. N., *The Electric Furnace* (London, 1921).
Py, Gaëtan, *Progrès de la métallurgie et leur influence sur l'aéronautique* (Paris, 1928).
Richards, J. W., *Aluminium: Its History, Occurrence, Properties, Metallurgy, and Applications, Including its Alloys* (3rd ed., rev.; Philadelphia, 1896).
——, "Conditions of Progress in Electrochemistry," presidential address, American Electrochemical Society, April 18, 1905.
Rosenhain, Walter, "Some Steps in Metallurgical Progress, 1908–1933," *Engineering,* CXXXVI, 725 (1933).
Schulz, Bruno, *Das Aluminium, seine Herstellung, Eigenschaften und Verwendung* (Berlin, 1926).
Slosson, E. E., *Creative Chemistry* (New York, 1919).
"Twenty Years in the Metal Industries," *Metal Industry,* XXI, 1 (1923).
Whyte, A. G., *The Electrical Industry* (London, 1904).
Wright, J., *Electric Furnaces and their Industrial Application* (new ed.; London, 1910).
Zeerleder, A. von, *Das Aluminium und seine Legierungen* (Zurich, 1927).

Industrial History and Economics

"The Aluminum Company of America," *Fortune,* March 1930.

"The Aluminum Company of America," *Fortune,* September 1934.

"Der Aluminiumzoll als Waffe im Internationalen Quotenkampf," *Magazin der Wirtschaft,* V, 1728 (1929); and other articles in this journal in 1929.

Anderson, R. J., "The Aluminum Industry," chap. I in *Representative Industries in the United States* (New York, 1928), ed. by H. T. Warshow.

Bannert, Hans, *Der Rohaluminiumweltmarkt und die deutsche Rohaluminiumindustrie* (Halle, 1927).

Barut, Victor, *L'Industrie de l'électro-chimie et de l'électro-métallurgie en France* (Paris, 1924).

Benni, A. S., and others, *Review of the Economic Aspects of Several International Industrial Agreements* (League of Nations, Industrial and Financial Section, Geneva, 1930).

Buschlinger, Heinrich, "Entwicklung und Aufbau der Aluminiumwirtschaft" (unpublished dissertation, Hamburg, 1924).

Chiati, M. B., *Les Ententes industrielles internationales* (Paris, 1928).

Clark, V. S., *History of Manufactures in the United States,* vols. II, III (New York, 1929).

Collins, H. C., and Loudon, D., "The Aluminum Industry," *Harvard Business Review,* VIII, 1 (1929).

Costa, J. L., *Le Rôle économique des unions internationales de producteurs* (Paris, 1932).

Czimatis, Albrecht, *Rohstoffprobleme der deutschen Aluminiumindustrie im Rahmen ihrer wirtschaftlichen Entwicklung* (Dresden, 1930).

Debar, Rudolph, *Die Aluminiumindustrie* (Braunschweig, 1925).

Dejean, Pierre, *Petite Histoire de la métallurgie dans le Sud-est* (Grenoble, 1929).

Dux, C., *Die Aluminium-Industrie-Aktiengesellschaft Neuhausen und ihre Konkurrenz-Gesellschaften* (Lucerne, 1913).

Ertel, Erich, *Internationale Kartelle und Konzerne der Industrie* (Stuttgart, 1930).

Escard, Jean, *L'Aluminium dans l'industrie* (2nd ed.; Paris, 1925).

Gautschi, Alfred, *Die Aluminiumindustrie* (Zurich, 1925).

Guillet, Léon, *L'Évolution de la métallurgie* (Paris, 1928).

Günther, Georg, *Die deutsche Rohaluminiumindustrie* (Leipzig, 1931).

Kossmann, Wilfried, *Über die wirtschaftliche Entwicklung der Aluminiumindustrie* (Frankfurt a.M., 1911).

Kupczyk, Edwin, "Zur Lage der Aluminiumindustrie," *Wirtschaftsdienst,* XVI, Heft 7, 281 (1931).

Marcus, Alfred, *Grundlagen der modernen Metallwirtschaft* (Berlin, 1928).

Minet, Adolphe, *The Production of Aluminum and its Industrial Use* (1st ed., rev.; New York, 1905), trans. by Leonard Waldo.

Pannell, E. V., "Aluminum: Its Present and Future Status," *Metal Industry,* XXVII, 72 (1929).

"Pittsburgh and the Pittsburgh Spirit." Addresses before Chamber of Commerce of Pittsburgh, 1927–1928.
Rousiers, Paul de, *Les Industries chimiques* (Paris, 1928), vol. V of *Les Grandes Industries modernes.*
Schoenebeck, *Das Aluminiumzollproblem* (Berlin, 1929).
Schulthess, M. von, "Die Entwicklung der Aluminiumindustrie in der Schweiz und ihre Beziehungen zur Wasserkraftnutzung," *Schweizerische Technische Zeitschrift,* Jahrgang 1926, no. 34/35.
Schwarzmann, R., "Die internationale Verflechtung der schweizerischen Aluminiumindustrie," *Weltwirtschaftliches Archiv,* XXXV, 585 (1932).
Sonnenberger, Erna, *Wirtschaftliche Bedeutung der Aluminium Herstellung in Deutschland* (Dissertation, Cologne, 1925).
Spurr, J. E., *Political and Commercial Geology and the World's Mineral Resources* (New York, 1920).
Tiedmann, H., "Die Standortsfrage in der Aluminiumindustrie," *Metallwirtschaft,* XIII, 36 (1934).

PERIODICALS

Scientific, Trade, and Financial

United States

Automotive Industries
Boston News Bureau
Cassier's Magazine
Chemical and Metallurgical Engineering (before 1902, *Electrochemical Industry*)
Commercial and Financial Chronicle
Engineering and Mining Journal
Engineering News
Foundry
Industrial and Engineering Chemistry
Iron Age
Journal of the Franklin Institute
Manufacturers' Record
Metal Industry (New York; before 1902, *Aluminum World*)
Mining and Metallurgy
Power
Wall Street Journal

Europe

Aluminium (Berlin)
Deutsche Oekonomist (Berlin)
Economist (London)
Engineering (London)
Hauszeitschrift der Vereinigte Aluminiumwerke und der Erftwerk A. G. für Aluminium (Berlin)
Journal du four électrique (Paris)

Kartellrundschau (Berlin)
Magazin der Wirtschaft (Leipzig)
Metall und Erz (Halle)
Metallwirtschaft (Berlin)
Mining Journal (London)
Nachrichten des Vereins deutscher Ingenieure (Berlin)
Revue de l'aluminium (Paris)
Weltwirtschaftliches Archiv (Jena)
Wirtschaftsdienst (Hamburg)

Miscellaneous

Aluminiumindustrie A. G., Neuhausen, Switzerland. Annual Reports.
Aluminium Limited, Toronto. Annual Reports.
Aluminum Company of America, Pittsburgh. Annual Reports.
American Bureau of Metal Statistics. Yearbook.
American Metal Market. Daily.
Les Assemblées générales.
The British Guiana Handbook.
Compagnie de Produits Chimiques et Électrométallurgiques, Alais, Froges, et Camargues, Lyon. Annual Reports.
Handbuch der deutschen Aktiengesellschaften.
Metallgesellschaft, *Statistische Zusammenstellungen über Aluminium, Blei, Kupfer, Nickel, Quecksilber, Zink und Zinn* (Frankfurt a.M.). Annual.
Metal Statistics (New York). Annual.
Minerais et Métaux, Société Anonyme, *Renseignements statistiques concernant les métaux cuivre, plomb, zinc, aluminium, argent, or* (Paris). Annual.
The Mineral Industry . . . in the United States (New York). Annual.
Moody's Manual.
Poor's Cumulative Index.
Vereinigte Aluminiumwerke A.G., Berlin. Annual Reports.
Vereinigte Industrie-Unternehmungen A.G., Berlin. Annual Reports.
Yearbook of the Bermudas, Bahamas, British Guiana, Etc.

GOVERNMENT DOCUMENTS

Canada
Department of Mines:
Economic Minerals and Mining Industries of Canada (Ottawa, 1913).
Dominion Bureau of Statistics:
Mineral Production of Canada. Annual Reports.
France
Ministère des Travaux Publics:
Statistique de l'industrie minérale. Annual.

Germany

Ausschuss zur Untersuchung der Erzeugungs- und Absatzbedingungen der deutschen Wirtschaft:

Die Versorgung der deutschen Wirtschaft mit Nicht-Eisen Metallen (Berlin, 1931).

Preussische Geologische Landesanstalt:

Weltmontanstatistik (2 vols., Stuttgart, 1929, 1932).

Great Britain

British Guiana, Combined Court:

Report on the Condition of the Colony of British Guiana during the Great European War (Georgetown, Demerara, 1919).

Report of the Land and Mines Department. Annual.

Imperial Institute:

Bauxite and Aluminium (London, 1925), by W. G. Rumbold.

Bulletin. Quarterly.

The Mineral Industry of the British Empire and Foreign Countries. Annual.

Switzerland

Statistisches Bureau des eidgenössichen Departments des Innern:

Betriebszählung vom 9 August 1905.

Schweizerische Fabrikstatistik vom 26 September 1923; Schweizerische Statistische Mitteilungen, VI Jahrgang, 1924, 6 Heft.

Betriebszählung vom 22 August 1929.

Gewerbebetrieben in den Kantonen; Statistische Quellenwerke der Schweiz, Heft 15, 1929.

United States

Congress

House

Committee on Military Affairs:

Hearings on the Tennessee Valley Authority, 74th Congress, 1st Session.

Committee on Ways and Means:

Tariff Hearings, 60th Congress, 2nd Session; House Document no. 1505, Schedule C., 1908–1909.

Tariff Hearings, 62nd Congress, 3rd Session; House Document no. 1447, vol. II, Schedule C., 1913.

Tariff Hearings, 67th Congress, 2nd Session, Part II, Schedule C., 1921.

Comparison of Tariff Acts of 1909, 1913, and 1922 (revised to June 1, 1924) prepared for Committee on Ways and Means under the direction of Clayton F. Moore (Washington, 1924).

Senate

Committee on Finance:

Replies to Tariff Inquiries, 53rd Congress, 2nd Session; Report no. 423, Bulletin no. 13.

Tariff Hearings, 62nd Congress, 2nd Session; Senate Document no. 5, Schedule C., 1921–1922.

Committee on Judiciary:

Hearings, 69th Congress, 1st Session, pursuant to Senate Resolution 109 directing an inquiry by the Committee on Judiciary as to whether due expedition has been observed by the Department of Justice in prosecuting the inquiry in the matter of the Aluminum Company of America (1926).

Report of Committee on Judiciary pursuant to Senate Resolution 109, Senate Report no. 177. February 15, 1926.

Select Committee on Investigation of Bureau of Internal Revenue:

Hearings, Part 10, 68th Congress, 2nd Session.

Department of Commerce

Bureau of Foreign and Domestic Commerce:

Commerce Reports.

Mineral Raw Materials (Washington, 1929), by J. W. Furness and L. M. Jones. Trade Promotion Series, no. 76.

Representative International Cartels, Combines and Trusts (Washington, 1929), by W. F. Notz. Trade Promotion Series, no. 81.

Bureau of Mines:

Mineral Resources of the United States, 1882–1931. (Published by the United States Geological Survey until 1925.)

Minerals Yearbook, 1932–1936. A continuation of *Mineral Resources.*

Department of the Interior

United States Geological Survey:

Political and Commercial Control of Mineral Resources of the World, no. 19, *Bauxite and Aluminum,* by J. M. Hill.

Department of Justice

Aluminum Company of America, Report of Special Assistant to the Attorney General, William R. Benham, concerning alleged violations by the Aluminum Company of America of the decree entered against it in the United States District Court of the western district of Pennsylvania on June 7, 1912 (February 22, 1926). Senate Document no. 67, 69th Congress, 1st Session, 1926.

Department of the Treasury

Tariff of 1897 on Imports into the United States to 1899; Document no. 2099.

Federal Trade Commission

Annual Reports.

Docket 248. Federal Trade Commission v. Aluminum Company of America. Complaint, answer, briefs, findings of fact, order to divest stock.

Docket 1335. In the Matter of the Aluminum Company of America. Complaint, answer, briefs, order, record of testimony, exhibits.

Report of the Federal Trade Commission on House Furnishings In-

dustries, vol. III, *Kitchen Furnishings and Domestic Appliances, October 6, 1924* (Washington, 1925).

National Recovery Administration

Report on the Aluminum Industry, submitted by Leon Henderson, Director of Research and Planning Division (1935).

Tariff Commission

Digest of Tariff Hearings before the Committee on Finance, United States Senate (1922)

Summary of Tariff Information . . . on Tariff Act of 1922, Schedule 3 (1929).

Tariff Information Survey, C–16, Aluminum, Magnesium, etc. (1921).

United States Court Papers

Aluminum Company of America v. Federal Trade Commission. 284 Fed. Rep. 401 (1922).

Petitioner's Brief, Petitioner's Reply Brief.

Re application by Federal Trade Commission for injunction and modification of order. 299 Fed. Rep. 361 (1924).

Brief on Behalf of Aluminum Company of America.

Reply of Aluminum Company of America.

Petitioner's Memorandum of Testimony on Special Points.

Baush Machine Tool Company, appellant v. Aluminum Company of America. 72 Fed. Rep. (2) 236 (1934).

Record of Pleadings, Testimony, Exhibits, Charge, and Exceptions.

Baush Machine Tool Company v. Aluminum Company of America, appellant. 79 Fed. Rep. (2) 217 (1935).

Record of Pleadings, Testimony, Exhibits, Charge, and Exceptions.

Haskell v. Perkins et al. 31 Fed. Rep. (2) 54 (1929).

Record of Pleadings, Testimony, Exhibits, Charge, and Exceptions.

United States v. Aluminum Company of America. United States District Court, western district of Pennsylvania, session of 1912.

Petition and Decree.

War Industries Board

History of Prices during the War; Prices of Ferroalloys, Nonferrous and Rare Metals (1919), Price Bulletin no. 34, by H. R. Aldrich and Jacob Schmuckler.

II. Economic Theory and Method

Beckerath, Herbert von, *Modern Industrial Organization* (New York, 1933).

Chamberlin, E. H., *The Theory of Monopolistic Competition* (Cambridge, 1933).

——, "Duopoly: Value Where Sellers are Few," *Quarterly Journal of Economics,* XLIV, 63 (1929).

Clark, J. M., *Studies in the Economics of Overhead Costs* (Chicago, 1923).

Crum, W. L., *Corporate Earning Power* (Stanford University, 1929).

Epstein, R. C., *Industrial Profits in the United States* (New York, 1934).

Harrod, R. F., "The Law of Decreasing Costs," *Economic Journal,* XLI, 566 (1931).

——, "Doctrines of Imperfect Competition," *Quarterly Journal of Economics,* XLVIII, 442 (1934).

Jewkes, John, "Factors in Industrial Integration," *Quarterly Journal of Economics,* XLIV, 621 (1930).

Lederer, Emil, "Monopole und Konjunktur," *Vierteljahrshefte zur Konjunkturforschung,* 2. Jahrgang (1927), Ergänzungsheft 2.

Pigou, A. C., *The Economics of Welfare* (3rd ed.; London, 1929).

Robertson, D. H., *The Control of Industry* (New York, 1923).

Robinson, Joan, *The Economics of Imperfect Competition* (London, 1933).

——, "What is Perfect Competition?" *Quarterly Journal of Economics,* XLIX, 104 (1934).

Schneider, Erich, *Reine Theorie monopolistischer Wirtschaftsformen* (Tübingen, 1932).

Shove, G. F., "The Representative Firm and Increasing Returns," *Economic Journal,* XL, 94 (1930).

Taussig, F. W., *Principles of Economics* (3rd ed.; New York, 1921).

Tschierschky, Siegfried, *Kartellpolitik* (Berlin, 1930).

Viner, Jacob, "Cost Curves and Supply Curves," *Zeitschrift für Nationalökonomie,* III, 26 (1931).

Wolfers, Arnold, *Das Kartellproblem im Lichte der deutschen Kartell-literatur* (Munich, Leipzig, 1931).

Zeuthen, Frederik, *Problems of Monopoly and Economic Warfare* (London, 1930).

INDEX

INDEX

Acetylene, 514–515
Acheson Graphite Company, 131
Advertising, 198, 211, 249, 256–257, 259
Agreements, 39, 93, 117, 125, 157–158, 163, 287, 299–300, 350–352, 355, 363; *see also* Cartels
Aircraft, consumption of aluminum, 45; influence on alloy development, 45–47; use of aluminum and aluminum alloys, 45–46, 50–52, 55, 57–59, 61–66, 146
Alclad, 53
Alcoa Power Company, 73, 74, 77, 234, 299
Allgemeine Elektrizitätsgesellschaft, *see* Deutsche Edison Gesellschaft
Alliance Aluminium Compagnie, *see* Cartels
Alloys, aluminum, casting, 48, 51–52, 54–58; development, 255–259; duralumin, 50–54, 219, 256–257, 375, 384, 389; early ignorance concerning, 9, 11; heat treatment, 22, 49–50, 54–55, 256; high strength, 48–54, 215, 256, 375–376, 390–393; influence of automobile on development, 19–23; influence of aviation on development, 45–47; influence of world war on development, 45–47; increasing multiplicity, 48–60; price discrimination, 218, 222, 390, 393; production in electric furnace, 509–513; research, *see* Research; wrought, 48–52;
types: copper, 5, 19–22, 48; manganese, 49, 55, 257, 375; silicon, 55, 256; zinc, 19–22, 48;
trade names: Aldrey, 51; Almasilium, 51; Almélec, 51; Alpax, 56; Aludur, 51; Anticorodal, 54; Duralumin, 22, 50; Hyblum, 51, 52, 386; J. L., 51; K. S. Seewasser, 54; Lautal, 51; Silumin, 56, 67; Y, 55, 257;
alloys of Aluminum Company of America: *4S*, 52; *17S*, 52, 384–385, 389; *C17S*, 52; *25S*, 50–51, 384; *51S*, 51, 384; *52S*, 52; *122*, 56; *132*, 57; *195*, 55; *196*, 55; *427*, 52
Alumilite process, 53
Alumina, 7
Alumina plants,
Canada: Arvida, 70, 74; France: Gardannes, 34; La Barasse, 121; Marseilles, 34; Salindres, 35, 193; Germany: Lauta, 84; Silesia, 34; Trotha, 34; Ireland: Larne, 36; Italy: Marghera, 92; United States: East St. Louis, 25, 194–195
Aluminio Español, S. A., 92, 96–97
Aluminium Company of Canada, 74, 96; *see also* Northern Aluminum Company
Aluminium Corporation, 89, 120–121, 125, 158, 164, 269
Aluminium Erz Bergbau und Industrie A. G., 84
Aluminium Industrie A. G. (AIAG), formation, 6, 513; acquisition of bauxite, 34, 119, 138; capacity, 33–34, 37, 40, 86–87, 97, 123, 292, 305; horizontal expansion, 33–34, 37, 265, 292; integration, 34, 86; foreign properties, 34, 86, 92, 97, 292; new markets, 291; participation in cartels, 36, 93–94, 119, 125, 158, 163; agreement with Northern Aluminum Company, 39, 547–548; products other than aluminum, 34;
finance: investment, 278, 282–283; capitalization, 266; earnings, 34, 266–267, 278, 282–283, 285, 319, 323; dividends, 34, 267; depreciation, 266–267, 282, 323; reinvestment, 34, 266; issue of preference shares to prevent outside control, 90–91;
plants: alumina, 34; power, 33–34, 37, 40; reduction, 40, 84, 97; rolling, 197
Aluminium Limited, formation, 74; ownership of bauxite, 70–71; capacity, 74, 96; foreign markets, 163, 168, 306–312; participation in cartel, 77, 94, 163; relations with Aluminum Com-

pany of America, 74–76; sales, 324; plants, 92, 96;
finance: investment, 324; capitalization, 75; earnings, 324; depreciation, 324; ownership of stock, 75–76
Aluminiumwerke, G.m.b.H., Bitterfeld, 84, 97, 153, 158
Aluminiumwerke Manfred Weiss A.G., 93, 97, 153
Aluminiumwerke Steeg, 97
Aluminum, discovery, 3, 505; relative abundance, 3; influence on development of electrochemistry, 508–516, 519–526; position among nonferrous metals, 46; physical properties, 11, 15; purity, 182, 216
Aluminum Castings Company, 27, 79–80, 396, 443–444
Aluminum Colors, Inc., 53
Aluminum Company of America, formation, 5, 529; acquisition of bauxite, 28, 69–72, 76, 103–110, 129–131, 139; acquisition of water power, 25–29, 72–74, 76, 110–111, 116, 136; capacity, 28, 41, 43, 77–78, 96, 110–111, 117, 247–248; horizontal expansion, 24–29, 44, 69–74, 77–81, 110–111, 117, 149–150; integration, 24–27, 185–186; acquisition of other companies, 26–27, 72–73, 79, 115–116, 131, 133, 136; merger with Canadian Manufacturing and Development Company, 73, 135–136; foreign properties, 69–74, 92, 129–131, 136, 248; relations with Aluminium Limited, 74–76, 481–483; costs, 143, 248–250; investment and price policy, 111–114, 238–263; price and output policy in depression, 316–320, 324–330; rationalization, 263; progressiveness, 50–54, 56–57, 59, 63, 256–257; sales data, 21, 251–252, 325, 470–471; reservation of home market, 39, 548; position in utensil branch, 397, 408–409; position in castings branch, 396–397, 443–444; scrap purchases, 248, 444–462, 473, 560–561; government investigation, 369–370, 398–402, 440; antitrust suits, 371–374, 384, 386, 398, 401, 480–484, 547–555 (*see also* Consent decree); complaint by Federal Trade Commission, 370, 401, 411–412, 444–447, 556–563, 568; answer to complaint, 564–567;
finance: investment, 28, 30–31, 69, 117, 225–237, 249–252, 258–263, 328–329, 538–546; earnings, 29–31, 43, 102, 112, 225–235, 249–252, 258–263, 319, 328–329, 389–391, 471, 538–546; capitalization, 30–31, 262, 540–541; dividends, 30–31, 261; depreciation, 227, 328; reinvestment of earnings, 30–31, 81, 111, 261–262, 540–543; ownership of stock, 75–76;
plants: alumina, 25, 194–195; power, 25–29, 41, 44, 73–74, 77–78, 149, 213, 299; reduction, 12, 25–28, 44, 72, 74, 77, 96, 111, 149; electrode, 25–29, 129; rolling, 11, 79, 196–197, 433; blooming, 54, 79, 197; foundries, 11, 27, 79–80, 396, 443; fabricating plants, 11, 13, 26–27, 79, 197; research laboratory, 80
Aluminum Company of America v. *Federal Trade Commission,* 370, 373–374
Aluminum Cooking Utensil Company, 24, 408, 414–415, 422–425, 430
Aluminum foil, 18, 67
Aluminum Goods Manufacturing Company, 370, 397, 408, 413–415, 417, 420–426, 431–437, 551
Aluminum Manufactures, Inc., 79–80, 149, 401, 444, 446, 470
Aluminum Ore Company, 79
Aluminum oxide, *see* Alumina
Aluminum paint, 18
Aluminum Products Company, 437
Aluminum Rolling Mill Company, 372–374
Aluminum ware industry, growth, 409
Alunite, 89, 148
American Bauxite Company, 106
American Cyanamid Company, 132, 134–135
American Magnesium Corporation, 81
American Nitrogen Company, 117
Ammonal, 18
Anaconda Copper Mining Company, 151
Antitrust laws, U. S., 109, 118, 136, 152, 369–374, 381, 394, 480–486; *see also* Consent decree; Department of Justice; Federal Trade Commission
Archer, R. S., 55
Association of Manufacturers in the Aluminum Industry, 474

A. S. Vigelands Brug, 120–121, 126
Automobile, consumption of aluminum, 20–21, 60–61, 65–66; influence on alloy development, 19–23, 254; use of aluminum and aluminum alloys, 19–23, 45, 55–58, 214–215, 254

Bankers, lack of interest in aluminum, 116, 133, 151–152
Baush Machine Tool Company, 52, 132, 375, 378, 480
Baush Machine Tool Company v. *Aluminum Company of America,* 384, 480–484
Bauxit Trust A. G., 71, 84, 139
Bauxite, composition, 7, 102–103, 138; imports into United States, 71; ownership, 34–36, 69–72, 88, 91–94, 103–110, 119–121, 129–131, 137–141, 154; refining process, 7; restrictions on export: France, 95; Hungary, 84, 92–93, 138–139; Italy, 91, 138–139; Jugo-Slavia, 92, 138–139; Rumania, 92, 138–139; restrictive agreements in the United States, 103–105, 108, 548–551;
Australia, 140; British Guiana, 69–71, 129–131, 137; Dalmatia, 138; Dutch Guiana, 70–71, 129–131; France, 31, 71, 95, 108, 119–121, 137; Germany, 86, 127; Gold Coast, 140; India, 140; Italy, 32, 71, 91, 120, 138–139; Istria, 71, 91, 138; Jugo-Slavia, 71, 138; Russia, 89; United States: Alabama, 24, 103; Arkansas, 25, 103, 106–107; Georgia, 24, 103
Bauxites du Midi, 71
Bayer alumina process, 7
Best utilization, 176, 177, 188, 205, 206, 208, 329, 340
Blanc process, 148
Blooming mill, 54, 79, 197
Bohn Aluminum and Brass Corporation, 57, 148, 470
Bohn Foundry Company, 444, 470–472
Bradley, Charles S., 516, 518, 531, 534, 535, 536, 537
Brass, competition with aluminum, 17–18
Bremer-Walz Corporation, 370
British Aluminium Company, formation, 6; acquisition of bauxite, 36, 70, 119, 139; capacity, 37–38, 41, 97, 123, 305; horizontal expansion, 37–38, 88–89, 265, 292; integration, 36; foreign properties, 38, 72, 88, 97, 121; government aid, 89, 155; participation in cartels, 36, 93–94, 119, 125, 158, 163;
finance: investment, 286; earnings, 269, 286, 319, 324; depreciation, 269; reorganization, 36, 269;
plants: alumina, 36; power, 36, 37, 38, 41; reduction, 97; fabricating, 36
Bunsen, Robert Wilhelm, 505, 506
Bus body, aluminum alloy, 62

Calcium carbide, 514–515, 521
Calorizing, 18
Canadian Manufacturing and Development Company, 73
Capacity,
power plants: Canada, 73–74, 111; Europe, 32–33, 265, 269–271, 276, 293; France, 121–123; Great Britain, 88–89; United States, 77–78; 110–111, 247–248; cartel members, 38, 123, 264;
reduction works: America, 41, 290–291, 307, 317; Austria, 98; Canada, 41, 73–74, 98; Europe, 38, 41, 95–98, 158–159, 271, 273, 287, 290–293, 306–307; France, 40–41, 98, 292; Germany, 83–86, 94–95, 98, 290, 292; Great Britain, 41, 88–89, 98, 292; Hungary, 98; Italy, 92, 98; Japan, 98, 291; Norway, 98, 290, 292; Russia, 89–90, 98, 291; Spain, 98; Sweden, 98; Switzerland, 40, 87, 98, 292; United States, 41, 98, 110–111, 247–248, 274, 307; new enterprises, 38, 40–41; relative changes among European companies, 304–305. *See also individual companies*
Carborundum, 519, 524
Carborundum Company, 131
Cartels, international,
first cartel, *1901–1908*: formation, 119; membership, 36–37; terms of agreement, 36–37, 264; price and output policy, 37–38, 265, 273–276; weaknesses, 124; dissolution, 38, 121, 124, 265, 273, 275; consequences, 271–276;
second cartel, *1912–1914*: formation, 125; membership, 39, 125–126; terms of agreement, 125–126; weaknesses, 126; dissolution, 39, 126;
third cartel, *1926–1931*: formation, 76, 93, 158; membership, 93, 158;

terms of agreement, 161–163; announced purposes, 161; price and output policy, 93–94, 287, 293, 296, 300; sales quotas, 300–304; change in relative outputs of members, 300–301; influence on market results, 304, 306; weaknesses, 304, 306;
Alliance Aluminium Compagnie (fourth cartel, *1931–*) formation, 94, 163; membership, 94, 163–164; terms of agreement, 164–165, 306; price and output policy, 321; control of stocks, 321; interference from nationalistic policies of governments, 165–166; weaknesses, 162–163
Castings, 18; die castings, 22, 48, 56–58; permanent-mold castings, 48, 56–57;
sand castings: 48, 56, 254; competitive methods, 443–473; sales data, 470
Centralstelle für wissenschaftliche und technische Forschungen, 49
Christianburg, Colony of, bauxite lands owned by, 69–70, 131, 137
Chute-à-Caron Power Company, 73; *see also* Alcoa Power Company
Cleveland Metal Products Company, 370–373
Cole, Romaine, 529
Compagnie de Produits Chimiques d'Alais et de la Camargue (Alais), entry into aluminum industry, 35; acquisition of bauxite, 35; capacity, 37, 41, 123; horizontal expansion, 37, 44, 87–88, 122, 126, 265; foreign properties, 87; participation in cartels, 36, 119, 125; products other than aluminum, 35; acquisition of Société Électrométallurgique Française, 87; earnings, 268; plants, 35, 37, 41
Compagnie de Produits Chimiques et Électrométallurgiques Alais, Froges, et Camargues (Compagnie AFC), formation, 87; acquisition of bauxite, 88, 139; capacity, 96, 292; horizontal expansion, 88, 292; foreign properties, 72, 92, 96, 292; participation in cartels, 158, 163;
finance: investment, 277–282; earnings, 277–282, 286, 319, 323; depreciation, 277–280, 323; reinvestment of earnings, 277–279; issue of preference shares to prevent outside control, 90;
plants: reduction, 87–88, 96; power, 87–88
Competition, acquisition of bauxite, 71, 91, 105, 138–139, 152, 298; acquisition of power, 91, 144, 152, 298; development of new alloys and new products, 259, 298, 345–347, 351; influence on price and quotas, 157, 265; investment, 90–91, 124, 156–158, 163, 165, 264, 290, 298, 304–308, 338–343, 350; imports into United States, 157; price, 38–39, 157, 163, 167, 236–237, 244, 246, 265, 275–276, 295, 304, 310–311, 316–320, 326–327, 332–333; quality, 310–311; sales expenditure, 310–311; between substitutes, 11–23, 33, 46, 48–49, 57–58, 62–63, 66–68, 198, 214–215, 218–219, 247, 253–257, 322, 390–391, 420, 430, 436, 475, 477; summary of competitive forces, 169.
See also Potential competition
Competitive methods, complaint of Federal Trade Commission, 401, 411, 556–563; supply of materials, 376, 405–407, 410–412, 421–436, 445–462, 473, 486–490, 496, 498, 562; attitude toward potential competition, 410, 437–441; materials-castings price differential, 445–447, 450–455, 461, 469, 473; ingot-sheet price differential, 381, 386, 390, 392–395, 437–440, 481–483; price differentials in general, 474–475, 478, 486–498; price discrimination, 405, 410–411, 417–421, 441, 447, 451, 486–490, 496, 498; pricing of castings, 447–455, 461–473, 561; scrap purchases, 444–462, 473, 560–561;
unfair: consent decree, 398, 405–407; criteria, 402–404.
See also Delivery delays
Connecting rods, aluminum alloy, 51, 62
Consent decree, *United States* v. *Aluminum Company of America,* 39, 105–109, 369, 377, 396–399, 404–407, 421–422, 426, 431, 433–434, 436–440; text, 547–555
Consolidated Stamping Company, 53
Consumers' coöperation, 353–355, 485–487, 499
Consumption, Europe, 33, 65–68; Germany, 65–67, 95; United States, 11, 21, 60–65; during World War, 44–45;

ingot, world, 127; relative growth in European countries, 302
Continentalen Bauxit Bergbau und Industrie A. G., 84
Cooking utensils, 11, 13; consumption of aluminum, 60, 64–66; competitive methods, 408–442
Copper, competition with aluminum, 14–17, 33, 214, 218–219, 255, 322, 436. *See also* Price ratios; Prices
Cost, constant, 208–209, 212, 334, 408, 443; diminishing, 206, 208–209, 212, 260, 334; European relative to American, 160; ingot, 270–271; labor, 269; power, 142–144, 289; reductions, 93, 112, 150, 237, 246, 248–250, 270, 281, 296, 324; sand castings, 443, 463–469; sheet, 387–390; standards for measuring cost at one stage, 463–465
Cowles, Alfred and Eugene, 508–514, 520–521, 528–532, 535, 537
Cowles Electric Smelting and Aluminum Company, 5, 530, 532
Cryolite, 8
Cylinder heads, aluminum alloy, 65

Davis, A. V., 135
Davis, E. K., 75
Davy, Sir Humphry, 503, 534
Defender, Herreschoff, 12
Delivery delays, 248, 376, 405, 411, 421–436, 441, 553, 562
Demand, characteristics, 214, 215; elasticity, 112, 146–147, 167, 214–215, 221, 235, 237, 253–261, 297–298, 320, 322, 329, 331–332, 390–391; ideal, 209–211, 258, 297; increases, 38, 43, 120, 156, 214, 235, 252, 253, 255–258, 272–273, 275, 456
Demerara Bauxite Company, 69, 130
Department of Justice, Benham report, 400, 410, 413; consent decree of *1912*, 106–109, 398; investigations, 399–401, 410–411; *re* acquisition of Southern Aluminium Company, 116
Det Norske Aktieselskab for Electrokemisk Industri, 72
Det Norske Nitridaktieselskab, 43–44, 72, 153; capacity, 96, 290, 292; plants, 87, 96
Deutsche Edison Gesellschaft, 6, 513
Deville, Henri St. Claire, 35, 505–506
Differentiation of product, *see* Monopoly elements
Diversification of markets, 61–68, 255, 297
Dixie Bauxite Company, 107
Doehler die-casting process, 22, 257
Duke, J. B., 73, 132–135
Duke-Price Power Company, 73–74, 135; *see also* Saguenay Power Company
Duralumin, *see* Alloys
Dürener Metallwerke A. G., 50
Dynamo, stimulus to development of electrochemistry, 506–508

Earnings, metal corporations, 231–233. *See also individual companies*
Economies, combination, 196–200; scale of plants, 189–196
Efficiency, factors affecting, 31–32; *see also* Scale of investment; Integration; Location
Electric cable, 14–16, 60–64, 66–67, 146, 214, 218–219
Electric Smelting and Aluminum Company, 521, 532–533
Electric Smelting and Aluminum Company v. *Pittsburgh Reduction Company*, 532–537
Electrochemistry, influence of aluminum on, 508–516, 519–526
Electrode plants, 25–26, 34, 74, 84, 129, 191
Emory, L. T., 130–132, 140
Entry into aluminum industry, affected by conditions of price making, 144–147; castings branch, 396, 443–445; costs, 144–147; difficulties, 28–29, 105–118, 129–136, 137–141, 142–148, 149–152, 155–156; government control in Germany, 153; new ventures, 29, 35, 38, 82, 111, 115–117, 120–124, 129–136, 148, 153, 225, 265, 269, 289; sheet, 374–381, 392–395; utensils branch, 396, 408, 415.
See also Potential competition
Erftwerk A. G., 83–84, 283
Exports from America, 163, 166, 309–311; Canada, 307–312, 317; cartel countries, 303, 310; Europe, 39, 159–160, 166, 243–244, 246, 266, 309–311, 378–379; France, 124, 303; Germany,

243, 246, 303, 310; Great Britain, 303, 311; Norway, 301–303, 308–309; Switzerland, 124, 301–303, 310–312; United States, 310
Extrusion, 58

Fairmont Manufacturing Company, 375, 378, 384
Faraday, Michael, 503–504
Federal Trade Commission, complaint against Aluminum Company of America, 370, 401, 411, 444–447; text of complaint, 556–563; answer to complaint, 564–567; dismissal of complaint, 401, 412, 473, 568; report on aluminum, 399, 409–412, 442; rolling mill case, 371–374
Ferroalloys, 515, 516, 522–523
Ford, Henry, 133, 151
Foundries, 11, 27, 79–80, 396–397, 443
Frontier Corporation, 80–81
Full utilization, 182, 199, 235–237, 258, 260, 271, 291, 293–294, 299, 313, 327, 332

Gebrüder Giulini, 51, 97, 120–121, 158
General Bauxite Company, 103–105
General Chemical Company, 103–105, 107, 549
General Electric Company, 80
Georgia Bauxite Company, 24
Goldschmidt, Hans, 18
Government control, competitive methods, United States, 370, 377, 398–402, 404–480; entry, Germany, 153; imports, Germany, 154; export of bauxite, various countries, 139; exports of aluminum and materials, France, 95; ownership of bauxite, Great Britain, 69–70.
See also Import duties; Antitrust laws
Government encouragement, 153–156; Austria, 153; Great Britain, 88–89, 155, 311; Italy, 91, 92, 139, 156; Spain, 92
Government enterprise, Germany, 83–86, 153–154; Russia, 89–90
Government policy, possibilities: establishment of oligopoly, 355–357; import duties, 488–489; regulation of investment, 357, 485, 491, 495–498; regulation of price, 357–360, 485, 492–498; taxes, 359–360; government competition, 360–365, 485–489, 498–499; public monopoly, 364–365
Great Britain, bauxite lands owned by, 69–70
Guillet, Léon, 50

Hall, Charles M., 5, 25, 118, 509, 516–518, 520, 527–537
Haskell, George, 73, 132–136
Haskell v. *Perkins et al.*, 136
Henderson, Leon, report on aluminum, *see* National Recovery Administration
Héroult, Paul L. T., 5, 6, 115, 512–514, 520, 523–524, 527, 532, 535
Hoopes, William, 16, 54
Horizontal expansion, balance in horizontal and vertical extension, 199–201; efficiency, *see* Scale of investment *and* Economies; relation to monopoly, 189–203; *see also individual companies*
Hunt, Alfred E., 529
Hybinette, Victor, 51
Hydroelectric plants, *see* Water power; Power plants

I. G. Farbenindustrie, 84, 153
Illinois Pure Aluminum Company, 424–425, 437–439
Import duties, France, 154; Germany, 154; Great Britain, 155; Hungary, 93; Italy, 94, 164; Spain, 92; Switzerland, 155; United States, 29, 39, 82, 102, 112, 118, 143, 159–160, 238–241, 318, 321, 379; relation to monopoly, 149, 159–160, 168, 355–356
Imports into America, 309; cartel countries, 303; France, 291; Germany, 127, 156, 301–303, 311; Great Britain, 301–303, 311; India, 310–311; Japan, 310–311; United States, 39, 81–82, 110, 114, 160, 166–167, 243–244, 246, 266, 274, 307–308, 317–318, 324–327, 378–379; government control, Germany, 154, 157
Innwerk, Bayerische Aluminium A. G., 83
Integration, definition, 176; advantages in aluminum production, 182–188; balance in horizontal and vertical extension, 199–201; problem of fitting scales, 177–179, 200; relation to mo-

nopoly, 177–179; relation to progressiveness, 181, 187; relation to research, 187; strategic advantages, 108–109, 179, 188; theoretical analysis of advantages, 180.
See also individual companies
Investment, appropriate to maximum monopoly revenue, 235–237, 260, 294–296, 331, 334; ideal, 174, 175, 204–210, 260, 266, 271–272, 274, 297, 314, 331, 357, 362.
See also individual companies
Iron, competition with aluminum, 17–18, 62, 254, 255, 436

Jadranski Bauxite Dioni'co Drus'tvo, 71
Japan Aluminum Reduction Company, 93, 97, 164
Jeffries, Zay, 55

Kewaskum Aluminum Company, 424, 437–439
Kiliani, Martin, 513, 535
Knoxville Power Company, 26
Kossmann, Wilfried, 270

L'Aluminium français, 39, 51, 87, 93–94, 115, 122, 154–155, 301, 305
L'Aluminium du Sud-Ouest, 121–122, 269
Large scale, *see* Scale of investment
Leucite, Italy, 92, 148
Location of plants, 31–32, 184
Low-grade ores, attempts to utilize, 148, 298
Lumpiness of equipment, 206, 212–213, 216, 222, 225

Management, balance in horizontal and vertical extension, 199–201
Marginal cost, 141, 145–147, 205, 209, 236, 314–316, 329, 340, 351, 454–455, 469, 472, 475, 490–491, 496
Materials required for aluminum production, 31
Mellon, Andrew, 398
Merica, P. D., 50
Metallgesellschaft, 39, 56, 84–85, 153
Minet, Adolphe, 118, 535
Monopoly elements, water power, 142–144;
bauxite: France, 154–155; Europe, 109; South America, 70–71; United States, 28, 103–110, 113–114, 548–551; world, 139–140;
aluminum: separation of markets, 36, 166, 264, 547–548; patents, 6, 101, 118, 537; price control, 36–37, 93–94, 126, 161–162, 165, 264, 273–276, 321, 332–333, 350–351; production control, 94, 165, 306, 321, 350–351; sales control, 39, 94, 125, 161, 300–304; control of stocks, 164–166, 321; tariffs, 118, 149, 159, 168, 355; difficulties of entry, *see* Entry;
oligopoly: and antitrust laws, 483; difficulties in policy, 298–300, 306, 346, 351; price policies, 157–161, 319–320; relation to efficiency, 202–203; relation to progressiveness, 345–349; restriction of investment, 300; rivalry in expansion, 157, 298, 304–306, 338–343; rivalry in variation of product, 345–347, 351; proportionate shares in the market, 337–343, 346–347, 350–351; sheet, 374–376, theoretical comparison with single-firm monopoly, 333–352; France, 87–88, 120–122, 154–155; Germany, 153–154, Great Britain, 120–121; international relations, 118–128, 156–161, 306–312. *See also* Agreements; Cartels;
single-firm monopoly: relation to efficiency, 202–203; (*see also* Scale of investment); relation to progressiveness, 345–349; continued existence in United States, 101–118, 149–152;
differentiation of product: alloys, 198, 259, 386; limited by tests of quality, 198, 259, 363; and price regulation, 495–496. *See also* Alloys
Montecatini, 91–92
Montreal Light, Heat, and Power Company, 111

Nantahala Power and Light Company, 80
National Physical Laboratory, 50, 54, 257
National Recovery Administration, code for aluminum industry, 474–479; report on aluminum, 401–402, 477
Nationalistic policies, 47, 138–139, 299, 311–312, 355; Austria, 86; France, 95;

Germany, 85–86, 94, 165, 299; Hungary, 92; Italy, 86, 91, 94, 163–164, 166, 299; Japan, 93; Jugo-Slavia, 92; Rumania, 92; Spain, 86, 92
New enterprises, *see* Entry
Niagara Falls Hydraulic Power and Manufacturing Company, 26
Niagara Hudson Power Company, 81
Normal earnings, definition, 204
Norsk Aluminium Company, 72, 133, 153; capacity, 96, 290
North British Aluminium Company, 89
Northern Aluminum Company, plants and capacity, 25, 96; participation in agreements and cartels, 37, 117, 126, 547–548; name changed, 74
Norton Company, 104–105, 107, 550, 552

Oersted, Hans C., 504–505
Oligopoly, *see* Monopoly elements
Ore, *see* Bauxite *and* Low-grade ores
Output, ideal, 209
Outsiders, 121–124, 158, 163–164, 273–275
Overinvestment, 174–175, 204–208, 212, 216, 222, 272–273, 332, 344, 351, 387, 389, 392–393, 470, 491–497

Pacz, Aladar, 55, 256–257
Patents, infringement suits, 5, 530–537; relation to monopoly, 6, 101, 118, 537; expiration, 6, 29, 101–102, 120; aluminum nitride, 117; Bradley, 5–6, 29, 101–102, 531–537; Cowles, 530; Hall, 5–6, 101, 530–531; Héroult, 118, 512
Pennsylvania Salt Manufacturing Company, 104, 107, 550–552
Pistons, aluminum alloy, 23, 45, 55–57, 62, 214; Nelson, 57
Pittsburgh Reduction Company, formation, 5, 529; acquisition of bauxite, 24–25; patent litigation, 530–537; name changed to Aluminum Company of America, 26
Pittsburgh Reduction Company v. *Cowles Electric Smelting and Aluminum Company,* 530–531, 536
Potential competition, 101–102, 120, 129–136, 140, 150–152, 236–237, 258–259, 289, 298, 401, 437–441.
See also Entry
Power plants
Austria: Lend, 34, 37, 40, 269; Rauris, 34, 40, 269;
Canada: Cedar Rapids, 111; Chute-à-Caron, 149, 299; Isle Maligne, 73–74, 77, 149–150; Shawinigan, 25, 41;
France: Auzat, 41, 121; Beyrède, 88, 121, 270; Calypso, 37, 41; Chedde, 41, 121; L'Argentière, 37, 40, 270; La Praz, 37, 40; La Saussaz, 37, 40; Les Clavaux, 88; Prémont, 41; Rioupéroux, 88; Sabart, 88; St. Auban, 88; St. Félix, 37, 41; St. Jean-de-Maurienne, 37, 41, 270; Venthon, 88, 121, 270;
Germany: Inn River, 83–84, 292; Lauta, 83–84; Rheinfelden, 34, 37, 40, 269;
Great Britain: Dolgarrog, 41, 120; Foyers, 36, 37, 40; Kinlochleven, 41, 269; Lochaber, 88, 155, 292, 302;
Italy: Bussi, 41; Cismon, 92; Mori, 92;
Norway: Eydehavn, 87; Glomfjord, 89; Höyanger, 72; Otterdal, 120; Stangfjord, 38, 41; Tyssedal, 87;
Russia: Dnieper, 89; Kamensk, 89; Rion, 89; Swanka, 89;
Switzerland: Borgne, 40; Martigny, 40, 120; Navizance, 37, 40; Neuhausen, 33, 37, 40, 269; Rhône, 40;
United States: Calderwood, Tenn., 77; Cheoah, Tenn., 29, 77, 213; High Rock, N. C., 77; Massena, N. Y., 26, 41; Niagara Falls, 12, 26, 41; Santeetlah, Tenn., 77, 149
Price control, *see* Monopoly elements
Price differentials, materials and products, 349, 379–395, 437–440, 445–447, 450–455, 461, 469, 473–475, 478, 481–483, 486–498; scrap and virgin, 446–451, 457–458, 461, 473
Price discrimination, alloys, 218, 222, 390–391, 393; competitive methods, 405, 417–421, 488, 498, 553; in depression, 315–316, 322, 327, 329; geographical, 221, 299, 304; government control, 357, 486–490, 496, 498; price differentials, 218–220, 380–381, 386, 389–394, 486–490; price structure, 217–224, 252, 297, 322, 327, 386, 389–394, 396, 477

Price ratios, aluminum and copper, 17, 244–246, 288; aluminum and tin, 17; aluminum and zinc, 17; aluminum and a nonferrous index, 244–246
Price stabilization, 38–39, 42, 161, 174, 313, 317–322, 326–327, 332
Price, Sir William, 132
Prices, aluminum
United States: Aluminum Company of America, 9, 13–17, 33, 43, 112, 159, 167, 220, 238–241, 244–247, 249–256, 316–321, 326–327, 329, 382–384, 388, 530; open market, 43, 112, 159, 239, 242, 244, 316–318, 326–327; foreign aluminum, 159, 239, 244, 317–318; scrap, 457–458; sheet, 382–384, 388–389;
Europe: 93–94, 158–159, 162, 167, 221, 240–241, 246, 265, 268, 270–276, 283–288, 291, 293, 296–300, 316, 318, 321–322, 329; cartel, 38–39, 42, 240–241, 265, 287; France, 281, 299, 322; Germany, 288, 299, 322; Great Britain, 288, 321; Russia, 322;
Orient, 221
Prices, copper, 15, 158, 245, 288
Primorske Bauxite Dioni'co Drus'tvo, 71
Processes of aluminum reduction, *see* Reduction of aluminum
Production control, *see* Monopoly elements
Profits, *see* Earnings
Progressiveness, alloys, 45–60, 255–259, 291, 297–298; cost reductions, 248–250; improvements, 52–54; 212, 255–259; new products, 56–68, 256–258, 291, 297–298, 344–347; relation to integration, 187; relation to size of firm, 201–202, 348; attempts to utilize low-grade ores, 148; sheet, 254
Pure competition, 173

Quebec Aluminium Company, 135
Quebec Development Company, 132–133

Railroads, use of aluminum, 63–65, 68
Rationalization, 174–176, 180, 208–209, 263, 273, 313, 330
Reduction of aluminum, chemical process, 4, 506; electrolytic processes, 4–5, 7–8, 190–191, 505, 512–513, 516–518, 527–537
Reduction plants
Austria: Lend, 97; Steeg, 97, 153;
Canada: Arvida, 74, 96, 149; Shawinigan Falls, 25, 96, 111;
France: Auzat, 96, 120, 123; Beyrède, 96, 120, 123; Chedde, 96, 120, 123; Calypso, 96; L'Argentière, 96; La Praz, 96; La Saussaz, 96; Les Clavaux, 97; Prémont, 97, 120, 123; Rioupéroux, 88, 96; Sabart, 88, 96; St. Auban, 88, 96; St. Jean, 96; St. Michel-de-Maurienne, 34; Venthon, 88, 97;
Germany: Bitterfeld, 84, 97; Erftwerk, 83–84, 96, 294; Innwerk, 96, 292; Lautawerk, 83–84, 96; Rheinfelden, 84, 97;
Great Britain: Dolgarrog, 97, 120, 123; Foyers, 36, 97; Kinlochleven, 97; Lochaber, 97;
Hungary: Csepel, 93, 97, 153;
Italy: Borgofranco, 92, 96; Bussi, 91, 120, 123; Marghera, 92, 97; Mori, 92, 97, 139;
Japan, 93, 97;
Norway: Eydehavn, 87, 96; Glomfjord, 97; Höyanger, 72, 96; Stangfjord, 97; Tysse, 87, 96; Vigelands, 97;
Russia, 97;
Spain: Sabinanigo, 96;
Sweden: Mansbo, 96;
Switzerland: Chippis, 37, 97, 292; Martigny, 97, 120, 123; Neuhausen, 97;
United States: Alcoa, Tenn., 28, 96, 111; Badin, N. C., 29, 96, 149; Massena, N. Y., 26, 28, 96, 111; Niagara Falls, 12, 25, 96
Refining aluminum, Hoopes process, 54, 217
Reinvestment of earnings, 175, 290. *See also individual companies*
Republic Carbon Company, 131
Republic Mining and Manufacturing Company, 81, 104–105
Research, alloys, 22, 47–60; auto bodies, 254–255; coloring, 53; corrosion resistance, 52–54; fabricating technique, 54; heat treatment, 49–50; use of low-grade ores, 86, 92–93, 148; pistons, 55–57; refining, 54; command of

funds, 347–349; company staffs, 47, 50–59, 256–257; government bureaus, 47, 49–50, 52, 54–55, 59, 256–257; influence of aviation, 46–47, 50; influence of World War, 46–47, 50; relation to integration, 187; relation to size of firm, 201–202, 348
Rheinisch-Westfälisches Elektrizitätswerk A. G., 83–84
Rolling mills, Germany, 196–197; Switzerland, 196–197; United States, 11, 79, 196–197, 370–371, 386, 433, 437
Rosenhain, Walter, 50, 54, 256
Russia, aluminum capacity, 89, 98; aluminum in second five-year plan, 89–90

Saguenay Power Company, 74, 77
St. Lawrence River Power Company, 26, 80
St. Lawrence Valley Power Corporation, 80–81
Sales control, *see* Monopoly elements
Scale of investment, definition, 176; best scale, 189; integration and fitting scales, 177–179, 200; relation to monopoly, 189–203, 485; relation to research and progressiveness, 201–202, 348; bauxite mining, 195; extraction of alumina, 193–195; power, 191–193; reduction of aluminum, 190–192; rolling, 195–196; castings, 443; utensils, 408
Schweizerische Metallurgische Gesellschaft, 118, 512
Scrap aluminum, 43, 326, 376; competitive methods, 444–462, 473
Secondary aluminum, 215, 247, 253, 319, 326–327, 376, 443, 450, 487, 572
Self-sufficiency, *see* Nationalistic policies
Shawinigan Water and Power Company, 74
Sheet, in automobiles, 254; cost, 387–390; sources of materials, 376–379; oligopoly, 374–376; new ventures, 375; potential competition, 374; ingot-sheet price differential, 379–395, 437–440; sales data, 251–252, 384–385
Sheet Aluminum Corporation, 51, 375, 378, 386, 480
Sherman Act, *see* Antitrust laws
Single-firm monopoly, *see* Monopoly elements
Size of firm, *see* Scale of investment
Societa Anonyma Mineraria Triestina, 71
Societa Anonyma Veneta dell'Alluminio, 92, 97
Societa dell'Alluminio Italiano, 73, 92, 96
Societa Italiano dell'Alluminio, 91, 97
Société Anonyme des Forces Motrices du Béarn, 73, 87
Société des Bauxites de France, 119
Société d'Électrochimie, d'Électrométallurgie et des Aciéries Électriques d'Ugine, 87–88, 97, 122
Société Électrométallurgique Française (Froges), formation, 6, 513; acquisition of bauxite, 34; capacity, 37, 40; horizontal expansion, 37, 265; integration, 34; participation in cartels, 36, 119, 125; merger with Compagnie Alais, 87;
finance: investment, 267–268; earnings, 267–268; dividends, 268; depreciation, 267–268;
plants: alumina, 34; power, 34, 37, 40; electrode, 34; reduction, 40
Société Électrométallurgique du Sud-Est, 121–122, 269
Société des Forces Motrices et Usines de l'Arve, 121–122
Société Générale des Nitrures, 117, 126
Société Industrielle de l'Aluminium, 34
Société des Produits Électrochimiques et Métallurgiques des Pyrénées, 121–122, 269
Southern Aluminium Company, 111–112, 115–117
Steam power, Germany, 83, 144
Steamship, aluminum alloy, 65
Steel, competition with aluminum, 48–49, 52, 62–63, 214–215, 254–255, 390–391, 436
Stern und Hafferl, 153, 158
Stocks, aluminum, 94, 163, 167, 272, 293, 300, 317, 321, 327; control, *see* Monopoly elements
Structure of industry, 176
Supply characteristics, 211–212
Surinaamsche Bauxite Maatschappij, 71, 74

Tallassee Power Company, 26

Tapolcza Mining Company, 84
Tariff, *see* Import duties
Tennessee Valley Authority, control of hydroelectric development, 78; experiments on use of low-grade ores, 148
Thermit, 19, 257
Tin, competition with aluminum, 17–18, 67
Transmission lines, 15–16, 63–64, 66–67; *see also* Electric cable

Uihlein venture, 129–131, 151
Uncertainties, demand, 300, 306, 336–339, 342; indirect effects of policies, 298
Underinvestment, 174, 204–205, 263, 271, 273, 296–297, 344
Underutilization, 174, 208, 212, 216, 225, 258, 260, 263, 265–266, 289–294, 314, 316–320, 323–324, 327–328, 340–341; reasons for, 205–209
Unfair methods of competition, *see* Competitive methods
Union des Bauxites de France, 119
Union Development Company, 26
United Smelting and Aluminum Company, 370–371, 375, 378, 384
United States v. *Aluminum Company of America, see* Consent decree
United States Aluminum Company, 24, 27, 80, 373, 397, 408, 444, 463
United States Bureau of Standards, 50, 52
Uses of Aluminum, 10–23, 33, 44–68, 214–215, 390, 512

Variations of product, 175, 210–211, 255–259, 344–347, 351
Vereinigte Aluminiumwerke A. G., formation, 83; acquisition of bauxite, 84, 139; capacity, 94–96, 290, 292, 305; horizontal expansion, 94–95, 292; integration, 84; costs, 85; foreign properties, 84, 91; government protection, 153–154; participation in cartels, 93–94, 158, 163; products other than aluminum, 85; progressiveness, 51, 53;
finance: investment, 279, 283–286; earnings, 279, 283–286, 323; depreciation, 284–285, 323; ownership of stock, 83–84;
plants: alumina, 84; reduction, 83–84, 96; electrode, 84
Vereinigte Industrieunternehmungen A. G., 84
Vereinigte Leichtmetallwerke, 53
Vickers Sons and Maxim, 50

Water power, costs, 142–144, 185–186, 191–192, 202, 269, 289; site characteristics, 191–192; ownership, 25–29, 33–34, 36–38, 72–74, 78–81, 132–137, 144; Austria, 32; Europe, 289; France, 31, 119; Germany, 32, 127; Great Britain, 32, 289; Norway, 32, 72, 119, 144, 186, 289; Switzerland, 32, 119; Canada, 25, 32, 73–74, 132–137, 143, 150, 185–186, 289; Niagara Falls, 12, 131, 185, 520; North Carolina, 150; St. Lawrence River, 26, 80–81, 186, 192; Tennessee, 26–27, 78–79, 150, 186, 213; Washington, 185.
See also Power plants
Wilm, Alfred, 22, 49, 256
Willson, Thomas L., 514, 520
Wire, *see* Electric cable; Transmission lines
Wöhler, Friedrich, 505

Zeppelinwerke, 50
Zinc, competition with aluminum, 17–18, 255